ADVANCES IN
MOLECULAR AND CELL BIOLOGY
VOLUME 37

Aspects of the Cytoskeleton

ADVANCES IN
MOLECULAR AND CELL BIOLOGY
VOLUME 37

Aspects of the Cytoskeleton

Series Editor

E. Edward Bittar
University of Wisconsin – Madison
Madison, Wisconsin

Volume Editor

Seema Khurana
University of Tennessee Health Science Center
Memphis, Tennessee

2006

ELSEVIER

AMSTERDAM • BOSTON • HEIDELBERG • LONDON • NEW YORK • OXFORD
PARIS • SAN DIEGO • SAN FRANCISCO • SINGAPORE • SYDNEY • TOKYO

Elsevier
525 B Street, Suite 1900, San Diego, California 92101-4495, USA
84 Theobald's Road, London WC1X 8RR, UK

This book is printed on acid-free paper. ♾

For all information on all Elsevier Academic Press publications visit our Web site at www.books.elsevier.com

ISBN-13: 978-0-4445-2868-1
ISBN-10: 0-444-52868-7
ISSN: 1569-2558

Transferred to digital print 2008
Printed and bound by CPI Antony Rowe, Eastbourne

Contents

Contributors

Patrik Asp *Department of Cell Biology, The Wenner-Gren Institute, Stockholm University, SE-106 91 Stockholm, Sweden*

Horacio F. Cantiello *Massachusetts General Hospital East and Harvard Medical School, Charlestown, Massachusetts*

Erica Cavellán *Department of Cell Biology, The Wenner-Gren Institute, Stockholm University, SE-106 91 Stockholm, Sweden*

Daniel R. Clayburgh *Department of Pathology, The University of Chicago, 5841 South Maryland, Chicago, Illinois 60637*

R. Brian Doctor *Department of Medicine, Division of Gastroenterology, University of Colorado Health Sciences Center, Denver, Colorado 80262*

Ronald R. Dubreuil *Department of Biological Sciences, University of Illinois, Chicago, Illinois 60607*

Martin Eigenthaler *Institute of Clinical Biochemistry and Pathobiochemistry, University of Wuerzburg, Germany*

John E. Eriksson *Turku Centre for Biotechnology, University of Turku and Åbo Akademi University, FIN-20521 Turku, Finland; Department of Biology, Åbo Akademi University, FI-20520 Turku, Finland*

Karl R. Fath *Biology Department, Queens College and Graduate Center of the City University of New York, 65-30 Kissena Boulevard, Flushing, New York 11367*

Pirta Hotulainen *Institute of Biotechnology, University of Helsinki, Finland*

Zhihong Hu *Department of Pathology, The University of Chicago, 5841 South Maryland, Chicago, Illinois 60637*

Roger Karlsson *Department of Cell Biology, The Wenner-Gren Institute, Stockholm University, SE-106 91 Stockholm, Sweden*

Seema Khurana *Department of Physiology, University of Tennessee Health Science Center, Memphis, Tennessee 38163*

Anna Kobsar *Institute of Clinical Biochemistry and Pathobiochemistry, University of Wuerzburg, Germany*

Pekka Lappalainen *Institute of Biotechnology, University of Helsinki, Finland*

Uno Lindberg *Department of Cell Biology, The Wenner-Gren Institute, Stockholm University, SE-106 91 Stockholm, Sweden*

Paul Martin *Departments of Physiology and Biochemistry, University of Bristol, University Walk, Bristol BS8 1TD, United Kingdom*

Verena Niggli *Department of Pathology, University of Bern, Murtenstr. 31, CH-3010 Bern, Switzerland*

Ann-Kristin Östlund Farrants *Department of Cell Biology, The Wenner-Gren Institute, Stockholm University, SE-106 91 Stockholm, Sweden*

Hanna-Mari Pallari *Turku Centre for Biotechnology, University of Turku and Åbo Akademi University, FIN-20521 Turku, Finland; Department of Biology, Åbo Akademi University, FI-20520 Turku, Finland*

Avner Priel *Department of Physics, University of Alberta Edmonton, AB, T6G 2J1, Canada*

Herwig Schüler *Max Delbrueck Center for Molecular Medicine Neuroproteomics, D-13125, Berlin-Buch, Germany*

Clarence E. Schutt *Department of Chemistry, Princeton University, Princeton, New Jersey 08544*

Aneta Skwarek-Maruszewska *Institute of Biotechnology, University of Helsinki, Finland*

Christos Stournaras *Department of Biochemistry, University of Crete Medical School, GR 71110 Heraklion Voutes, Greece*

Jerrold R. Turner *Department of Pathology, The University of Chicago, 5841 South Maryland, Chicago, Illinois 60637*

Jack A. Tuszynski *Division of Experimental Oncology, Cross Cancer Institute, Edmonton, Alberta, T6G 1Z2, Canada*

Sarah Woolner *Departments of Physiology and Biochemistry, University of Bristol, University Walk, Bristol BS8 1TD, United Kingdom*

Preface

While the research in the field of cytoskeleton continues with its original intent, namely the identification of new cytoskeletal proteins, significant advances have been made in our understanding of the structure and function of the cytoskeleton. As is evident from the chapters included in this volume, in the last two decades a more integrated view has emerged which provides compelling evidence of the intimate association between cytoskeletal elements and signal transduction pathways which also demonstrates the significance of the cytoskeleton in organizing, integrating, and regulating multiple biological processes such as cell structure, cell motility, vesicular/organelle traffic, cell death, and cell proliferation among others. To perform these diverse biological functions, the actin cytoskeleton must be tightly regulated both spatially and temporally and must be responsive to extracellular stimuli. Research in the last two decades has clearly demonstrated how many proteins associated with the cytoskeleton are targets of signaling pathways or alternatively regulate the signaling pathways themselves.

While the major emphasis in this volume remains on the microfilament structure, some discussion has been included to illustrate the similarities and differences between the three cytoskeletal elements, namely the actin microfilament, the intermediate filaments, and the microtubules. For instance, Pallari and Eriksson (Chapter 14) describe how intermediate filaments are targets of protein kinases and how phosphorylation of intermediate filaments regulates cell structure and function. Kobsar and Eigenthaler (Chapter 1) describe the cytoskeleton (the microfilament and the microtubule) in platlet cells and this chapter serves to provide an overview of the central focus of this book namely to offer a comprehensive account of the relevance of the cytoskeleton to cell structure and function as well as to highlight the dynamic nature of the cytoskeleton by describing the regulation of the cytoskeleton and its associated proteins by signaling pathways. Polarized epithelial cells display a cell surface organization uniquely adapted to their specialized functions of vectorial transport. Likewise the apical surface of epithelial cells demonstrate three unique structures consisting of actin and actin-binding proteins. The specialized cytoskeleton of epithelial cells is discussed in Chapter 2 by Brain Doctor. In the next chapter (Chapter 3), the regulation of actin kinetics by nucleotides is examined. This chapter illustrates how actin ATPase activity determines the free energy changes that drive actin polymerization and depolymerization. This study can also be read with interest by scientists exploring tubulin GTPase and its role in assembly of the microtubular structure, thus highlighting the conserved regulatory mechanisms between the various cytoskeletal elements of the cell.

Ronald Dubreuil (Chapter 4) describes the spectrin structure in erythrocytes, in nonerythroid cells as well as in *Drosophila*. This chapter illustrates how the structure and function of actin-binding proteins is conserved from invertebrates to humans. It also highlights the significant progress that has been made in recent years in understanding the role of the cytoskeleton using genetic approaches. Three additional actin-binding proteins are discussed in greater detail and each exemplifies unique actin-modifying functions and biological roles these proteins play. Villin is unique among the actin regulatory proteins in that demonstrate actin-polymerizing and -depolymerizing activities. Villin belongs to a large family of proteins that share structural and/or functional homology and include gelsolin, adseverin, fragmin, flightless I, advillin, protovillin, supervillin, coronins. This chapter describes the ligand-binding properties of villin and how these are mechanistically important to villin's function in regulating actin reorganization, cell shape, and cell migration. Villin interacts with several second messengers including phosphatidylinositol 4,5-bisphosphate, phospholipase C-γ_1, calcium, and is the substrate for tyrosine kinases and phosphatases, thus it also illustrates the synergy between actin-binding proteins and signal transduction pathways in the regulation of biological processes. Karl Fath (Chapter 6) describes the microfilament-based motor proteins and their role on intracellular trafficking. Studies have clearly demonstrated that actin assembly also regulates the movement of intracellular vesicles, suggesting that this may be the mechanism underlying endomembrane trafficking. A little over a decade ago, most intracellular vesicular transport was attributed to the microtubule-dependent motors. The identification of myosin isoforms has illustrated the equally significant role the actin cytoskeleton plays in cytoplasmic trafficking. Karl Fath illustrates this by describing the role of the microfilament in the spatial organization of the Golgi complex and describes the role of specific myosin isoforms in actin-dependent transport of cargo. In the same vein, Östlund Farrants and coauthors (Chapter 7) describe the role of the SWI/SNF complexes in chromatin remodeling. Interestingly, while actin was often seen in isolated nuclei preparation, it was most often attributed to cytoplasmic contamination. However, very convincing studies in recent years have identified not only actin in the nuclei but also actin-binding proteins. The microfilament structure has been implicated in mRNA export, transcription, DNA repair, and chromatin remodeling. Östlund Farrants and coauthors describe the exciting new findings in this emerging field. Priel, Tuszynski, and Cantiello (Chapter 8) describe the synergy between ion transport and the microfilament/microtubule structure. They propose an appealing new model for the propagation of signals in cells, such as neurons, and the role of the cytoskeleton in regulating membrane properties in such cells using its capability of ion conductivity.

Christos Stournaras (Chapter 9) provides a very comprehensive analysis of the role of the microfilament structure in communicating cell surface signals to the structural machinery of the cell resulting in rapid modification of the microfilament structure in response to extracellular signals. The author also describes the significance of these studies in our understanding of malignant cells. It is worth noting that despite the significance of the microfilament structure in diverse pathophysiological conditions, no cytoskeleton-based strategies are available to treat or prevent any human disease. The author describes potential applications of regulating the actin dynamics in the diagnosis and treatment of cancer. In the two following chapters, Niggli (Chapter 11) and Skwarek-Maruszewska (Chapter 10) describe the interaction of actin-binding proteins with phospholipids, specifically phosphoinositides and the significance of these interactions in regulating membrane events. These chapters not only elucidate the role of the microfilament structure in converging signals at the cell surface in the regulation of cell migration, phagocytosis, and endocytosis but also intracellularly in the regulation of cytokinesis. Other examples of cell function regulated by the microfilament structure are provided by Woolner and Martin in Chapter 12, where they describe the role of the actin cytoskeleton in embryo morphogenesis and by Hu and coauthors in Chapter 13, where they describe the regulation of ion transport proteins by actin and actin-regulatory proteins.

In this volume, we have made an attempt to provide an overview of the cytoskeleton rather than provide a comprehensive description. The fundamental role of the cytoskeleton in the regulation of cell structure and function has been very ably articulated by various contributors to this book. This is an exciting period for biological research in general and the cytoskeletal field more specifically and we hope to follow these new discoveries which will only establish what is manifested from the discussions included in this book, namely the significance of the cytoskeleton in integrating cellular functions.

Seema Khurana

Chapter 1

The Cytoskeleton of the Platelet

Anna Kobsar and Martin Eigenthaler

Institute of Clinical Biochemistry and Pathobiochemistry, University of Wuerzburg, Germany

Platelets contain a membrane cytoskeleton, a cytoplasmic core cytoskeleton, and a microtubule system that are closely connected and act in concert to maintain the resting platelet discoid shape. A system of intracellular activatory and inhibitory signaling molecules interacts directly or indirectly with these cytoskeletal structures and tightly controls the cytoskeletal function. During platelet adhesion, activation, and aggregation, these cytoskeletal structures are rearranged resulting in the formation of a variety of F-actin-based structures, including filopodia and lamellipodia. Cytoskeletal remodeling is initiated by membrane receptors and signaling proteins that are in close contact with the platelet cytoskeleton. In turn, the cytoskeleton serves as a scaffold that binds and localizes signaling molecules at the right place and the right time to warrant platelet hemostatic function.

Advances in Molecular and Cell Biology, Vol. 37

ISBN: 0-444-52868-7
DOI: 10.1016/S1569-2558(06)37001-4

I. INTRODUCTION

Platelets play a key role in physiologic hemostasis as well as in thrombotic diseases and bleeding disorders. Platelets are activated by a number of stimuli resulting in platelet shape change, expression and/or activation of surface receptors, secretion of vasoactive substances, adhesion, aggregation, and finally thrombus formation. Vasoconstrictors (TxA_2, vasopressin) and endothelium-dependent vasoactive substances (ADP, thrombin, serotonin) are potent physiologic platelet activators. The initial event in platelet activation is binding of such agonists to specific membrane-spanning G-protein–linked receptors on the platelet surface, which transmit the signal via heterotrimeric and small GTP-binding proteins and multiple protein kinases into an ordered cascade of intracellular signaling pathways. All these pathways induce remodeling of the platelet cytoskeleton resulting in the appearance of actin filament-based structures, including filopodia and lamellipodia. Platelet adhesion and activation are strictly controlled through complex signaling interactions between extracellular ligands, adhesion receptors in the plasma membrane, and cytoskeletal structures.

This chapter discusses the structure of the platelet cytoskeleton and its interaction with platelet adhesion receptors. Furthermore, key players in these complex interactions (e.g., tyrosine kinases, protein kinases) as well as inhibitory signaling pathways regulated by cyclic nucleotides will be discussed in detail. A summary of the functions of proteins associated with the platelet cytoskeleton is provided in Table I. The reader is further referred to excellent reviews on these topics for additional information which cannot be presented in such detail in the context of this chapter (Canobbio *et al.*, 2004; Gibbins, 2004; Abrams, 2005; Ozaki *et al.*, 2005; Watson *et al.*, 2005; Hartwig, 2006; Larsson, 2006).

II. CYTOSKELETON OF THE RESTING PLATELET

The platelet cytoskeleton is composed of three major components: the actin filament system consisting of a spectrin-based membrane skeleton, the cytoplasmic "core" cytoskeleton, and a microtubule-formed ring structure around the platelet plasma membrane edge (Hartwig, 2006).

A. Actin Cytoskeleton

The actin filament system of a resting platelet consists of about 2000–5000 F-actin filaments that are distributed mainly in two locations: a cytoplasmic "core" cytoskeleton and a membrane cytoskeleton that attaches to the plasma membrane (Barkalow *et al.*, 2003). The platelet actin cytoskeleton fulfills manifold functions: it is responsible for the disk shape of the resting platelet and regulates shape change, assembly of filopodia and lamellipodia as well as the movement of subcellular components during platelet adhesion and activation (Hartwig, 2006). The role of the microfilament system in these processes is still not fully elucidated and will be dicussed later.

Table I
Functions of Proteins Associated with the Cytoskeleton

Protein	Function	Knockout mouse phenotype	References
α-Actinin	Bundles F-actin filaments	Increased lymphocyte cell motility	Gluck and Ben-Ze'ev, 1994; Kos *et al.*, 2003
β-Adducin	Targets barbed actin filament ends to spectrin; caps barbed ends	Actin levels were diminished; elliptocytes, ovalocytes, and occasionally spherocytes were found in the blood; mice suffer from mild anemia with compensated hemolysis	Muro *et al.*, 2000
α-Spectrin	Assembly of membrane skeleton; binds F-actin and adducin	Hereditary spherocytosis; thrombosis and infarction in neonatal animals	Wandersee *et al.*, 2001; Hartwig, 2006
Cofilin	Binds G-actin and severs filaments		Thirion *et al.*, 2001
Filamin	Attaches F-actin and signaling molecules to GPIbα; crosslinks F-actin	Enlarged and fragile platelets, reduced platelet counts	Nakamura *et al.*, 2006
Gelsolin	Severs actin filaments; increases actin-barbed ends	Altered platelet shape change	Kiuru *et al.*, 2000
Profilin	Binds G-actin		Mahoney *et al.*, 1997
Tubulin	Essential part of microtubules	Discontinous microtubule ring in platelets; spheroid platelets	Schwer *et al.*, 2001
PKCβ	Intracellular signal transduction	Platelets spread poorly on fibrinogen, however, normal agonist-induced fibrinogen binding	Buensuceso *et al.*, 2005
p85α Subunit of PI3-kinase	Intracellular signal transduction; regulatory subunit of heterodimeric PI3-kinase	Aggregation induced by ADP, thrombin, U46619, or PMA was normal; aggregation induced by collagen and collagen-related peptide (CRP) was partially impaired; both P-selectin expression and fibrinogen binding in response to CRP were decreased; poor spreading over a CRP-coated surface with intact filopodial protrusions	Watanabe *et al.*, 2003

(Continued)

Table I (*Continued*)

Protein	Function	Knockout mouse phenotype	References
PLCγ2	Cleaves phosphatidyl-inositol (4,5)-bisphosphate to generate inositol (1,4,5)-trisphosphate and diacylglycerol	Defective thrombus formation on a collagen surface; *in vivo* defective thrombus formation at superficial lesions, however, productive thrombosis after a more severe injury; only a minimal effect on platelet adhesion to immobilized vWF; impaired spreading and thrombus growth on vWF	Rathore *et al.*, 2004; Nonne *et al.*, 2005
SLP-76	Multidomain SH2 domain-containing adapter protein lacking intrinsic catalytic activity	Fetal hemorrhage and increased perinatal mortality; although megakaryocyte and platelet development proceeds normally, collagen-induced platelet aggregation and granule release is markedly impaired; treatment of SLP-76-deficient platelets owith collagen fails to elicit tyrosine phosphorylation of PLCγ2	Clements *et al.*, 1999
Syk	Tyrosine kinase	Tyrosine phosphorylation of PLCγ2 after collagen stimulation is absent	Poole *et al.*, 1997
VASP	Cytoskeleton- and focal adhesion-associated protein; major substrate of both cAMP- and cGMP-dependent protein kinases	Reduced cAMP- and cGMP-mediated inhibition of platelet aggregation; hyperplasia of megakaryocytes in bone marrow and spleen; activation of platelets with thrombin induced an increased surface expression of P-selectin and fibrinogen binding	Aszodi *et al.*, 1999; Hauser *et al.*, 1999
Vav3	Cytoplasmic guanine nucleotide exchange factor		Pearce *et al.*, 2004
Vav1/2 and Vav1/2/3 knockout	Cytoplasmic guanine nucleotide exchange factors	Marked inhibition of aggregation and spreading on activation of GPVI, which is associated with a reduction in tyrosine phosphorylation of PLCγ2	Pearce *et al.*, 2004

The cytoplasmic core cytoskeleton of the resting platelet is an actin filament network containing actin-binding proteins like α-actinin and, probably the major functional component, filamin. Filamin functions as a scaffolding molecule that collects and targets the corresponding binding partners toward the plasma membrane, thereby creating a tight link of the core cytoskeleton to the membrane cytoskeleton (Hartwig, 2006). Binding to filamin occurs mostly in its C-terminal part and includes binding partners like the small GTPases (Rac, Rho, Cdc42, RalA), the exchange factors Trio and Toll, adhesion molecules (β-integrins, glycoprotein Ibα: GPIbα), and various other signaling molecules. The Rho family of small GTPases (Rac, Rho, Cdc42) plays a distinct role in the regulation of actin assembly and cell motility. These proteins cycle between an active (GTP bound) and an inactive (GDP bound) state and interact with a variety of effector proteins.

The membrane cytoskeleton resembles a tightly woven planar sheet which lies at the cytoplasmic side of the platelet plasma membrane and the cytoplasmic side of the open canalicular system (Hartwig, 2006). This membrane skeleton consists of elongated spectrin strands that are connected to the central cytoplasmic core network through radial actin filaments. These filaments extend from the platelet center toward the plasma membrane, turn at this point, follow the membrane, and locate the actin-barbed ends close to the membrane. Besides actin and spectrin, major components of this network are adducin, filamin, and the von Willebrand factor receptor, glycoprotein GPIb-IX-V complex. The membrane cytoskeleton is further associated and interacts with a variety of cytoskeletal proteins and signaling molecules which will be discussed in detail later.

1. Spectrin and Adducin

Spectrin molecules are head-to-head assemblies of two $\alpha\beta$-heterodimers associating lengthwise in an antiparallel conformation (Hartwig, 2006). The dimerization site of spectrin is located at the C-terminus of the β-chain and the N-terminus of the α-chain. The N-terminal β-chain of spectrin exposes both an actin- and adducin-binding site which allows spectrin to bind to and support this actin structure.

Adducin is a membrane-skeletal protein with a tetrameric structure of either α/β- or α/γ-heterodimers. Adducin preferentially recruits spectrin to the fast-growing ends of actin filaments, bundles actin filaments, and caps the majority of the actin-barbed ends (Barkalow *et al.*, 2003). The C-terminal domains of all adducin subunits contain a highly conserved, 22 amino acid residue of myristoylated alanin-rich C kinase substrate that has high homology to MARCKS protein, an important substrate for protein kinase C (PKC) in human platelets (Matsuoka *et al.*, 2000). This MARCKS-related domain is essential for adducin function and represents a substrate for PKC and Rho-associated kinase. Furthermore, adducin binds calmodulin. Both adducin functions, spectrin recruitment and actin capping, are inhibited by PKC phosphorylation and calmodulin binding.

2. Filamin A and Filamin B

Another central element in membrane skeleton is filamin, present in platelets as Filamin A and B (Gorlin *et al.*, 1990; Takafuta *et al.*, 1998). Similar to spectrin, filamin assembles into bipolar homodimers with the subunits assembled in a V-shaped form and actin filament-binding sites on each strand at their free N-terminal ends. Each Filamin A/B molecule has two binding sites for the GPIbα chain of the von Willebrand receptor (GPIb-IX-V complex). Interaction of filamin with this platelet adhesion receptor complex is of great importance for platelet function. This interaction not only links a variety of signaling molecules to the von Willebrand factor (vWF) receptor, it also arranges the alignment of vWF receptor on the platelet membrane and strengthens the linkage of the membrane cytoskeleton to the plasma membrane (Hartwig, 2006). Platelets from mice deficient in Filamin A are enlarged and fragile, platelet counts are reduced (Ware *et al.*, 2000). A similar phenotype is observed in mice lacking GPIbα as well as in the corresponding human disease, the Bernard-Soulier syndrome.

B. Platelet Microtubule Coil

Besides the main structural component of platelet shape, the internal actin filament cytoskeleton, the microtubule structure of platelets is an essential component for the structure of the resting platelet. Platelet shape is maintained by an internal microtubule coil that is concentrated closely at the plasma membrane edge of the platelet disk (Hartwig, 2006). A microtubule is a hollow cylinder with a basic structure formed by α-β-tubulin heterodimers which do not come apart after formation. Along the microtubule axis, tubulins are arranged in head-to-tail order to form protofilaments with alternating α- and β-subunits. Platelets express four different β-tubulins (β_1, β_2, β_4, and β_5), however, β_1 is the dominant isoform in human and mouse platelets (Schwer *et al.*, 2001). Mouse models with deficiency in β_1-tubulin still polymerize some amount of tubulin, however, the overall tubulin content was strongly reduced and a discontinuous microtubule ring is formed in such platelets. Studies in megakaryocytes of these mice showed that β-tubulin is essential for the normal proplatelet formation and that circulating platelets failed to form the characteristic disk shape (Schwer *et al.*, 2001). Recently, a P for Q substitution in β_1-tubulin was found in humans. Heterozygous carriers of the Q43P variant showed a reduced platelet β_1-tubulin expression and in their blood a mixture of two different platelet populations appeared: normal discoid-shaped platelets and enlarged spheroid platelets (Freson *et al.*, 2005). In contrast to the β_1-tubulin-deficient mice that had only minor platelet abnormalities, the human heterozygous Q43P β_1-tubulin substitution resulted in a defective platelet function with reduced platelet adhesion, secretion, and aggregation, suggesting that these carriers might even be protected against cardiovascular diseases.

Microtubule-based dynamics might play a role in many aspects of cytoplasmic motility, especially since platelets express cytoplasmic dynein and kinesin (Hartwig, 2006).

Both proteins are certainly involved in proplatelet formation in megakaryocytes and in vesicular trafficking. However, the role of microtubule-based movement of organelles and granules in physiologic platelet function still needs to be elucidated in detail.

III. REARRANGEMENT OF THE CYTOSKELETON DURING PLATELET ACTIVATION

The morphologic changes in platelet activation are well characterized in platelets stimulated by agonists in suspension (e.g., thrombin, ADP, collagen, thromboxane). Cytoskeletal reorganization occurs in two phases: the early phase is independent of platelet aggregation, starts immediately, and is completed within minutes. Platelets round up, change their shape from discoid to spherical, and form actin filament networks in developing filopodia and at the cell periphery. The second phase requires the binding of ligands (fibrinogen or vWF) to cell surface receptors, close cell–cell contact, and formation of platelet aggregates. Shape change occurs rapidly within seconds and platelet aggregation is completed within minutes, depending on the type and concentration of the agonists used.

On activation by soluble ligands via their specific platelet membrane receptors, platelet cytoskeleton undergoes a complex remodeling, and F-actin content increases rapidly to about 70% of total actin. This increase in F-actin is primarily due to the generation of actin filament-free barbed ends which then serve as centers to promote actin polymerization. This occurs through a combination of filament fragmentation and the release of barbed-end-capping molecules. Key events in these processes are the dissociation of adducin and the activation of gelsolin and cofilin.

1. Dissociation of Adducin from Spectrin

Following platelet activation, adducin dissociates from the barbed ends of actin filaments and also from spectrin (Gilligan *et al.*, 2002). This reorganizes and dislocates the membrane skeleton from the plasma membrane localization. This inactivation of adducin might occur by various signaling pathways inclucing PKC-induced phsophorylation, activation of phosphoinositide 3-kinase (PI3-kinases) and calcium–calmodulin binding as discussed later (Gilligan *et al.*, 2002).

2. Gelsolin

Gelsolin severs actin filaments and is the main generator of pointed filament ends. During platelet activation, gelsolin undergoes a calcium-dependent conformational change that allows it to bind to actin filaments, fragment them, and then stay bound to the barbed ends of these fragments (Sun *et al.*, 1999; Hartwig, 2006). It is of interest that gelsolin-deficient platelets turn over actin filaments much slower and can still form filopodia, but not lamellae (Witke *et al.*, 1995).

3. Cofilin

Cofilin is found in high concentration in platelets. In unstimulated platelets, the major fraction of cofilin is phosphorylated and therefore inactive. Cofilin becomes rapidly dephosphorylated during platelet activation and stays dephosphorylated for some time as long as platelet integrin receptors are ligated (Falet *et al.*, 2005). In its dephosphorylated state, cofilin binds to and induces the disassembly of actin filaments. This reaction follows a slower time kinetic than the activation of gelsolin (Hartwig, 2006). The combined action of cofilin and gelsolin is enhanced by the dissociation of adducin from spectrin and from the barbed ends of actin that further disassembles the membrane cytoskeleton. Very recently, cofilin dephosphorylation by an okadaic acid-insensitive phosphatase at the beginning of aggregation and, later in aggregation, slow rephosphorylation in a Rho-kinase dependent manner was described (Pandey *et al.*, 2006). However, this cofilin phosphorylation/rephosphorylation was independent of integrin $\alpha_{\mathrm{IIb}}\beta_3$-ligation and did not regulate cofilin association with F-actin.

A. Formation of Filopodia

On activation with soluble platelet agonists, platelets rapidly form filopodia that represent thin elongations of the plasma membrane with actin filament bundles in their center. Although filopodia formation has been extensively studied in other cell systems, the molecular mechanisms of platelet filopodia formation are still far from being elucidated. Members of the protein family of Wiskott–Aldrich syndrome protein (WASP) including WASP, neuronal WASP (N-WASP), and WAVE (WASP family verprolin homologous proteins) are certainly involved in this process. In fibroblast, Cdc42 activation leads to extensive filopodia formation (Nobes and Hall, 1995). In a cell-free system it was shown that active Cdc42 stimulated the actin-depolymerizing activity of N-WASP. WASP and N-WASP then mediate the signaling toward the Arp2/3 complex and create free barbed ends from which actin polymerization can then take place (Miki *et al.*, 1998). In contrast to these findings, platelets from WASP/N-WASP-deficient mice have normal actin assembly and produce normal or even an enhanced number of filopodia after stimulation. Due to the functional homology of the WASP family members, homologous proteins might be involved in filopodia formation and take over the function after the loss of other family members. Other proteins have been discussed to regulate filopodia formation: CapZ, a barbed end capping protein that is able to terminate actin filament assembly, and the cytoskeletal proteins Ena and VASP, members of the Ena/VASP protein family (Reinhard *et al.*, 2001). Depletion of the capping protein caused the loss of lamellipodia and strongly enhanced formation of filopodia. In this system, gelsolin overexpression did not restore the original phenotype, indicating that this phenotype was specific for CapZ. In Ena/VASP-deficient cells, CapZ depletion resulted in ruffling instead of filopodia formation. Therefore it was proposed that CapZ is a negative regulator of filopodia formation and that, in addition to their previously suggested anticapping and antibranching activity, Ena/VASP have activating functions downstream of actin filament elongation (Krause *et al.*, 2003).

B. Signaling Pathways Regulating the Cytoskeleton of the Platelet

A variety of activatory platelet-signaling pathways including increase in phosphatidylinositides, activation of calcium-calmodulin–regulated pathways, activation of PKC, and tyrosine kinases are involved in remodeling of the cytoskeleton. Figure 1 shows an overview of signaling events during platelet activation.

1. Phospholipase C and Calcium Signaling

Platelet activation by many stimulatory agonists involves activation of tyrosine kinases and/or binding to G-protein–linked receptors followed by the dissociation of G-protein subunits. Both events lead to activation of phospholipase C (PLC) isoforms: PLCβ is activated by heterotrimeric G-proteins and activation of PLCγ2 is regulated by tyrosine phosphorylation through Src kinase. Real time phase-contrast microscopy showed a strong reduction of filopodia and lamellapodia formation in murine platelets on a fibrinogen surface in the absence of PLCγ2 (Wonerow *et al.*, 2003). Once activated, PLC generates 1,2-diacylglycerol (DAG) and IP_3 through hydrolysis of phosphatidylinositol 4,5-bisphosphate (PIP_2). DAG then activates PKC, whereas IP_3 binds to its receptors at the plasma membrane and the dense tubular system and elevates cytosolic calcium by release from intracellular stores as well as by stimulation of the extracellular calcium entry. This subsequent increase of intracellular free calcium plays the key role in platelet activation, since it regulates multiple calcium-dependent enzymes, like the calcium-dependent PKC isoforms and the calcium/calmodulin-dependent myosin light chain (MLC) kinase. Major target proteins in platelets are further myosin, gelsolin, and filamin. These pathways are key players in controlling actin filament assembly and disassembly.

2. Nonreceptor Tyrosine Kinases

Src tyrosine kinases form the largest class of nonreceptor tyrosine kinases. Activity of Src kinase is regulated by tyrosin phosphorylation at two different sites: phosphorylation at tyrosine residue 416 leads to enzyme activation, whereas phosphorylation at Tyr-527 inhibits it. Src kinase and members of its family are membrane-associated proteins, however, unlike Src, the Src-related tyrosine kinases are associated with electron-dense cytoplasmic and plasma membrane structures that correspond to endocytotic vesicles and coated pits (Stenberg *et al.*, 1997). In contrast to Lyn and Fyn kinase, Src binds constitutively and selectively to β_3-integrins

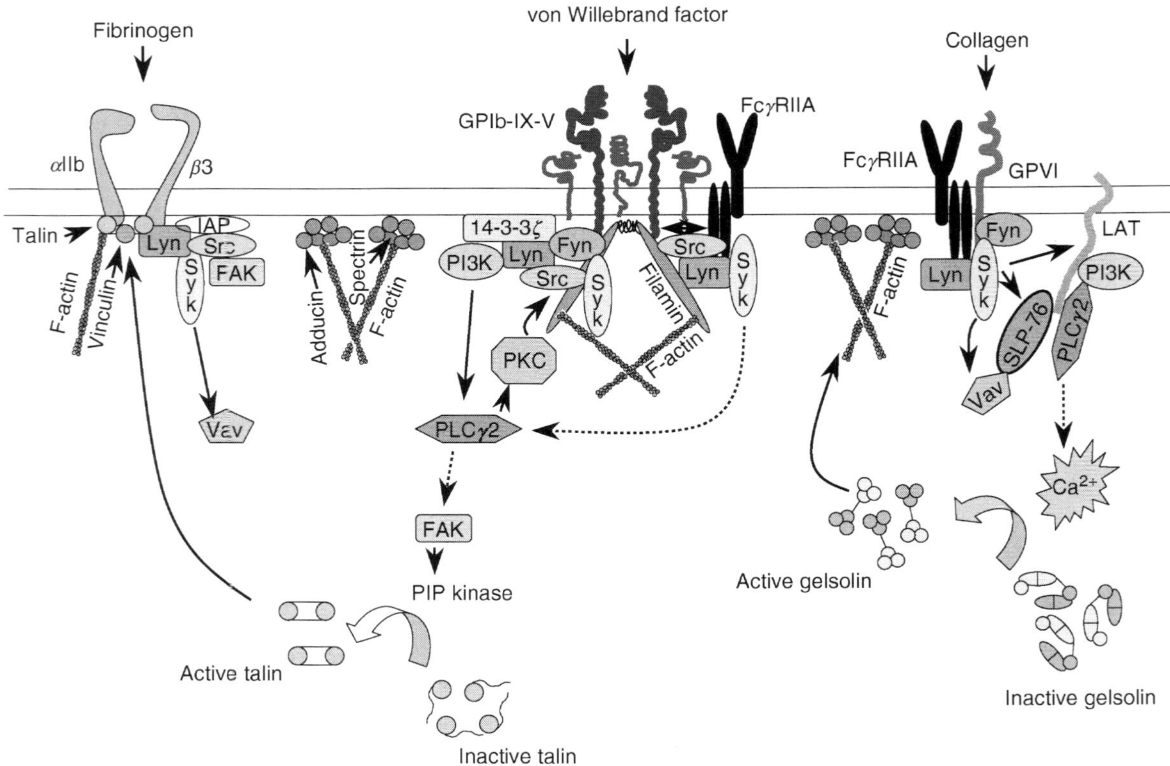

Fibrinogen
αIIb
β3
Talin
F-actin
Vinculin
Lyn
IAP
Src
FAK
Syk
Vav
Adducin
Spectrin
F-actin
von Willebrand factor
GPIb-IX-V
FcγRIIA
14-3-3ζ
PI3K
Lyn
Fyn
Src
Syk
Filamin
F-actin
Src
Lyn
Syk
PKC
PLCγ2
FAK
PIP kinase
Active talin
Inactive talin
Collagen
FcγRIIA
GPVI
F-actin
Lyn
Fyn
Syk
LAT
PI3K
SLP-76
PLCγ2
Vav
Ca2+
Active gelsolin
Inactive gelsolin

through an interaction involving the c-Src SH3 (SH3: Src homolgy 3) domain and the C-terminal region of the β_3 cytoplasmic tail (Kralisz and Cierniewski, 1998a, Arias-Salgado *et al.*, 2003; Shattil, 2005). Clustering of β_3-integrins *in vivo* activates c-Src by inducing intermolecular autophosphorylation (Arias-Salgado *et al.*, 2003; Shattil, 2005). Src kinases play an important role downstream of integrin adhesion receptors and are necessary for the generation of "outside-in signals" that regulate the cytoskeletal reorganization (Kralisz and Cierniewski, 1998a; Shattil, 2005). In activated, nonaggregated platelets the tyrosine kinases Src, Fyn, and Lyn are specifically associated with the actin cytoskeleton of platelets (Bertagnolli *et al.*, 1999). Analysis of Src phopsphorylation in collagen-stimulated platelets revealed a rapid increase in Src kinase phosphorylation, which was important for platelet shape change (Kralisz and Cierniewski, 1998b). Investigations show that activation of Src family kinases is also involved in GPIb-mediated platelet activation. Src and Lyn are dynamically associated with GPIb on platelet vWF receptor ligation, and ligand binding to GPIb-IX-V is sufficient to activate Src even without receptor clustering. Immunoprecipitation studies demonstrated that Src, PI3-kinase, and GPIb form a complex in agonist-stimulated platelets. The p85 subunit of PI3-kinase mediates GPIb-related activation signals and activates Src independently of the enzymatic activity of PI3-kinase (Wu *et al.*, 2003). Src kinase is further involved in the process of store-mediated calcium entry. In addition, depletion of intracellular calcium stores stimulates association of Src with cytoskeleton (Rosado *et al.*, 2000).

Syk, a protein tyrosine kinase with two Src homolgy 2 (SH2) domains also plays an important role in the early phase of platelet activation (Ozaki, 1999). Platelet stimulation with thrombin resulted in strong Syk phosphorylation, translocation to the cytoskeleton, and association with the actin filament network. Dissociation of the $\alpha_{IIb}\beta_3$-complex *in vitro* activated Syk kinase and these changes were insensitive to cyclic nucleotide inhibitory pathways and only in part dependent on cytosolic calcium (Negrescu and Siess, 1996). Platelet adhesion to fibrinogen resulted in both Src and Syk kinases activation. In contrast to Src, Syk kinase was associated with integrin $\alpha_{IIb}\beta_3$ only after fibrinogen binding. Inhibition of Src kinases blocked Syk activation and inhibited phosphorylation of Syk substrates (Vav1, Vav3, SLP-76) which are

Figure 1. Model of platelet receptor signaling and cytoskeletal organization. Filamin attaches the GPIb-IX-V complex to actin filaments via the cytoplasmic domain of GPIbα. Binding of vWF to platelet GPIb-IX-V recruits a signaling complex including tyrosine kinases and the FcγRIIA receptor that leads to PI3-kinase-mediated activation of phospholipase Cγ2 (PLCγ2). PLCγ2 generates inositol (1,4,5)-trisphosphate and diacylglycerol, which lead to the mobilization of calcium from intracellular stores and activation of PKC, respectively. In platelets, the FcγRIIA receptor may associate with GPVI. Collagen binding leads to GPVI clustering, phosphorylation of FcγRIIA by tyrosine kinases as well as phosphorylation of the transmembrane adaptor protein LAT, which assembles a complex of signaling proteins including PI3-kinase, PLCγ2 and the adapter proteins SLP-76 and Vav. Increased cytoplasmic concentration of calcium induces a conformational change of gelsolin to expose its actin-binding sites. Active gelsolin severs actin filaments via a unique tail latch mechanism. Adductin caps the fast-growing ends of actin filaments and recruits spectrin to the filaments. Both spectrin and adductin provide the connection of the membrane skeleton to the plasma membrane. (See Color Insert.)

implicated in cytoskeletal regulation. Syk-deficient platelets exhibited Src activation on adhesion to fibrinogen, but no spreading or phosphorylation of Vav1, Vav3, and SLP-76 (Obergfell *et al.*, 2002). Syk is further associated with FAK kinase and this association correlates with the state of actin polymerization (Sada *et al.*, 1997). Besides being part of the integrin α_{IIb}/β_3 signaling pathway, Syk also becomes tyrosine phosphorylated and activated downstream of GPIb. Via its SH2 domains, Syk is further associated with the FcγRIIA receptor (Falati *et al.*, 1999; Ozaki *et al.*, 2000; Satoh *et al.*, 2000).

Focal adhesion kinase (FAK) is a nonreceptor tyrosine kinase involved in integrin-regulated signaling pathways (Cobb *et al.*, 1994; Schaller *et al.*, 1994). Cell adhesion and integrin ligand binding activate FAK via phosphorylation on tyrosine residues (Lyman *et al.*, 1997). PKC plays an essential role in the transduction of intracellular signals downstream of α_{IIb}/β_3 and regulates FAK phosphorylation in platelets immobilized to fibrinogen (Haimovich *et al.*, 1996). In collagen-stimulated human platelets both intracellular calcium and PKC activity are essential for FAK activation (Achison *et al.*, 2001). Stimulation of human platelets with vWF results in a rapid tyrosine phosphorylation of FAK. This was dependent on thromboxane A2 production and was inhibited by RGDS peptide that blocks the integrin α_{IIb}/β_3 signaling (Canobbio *et al.*, 2002). Moreover, chelation of intracellular calcium or inhibition of PKC totally blocked vWF-induced tyrosine phosphorylation of FAK, indicating that this event is downstream of PLC activation (Canobbio *et al.*, 2002). Immunoprecipitation techniques showed a close complex formation between α_{IIb}/β_3-integrin subunits, Src and FAK. An FAK mutant with substitutions destroying its SH3-binding site showed reduced binding to Src *in vivo* (Chung *et al.*, 1997). This mutation also reduced Src-dependent tyrosine phosphorylation of the FAK mutant itself and of downstream substrates, such as paxillin (Thomas *et al.*, 1998).

3. Phosphatidylinositol 3-Kinases

PI3-kinases are a group of enzymes that phosphorylate phosphatidylinositol to generate phosphatidylinositol 3-phosphate (PIP_3). Four different isoforms, classified according to their catalytic subunits (p110α, p110β, p110γ, and p110δ), have been described in human platelets until now, and specific inhibitors as well as transgenic mouse models help to elucidate their function in platelets: mice lacking the p110γ subunit have a defective ADP response and an inhibition of p110β inhibited platelet adhesion in mice. Furthermore, mice lacking the regulatory p85α-subunit of PI3-kinase have defective glycoprotein VI (GPVI) signaling (Watanabe *et al.*, 2003). The p85 subunit of PI3-kinase also mediates GPIb-related activation signals and activates Src independently of the enzymatic activity of PI3-kinase. Changes in GPIb-XI-V signaling in p85-deficient animals should be interesting to investigate (Wu *et al.*, 2003).

4. Protein Kinase C

Platelets express members of all three classes of PKC isoforms. Until now, PKCα, PKCβ, PKCζ, PKCδ, PKCη, PKCθ, and PKCϵ have been detected and differ in their subcellular distribution (Crabos *et al.*, 1991). PKC is involved in the regulation of calcium-induced platelet aggregation (Tabuchi *et al.*, 2003), platelet shape change (Ozaki *et al.*, 1993; Dandona *et al.*, 1996), granula secretion (Murphy and Westwick, 1992; Murugappan *et al.*, 2004), and vesicle trafficking (Fitzgerald and Reed, 1999). Analysis of signaling events in activated platelets demonstrated that Syk and PKCα associated in a stimulation-dependent manner, requiring Syk activation but not PKC activity. Subsequently, PKCα and Syk cotranslocated from the cytosol to the plasma membrane. PKCα also associated with Src and PKCα activity was required for the association of these kinases but not for the stimulation-induced translocation of Src to the cell membrane. Src activity was negatively regulated by PKC (Pula *et al.*, 2005). Furthermore, it was found that in platelets thrombin caused tyrosine phosphorylation of PKCδ, which was dependent on calcium and Src family tyrosine kinases (Murugappan *et al.*, 2005). PKCδ and the classical PKC isoforms play a differential role in platelet dense granule release mediated by protease-activated receptors and GPVI. PKCδ played an enhancing role in protease-activated receptor-mediated dense granule secretion, whereas it acted as a negative regulator downstream of GPVI signaling (Murugappan *et al.*, 2004).

5. Vav Proteins

Vav proteins belong to the family of guanine nucleotide exchange factors activating Rho GTPases (Bustelo, 2000). All members of Vav family proteins contain various functional domains: an *abl* homology (DH) region which exhibits a nucleotide exchange factor activity, a pleckstrin homology (PH) domain which interacts with polyphosphoinositides, an SH2 and two SH3 domains which mediate protein–protein interactions, a proline-rich motif, and a calponin-homology region which acts as an actin-binding domain (Billadeau *et al.*, 1998; Schuebel *et al.*, 1998; Abe *et al.*, 2000). Potent platelet agonists like thrombin and collagen stimulate tyrosine phosphorylation of Vav, whereas weaker platelet agonists (ADP, epinephrine, and thromboxane A_2) failed to do so. Vav phosphorylation was also observed when platelets adhere to immobilized collagen, fibronectin, or fibrinogen (Cichowski *et al.*, 1996). Vav phosphorylation was stimulated by an increase of intracellular calcium or by PKC stimulation, a mechanism of Vav phosphorylation that is unique to platelets (Miyakawa *et al.*, 1997). In human platelets, the GPVI-specific agonist collagen-related peptide (CRP) and thrombin stimulate tyrosine phosphorylation of Vav1 but not Vav2. Activation of PLC by GPVI and thrombin was unaltered in Vav1-, Vav2-, and Vav1/Vav2-deficient platelets. A weak inhibition of late-stage aggregation to CRP and thrombin was observed in platelets deficient in Vav1 but not Vav2, whereas spreading on fibrinogen was not changed. Inhibitors or Src in wild-type platelets inhibited adhesion-dependent tyrosine phosphorylation of both Syk and Vav1, therefore it appears likely that Syk is downstream of Src and upstream of Vav1 in an α_{IIb}/β_3 signaling pathway that regulates the platelet actin cytoskeleton. These data allow the conclusion that Vav1 but not Vav2 is essential in the later stages of platelet aggregation (Pearce *et al.*, 2002). Furthermore, platelets deficient in both Vav1 and Vav3 show a marked inhibition of aggregation and spreading on activation of GPVI, which is associated with a reduction in tyrosine phosphorylation of PLCγ2. Thus, Vav3 and Vav1 play crucial but redundant roles in the activation of PLCγ2 by GPVI (Jay *et al.*, 2004).

IV. REARRANGEMENT OF THE CYTOSKELETON DURING PLATELET ADHESION

Platelets that are activated in suspension by soluble agonists round up, create elongated filopodia, and finally form large stable aggregates. The adhesion of platelets to immobilized ligands leads to a different type of cytoskeletal rearrangement.

A. Formation of Lamellipodia

On attachment to surfaces, platelets form broad, actin filament-containing lamellae that can cover an area of several μm^2 in size (Hartwig, 2006). Consequently, the platelet F-actin content nearly doubles due to filament growth, however, unlike leukocytes, platelets do not use these structures for directed movement. Later, in adhesion also some filopodia may appear on the apical surface of the platelet and elongate around the cell margin (Hartwig, 2006). Two major molecular mechanisms are used for lamellipodia formation: actin filament severing/uncapping and activation of the actin-related Arp2/3 complex that is highly expressed close to the lamellar actin filaments and generates the branched actin network. In platelets, new barbed ends are provided by the activity of gelsolin and cofilin. Transgenic mice lacking gelsolin have a strongly reduced ability to assemble actin and, compared to wild-type platelets, a reduced actin filament content after stimulation with platelet agonists. Cofilin contributes to a lower degree to the generation of barbed ends, however, both gelsolin and Arp2/3 functions seem to depend on cofilin activity. After the initial increase in actin-barbed ends, actin assembly is continued by the Arp2/3 complex which binds to actin filaments close to their barbed ends. Arp2/3 activity is regulated by cortactin and the protein family of WASP, including WASP, neuronal WASP, and WAVE. WASP-deficient mice reveal normal actin assembly, most certainly due to functional homology among the WASP proteins and takeover of WASP function by other family members. Activity and localization of the Arp2/3 complex strongly depends on the presence of functional gelsolin. The essential role of Rac1, a member of the Rho family of small GTPases, for lamellipodia formation has been demonstrated in Rac1-deficient mice. Loss of Rac1 did not influence filopodia formation and thrombin-induced platelet aggregation. In contrast, Rac1-deficient platelets showed strongly reduced spreading on fibrinogen, collagen, and laminin with loss of lamellipodia formation (McCarty *et al.*, 2005). Furthermore, Rac1 was essential for the formation of stable platelet aggregates under flow conditions.

V. CONTRACTILE ELEMENTS OF THE PLATELET CYTOSKELETON

Platelet contraction is a key element of clot retraction in hemostasis. During platelet activation, the cytoplasmic domains of the fibrinogen receptor integrin $\alpha_{IIb}\beta_3$ not only tightly associate with the underlying actin filaments, they also bind a variety of intracellular signaling molecules including talin, vinculin, zyxin, paxillin, filamin, and α-actinin, just to name a few of them. Formation of this large complex is regulated by a number of protein kinases, including tyrosine kinases and PI3-kinase, as discussed earlier. Clot retraction then occurs through the action of myosin that is attached to actin and contracts the actin filaments. Platelets express two isoforms of myosin (nonmuscle myosin IIA and IIB) which have a hexameric structure with two heavy chains, two 20-kDa light chains, and two 15-kDa light chains (Hartwig, 2006). Myosin activity depends on phosphorylation of the 20-kDa light chains which then undergo a conformational change and promote actin filament assembly. Two main pathways lead to MLC phosphorylation: calcium/calmodulin-dependent activation of MLC kinase and a second pathway depending on the small GTPase RhoA that activates Rho-kinase which in turn phosphorylates and inhibits MLC phosphatase. Since calcium increase is one of the first steps in platelet activation, calcium/calmodulin-dependent MLC kinase activation certainly occurs already in the early phase of platelet activation. This indicates that MLC activation not only leads to late-phase clot retraction but also is involved in early phase reactions like granule secretion or membrane receptor organization.

VI. INHIBITION OF THE PLATELET CYTOSKELETON BY CYCLIC NUCLEOTIDES

In blood vessels, platelets are continuously exposed to high shear rates that might induce platelet activation as shown by many investigators. Therefore, a tightly controlled system that prevents platelet activation at the wrong place is essential for hemostasis. The vascular endothelium provides such a system. Endothelial cells produce large quantities of mainly two endogenous vasodilators that strongly inhibit platelet activation: endothelium-derived relaxing factor (EDRF) and Prostaglandin I_2 (prostacyclin, PG-I_2) that elevate platelet cGMP and cAMP, respectively. The major target proteins for cyclic nucleotides in platelets are cGMP-dependent protein kinase (cGMP-PK) and cAMP-dependent protein kinase (cAMP-PK).

1. Inhibition of Intracellular Calcium Increase

As described earlier, platelet activation by most stimulatory agonists involves PLC activation and elevation of cytosolic calcium levels. Increase of intracellular cAMP or cGMP leads to inhibition of PLC activation, however, direct inhibition of PLC by cAMP-PK or cGMP-PK has not been shown. PLC catalyzes the conversion of PIP_2 into IP_3 and DAG. It is likely that cAMP and cGMP inhibit the PLC reaction by reduction of PIP_2 resynthesis (Ryningen *et al.*, 1998). Additional inhibition of intracellular calcium release occurs downstream of PLC. IP_3 receptors, which mediate the release of calcium from the dense tubular system and probably also the secondary store-related influx, are directly phosphorylated by cAMP-PK and cGMP-PK in human platelets, however, the role of this phosphorylation for regulation of intracellular calcium elevation is still not clear (Cavallini *et al.*, 1996; El-Daher *et al.*, 2000).

2. Inhibition of Cytoskeletal Reorganization

One of the major substrates of cAMP-PK and cGMP-PK is the cytoskeleton-associated protein vasodilator-stimulated phosphoprotein (VASP). VASP was found in most cell types to be located in focal adhesions, stress fibers, cell–cell contacts, and highly dynamic membrane regions (Reinhard *et al.*, 2001). In platelets it is present in very high concentrations of about 80,000 molecules/platelet (Eigenthaler *et al.*, 1992). Depending upon its phosphorylation status and the system investigated (*in vitro*, listeria, or mammalian cells), VASP is able to regulate actin polymerization and actin filament bundling (Reinhard *et al.*, 2001). *In vitro* experiments with VASP phosphorylated at Ser-157 diminished VASP binding to F-actin, and suppressed actin polymerization and actin filament bundling (Harbeck *et al.*, 2000). Platelets from VASP-deficient mice also showed reduced aggregation time in response to collagen, increased thrombin-induced activation of platelet fibrinogen receptor, and enhanced surface expression of P-selectin (Aszodi *et al.*, 1999; Hauser *et al.*, 1999). These observations suggest that phosphorylation downregulates the enhancing functions of VASP on actin polymerization. Still, the role of VASP in cytoskeletal reorganization and integrin activation is not completely understood and VASP phosphorylation may be only one of the effects of cAMP-PK and cGMP-PK action on the platelet cytoskeletal system.

Other known substrate proteins that might be involved in regulation of the cytoskeleton are filamin, αGPIbα, heat shock protein 27 (Hsp27) (Butt *et al.*, 2001), LIM and SH3 domain protein (LASP), and MLC kinase (Hathaway *et al.*, 1981; Nishikawa *et al.*, 1984; Butt *et al.*, 2003).

Filamin stabilizes the actin filaments and links them to the vWF receptor. Filamin possesses two phosphorylation sites for cAMP-PK (Jay *et al.*, 2004), suggesting that its phosphorylation by cAMP-PK in human platelets might stabilize the protein against proteolysis, thereby stabilizing the actin filament network and making it resistant to cytoskeletal reorganization (Chen and Stracher, 1989).

GPIbα, a subunit of the GPIb-IX-V complex, mediates adhesion to the subendothelium of damaged vessel walls. As described earlier, this complex is linked to underlying actin filaments by filamin (Fox, 1985). PG-E_1-induced phosphorylation of GPIbα led to inhibition of collagen-induced actin polymerization, an effect that was absent in platelets from patients with Bernard-Soulier syndrome, who lack the GPIb-IX-V complex (Fox and Berndt, 1989). Therefore, cAMP-PK-mediated GPIbα phosphorylation may also participate in inhibition of the agonist-induced cytoskeletal reorganization.

Another protein involved in the regulation of actin polymerization is Hsp27. Phosphorylation of this protein by the p38 MAPK-activated MAPKAPK-2 leads to stimulation of actin polymerization. In human platelets, Hsp27 is additionally phosphorylated by cGMP-PK and probably also by cAMP-PK (Butt *et al.*, 2001). *In vitro* experiments showed that cGMP-PK-phosphorylated Hsp27 reduced the stimulation of actin polymerization (Butt *et al.*, 2001).

The LASP that was cloned from human breast cancer cells was identified as a novel substrate of cGMP-PK and cAMP-PK in human platelets. LASP is an actin-binding protein, and Ser-146 phospho-LASP showed reduced binding affinity for F-actin. Although LASP function in human platelets is still not resolved, these data suggest that phosphorylation of LASP by cyclic nucleotide-depending protein kinases may be involved in organization of the cytoskeleton and platelet motility (Butt *et al.*, 2001).

Another effect of cyclic nucleotides in platelets is reduction of MLC phosphorylation. MLC phosphorylation promotes myosin filament formation and contractility, resulting in actin stress fiber formation and clustering of integrins into focal adhesions (Schoenwaelder and Burridge, 1999). The calcium/calmodulin-dependent MLC kinase is directly phosphorylated by cAMP-PK and cGMP-PK *in vitro* (Hathaway *et al.*, 1981; Nishikawa *et al.*, 1984), leading to a decreased affinity of this enzyme for calmodulin and subsequently decreased MLC phosphorylation. The involvement of cyclic nucleotide-regulated protein kinases in the regulation of myosin phosphatase in platelets has to be considered. In smooth muscle cells, direct interaction of cGMP-PK with myosin phosphatase was essential for MLC dephosphorylation (Surks *et al.*, 1999). Another possible mechanism for the regulation of MLC phosphorylation is through phosphorylation of the GTPase RhoA. RhoA stimulates Rho-kinase, which in turn phosphorylates myosin phosphatase, decreases its activity, and thereby leads to increase in MLC phosphorylation (Schoenwaelder and Burridge, 1999). In smooth muscle cells, cGMP-PK phosphorylates RhoA which inhibits RhoA-induced calcium sensitization of contraction and actin cytoskeleton organization (Sauzeau *et al.*, 2000). In platelets, these mechanisms of regulating MLC phosphorylation may also play a role in cyclic nucleotide-mediated inhibition of cytoskeletal rearrangement.

VII. SUMMARY

Platelets contain a complex cytoskeletal system consisting of a membrane cytoskeleton, a cytoplasmic core cytoskeleton, a microtubule system, and multiple regulatory signaling molecules that all together control the resting platelet discoid shape. During platelet adhesion, activation, and aggregation, these cytoskeletal structures are rearranged resulting in the formation of a variety of F-actin-based structures including filopodia and lamellipodia. Cytoskeletal remodeling is mediated through proteins that are part of the platelet cytoskeleton or in close contact to it. In turn, the cytoskeleton serves as a scaffold that binds and localizes signaling molecules at the right place and the right time to warrant platelet hemostatic function. Activatory signaling pathways initiate and control cytoskeletal stability and rearrangements, and include lipid kinases, G-protein–linked surface receptors, heterotrimeric and small GTP-binding proteins, and protein kinases. Still, many aspects and details of the platelet cytoskeleton are far from being understood. The availability of a variety of transgenic animal models as well as new approaches toward identifying the platelet proteom will open up chances to understand these complex interactions in the near future.

REFERENCES

Abe, K., Rossman, K. L., Liu, B., Ritola, K. D., Chiang, D., Campbell, S. L., Burridge, K., and Der, C. J. (2000). Vav2 is an activator of Cdc42, Rac1, and RhoA. *J. Biol. Chem.* **275,** 10141–10149.

Abrams, C. S. (2005). Intracellular signalling in platelets. *Curr. Opin. Hematol.* **12,** 401–405.

Achison, M., Elton, C. M., Hargreaves, P. G., Knight, C. G., Barnes, M. J., and Farndale, R. W. (2001). Integrin-independent tyrosine phosphorylation of p125(fak) in human platelets stimulated by collagen. *J. Biol. Chem.* **276,** 3167–3174.

Arias-Salgado, E. G., Lizano, S., Sarkar, S., Brugge, J. S., Ginsberg, M. H., and Shattil, S. J. (2003). Src kinase activation by direct interaction with the integrin beta cytoplasmic domain. *Proc. Natl. Acad. Sci. USA* **100,** 13298–13302.

Aszodi, A., Pfeifer, A., Ahmad, M., Glauner, M., Zhou, X. H., Ny, L., Andersson, K. E., Kehrel, B., Offermanns, S., and Fassler, R. (1999). The vasodilator-stimulated phosphoprotein (VASP) is involved in cGMP- and cAMP-mediated inhibition of agonist-induced platelet aggregation, but is dispensable for smooth muscle function. *EMBO J.* **18,** 37–48.

Barkalow, K. L., Italiano, J. E., Jr., Chou, D. E., Matsuoka, Y., Bennett, V., and Hartwig, J. H. (2003). Alpha-adducin dissociates from F-actin and spectrin during platelet activation. *J. Cell. Biol.* **161,** 557–570.

Bertagnolli, M. E., Hudson, L. A., and Stetsenko, G. Y. (1999). Selective association of the tyrosine kinases Src, Fyn, and Lyn with integrin-rich actin cytoskeletons of activated, nonaggregated platelets. *Biochem. Biophys. Res. Commun.* **260,** 790–798.

Billadeau, D. D., Brumbaugh, K. M., Dick, C. J., Schoon, R. A., Bustelo, X. R., and Leibson, P. J. (1998). The Vav-Rac1 pathway in cytotoxic lymphocytes regulates the generation of cell-mediated killing. *J. Exp. Med.* **188,** 549–559.

Buensuceso, C. S., Obergfell, A., Soriani, A., Eto, K., Kiosses, W. B., Arias-Salgado, E. G., Kawakami, T., and Shattil, S. J. (2005). Regulation of outside-in signalling in platelets by integrin-associated protein kinase C beta. *J. Biol. Chem.* **280,** 644–653.

Bustelo, X. R. (2000). Regulatory and signalling properties of the Vav family. *Mol Cell Biol.* **20,** 1461–1477.

Butt, E., Immler, D., Meyer, H. E., Kotlyarov, A., Laass, K., and Gaestel, M. (2001). Heat shock protein 27 is a substrate of cGMP-dependent protein kinase in intact human platelets: Phosphorylation-induced actin polymerization caused by HSP27 mutants. *J. Biol. Chem.* **276,** 7108–7113.

Butt, E., Gambaryan, S., Gottfert, N., Galler, A., Marcus, K., and Meyer, H. E. (2003). Actin binding of human LIM and SH3 protein is regulated by cGMP- and cAMP-dependent protein kinase phosphorylation on serine 146. *J. Biol. Chem.* **278,** 15601–15607.

Canobbio, I., Lova, P., Sinigaglia, F., Balduini, C., and Torti, M. (2002). Proline-rich tyrosine kinase 2 and focal adhesion kinase are involved in different phases of platelet activation by vWF. *Thromb. Haemost.* **87,** 509–517.

Canobbio, I., Balduini, C., and Torti, M. (2004). Signalling through the platelet glycoprotein Ib-V-IX complex. *Cell Signal* **16,** 1329–1344.

Cavallini, L., Coassin, M., Borean, A., and Alexandre, A. (1996). Prostacyclin and sodium nitroprusside inhibit the activity of the platelet inositol 1,4,5-trisphosphate receptor and promote its phosphorylation. *J. Biol. Chem.* **271,** 5545–5551.

Chen, M. and Stracher, A. (1989). In situ phosphorylation of platelet actin-binding protein by cAMP-dependent protein kinase stabilizes it against proteolysis by calpain. *J. Biol. Chem.* **264,** 14282–14289.

Chung, J., Gao, A. G., and Frazier, W. A. (1997). Thrombspondin acts via integrin-associated protein to activate the platelet integrin alphaIIbbeta3. *J. Biol. Chem.* **272,** 14740–14746.

Cichowski, K., Brugge, J. S., and Brass, L. F. (1996). Thrombin receptor activation and integrin engagement stimulate tyrosine phosphorylation of the proto-oncogene product, p95vav, in platelets. *J. Biol. Chem.* **271,** 7544–7550.

Clements, J. L., Lee, J. R., Gross, B., Yang, B., Olson, J. D., Sandra, A., Watson, S. P., Lentz, S. R., and Koretzky, G. A. (1999). Fetal hemorrhage and platelet dysfunction in SLP-76-deficient mice. *J. Clin. Invest.* **103,** 19–25.

Cobb, B. S., Schaller, M. D., Leu, T. H., and Parsons, J. T. (1994). Stable association of pp60src and pp59fyn with the focal adhesion-associated protein tyrosine kinase, pp125FAK. *Mol. Cell Biol.* **14,** 147–155.

Crabos, M., Imber, R., Woodtli, T., Fabbro, D., and Erne, P. (1991). Different translocation of three distinct PKC isoforms with tumor-promoting phorbol ester in human platelets. *Biochem. Biophys. Res. Commun.* **178,** 878–883.

Dandona, P., Thusu, K., Khurana, U., Love, J., Aljada, A., and Mousa, S. (1996). Calcium, calmodulin and protein kinase C dependence of platelet shape change. *Thromb. Res.* **81,** 163–175.

Eigenthaler, M., Nolte, C., Halbrugge, M., and Walter, U. (1992). Concentration and regulation of cyclic nucleotides, cyclic-nucleotide-dependent protein kinases and one of their major substrates in human platelets. Estimating the rate of cAMP-regulated and cGMP-regulated protein phosphorylation in intact cells. *Eur. J. Biochem.* **205,** 471–481.

El-Daher, S. S., Patel, Y., Siddiqua, A., Hassock, S., Edmunds, S., Maddison, B., Patel, G., Goulding, D., Lupu, F., Wojcikiewicz, R. J., and Authi, K. S. (2000). Distinct localization and function of (1,4,5)IP(3) receptor subtypes and the (1,3,4,5)IP(4) receptor GAP1(IP4BP) in highly purified human platelet membranes. *Blood* **95,** 3412–3422.

Falati, S., Edmead, C. E., and Poole, A. W. (1999). Glycoprotein Ib-V-IX, a receptor for von Willebrand factor, couples physically and functionally to the Fc receptor gamma-chain, Fyn, and Lyn to activate human platelets. *Blood* **94,** 1648–1656.

Falet, H., Chang, G., Brohard-Bohn, B., Rendu, F., and Hartwig, J. H. (2005). Integrin alpha(IIb)beta3 signals lead cofilin to accelerate platelet actin dynamics. *Am. J. Physiol. Cell. Physiol.* **289,** C819–C825.

Fitzgerald, M. L. and Reed, G. L. (1999). Rab6 is phosphorylated in thrombin-activated platelets by a protein kinase C-dependent mechanism: Effects on GTP/GDP binding and cellular distribution. *Biochem. J.* **342**(Pt. 2), 353–360.

Fox, J. E. (1985). Identification of actin-binding protein as the protein linking the membrane skeleton to glycoproteins on platelet plasma membranes. *J. Biol. Chem.* **260,** 11970–11977.

Fox, J. E. and Berndt, M. C. (1989). Cyclic AMP-dependent phosphorylation of glycoprotein Ib inhibits collagen-induced polymerization of actin in platelets. *J. Biol. Chem.* **264,** 9520–9526.

Freson, K., De Vos, R., Wittevrongel, C., Thys, C., Defoor, J., Vanhees, L., Vermylen, J., Peerlinck, K., and Van Geet, C. (2005). The TUBB1 Q43P functional polymorphism reduces the risk of cardiovascular disease in men by modulating platelet function and structure. *Blood* **106,** 2356–2362.

Gibbins, J. M. (2004). Platelet adhesion signalling and the regulation of thrombus formation. *J. Cell. Sci.* **117,** 3415–3425.

Gilligan, D. M., Sarid, R., and Weese, J. (2002). Adducin in platelets: Activation-induced phosphorylation by PKC and proteolysis by calpain. *Blood* **99,** 2418–2426.

Gluck, U. and Ben-Ze'ev, A. (1994). Modulation of alpha-actinin levels affects cell motility and confers tumorigenicity on 3T3 cells. *J. Cell. Sci.* **107**(Pt. 7), 1773–1782.

Gorlin, J. B., Yamin, R., Egan, S., Stewart, M., Stossel, T. P., Kwiatkowski, D. J., and Hartwig, J. H. (1990). Human endothelial actin-binding protein (ABP-280, nonmuscle filamin): A molecular leaf spring. *J. Cell. Biol.* **111,** 1089–1105.

Haimovich, B., Kaneshiki, N., and Ji, P. (1996). Protein kinase C regulates tyrosine phosphorylation of pp125FAK in platelets adherent to fibrinogen. *Blood* **87,** 152–161.

Harbeck, B., Huttelmaier, S., Schluter, K., Jockusch, B. M., and Illenberger, S. (2000). Phosphorylation of the vasodilator-stimulated phosphoprotein regulates its interaction with actin. *J. Biol. Chem.* **275,** 30817–30825.

Hartwig, J. H. (2006). The platelet: Form and function. *Semin. Hematol.* **43,** S94–S100.

Hathaway, D. R., Eaton, C. R., and Adelstein, R. S. (1981). Regulation of human platelet myosin light chain kinase by the catalytic subunit of cyclic AMP-dependent protein kinase. *Nature* **291,** 252–256.

Hauser, W., Knobeloch, K. P., Eigenthaler, M., Gambaryan, S., Krenn, V., Geiger, J., Glazova, M., Rohde, E., Horak, I., Walter, U., and Zimmer, M. (1999). Megakaryocyte hyperplasia and enhanced agonist-induced platelet activation in vasodilator-stimulated phosphoprotein knockout mice. *Proc. Natl. Acad. Sci. USA* **96,** 8120–8125.

Jay, D., Garcia, E. J., and de la Luz Ibarra, M. (2004). In situ determination of a PKA phosphorylation site in the C-terminal region of filamin. *Mol. Cell. Biochem.* **260,** 49–53.

Kiuru, S., Javela, K., Somer, H., and Kekomaki, R. (2000). Altered platelet shape change in hereditary gelsolin Asp187Asn-related amyloidosis. *Thromb. Haemost.* **83,** 491–495.

Kos, C. H., Le, T. C., Sinha, S., Henderson, J. M., Kim, S. H., Sugimoto, H., Kalluri, R., Gerszten, R. E., and Pollak, M. R. (2003). Mice deficient in alpha-actinin-4 have severe glomerular disease. *J. Clin. Invest.* **111,** 1683–1690.

Kralisz, U. and Cierniewski, C. S. (1998a). Association of pp60c-src with alpha IIb beta 3 in resting platelets. *Biochem. Mol. Biol. Int.* **45,** 735–743.

Kralisz, U. and Cierniewski, C. S. (1998b). Tyrosine phosphorylation events during different stages of collagen-platelet activation. *Biochim. Biophys. Acta* **1405,** 128–138.

Krause, M., Dent, E. W., Bear, J. E., Loureiro, J. J., and Gertler, F. B. (2003). Ena/VASP proteins: Regulators of the actin cytoskeleton and cell migration. *Annu. Rev. Cell Dev. Biol.* **19,** 541–564.

Larsson, C. (2006). Protein kinase C and the regulation of the actin cytoskeleton. *Cell Signal* **18,** 276–284.

Lyman, S., Gilmore, A., Burridge, K., Gidwitz, S., and White, G. C., II (1997). Integrin-mediated activation of focal adhesion kinase is independent of focal adhesion formation or integrin activation. Studies with activated and inhibitory beta3 cytoplasmic domain mutants. *J. Biol. Chem.* **272,** 22538–22547.

Mahoney, N. M., Janmey, P. A., and Almo, S. C. (1997). Structure of the profilin-poly-L-proline complex involved in morphogenesis and cytoskeletal regulation. *Nat. Struct. Biol.* **4,** 953–960.

Matsuoka, Y., Li, X., and Bennett, V. (2000). Adducin: Structure, function and regulation. *Cell. Mol. Life Sci.* **57,** 884–895.

McCarty, O. J., Larson, M. K., Auger, J. M., Kalia, N., Atkinson, B. T., Pearce, A. C., Ruf, S., Henderson, R. B., Tybulewicz, V. L., Machesky, L. M., and Watson, S. P. (2005). Rac1 is essential for platelet lamellipodia formation and aggregate stability under flow. *J. Biol. Chem.* **280,** 39474–39484.

Miki, H., Sasaki, T., Takai, Y., and Takenawa, T. (1998). Induction of filopodium formation by a WASP-related actin-depolymerizing protein N-WASP. *Nature* **391,** 93–96.

Miyakawa, Y., Oda, A., Druker, B. J., Ozaki, K., Handa, M., Ohashi, H., and Ikeda, Y. (1997). Thrombopoietin and thrombin induce tyrosine phosphorylation of Vav in human blood platelets. *Blood* **89,** 2789–2798.

Muro, A. F., Marro, M. L., Gajovic, S., Porro, F., Luzzatto, L., and Baralle, F. E. (2000). Mild spherocytic hereditary elliptocytosis and altered levels of alpha- and gamma-adducins in beta-adducin-deficient mice. *Blood* **95,** 3978–3985.

Murphy, C. T. and Westwick, J. (1992). Selective inhibition of protein kinase C. Effect on platelet-activating-factor-induced platelet functional responses. *Biochem. J.* **283**(Pt. 1), 159–164.

Murugappan, S., Tuluc, F., Dorsam, R. T., Shankar, H., and Kunapuli, S. P. (2004). Differential role of protein kinase C delta isoform in agonist-induced dense granule secretion in human platelets. *J. Biol. Chem.* **279,** 2360–2367.

Murugappan, S., Shankar, H., Bhamidipati, S., Dorsam, R. T., Jin, J., and Kunapuli, S. P. (2005). Molecular mechanism and functional implications of thrombin-mediated tyrosine phosphorylation of PKCdelta in platelets. *Blood* **106,** 550–557.

Nakamura, F., Pudas, R., Heikkinen, O., Permi, P., Kilpelainen, I., Munday, A. D., Hartwig, J. H., Stossel, T. P., and Ylanne, J. (2006). The structure of the GPIb-filamin A complex. *Blood* **107,** 1925–1932.

Negrescu, E. V. and Siess, W. (1996). Dissociation of the alphaIIbbeta3-integrin by EGTA stimulates the tyrosine kinase pp72(syk) without inducing platelet activation. *J. Biol. Chem.* **271,** 26547–26553.

Nishikawa, M., de Lanerolle, P., Lincoln, T. M., and Adelstein, R. S. (1984). Phosphorylation of mammalian myosin light chain kinases by the catalytic subunit of cyclic AMP-dependent protein kinase and by cyclic GMP-dependent protein kinase. *J. Biol. Chem.* **259,** 8429–8436.

Nobes, C. D. and Hall, A. (1995). Rho, rac, and cdc42 GTPases regulate the assembly of multimolecular focal complexes associated with actin stress fibers, lamellipodia, and filopodia. *Cell* **81,** 53–62.

Nonne, C., Lenain, N., Hechler, B., Mangin, P., Cazenave, J. P., Gachet, C., and Lanza, F. (2005). Importance of platelet phospholipase Cgamma2 signalling in arterial thrombosis as a function of lesion severity. *Arterioscler. Thromb. Vasc. Biol.* **25,** 1293–1298.

Obergfell, A., Eto, K., Mocsai, A., Buensuceso, C., Moores, S. L., Brugge, J. S., Lowell, C. A., and Shattil, S. J. (2002). Coordinate interactions of Csk, Src, and Syk kinases with [alpha]IIb[beta]3 initiate integrin signalling to the cytoskeleton. *J. Cell. Biol.* **157,** 265–275.

Ozaki, Y. (1999). Platelet activation mediated through membrane glycoproteins: Involvement of tyrosine kinases. *Rinsho. Byori.* **47,** 799–808.

Ozaki, Y., Jinnai, Y., Yatomi, Y., and Kume, S. (1993). Protein kinase C inhibitors suppress disc-sphere changes of human platelets, as assessed with the shape-change parameter. *Eur. J. Pharmacol.* **235,** 255–265.

Ozaki, Y., Qi, R., Satoh, K., Asazuma, N., and Yatomi, Y. (2000). Platelet activation mediated through membrane glycoproteins: Involvement of tyrosine kinases. *Semin. Thromb. Hemost.* **26,** 47–51.

Ozaki, Y., Asazuma, N., Suzuki-Inoue, K., and Berndt, M. C. (2005). Platelet GPIb-IX-V-dependent signalling. *J. Thromb. Haemost.* **3,** 1745–1751.

Pandey, D., Goyal, P., Bamburg, J. R., and Siess, W. (2006). Regulation of LIM-kinase 1 and cofilin in thrombin-stimulated platelets. *Blood* **107,** 575–583.

Pearce, A. C., Wilde, J. I., Doody, G. M., Best, D., Inoue, O., Vigorito, E., Tybulewicz, V. L., Turner, M., and Watson, S. P. (2002). Vav1, but not Vav2, contributes to platelet aggregation by CRP and thrombin, but neither is required for regulation of phospholipase C. *Blood* **100,** 3561–3569.

Pearce, A. C., Senis, Y. A., Billadeau, D. D., Turner, M., Watson, S. P., and Vigorito, E. (2004). Vav1 and vav3 have critical but redundant roles in mediating platelet activation by collagen. *J. Biol. Chem.* **279,** 53955–53962.

Poole, A., Gibbins, J. M., Turner, M., van Vugt, M. J., van de Winkel, J. G., Saito, T., Tybulewicz, V. L., and Watson, S. P. (1997). The Fc receptor gamma-chain and the tyrosine kinase Syk are essential for activation of mouse platelets by collagen. *EMBO J.* **16,** 2333–2341.

Pula, G., Crosby, D., Baker, J., and Poole, A. W. (2005). Functional interaction of protein kinase Calpha with the tyrosine kinases Syk and Src in human platelets. *J. Biol. Chem.* **280,** 7194–7205.

Rathore, V., Wang, D., Newman, D. K., and Newman, P. J. (2004). Phospholipase Cgamma2 contributes to stable thrombus formation on VWF. *FEBS Lett.* **573,** 26–30.

Reinhard, M., Jarchau, T., and Walter, U. (2001). Actin-based motility: Stop and go with Ena/VASP proteins. *Trends Biochem. Sci.* **26,** 243–249.

Rosado, J. A., Graves, D., and Sage, S. O. (2000). Tyrosine kinases activate store-mediated Ca^{2+} entry in human platelets through the reorganization of the actin cytoskeleton. *Biochem. J.* **351**(Pt. 2), 429–437.

Ryningen, A., Olav Jensen, B., and Holmsen, H. (1998). Elevation of cyclic AMP decreases phosphoinositide turnover and inhibits thrombin-induced secretion in human platelets. *Biochim. Biophys. Acta* **1394,** 235–248.

Sada, K., Minami, Y., and Yamamura, H. (1997). Relocation of Syk protein-tyrosine kinase to the actin filament network and subsequent association with Fak. *Eur. J. Biochem.* **248,** 827–833.

Satoh, K., Asazuma, N., Yatomi, Y., Fujimura, Y., Miura, S., Titani, K., and Ozaki, Y. (2000). Activation of protein-tyrosine kinase pathways in human platelets stimulated with the A1 domain of von Willebrand factor. *Platelets* **11,** 171–176.

Sauzeau, V., Le Jeune, H., Cario-Toumaniantz, C., Smolenski, A., Lohmann, S. M., Bertoglio, J., Chardin, P., Pacaud, P., and Loirand, G. (2000). Cyclic GMP-dependent protein kinase signalling pathway inhibits RhoA-induced Ca2+ sensitization of contraction in vascular smooth muscle. *J. Biol. Chem.* **275,** 21722–21729.

Schaller, M. D., Hildebrand, J. D., Shannon, J. D., Fox, J. W., Vines, R. R., and Parsons, J. T. (1994). Autophosphorylation of the focal adhesion kinase, pp125FAK, directs SH2-dependent binding of pp60src. *Mol. Cell. Biol.* **14,** 1680–1688.

Schoenwaelder, S. M. and Burridge, K. (1999). Bidirectional signalling between the cytoskeleton and integrins. *Curr. Opin. Cell. Biol.* **11,** 274–286.

Schuebel, K. E., Movilla, N., Rosa, J. L., and Bustelo, X. R. (1998). Phosphorylation-dependent and constitutive activation of Rho proteins by wild-type and oncogenic Vav-2. *EMBO J.* **17,** 6608–6621.

Schwer, H. D., Lecine, P., Tiwari, S., Italiano, J. E., Jr., Hartwig, J. H., and Shivdasani, R. A. (2001). A lineage-restricted and divergent beta-tubulin isoform is essential for the biogenesis, structure and function of blood platelets. *Curr. Biol.* **11,** 579–586.

Shattil, S. J. (2005). Integrins and Src: Dynamic duo of adhesion signalling. *Trends Cell Biol.* **15,** 399–403.

Stenberg, P. E., Pestina, T. I., Barrie, R. J., and Jackson, C. W. (1997). The Src family kinases, Fgr, Fyn, Lck, and Lyn, colocalize with coated membranes in platelets. *Blood* **89,** 2384–2393.

Sun, H. Q., Yamamoto, M., Mejillano, M., and Yin, H. L. (1999). Gelsolin, a multifunctional actin regulatory protein. *J. Biol. Chem.* **274,** 33179–33182.

Surks, H. K., Mochizuki, N., Kasai, Y., Georgescu, S. P., Tang, K. M., Ito, M., Lincoln, T. M., and Mendelsohn, M. E. (1999). Regulation of myosin phosphatase by a specific interaction with cGMP-dependent protein kinase Ialpha. *Science* **286,** 1583–1587.

Tabuchi, A., Yoshioka, A., Higashi, T., Shirakawa, R., Nishioka, H., Kita, T., and Horiuchi, H. (2003). Direct demonstration of involvement of protein kinase Calpha in the Ca^{2+}-induced platelet aggregation. *J. Biol. Chem.* **278,** 26374–26379.

Takafuta, T., Wu, G., Murphy, G. F., and Shapiro, S. S. (1998). Human beta-filamin is a new protein that interacts with the cytoplasmic tail of glycoprotein Ibalpha. *J. Biol. Chem.* **273,** 17531–17538.

Thirion, C., Stucka, R., Mendel, B., Gruhler, A., Jaksch, M., Nowak, K. J., Binz, N., Laing, N. G., and Lochmuller, H. (2001). Characterization of human muscle type cofilin (CFL2) in normal and regenerating muscle. *Eur. J. Biochem.* **268,** 3473–3482.

Thomas, J. W., Ellis, B., Boerner, R. J., Knight, W. B., White, G. C., II, and Schaller, M. D. (1998). SH2- and SH3-mediated interactions between focal adhesion kinase and Src. *J. Biol. Chem.* **273,** 577–583.

Wandersee, N. J., Roesch, A. N., Hamblen, N. R., de Moes, J., van der Valk, M. A., Bronson, R. T., Gimm, J. A., Mohandas, N., Demant, P., and Barker, J. E. (2001). Defective spectrin integrity and neonatal thrombosis in the first mouse model for severe hereditary elliptocytosis. *Blood* **97,** 543–550.

Ware, J., Russell, S., and Ruggeri, Z. M. (2000). Generation and rescue of a murine model of platelet dysfunction: The Bernard-Soulier syndrome. *Proc. Natl. Acad. Sci. USA* **97,** 2803–2808.

Watanabe, N., Nakajima, H., Suzuki, H., Oda, A., Matsubara, Y., Moroi, M., Terauchi, Y., Kadowaki, T., Koyasu, S., Ikeda, Y., and Handa, M. (2003). Functional phenotype of phosphoinositide 3-kinase p85alpha-null platelets characterized by an impaired response to GP VI stimulation. *Blood* **102,** 541–548.

Watson, S. P., Auger, J. M., McCarty, O. J., and Pearce, A. C. (2005). GPVI and integrin alphaIIb beta3 signalling in platelets. *J. Thromb. Haemost.* **3,** 1752–1762.

Witke, W., Sharpe, A. H., Hartwig, J. H., Azuma, T., Stossel, T. P., and Kwiatkowski, D. J. (1995). Hemostatic, inflammatory, and fibroblast responses are blunted in mice lacking gelsolin. *Cell* **81,** 41–51.

Wonerow, P., Pearce, A. C., Vaux, D. J., and Watson, S. P. (2003). A critical role for phospholipase Cgamma2 in alphaIIbbeta3-mediated platelet spreading. *J. Biol. Chem.* **278,** 37520–37529.

Wu, Y., Asazuma, N., Satoh, K., Yatomi, Y., Takafuta, T., Berndt, M. C., and Ozaki, Y. (2003). Interaction between von Willebrand factor and glycoprotein Ib activates Src kinase in human platelets: Role of phosphoinositide 3-kinase. *Blood* **101,** 3469–3476.

Chapter 2

The Actin Cytoskeleton in the Apical Domain of Epithelial Cells

R. Brian Doctor

Department of Medicine, Division of Gastroenterology, University of Colorado Health Sciences Center, Denver, Colorado 80262

Advances in Molecular and Cell Biology, Vol. 37

ISBN: 0-444-52868-7
DOI: 10.1016/S1569-2558(06)37002-6

I. INTRODUCTION

Epithelial cells line the cavities and surfaces of the body. In organs such as the skin, epithelial cells serve primarily as a barrier with comparatively little movement of water, ions, or solutes across the epithelial layer. In contrast, in organs such as the intestine, kidney, and liver, epithelial cells orchestrate the vectorial flux of specific ions, solutes, or water across the epithelial layer. For the bulk movement of water and ions, epithelia such as renal proximal tubules draw water across a "leaky" tight junction and between cells while dragging other ions and solutes through this paracellular pathway. The more selective movement of specific ions or solutes generally occurs in a transcellular fashion. Specific transport or channel proteins are localized in the apical and basolateral membranes to permit this vectorial movement through epithelial cells. Epithelial cells have evolved a number of distinct molecular mechanisms to regulate the distribution and activities of the transport and channel proteins and the actin cytoskeleton is at the core of many of these molecular mechanisms.

The goal of this chapter is to detail how the actin cytoskeleton impacts transport functions at the apical membrane of epithelial cells and contributes to specific disease states. Within epithelial cells, actin filaments are distributed predominantly at the plasma membrane and are highest in concentration at the apical membrane domain (Fig. 1). The last decade has seen the discovery of how the fundamental properties of the actin cytoskeleton are harnessed by cells to regulate cellular functions. This chapter is presented in four sections. The first section briefly highlights poignant molecular and biochemical features of actin and its associated proteins. The second section reviews the structural organization of the actin cytoskeleton in the apical domain of epithelial cells. The third section describes five distinct roles the actin cytoskeleton plays in moderating transport activities at the apical membrane. Finally, the fourth section highlights the pivotal role that the actin cytoskeleton plays in the molecular pathophysiology of specific epithelial disease states.

II. PROPERTIES OF ACTIN AND ACTIN-ASSOCIATED PROTEINS

A. Cellular Cytoskeleton

The eukaryotic cytoskeleton is composed of three distinct types of filaments: intermediate filaments, microtubules, and microfilaments. While these three filamentous systems are interactive, each has discernible roles in eukaryotic cells. In epithelial cells, intermediate filaments form stable, elastic cables that course through the body of cells while binding at sites of cell–cell (i.e., desmosome) and cell–matrix (i.e., hemidesmosome) contact. The intermediate filament system confers mechanical stability both to the individual cells and to epithelial layers. Microtubules are dynamic polymers capable of undergoing rapid polymerization and depolymerization. Emanating from the supranuclear microtubule-organizing center, microtubules, microtubule-associated

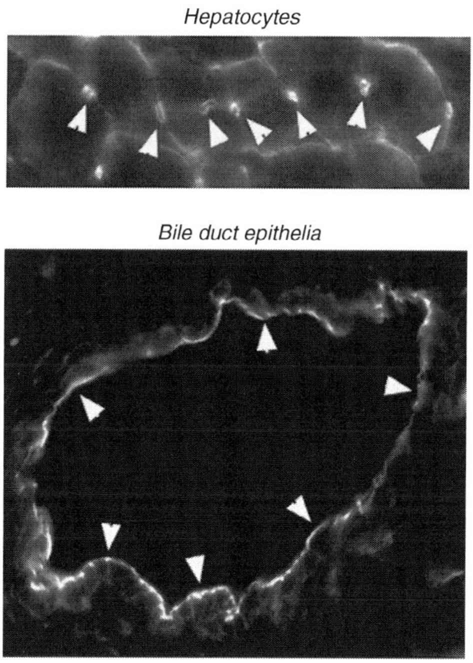

Figure 1. Filamentous actin concentrates at the plasma membrane domains. Rhodamine-phalloidin staining (binds filamentous actin) of hepatocytes and cholangiocytes in liver sections shows filamentous actin at the cell periphery and is concentrated at the apical domain (upper panel). In cords of hepatocytes, the sinusoidal (basal), lateral, and apical membranes are stained; staining of the canalicular (apical) membranes is particularly intense (arrowheads) (lower panel). Similarly, in cholangiocytes lining intrahepatic bile duct, modest staining is observed along the basolateral membranes and intense staining is seen along the apical domain (arrowheads).

proteins, and microtubular motor proteins allow for the targeted delivery of vesicles, organelles, and proteins throughout the cell.

In contrast to intermediate filaments and microtubules, actin filaments in epithelial cells are concentrated at the plasma membrane where they provide structural support and shape to the membrane, drive vesicle movement to and from the membrane, restrict and retain proteins within the membrane, and regulate the function of integral membrane proteins. These actin-dependent activities are pivotal to regulating transport activities at the apical membrane of epithelial cells and will be exemplified later (Section III).

B. Individual Epithelial Cells Can Coexpress *β*- and *γ*-Actin Genes

In mammals, distinct genes encode for α-, β-, and γ-actin. α-Actin genes are expressed primarily within smooth, cardiac and striated muscle cells while β- and γ-actin are more broadly expressed, including expression in epithelial cells. β-Actin and γ-actin share 98% amino acid sequence identity and have *in vitro* biochemical characteristics that are markedly similar. Despite these similarities, specific epithelial cell types will express both β- and γ-actin and likely utilize the two forms of actin for distinct functions. In support of this hypothesis, β- and γ-actin are concurrently expressed in gastric parietal cells but γ-actin is distributed predominantly along the basolateral membranes while β-actin is concentrated around the canalicular membranes (Yao *et al.*, 1995). Gastric parietal cells sequester H^+-ATPase proton pump proteins in subapical vesicles and, in response to feeding, translocate and fuse the H^+-ATPase–laden vesicles with the apical membrane. While it remains to be definitively demonstrated, β-actin is distributed in a pattern to facilitate H^+-ATPase distribution and activity.

C. Actin-Associated Proteins Harness the Potential of Actin Filaments

The utility of actin stems first from its innate capacity to form polymers and second from diverse actin-associated proteins that harness the potential of actin polymerization to perform useful work within cells. Actin polymerizes in a head-to-tail fashion, binds and hydrolyzes ATP, and has a greater ATPase activity when it is within an actin filament. Further, the binding affinity of ATP-actin monomers is greater for actin polymers with ATP bound rather than ADP and the on–off rates are greater at the head (i.e., plus end) versus the tail (i.e., minus end) of the filament. As a consequence of these characteristics, under physiologic conditions, actin polymerizes in a polarized fashion with monomer addition occurring predominantly at the "barbed" or (+)-end of actin filaments.

An individual actin filament, however, has little functional significance. To use the potential of actin polymerization and actin filaments, cells have evolved specific actin-associated proteins that nucleate *de novo* filaments, moderate the rates of polymerization and depolymerization, transduce filament polymerization into mechanical work, stabilize the filaments, utilize stabilized filaments as structural elements, and move along the length of filaments (Table I). Described later, many of these actin-associated proteins are present within the apical domain of epithelial cells.

D. Arp2/3 Complex Nucleates Actin Filaments

The rate-limiting step in actin polymerization *in vitro* is the nucleation of three actin monomers to seed filament growth. The molecular mechanisms permitting and directing actin nucleation within cells was an enduring mystery until a complex of seven proteins, termed the Arp2/3 complex, was discovered (Machesky *et al.*, 1994; Millard *et al.*, 2004). The Arp2/3 complex utilizes actin-related proteins (Arps) to mimic a nucleated actin site and initiate the formation of new actin filaments (Mullins and Pollard, 1999). Not surprisingly, filament nucleation is a tightly regulated process and cells have evolved a number of mechanisms to limit, direct, and coordinate the

Table I
Actin-Associated Proteins and Their Activities

Activity	Protein	Comments
Nucleation	Arp2/3 complex	Critical for *de novo* filament formation
Monomer binding	b-Thymosin	Actin monomer buffer
	Profilin	Induces accelerated polymerization
Capping	CapZ	Regulates availability of (+) ends
	Adducin	Recruits spectrin to capped ends
Severing	ADF/cofilin	ADP-actin severing
	Gelsolin	Actin severing and capping
Lateral binding	Tropomyosin	Filament stabilization
Cross-linking	Villin	Actin-bundling and actin-severing activity
	a-Actinin	Parallel actin filament bundling
	Spectrin	Orthogonal actin filament cross-linking
	Filamin	Actin gelation and membrane attachment
Tethering	ERM proteins	Links actin, membrane, and regulatory proteins
	Ankyrin	Site-specific localization of membrane proteins
Motor	Myosins	Family of actin mechanoenzymes

de novo nucleation of new actin filaments. Among the best understood is the Wiskott–Aldrich syndrome protein (WASP). When activated to unfold at specific membrane sites, WASP then recruits and activates the Arp2/3 complex to initiate actin polymerization. Directed initiation of actin filament formation is important in a number of cellular processes, including Arp2/3-dependent nucleation that drives the motility of migrating cells. Nonmotile epithelial cells use the same Arp2/3-dependent nucleation and polymerization to drive the budding and internalization of endosomes from the apical membrane of epithelial cells (Section III). Further, specific bacteria can hijack this actin nucleation/polymerization machinery to assist in driving their internalization, cytoplasmic motility, and invasion of neighboring cells (Section IV).

E. Disassembly of Filaments Is an Essential Part of Actin Dynamics

Equally important as the regulated polymerization of actin filaments is the coordinated disassembly of actin filaments. Filament disassembly can occur via severing filaments along the length of a filament or by accelerating the removal of monomers from the "minus" end of filaments. Disassembly is important in remodeling of filament networks that are no longer required or are an impediment to cellular events. Their disassembly also supplies monomers needed for actin polymerization at other cellular sites.

The gelsolin and actin-depolymerizing factor (ADF)/cofilin families of proteins are two well-described actin filament-severing proteins with distinct modes of severing and regulation. Gelsolin-related proteins are Ca^{2+}-regulated actin-severing proteins that bind along the lateral aspect of filaments, cleaves the filament, and caps the "plus" end of the resultant product. Proteins in the gelsolin family play a pivotal role in providing access to or from the apical membrane of epithelial cells and remodeling of microvillar actin filaments in disease states (Sections III and IV).

Distinct from the gelsolin family, ADF/cofilin is a family of actin filament disassembly proteins with an essential role in cellular physiology (Bamburg, 1999). ADF/cofilin binds either laterally onto actin filaments or to the "minus" end of filaments. When activated, ADF/cofilin "twists" the actin filament to increase the molecular strain and facilitate filament severing (McGough *et al.*, 1997). The severing activity of ADF/cofilin is regulated by multiple means including inhibition by phosphorylation on Ser-3, low intracellular pH, and elevated levels of $PI(4,5)P_2$. ADF/cofilin plays a key role in degrading outdated or inefficient actin filaments and contributes to the remodeling of microvillar actin filaments in ischemic epithelial cells (Sections III and IV).

F. Filaments Can be Stabilized from Dynamic Turnover

Actin-associated proteins that moderate actin nucleation, polymerization, depolymerization, or severing moderate the dynamic aspects of actin filaments to perform cellular functions. There are also a number of cellular functions that require stabilized populations of actin filaments. These, in general, consist of "capping" proteins to block the addition of monomers to the (+)-end or removal of monomers from the (−)-end and lateral binding proteins to sterically hinder access of severing proteins along the length of actin filaments. Tropomyosin, the most noted lateral binding protein, spans across seven adjacent actin subunits and impedes the access and activities of gelsolin, ADF/cofilin, and Arp2/3 (Sections II and IV).

G. Stabilized Filaments Are Cross-Linked into Structural Lattices

To enhance the structural characteristics of individual actin filaments, "structural" filaments are cross-linked into an orthogonal meshwork or bound together into parallel bundles. Cross-linking actin filaments into a cross-linked mesh transforms intersecting filaments into a cohesive gel. This meshwork lies adjacent to the plasma membrane and is termed "cortical actin." Two distinct orthogonal cross-linking proteins include filamin and spectrin. Filamin is an extended protein dimer with an actin-binding motif on its N-terminus, an extended rod domain composed of multiple repeats, and a C-terminus containing a dimerization motif. Filamin extends to 160 nm in a flexible V-shaped dimer with ends of the V interacting with actin filaments. Found in the terminal web below the apical membrane of epithelial cells, α- and β-spectrin chains form antiparallel dimers and dimer pairs associate in a head-to-head direction to form spectrin tetramers that extend up to 200 nm in length. The tails of the spectrin tetramers associate with adducin, an actin-binding protein that mediates the spectrin tetramer cross-linking of actin filaments.

In contrast to the orthogonally linked cortical actin meshwork, actin filaments are also cross-linked into parallel actin filament bundles. "Looser" parallel actin bundles are generally cross-linked by dimers that contain a single actin-binding domain and a rod domain that is capable of self-association. α-Actinin, a member of the spectrin family, is the most noted example of an actin-bundling dimer. "Tighter" actin bundles are formed by single proteins with two actin-binding motifs. Villin, fimbrin, and espin are noted examples of "tight" actin-bundling proteins and are distributed within the apical microvilli of epithelial cells.

H. Actin Filaments Are Tethered to the Plasma Membrane

While bundled and orthogonal arrays of actin filaments are important structural elements, the plasma membrane must be physically linked to these arrays to confer their structural properties to the membrane. The predominant means of linking the plasma membrane to the underlying actin cytoskeleton is for actin-associated proteins to bind, either directly or indirectly, to specific integral membrane proteins. Orthogonal cross-linking proteins, including filamin and spectrin, tether to integral membrane proteins. The significance of the actin–filamin–membrane protein linkage is seen in a spontaneous filamin knockout melanoma cell line (Cunningham *et al.*, 1992). These cells have lowered membrane tensions and undergo dramatic, continuous blebbing of their plasma membrane. Spectrin also links the membrane to the underlying actin cytoskeleton. Along the basolateral membrane of many epithelial cells, actin–spectrin tethers to Na,K-ATPase through the linking protein ankyrin (Nelson and Veshnock, 1987; Nelson and Hammerton, 1989). Linkage of Na,K-ATPase to the actin–spectrin–ankyrin complex markedly increases the polarized retention of Na,K-ATPase within the basolateral membrane.

Within apical microvilli of epithelial cells, early electron microscopy techniques showed a series of molecular bridges extending from the actin cores to the microvillar membrane (Mukherjee and Staehelin, 1971; Mooseker and Tilney, 1975). Subsequently, this actin–membrane linkage was determined to be a myosin I protein (Mooseker *et al.*, 1978). More recently, proteins of the protein 4.1 superfamily have been found to form important actin–membrane linkages. Within this superfamily, ezrin-radixin-moesin (ERM) family members are of particular importance in the microvilli of epithelial cells. ERM proteins have multiple protein-binding domains including an actin–binding domain, an A-kinase–binding domain, and a protein-binding domain that allows it to bind directly or indirectly to integral membrane proteins. The A-kinase–binding domain allows for the targeted distribution of the activated protein kinase A (PKA) subunits at specific membrane microdomains.

A pivotal linkage of ERM proteins to microvillar membrane proteins in epithelial cells occurs through ERM-binding phosphoprotein 50 (EBP50). EBP50 is composed of two PSD95-Dlg-ZO1 (PDZ) domains and an ezrin-binding region whose interactions are regulated through phosphorylation (Weinman *et al.*, 1995; Reczek *et al.*, 1997; Hall *et al.*, 1999; Fouassier *et al.*, 2005). Expressed in hundreds of different proteins, PDZ domains are often concentrated in submembranous regions and bind the cytoplasmic C-terminal tails of specific integral membrane receptors, transporters, and channels. By expressing multiple PDZ domains and other protein-binding domains, PDZ domain-containing proteins form critical scaffolds that cluster and integrate functionally related proteins within membrane microdomains. The actin cytoskeleton plays a central role in the function of many PDZ proteins.

I. Actin Cytoskeleton Is Used to Generate Force and Movement

Since its discovery as the thin filament in muscle sarcomeres, actin filaments have been associated with force generation and movement. Antiparallel bundles of the myosin II mechanoenzyme make up the thick filaments of sarcomeres and use their mechanical characteristics to ratchet along the actin thin filaments. In humans, about 40 distinct myosin genes have been identified. Each myosin family member has a highly conserved head domain that utilizes ATP hydrolysis to drive its motor function. In contrast, the tail domain is highly variable and specifies the cargo that the myosin will carry and is largely responsible for specifying the function of the individual myosin proteins. Most myosins are (+)-end–directed motors but there are (−)-end–directed motors including myosin VI. Myosins I, II, V, and VI are distributed within the apical domain of epithelial cells (Section III) and contribute to a disparate functions within that domain (Raposo *et al.*, 1999).

More recently, actin polymerization itself is transduced by cells into a motive force. The most striking example resides at the leading edge of motile cells where actin polymerization drives membrane protrusion forward. Nonmotile cells also utilize actin polymerization. In phagocytic cells, such as Kupfer cells in the liver, actin polymerization drives plasma membrane up and around a bound particle to allow its engulfment. In the apical domain of epithelial cells, the force generated from actin polymerization is utilized to drive the formation and internalization of endosomes from the plasma membrane into the cell interior (Section III).

III. STRUCTURAL ORGANIZATION OF ACTIN WITHIN EPITHELIAL CELLS

The actin cytoskeleton is concentrated at the plasma membrane of epithelial cells. The organization of the actin cytoskeleton, however, is markedly different within the distinct regions of these cells. Along the basal membrane of cultured epithelial cells, parallel actin cables are readily observed extending to and from focal adhesion sites that anchor the cell to the underlying extracellular matrix. These anchorage sites permit actin and myosin II to impart tension on the underlying substratum. Further, the actin cytoskeleton emanating from these sites transduces information about the cell exterior to the cell interior. Other regions of the basolateral membrane are underlaid with a meshwork of actin filaments. The meshwork of actin filaments is cross-linked by proteins such as spectrin and filamin. Further, these proteins bind directly or indirectly with integral membrane proteins to tether the plasma membrane to the underlying actin cytoskeleton. The membrane-cytoskeleton imparts tension into the plane of the membrane and allows the membrane to conform to the shape of the cell. In renal proximal tubules, disruption of the actin cytoskeleton profoundly decreases membrane tension to the point where the intracellular hydrostatic pressure drives the plasma membrane from the cell and into membrane blebs (Doctor *et al.*, 1997).

Interposed at the apical-basolateral border is the epithelial junctional complex that includes tight junctions (aka zonula occludens) and cell–cell adhesion junctions (aka zonula adherens). Associated with these adhesion sites and circumscribing each cell is an actinomyosin ring. Regulated contraction of the actinomyosin ring can modify the junctional properties and paracellular permeability of the epithelium (Nusrat *et al.*, 1995). While the physiologic interplay between the junctional proteins and the actin cytoskeleton remains under study (Peifer and Gates, 2005), the interaction of the actin cytoskeleton and the junctional complexes is important in establishing cell polarity and modulating the function of the junctions.

The actin cytoskeleton within the apical domain of most transporting epithelial cells has two prominent areas: the terminal web and microvilli. The electron dense terminal web extends from the junctional complexes and lies just below the apical membrane. The actin-associated proteins within the terminal web include spectrin, α-actinin, tropomyosin, and distinct myosins. Actin within the terminal web is largely organized as a meshwork of filaments; spectrin likely serves to cross-link this meshwork. α-Actinin is concentrated at the lateral edges of the terminal web and is likely bundling filaments associated with the junctional complexes. Tropomyosin is concentrated along the rootlets of the microvillar bundles that are embedded within the terminal web. Tropomyosin is likely serving to stabilize the (−)-end or root of the microvillar core filaments (Bretscher and Weber, 1978; Burgess *et al.*, 1987; Ashworth *et al.*, 2004). In addition, while filament-severing proteins may reside within the terminal web under physiologic conditions, they do not appear to be concentrated within this domain. This balance of severing and stabilizing proteins would enhance the stability of actin filaments within the terminal web and microvilli.

Extending from the terminal web and toward the membrane, bundled actin filaments form the structural core of microvilli. These filaments are polarized with the (−)-end anchored in the terminal web region and the (+)-end residing at the microvillar tip. The microvillar filaments can extend several microns in length and are bundled together by specific actin-bundling proteins. A single microvillus may have multiple bundling proteins. For example, microvilli in enterocytes contain at least three different bundling proteins: fimbrin, villin, and espin. While there may be some overlap in function, the unique characteristics of the individual bundling proteins allows for the orderly genesis, stabilization, and degradation of the microvilli. Introduction of villin into fibroblasts results in microvilli (Friederich *et al.*, 1989), suggesting villin is capable of organizing microvillar formation. However, epithelial cells in villin knockout mice still form microvilli, indicating there is at least redundancy in the process (Pinson *et al.*, 1998). Villin is also a member of the gelsolin family and possesses Ca^{2+}-activated severing activity and is important in microvillar remodeling in disease states (Ferrary *et al.*, 1999). The role of fimbrin is not fully defined but is suggested to contribute to the ordering and secondary characteristics of microvilli (Fath and Burgess, 1995). The small espin isoform expressed in epithelial cells is present in lower abundance than fimbrin or villin but it has a Ca^{2+}-insensitive, high-affinity filament-bundling capacity and contributes in conferring stability and dictating the length of the microvillar actin filaments (Loomis *et al.*, 2005).

Membrane linking proteins are needed in order to confer the structural properties of the microvillar core filaments to the membrane. As described earlier, microvillar membranes are tethered to the underlying actin core by different classes of proteins including myosin I and ERM proteins. With its (+)-end–directed motor activity, it is possible that myosin I-linked proteins could be distributed from the cell interior to and along the microvillus by its actin motor-binding partner.

The crypts at the base of microvilli are also specialized regions within the apical domain. This is especially true in epithelial cells with densely packed microvilli such as renal proximal tubule cells. In these cells, the vesicle trafficking to the apical membrane and endosome formation at the apical membrane occurs at the microvillar crypts. In addition to constraining the access of vesicles to and from the crypt, the actin cytoskeleton plays a key role in driving the formation and initial internalization of endosomes (Section III).

IV. FUNCTIONAL CONTRIBUTIONS OF THE ACTIN CYTOSKELETON IN THE APICAL DOMAIN

Over the last decade, the actin cytoskeleton has been found to contribute significantly to the physiology of epithelial cells. Concentrated at the plasma membrane in the different regions of the cell, the actin cytoskeleton interacts with the plasma membrane, imparts tension into the membrane, and provides the membrane with structural support. Further, ion channel and transporter activities can be regulated by moderating their abundance in the membrane or their open probability/turnover rates while in the membrane. At the apical membrane, the actin cytoskeleton contributes significantly to: (1) trafficking of exocytic vesicles to the apical membrane, (2) controlling access to the plasma membrane, (3) retaining specific proteins in the membrane, (4) moderating protein activities while in the membrane, and (5) recovering proteins from the membrane (Fig. 2).

A. Trafficking of Exocytic Vesicles to the Apical Membrane

All epithelial cells traffic vesicles to the apical membrane. This is most apparent in "professional" secretory cells, such as pancreatic acinar cells, whose primary function is to synthesize and secrete digestive enzymes into the pancreatic duct. However,

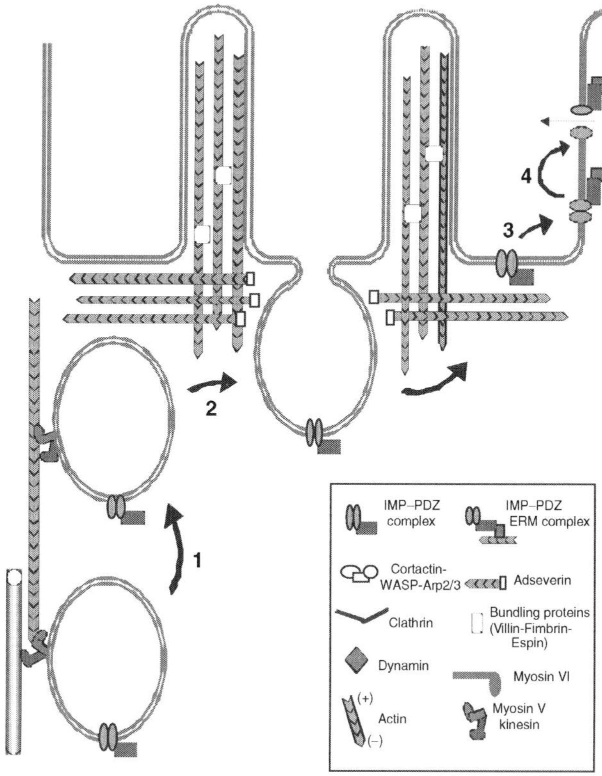

Figure 2. Actin-associated events in the apical domain of epithelial cells. The actin cytoskeleton impacts a number of functional activities at the apical membrane of epithelial cells. These include: (1) trafficking exocytic vesicles to the apical membrane, (2) accessing the plasma membrane, (3) retaining specific proteins in the membrane, (4) moderating the activities of the integral membrane proteins, and (5) formation and internalization of endosomes. See text for details.

"nonprofessional" secretory cells, such as those found in the kidney and liver, also traffic vesicles to the apical membrane in a regulated fashion. The significance of vesicular trafficking is reflected in the high rates of plasma membrane turnover (Kilic *et al.*, 2001; Doctor *et al.*, 2002). In cultured hepatocytes and cholangiocytes, membrane areas equivalent to approximately 1.5% of the total plasma membrane per minute are continually delivered and retrieved from the plasma membrane. These rates are moderated through various signaling pathways. Further, manipulating the integrity of the actin cytoskeleton impacts the rates of delivery and retrieval. The actin cytoskeleton contributes to regulated trafficking of exocytic vesicles to the apical membrane both by controlling the movement of vesicles from the cell interior to the membrane and by controlling final access of the vesicles to the membrane.

The temporal and targeted delivery of vesicles to the apical membrane is an active and regulated process. The actin cytoskeleton at the plasma membrane of cells is positioned to permit myosin motor proteins to coordinate the local delivery of vesicles. Accordingly, several myosin proteins, including myosins I, II, V, and VI, localize to the apical domain of epithelial cells (Heintzelman *et al.*, 1994). In pancreatic acinar cells, myosin I colocalizes with zymogen granules and inhibition of myosin ATPase activity inhibits zymogen granule release, suggesting myosin I participates in their regulated delivery to the apical membrane (Poucell-Hatton *et al.*, 1997). Myosin V has molecular features that promote its role in the targeted delivery of vesicles to the apical membrane. Kinesin is an anterograde microtubule motor protein that shuttles vesicles along microtubules from the cell interior to the plasma membrane. The tail domain of myosin V is capable of directly binding to conventional kinesin and myosin V and kinesin coprecipitate in rat brain extracts (Huang *et al.*, 1999). Further, the neck region of myosin V is capable of binding proteins within the SNARE complex, a group of proteins that facilitate specific docking and fusion of vesicles with plasma membranes. Thus, the design of myosin V enables it to take possession of vesicles that have been delivered to the apical domain by kinesin, direct the local delivery of these vesicles to the plasma membrane, and then interact with the SNARE complex to facilitate the docking and fusion of vesicles with the plasma membrane.

B. Moderating Access to the Plasma Membrane

The cytoskeletal matrix of the terminal web imposes an additional barrier to vesicle delivery to the plasma membrane. This is highlighted in studies that disrupted the actin cytoskeleton and observed increased rates of exocytosis (Muallem *et al.*, 1995). In neuroendocrine chromaffin cells, a "professional" secretory cell, the Ca^{2+}-activated severing protein adseverin (aka scinderin) regulates the spatial and temporal release of exocytic vesicles. Expression of adseverin in epithelial cells is cell-type specific but is present in kidney, intestine, and lung (Lueck *et al.*, 1998). In airway goblet cells, which express adseverin, the apical actin barrier is disassembled when stimulated to release mucin-containing granules and the release is blocked by the dominant-negative introduction of the actin-binding domains of adseverin to the cells (Ehre *et al.*, 2005). While a greater understanding is needed, the actin filament barrier and regulated actin-severing protein activity are positioned to moderate the targeted delivery of vesicles to the apical membrane.

C. Retaining Specific Proteins in the Apical Membrane

The cytoplasmic surface of the plasma membrane is a crowded space with protein concentrations approaching 1 g/ml (Sheetz, 1993). The actin cytoskeleton assists in the organization of this domain by sequestering integral membrane proteins with appreciable cytoplasmic tails into molecular corrals in the plane of the membrane. These actin-based corrals restrict lateral diffusion of proteins into approximately 0.6 μm^2 microdomains of the membrane (Edidin *et al.*, 1991). The actin cytoskeleton also contributes to retaining proteins in the apical membrane. In WIF-B cells, a polarized hepatocyte cell culture model, disruption of the actin cytoskeleton by cytochalasin D (CD) greatly diminished the dwell times of 5′-nucleotidase and aminopeptidase N in the canalicular (i.e., apical) membrane (Tuma *et al.*, 2002).

The apical concentration of multidrug resistance protein 2 (Mrp2) in WIF-B cells was not affected by this same treatment. This suggests Mrp2 has additional mechanisms that contribute to its retention in the apical membrane. One possibility is that Mrp2 is directly bound to CD-insensitive actin filaments. Parenthetically, CD acts by capping the (+) ends of filaments to block further monomer addition; it is largely ineffective against stabilized actin filaments. Concentrated in the canalicular (i.e., apical) membrane of hepatocytes, Mrp2 is linked to the actin cytoskeleton in a linear actin–radixin–EBP50–Mrp2 complex (Fig. 3). The physiologic significance of this linkage is highlighted in radixin knockout (*Rad–/–*) mice (Kikuchi *et al.*, 2002). Normally, hepatocytes conjugate and eliminate the conjugated bilirubin (a by-product of hemoglobin degradation) by secreting it into the bile. The conjugated bilirubin crosses the canalicular membrane through Mrp2. In humans, mutations in Mrp2 result in the autosomal recessive disorder Dubin–Johnson syndrome (DJS) that is

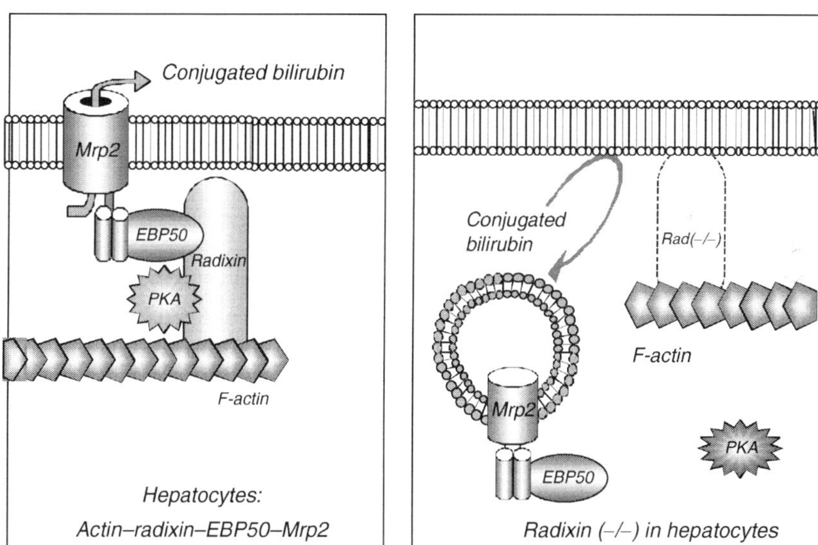

Figure 3. ABC transporters are linked to the actin cytoskeleton through an ERM–PDZ protein complex. Mrp2, which transports conjugated bilirubin, is sequestered into a signaling-effector complex in the apical domain of hepatocytes (left panel). Under physiologic conditions, the C-terminal tail of Mrp2 can bind to the PDZ domains of EBP50, which is in turn linked to radixin, an ERM protein. Radixin and EBP50 serve to anchor Mrp2 and PKA into a signaling-effector complex (right panel). The physiologic significance of the actin–radixin–EBP50–Mrp2 linkage is seen in *Rad(–/–)* mice. These mice have significantly diminished levels of Mrp2 in their canalicular membranes and elevated serum levels of bilirubin.

characterized by chronically elevated conjugated hyperbilirubinemia in blood. In *Rdx*$^{-/-}$ mice, serum concentrations of conjugated bilirubin increase soon after birth and mild liver injury develops, a pattern of congenital hyperbilirubinemia that parallels DJS in humans. The observed hyperbilirubinemia and decrease in Mrp2 protein within the bile cannalicular membrane of *Rad(−/−)* mice indicates the actin-associated radixin complex contributes to the canalicular localization and activity of Mrp2.

When the radixin-to-Mrp2-linking protein ERM-binding phosphoprotein 50 (EBP50; aka NHERF1) was knocked out in mice, hyperbilirubinemia was not reported. This may be due to the expression of proteins, such as NHE3 kinase A regulatory protein (E3KARP; aka NHERF2), or other functionally related proteins that are coexpressed in hepatocytes and can partially or completely substitute for EBP50 function in hepatocytes. *EBP50 (−/−)* mice were, however, hyperphosphouremic (Shenolikar *et al.*, 2002; Weinman *et al.*, 2003). Serum phosphate levels are largely and acutely regulated by the kidney through selective regulation of the amount of phosphate that is absorbed from the glomerular filtrate by the proximal tubule. Under low serum phosphate conditions, proximal tubule cells maximize the number of sodium-phosphate cotransport proteins (NaPiIIa) in the apical membrane and minimize the endocytic recovery of NaPiIIa. Conversely, under high serum phosphate conditions, proximal tubule cells diminish their retention and increase their endocytic recovery and lysosomal degradation of NaPiIIa. Regulation of NaPiIIa activity is modestly influenced by rates of synthesis, is not moderated by posttranslation modification (i.e., phosphorylation), and does not undergo significant recycling to the plasma membrane. The C-terminal tail of NaPiIIa is bound by specific PDZ domain proteins, including EBP50. In *EBP50(−/−)* mice, the absence of EBP50 led to diminished retention of NaPiIIa in the apical membrane, decreased phosphate recovery by the kidneys, and phosphate wasting in the urine.

In addition to Mrp2 and NaPiIIa, the actin–ERM–EBP50 linkage binds a number of additional integral membrane proteins in a variety of epithelial cell types (Brone and Eggermont, 2005). Another physiologically intriguing example is the interaction of actin–ezrin–EBP50 with the cystic fibrosis transmembrane conductance regulator (CFTR) protein (Hall *et al.*, 1998; Short *et al.*, 1998). CFTR is a cAMP-regulated Cl^- channel that normally promotes fluid secretion apically across CFTR-expressing epithelia. In cultured cholangiocytes, dominant-negative disruption of the CFTR–EBP50 interaction results in the complete ablation of the cAMP-mediated Cl^- secretory response in cholangiocytes (Fouassier *et al.*, 2001). While the specifics continue to be debated about the *in vivo* consequences, the binding of CFTR to actin–ezrin–EBP50 may enhance the apical polarization, access to PKA, and dimerization of the CFTR Cl^- channel (Moyer *et al.*, 1999; Fouassier *et al.*, 2001; Raghuram *et al.*, 2001; Benharouga *et al.*, 2003; Haggie *et al.*, 2004). Dimerization of CFTR enhances the open probability of its channel. Cystic fibrosis (CF) is a lethal autosomal recessive disorder that stems from mutations in the CFTR gene. While generally of a milder phenotype, approximately 10% of CFTR mutations in humans result in truncated C-terminus, the domain that binds EBP50.

D. Moderating Activities of Proteins While in the Membrane

As noted earlier, with nearly 400 PDZ domain proteins identified in the human genome, PDZ domain proteins play a pivotal role in forming supramolecular signaling-effector complexes (Sheng and Sala, 2001). Within the apical domain of epithelial cells, different PDZ domain proteins utilize their association with the actin cytoskeleton to influence the polarized distribution, molecular organization, trafficking, and regulated activity of specific integral membrane proteins. Multiple PDZ proteins may bind a single integral membrane protein. For example, EBP50 and Shank2 are two distinct PDZ proteins, each capable of binding the same transport proteins (e.g., NaPiIIa, NHE3, CFTR) but imparting disparate rather than redundant functions (McWilliams *et al.*, 2004, 2005). EBP50 appears to anchor these proteins to the microvillar actin filaments and place the protein in close apposition to regulatory proteins such as PKA. In contrast, Shank2 coassociates with proteins such as cortactin (activator of Arp2/3 and actin nucleation; Du *et al.*, 1998) and dynamin (vesicle "pinchase"; Okamoto *et al.*, 2001). Given their role in endocytosis, it suggests Shank2 may contribute to protein recovery, vesicle formation, or vesicle trafficking. Importantly, EBP50 and Shank2 are each associated with the actin cytoskeleton but likely utilize the actin cytoskeleton in disparate manners.

E. Recovering Proteins from the Membrane

Just as insertion of receptors, channels, or transporters into the plasma membrane is a potent means of upregulating their activity, removal of the protein from the membrane serves as a potent mechanism of downregulating their activity. The actin cytoskeleton plays a central role in regulated endocytosis. There are multiple means by which different cell types internalize membranes, including phagocytosis, macropinocytosis, clathrin-mediated endocytosis (CME), caveolae-mediated endocytosis, and endocytosis by mechanisms that are independent of either clathrin or caveolae. Each of these distinct modes of membrane internalization includes actin-dependent steps for internalization. CME is a well-characterized process that includes invagination of membrane into clathrin-coated pits, dynamin-dependent fission of the vesicle from the plasma membrane, and translocation of the vesicle away from the membrane. While not essential in all cells for CME to occur, the actin cytoskeleton enhances the efficacy of the endocytic steps and CME highlights how the actin cytoskeleton contributes to endocytosis.

In epithelial cells, endocytic vesicles repeatedly emerge from the same microdomain of the plasma membrane (Gaidarov *et al.*, 1999). The actin cytoskeleton contributes significantly to the "corraling" of proteins in clathrin-coated membranes; disruption of actin filaments results in a sevenfold increase in the diffusion area of proteins within coated pit membranes. The invagination of clathrin-coated membranes is coordinated by intersectin, a protein capable of scaffolding clathrin, AP2, dynamin, WASP, and Cdc42. WASP and Cdc42 are positive regulators of Arp2/3-dependent actin nucleation. Spatial and temporal analysis demonstrates that the appearance of filamentous actin occurs coincidently with the recruitment of dynamin to the neck region and suggests actin polymerization is important in driving the formation and internalization of CME vesicles (Merrifield *et al.*, 2002). Consistent with this hypothesis, disruption of the actin cytoskeleton with CD results in the accumulation of clathrin-coated vesicles along the apical membrane (Shurety *et al.*, 1996). In addition to nucleation-dependent formation of endosomes, myosin motor proteins also contribute to vesicle formation and internalization. CFTR is endocytosed from apical membranes by CME. Myosin VI, a (−)-end–directed motor, resides at sites of endocytosis, associates with CME proteins, and is needed for the efficient endocytosis of CFTR from the apical membrane of epithelial cells (Swiatecka-Urban *et al.*, 2004).

V. CONTRIBUTION OF THE ACTIN CYTOSKELETON IN EPITHELIAL DISEASE STATES

A. Infectious Pathogens Hijack the Host Actin Cytoskeleton

A variety of bacterial and viral pathogens "hijack" the host actin cytoskeleton to promote cellular uptake, intracellular motility, and infection of adjacent cells (Cossart and Sansonetti, 2004; Stevens *et al.*, 2006). Their utilization of the host's actin cytoskeleton is essential to their virulence and pathogenicity. An excellent example of "actin hijacking" is found in *Listeria monocytogenes. Listeria* expresses the protein ActA at an extracellular pole. ActA induces actin filament nucleation and polymerization through the recruitment and activation of the host's Arp2/3 complex. Similar to the extension of the plasma membrane during phagocytosis or driving the intracellular movement of endocytic vesicles, this polymerization propels the bacteria through the host's cytoplasm. When stained for F-actin, infected cells show a "comet tail" of F-actin behind the *Listeria*.

Helicobacter pylori, an invasive human pathogen that causes inflammation of the gastric epithelial mucosa, induces effacement of microvilli and reorganization of the actin cytoskeleton directly below the site of bacterial attachment (Segal *et al.*, 1996). Several *H. pylori* virulence factors, including vacuolating cytotoxin and a cytotoxin-associated antigen, have been implicated in this reorganization. Enteropathogenic *Escherichia coli* (EPEC) induces a similar loss of brush border microvilli and reorganizes the host cell actin cytoskeleton directly beneath the extracellular bacterium to form an actin-based pedestal (Campellone and Leong, 2003). The pathogenicity of enterotoxigenic *Bacteroides fragilis*, a noninvasive bacterium associated with diarrheal disease, induces toxin-mediated alterations of the apical F-actin structure at the level of the tight junctions in cultured human intestinal epithelial cells (Koshy *et al.*, 1996). This disruption of the actin cytoskeleton diminishes the paracellular permeability barrier function and induces the diarrheal disease that is associated with *B. fragilis*.

Chronic hepatitis B virus (HBV) infection and progression to cirrhosis are well-recognized risk factors for the development of hepatocellular carcinoma. HBV infection modifies the actin cytoskeleton and increases the metastasis of infected hepatocytes. Hepatitis B virus X protein (HBx), a 17-kD protein encoded by the HBV genome, induces rearrangement of the actin cytoskeleton (Lara-Pezzi *et al.*, 2001). This includes the formation of pseudopodia with F-actin–moesin–CD44 protein complexes clustered at the tips of the pseudopodia. CD44 is a surface glycoprotein known to moderate tumor cell growth and metastasis (Sherman *et al.*, 1994).

B. Apical Membrane Structure Is Specifically Altered in Epithelial Diseases

Disruption of the apical actin cytoskeleton can result in significant epithelial dysfunction. Microvillus inclusion disease is an autosomal recessive disorder characterized by refractory diarrhea from birth and is usually fatal. Enterocytes from individuals with microvillus inclusion disease have decreased levels of actin, a loss of apical microvilli, and an increase in apical cytoplasmic vesicles containing apically targeted proteins. Similarly, experimental models of cholestatic liver disease display marked alterations in the organization of the canalicular domain in hepatocytes. Hepatocytes from obstructive cholestatic livers have significantly decreased numbers of motile cytoplasmic vesicles and large intracellular pseudocanaliculi (Torok *et al.*, 2001). In general, these hepatocytes have impaired transcytotic trafficking and a loss of domain specificity of canalicular membrane proteins (Steiger *et al.*, 1994). The psuedocanaliculi are associated with a cortical actin meshwork, have luminal microvilli, and retain the ability to actively transport charged molecules into their lumen.

C. Ischemia Results in Profound Structural and Functional Alterations

Ischemia, the cessation of blood flow, has severe effects on organ function and occurs both in disease states (e.g., cardiac arrest) and in clinical protocols (e.g., surgery and organ transplantation). Ischemia is a well-defined example of a disease process in which the actin cytoskeleton is profoundly disrupted and the genesis of the alterations is understood at a molecular level (Fig. 4).

Cellular consequences of ischemia include the profound loss of cellular ATP, decreased pHi, and increased Ca^{2+} concentrations within the apical domain. Paradoxically, despite a profound decrease in the cellular ATP/ADP ratio, there is a relative increase in the amount of F-actin. This is likely attributed to the dysregulation of actin-binding proteins and resultant increase in available monomers for polymerization. Normally concentrated at the cell periphery, F-actin in ischemic or ATP-depleted cells redistributes from the cell periphery to a cytoplasmic/perinuclear localization (White *et al.*, 2000). These changes in F-actin coincide with significant structural and functional alterations. In ischemic or ATP-depleted epithelial cells, the most striking change occurs within the apical microvilli. In renal proximal tubule cells, ATP depletion leads to the blebbing, clubbing, and eventual loss of apical microvilli. Similarly, ATP depletion in cholangiocytes, which have short and less dense microvilli, leads to

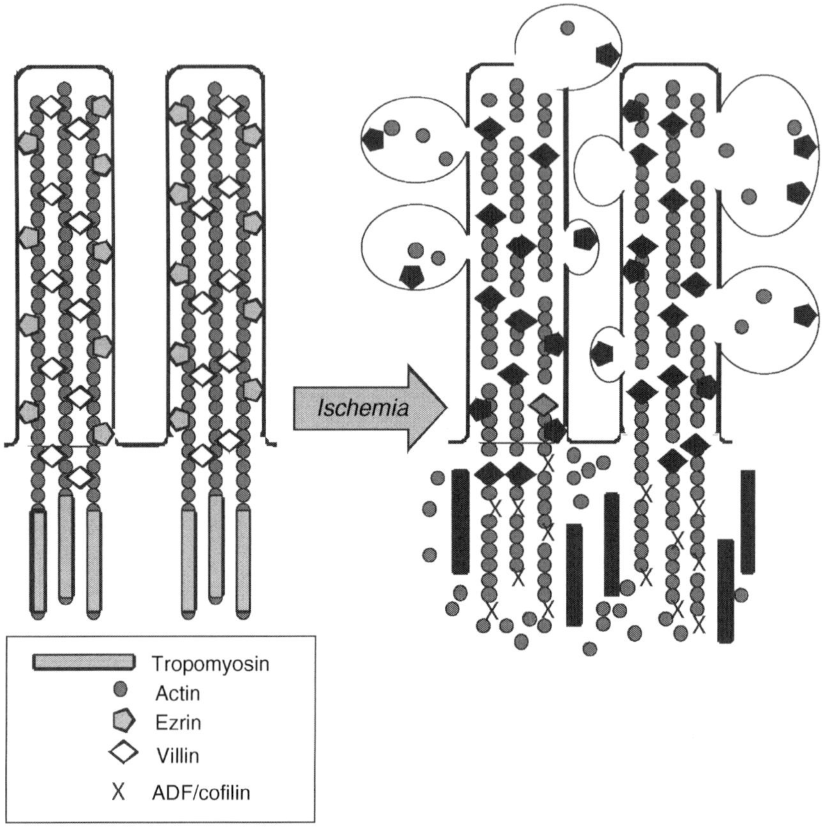

Figure 4. Actin-associated proteins remodel microvillar actin in ischemic epithelial cells. Under physiologic conditions (left panel), the microvillar core rootlets are stabilized by tropomyosin, the microvillar core filaments are bundled by multiple bundling proteins, including villin, and the membrane is tethered to the microvillar core filaments by multiple linking proteins, including by the ezrin–EBP50 complex. Villin also has Ca^{2+}-activated actin-severing activity. Ezrin-mediated protein interactions are positively moderated by phosphorylation. ADF/cofilin is not concentrated within this domain. Under ischemic conditions (right panel), ezrin is dephosphorylated and detaches from the microvillar filaments. Villin is activated and positioned to sever microvillar filaments. ADF/cofilin redistributes to the apical domain and appears to outcompete tropomyosin to access and sever the microvillar rootlet filaments. Each of these activities will impact the structure and organization of the microvillar membranes in ischemic epithelial cells.

an initial elongation of microvilli followed by loss of microvillar structure and density (Doctor *et al.*, 1999). Specific alterations to four microvillar proteins have been directly implicated in contributing to the changes in the microvilli. First, the protein-binding activities of ezrin are regulated by ezrin phosphorylation. Under ischemic conditions, ezrin is dephosphorylated, dissociates from the microvillar cytoskeleton, and redistributes from the microvilli (Chen *et al.*, 1994, 1995). This loss of ezrin activity would lead to diminished membrane tethering to the actin cytoskeleton and facilitate membrane blebbing. Second, the actin-severing activity of villin is positively regulated by elevated Ca^{2+} levels and microvillar Ca^{2+} concentrations are elevated during ischemia and ATP depletion (Spencer *et al.*, 1991). The actin-severing activity of villin would result in cleaving of the actin filaments along the length of the microvilli. Unlike actin along the length of the microvilli, actin within the microvillar rootlet is stabilized by tropomyosin, which hinders the access of severing proteins, including ADF/cofilin, to those filaments. Immunohistologic studies suggest tropomyosin partially redistributes from the apical domain of ischemic epithelial cells (Ashworth *et al.*, 2004). ADF/cofilin is activated by dephosphorylation and low pH (Bamburg, 1999), conditions that prevail in ischemic or metabolically inhibited epithelial cells. In ischemic proximal tubule cells, ADF/cofilin redistributes from the cell interior to the apical domain where it is positioned to compete with tropomyosin for access to the microvillar core filaments (Ashworth *et al.*, 2001).

The structural alterations in microvilli are paralleled by functional deficits at the apical membrane. In ATP-depleted cholangiocytes, the concentration of Na^+-glucose cotransporter 1 and activity of γ-glutamyl transpeptidase in the apical membrane are significantly decreased (Doctor *et al.*, 2000). *In vivo* studies of ischemic cholangiocytes also show significant internalization of the apical membrane protein leucine aminopeptidase. Together these observations suggest that membrane protein internalization is a result of ischemia-induced alterations in the actin cytoskeleton and its interaction with the plasma membrane.

VI. CONCLUDING REMARKS

For a number of years, there have been significant strides taken in the molecular understanding of actin-associated proteins and their functions in governing the biochemistry of actin filaments (Section I). Within a cellular context, the actin cytoskeleton was first appreciated for its structural contributions, including the formation and organization of the apical membrane of epithelial cells (Section II). Cell biologists have furthered this and made enormous strides over the last decade in unraveling the mysteries underlying how the actin cytoskeleton dictates cellular functions. This is especially apparent within the apical domain of epithelial cells where the actin cytoskeleton is central to regulating multiple facets of transport across the apical membrane (Section III). These functions include directing the movement of vesicles to the apical membrane, controlling vesicle access to the membrane, retaining and regulating proteins within the membrane, and recovering these proteins from the membrane. Given its central role in cellular physiology, it has not been surprising that the actin cytoskeleton plays a pivotal role in the pathogenesis of specific epithelial diseases (Section IV). The following decade should continue to see enormous progress in defining the molecular and cellular physiology of the actin cytoskeleton.

REFERENCES

Ashworth, S., Sandoval, R., Hosford, M., Bamburg, J., and Molitoris, B. (2001). Ischemic injury induces ADF relocalization to the apical domain of rat proximal tubule cells. *Am. J. Physiol.* **280,** F886–F894.

Ashworth, S. L., Wean, S. E., Campos, S. B., Temm-Grove, C. J., Southgate, E. L., Vrhovski, B., Gunning, P., Weinberger, R. P., and Molitoris, B. A. (2004). Renal ischemia induces tropomyosin dissociation-destabilizing microvilli microfilaments. *Am. J. Physiol.* **286,** F988–F996.

Bamburg, J. (1999). Proteins of the ADF/cofilin family: Essential regulators of actin dynamics. *Ann. Rev. Cell. Dev. Biol.* **15,** 185–230.

Benharouga, M., Sharma, M., So, J., Haardt, M., Dryzmala, L., Popov, M., Schwappach, B., Grinstein, S., Du, K., and Lukacs, G. (2003). The role of the C-terminus and NHERF in the functional expression of CFTR in nonpolarized cells and epithelia. *J. Biol. Chem.* **278,** 22079–22089.

Bretscher, A. and Weber, K. (1978). Localization of actin and microfilament-associated proteins in the microvilli and terminal web of the intestinal brush border by immunofluorescence microscopy. *J. Cell. Biol.* **79,** 839–845.

Brone, B. and Eggermont, J. (2005). PDZ proteins retain and regulate membrane transporters in polarized epithelial cell membranes. *Am. J. Physiol.* **288,** C20–C29.

Burgess, D. R., Broschat, K. O., and Hayden, J. M. (1987). Tropomyosin distinguishes between the two actin-binding sites of villin and affects actin binding properties of other brush border proteins. *J. Cell. Biol.* **104,** 29–40.

Campellone, K. G. and Leong, J. M. (2003). Tails of two Tirs: Actin pedestal formation by enteropathogenic *E. coli* and enterohemorrhagic *E. coli* O157:H7. *Curr. Opin. Micro.* **6,** 82–90.

Chen, J., Doctor, R. B., and Mandel, L. J. (1994). Cytoskeletal dissociation of ezrin during renal anoxia: Role in microvillar injury. *Am. J. Physiol.* **267,** C784–C795.

Chen, J., Cohn, J., and Mandel, L. J. (1995). Dephosphorylation of ezrin as an early event in renal microvillar breakdown and anoxic injury. *Proc. Natl. Acad. Sci. USA* **92,** 7495–7499.

Cossart, P. and Sansonetti, P. J. (2004). Bacterial invasion: The paradigms of enteroinvasive pathogens. *Science* **304,** 242–248.

Cunningham, C. C., Gorlin, J. B., Kwiatkowski, D. J., Hartwig, J. H., Janmey, P. A., Byers, H. R., and Stossel, T. P. (1992). Actin-binding protein requirement for cortical stability and efficient locomotion. *Science* **255,** 325–327.

Doctor, R. B., Zhelev, D. V., and Mandel, L. J. (1997). Loss of plasma membrane structural support in ATP-depleted renal epithelia. *Am. J. Physiol.* **272,** C439–C449.

Doctor, R. B., Dahl, R. H., Salter, K. D., and Fitz, J. G. (1999). Reorganization of cholangiocyte membrane domains represents an early event in rat liver ischemia. *Hepatology* **29,** 1364–1374.

Doctor, R. B., Dahl, R. H., Salter, K. D., Fouassier, L., Chen, J., and Fitz, J. G. (2000). ATP depletion in cholangiocytes leads to marked internalization of membrane proteins. *Hepatology* **31,** 1045–1054.

Doctor, R. B., Dahl, R. H., Fouassier, L., Kilic, G., and Fitz, J. G. (2002). Cholangiocytes exhibit dynamic, actin-dependent apical membrane turnover. *Am. J. Physiol.* **282,** C1042–C1053.

Du, Y., Weed, S. A., Xiong, W., Marshall, T. D., and Parsons, J. T. (1998). Identification of a novel cortactin SH3 domain-binding protein and its localization to growth cones of cultured neurons. *Mol. Cell. Biol.* **18,** 5838–5851.

Edidin, M., Kuo, S. C., and Sheetz, M. P. (1991). Lateral movements of membrane glycoproteins restricted by dynamic cytoplasmic barriers. *Science* **254,** 1379–1382.

Ehre, C., Rossi, A. H., Abdullah, L. H., DePestel, K., Hill, S., Olsen, J. C., and Davis, C. W. (2005). Barrier role of actin filaments in regulated mucin secretion from airway goblet cells. *Am. J. Physiol.* **288,** C46–C56.

Fath, K. and Burgess, D. (1995). Microvillus assembly. Not actin alone. *Curr. Biol.* **5,** 591–593.

Ferrary, E., Cohen-Tannoudji, M., Pehau-Arnaudet, G., Lapillonne, A., Athman, R., Ruiz, T., Boulouha, L., El Marjou, F., Doye, A., Fontaine, J. J., Antony, C., Babinet, C., *et al.* (1999). *In vivo*, villin is required for Ca(2+)-dependent F-actin disruption in intestinal brush borders. *J. Cell Biol.* **146,** 819–830.

Fouassier, L., Duan, C. Y., Feranchak, A., Yun, C. C., Sutherland, E., Simon, F., Fitz, J. G., and Doctor, R. B. (2001). ERM binding phosphoprotein 50 is expressed at the apical membrane of rat liver epithelia. *Hepatology* **33,** 166–176.

Fouassier, L., Nichols, M. T., Gidey, E., McWilliams, R. R., Robin, H., Finnigan, C., Howell, K. E., Housset, C., and Doctor, R. B. (2005). PKC regulates the phosphorylation and oligomerization of EBP50. *Exp. Cell Res.* **306,** 264–273.

Friederich, E., Huet, C., Arpin, M., and Louvard, D. (1989). Villin induces microvilli growth and actin redistribution in transfected fibroblasts. *Cell* **59,** 461–475.

Gaidarov, I., Santinin, F., Warren, R., and Keen, J. H. (1999). Spatial control of coated-pit dynamics in living cells. *Nat. Cell Biol.* **1,** 1–7.

Haggie, P. M., Stanton, B. A., and Verkman, A. S. (2004). Increased diffusional mobility of CFTR at the plasma membrane after deletion of its C-terminal PDZ binding motif. *J. Biol. Chem.* **279,** 5494–5500.

Hall, R., Ostegaard, L., Premont, R., Blitzer, J., Rahman, N., Welsh, M., and Lefkowitz, R. (1998). A C-terminal motif found in the b2-adrenergic receptor, P2Y1 receptor and CFTR determines binding to the Na+/H+ exchanger regulatory factor family of PDZ proteins. *Proc. Natl. Acad. Sci. USA* **95,** 8496–8501.

Hall, R. A., Spurney, R. F., Premont, R. T., Rahman, N., Blitzer, J. T., Pitcher, J. A., and Lefkowitz, R. J. (1999). G protein-coupled receptor kinase 6A phosphorylates the NHE regulatory factor via a PDZ domain-mediated interaction. *J. Biol. Chem.* **274,** 24328–24334.

Heintzelman, M., Hasson, T., and Mooseker, M. (1994). Multiple unconventional myosin domains of the intestinal brush border cytoskeleton. *J. Cell. Sci.* **107,** 3535–3543.

Huang, J., Brady, S., Richards, B., Stenoien, D., Resau, J., Copeland, N., and Jenkins, N. (1999). Direct interaction of microtubule- and actin-based transport motors. *Nature* **397,** 267–270.

Kikuchi, S., Hata, M., Fukumoto, K., Yamane, Y., Matsui, T., Tamura, A., Yonemura, S., Yamagishi, H., Kepple, D., Tsukita, S., and Tsukita, S. (2002). Radixin deficiency causes conjugated hyperbilirubinemia with loss of Mrp2 from bile canalicular membranes. *Nat. Genet.* **31,** 320–325.

Kilic, G., Doctor, R. B., and Fitz, J. G. (2001). Insulin stimulates membrane conductance in a liver cell line: Evidence for insertion of ion channels through a PI3 kinase-dependent mechanism. *J. Biol. Chem.* **276,** 26762–26768.

Koshy, S. S., Montrose, M. H., and Sears, C. L. (1996). Human intestinal epithelial cells swell and demonstrate actin rearrangement in response to the metalloprotease toxin of *Bacteroides fragilis*. *Infect. Immun.* **64,** 5022–5028.

Lara-Pezzi, E., Serrador, A., Montoya, M. C., Zamora, D., Yanez-Mo, M., Carretero, M., Furthmayr, H., Sanchez-Madrid, F., and Lopez-Cabrera, M. (2001). The hepatitis B virus X protein induces a migratory phenotype in a CD44-dependent manner. *Hepatology* **33,** 1270–1281.

Loomis, P. A., Zheng, L., Sekerkova, G., Changyaleket, B., Mugnaini, E., and Bartles, J. R. (2005). Espin cross-links cause the elongation of microvillus-type parallel actin bundles *in vivo*. *J. Cell. Biol.* **163,** 1045–1055.

Lueck, A., Brown, D., and Kwaitkowski, D. J. (1998). The actin-binding proteins adseverin and gelsolin are both highly expressed but differentially localized in kidney and intestine. *J. Cell. Sci.* **111,** 3633–3643.

Machesky, L. M., Atkinson, S. J., Ampe, C., Vandekerckhove, J., and Pollard, T. D. (1994). Purification of a cortical complex containing two unconventional actins from Acanthamoeba by affinity chromatography on profilin-agarose. *J. Cell. Biol.* **127,** 107–115.

McGough, A., Pope, B., Chiu, W., and Weeds, A. (1997). Cofilin changes the twist of F-actin: Implications for actin filament dynamics and cellular function. *J. Cell. Biol.* **138,** 771–781.

McWilliams, R. R., Gidey, E., Fouassier, L., Weed, S. A., and Doctor, R. B. (2004). Characterization of an ankyrin repeat-containing Shank2 isoform (Shank2E) in liver epithelial cells. *Biochem. J.* **380,** 181–191.

McWilliams, R. R., Breusegem, S. Y., Brodsky, K. F., Kim, E., Levi, M., and Doctor, R. B. (2005). Shank2E binds Na/Pi co-transporter at the apical membrane of proximal tubule cells. *Am. J. Physiol.* **289,** C1042–1051.

Merrifield, C., Feldman, M., Wan, L., and Almer, W. (2002). Imaging actin and dynamin recruitment during invagination of single clathrin coated pits. *Nat. Cell. Biol.* **4,** 691–698.

Millard, T. H., Sharp, S. J., and Machesky, L. M. (2004). Signalling to actin assembly via the WASP-family proteins and the Arp2/3 complex. *Biochem. J.* **380,** 1–17.

Mooseker, M. and Tilney, L. (1975). The organization of an actin filament-membrane complex: Filament polarity and membrane attachment in the microvilli of intestinal epithelial cells. *J. Cell. Biol.* **67,** 725–743.

Mooseker, M., Pollard, T., and Fujiwara, K. (1978). Characterization and localization of myosin in the brush border of intestinal epithelial cells. *J. Cell. Biol.* **79,** 444–453.

Moyer, B. D., Denton, J., Karlson, K. H., Reynolds, D., Wang, S., Mickle, J. E., Milewski, M., Cutting, G. R., Guggino, W. B., Li, M., and Stanton, B. A. (1999). A PDZ-interacting domain in CFTR is an apical membrane polarization signal. *J. Clin. Invest.* **104,** 1353–1361.

Muallem, S., Kwiatkowska, K., Xu, X., and Yin, H. (1995). Actin filament disassembly is a sufficient trigger for exocytosis in non-excitable cells. *J. Cell. Biol.* **128,** 589–598.

Mukherjee, T. and Staehelin, L. (1971). The fine structural organization of the brush border of intestinal epithelial cells. *J. Cell. Sci.* **8,** 573–599.

Mullins, R. and Pollard, T. (1999). Structure and function of the Arp2/3 complex. *Curr. Opin. Struct. Biol.* **9,** 244–249.

Nelson, W. J. and Veshnock, P. J. (1987). Ankyrin binding to (Na++ K+)ATPase and implications for the organization of membrane domains in polarized cells. *Nature* **328,** 533–536.

Nelson, W. J. and Hammerton, R. W. (1989). A membrane-cytoskeletal complex containing Na+,K+ ATPase, ankyrin, and fodrin in Madin-Darby canine kidney (MDCK) cells: Implications for the biogenesis of epithelial cell polarity. *J. Cell. Biol.* **108,** 893–902.

Nusrat, A., Giry, M., Turner, J. R., Colgan, S. P., Parkos, C. A., Carnes, K., Lemichez, E., Boquet, P., and Madara, J. L. (1995). Rho protein regulates tight junctions and perijunctional organization in polarized epithelial cells. *Proc. Natl. Acad. Sci. USA* **92,** 10629–10633.

Okamoto, P. M., Gamby, C., Wells, D., Fallon, J., and Vallee, R. B. (2001). Dynamin isoform-specific interaction with shank/ProSAP scaffolding proteins of the postsynaptic density and actin cytoskeleton. *J. Biol. Chem.* **276,** 48458–48465.

Peifer, M. and Gates, J. (2005). Can 1000 reviews be wrong? Actin, a-catenin and adherens junctions. *Cell* **123,** 769–772.

Pinson, K., Dunbar, L., Samuelson, L., and Gumucio, D. (1998). Targeted disruption of the mouse villin gene does not impair the morphogenesis of microvilli. *Develop. Dynam.* **211,** 109–121.

Poucell-Hatton, S., Perkins, P. S., Deerink, T. J., Ellisman, M. H., Hardison, W. G., and Pandol, S. J. (1997). Myosin I is associated with zymogen granule membranes in the rat pancreatic acinar cell. *Gastroenterology* **113,** 649–658.

Raghuram, V., Mak, D., and Foskett, J. (2001). Regulation of CFTR single-channel gating by bivalent PDZ domain-mediated interaction. *Proc. Natl. Acad. Sci. USA* **98,** 1300–1305.

Raposo, G., Cordonnier, M., Tenza, D., Menichi, B., Durbach, A., Louvard, D., and Coudrier, E. (1999). Association of myosin Ia with endosomes and lysosomes in mammalian cells. *Mol. Biol. Cell.* **10,** 1477–1494.

Reczek, D., Berryman, M., and Bretscher, A. (1997). Identification of EBP50: A PDZ-containing phosphoprotein that associates with members of the ezrin-radixin-moesin family. *J. Cell. Biol.* **139,** 169–176.

Segal, E., Falkow, S., and Tomkins, L. (1996). *H. pylori* attachment to gastric cells induces cytoskeletal rearrangements and tyrosine phosphorylation. *Proc. Natl. Acad. Sci. USA* **93,** 1259–1264.

Sheetz, M. P. (1993). Glycoprotein motility and dynamic domains in fluid plasma membranes. *Ann. Rev. Biophys. Biomol. Struct.* **22,** 417–431.

Sheng, M. and Sala, C. (2001). PDZ domains and the organization of supramolecular complexes. *Ann. Rev. Neurosci.* **24,** 1–29.

Shenolikar, S., Voltz, J., Minkoff, C. M., Wade, J., and Weinman, E. J. (2002). Targeted disruption of the mouse gene encoding a PDZ domain-containing protein adaptor, NHERF-1, promotes Npt2 internalization and renal phosphate wasting. *Proc. Natl. Acad. Sci. USA* **99,** 11470–11475.

Sherman, L., Sleeman, J., Herrlich, P., and Ponta, H. (1994). Hyaluronate receptors: Key players in growth, differentiation, migration and tumor progression. *Curr. Opin. Cell. Biol.* **6,** 726–733.

Short, D. B., Trotter, K. W., Reczek, D., Kreda, S. M., Bretscher, A., Boucher, R. C., Stutts, M. J., and Milgram, S. L. (1998). An apical PDZ protein anchors the cystic fibrosis transmembrane conductance regulator to the cytoskeleton. *J. Biol. Chem.* **273,** 19797–19801.

Shurety, W., Bright, N., and Luzio, J. (1996). The effects of cytochalasin D and PMA on the apical endocytosis of ricin in polarized Caco-2 cells. *J. Cell. Sci.* **109,** 2927–2935.

Spencer, A. J., LeFurgey, A., Ingram, P., and Mandel, L. J. (1991). Elemental microanalysis of organelles in proximal tubules. II. Effects of oxygen deprivation. *J. Am. Soc. Nephrol.* **1,** 1321–1333.

Steiger, B., Meier, P., and Landmann, L. (1994). Effects of obstructive cholestasis on membrane traffic and domain specific expression of plasma membrane proteins in rat liver parenchymal cells. *Hepatology* **20,** 201–213.

Stevens, J. M., Galyov, E. E., and Stevens, M. P. (2006). Actin-dependent movement of bacterial pathogens. *Nat. Rev. Microbiol.* **4,** 91–101.

Swiatecka-Urban, A., Boyd, C., Coutermarsh, B., Karlson, K. H., Barnaby, R., Aschenbrenner, L., Langford, G. M., Hasson, T., and Stanton, B. A. (2004). Myosin VI regulates endocytosis of the cystic fibrosis transmembrane conductance regulator. *Am. J. Physiol.* **279,** 38025–38031.

Torok, N., Larusso, E., and McNiven, M. (2001). Alterations in vesicle transport and cell polarity in rat hepatocytes subjected to mechanical or chemical cholestasis. *Gastroenterology* **121,** 1176–1184.

Tuma, P., Nyasae, L., and Hubbard, A. (2002). Nonpolarized cells selectively sort apical proteins from cell surface to a novel compartment but lack apical retention mechanisms. *Mol. Biol. Cell.* **13,** 3400–3415.

Weinman, E. J., Steplock, D., Wang, Y., and Shenolikar, S. (1995). Characterization of a protein cofactor that mediates protein kinase A regulation of the renal brush border membrane Na^{+}-H^{+} exchanger. *J. Clin. Invest.* **95,** 2143–2149.

Weinman, E. J., Boddeti, A., Cunningham, R., Akom, M., Wang, F., Wang, Y., Liu, J., Steplock, D., Shenolikar, S., and Wade, J. B. (2003). NHERF-1 is required for renal adaptation to a low-phosphate diet. *Am. J. Physiol.* **285,** F1225–F1232.

White, P., Doctor, R. B., Dahl, R., and Chen, J. (2000). Coincident microvillar actin bundle disruption and perinuclear actin sequestration in anoxic proximal tubules. *Am. J. Physiol.* **278,** F886–F893.

Yao, X., Chaponnier, C., Gabbiani, G., and Forte, J. G. (1995). Polarized distribution of actin isoforms in gastric parietal cells. *Mol. Biol. Cell.* **6,** 541–557.

Chapter 3

The Connection Between Actin ATPase and Polymerization

Herwig Schüler,[1] Roger Karlsson,[2] Clarence E. Schutt,[3] and Uno Lindberg[2]

[1] *Max Delbrueck Center for Molecular Medicine Neuroproteomics, D-13125, Berlin-Buch, Germany*
[2] *Department of Cell Biology, The Wenner-Gren Institute, Stockholm University, SE-106 91 Stockholm, Sweden*
[3] *Department of Chemistry, Princeton University, Princeton, New Jersey 08544*

Remodeling of the actin filament system in cells results from strictly regulated polymerization and depolymerization of actin, where hydrolysis of actin-bound ATP is crucial. Actin–actin interactions are influenced by the state of the bound nucleotide, and many microfilament regulators influence the actin ATPase by binding preferentially either to ATP/ADP·Pi- or ADP-bound actin. This chapter summarizes observations made concerning the actin ATPase and its role in the biological activity of actin and actin filaments.

I. ACTIN MICROFILAMENT SYSTEM

Actin and myosin, organized into supramolecular structures, cooperate to generate the force necessary for many types of dynamic cellular transport processes. As part of the energy-transducing mechanism in muscle cells, they generate large-scale movements.

Advances in Molecular and Cell Biology, Vol. 37

ISBN: 0-444-52868-7

DOI: 10.1016/S1569-2558(06)37003-8

In nonmuscle cells, the actin microfilament system (MFS) drives cell motility, cytokinesis, and vesicular movements. A weave of actin polymers (filaments) found in juxtaposition to the inner surface of the plasma membrane of all cells is intimately coupled to signal transduction controlling the formation of the actin filaments and their involvement in force-generating processes. The organization of the MFS is constantly being remodeled in response to transmembrane signals generated in the interactions between cells and between cells and extracellular matrices or soluble molecules like growth factors and hormones binding to cell surface receptors. All these processes ultimately depend on the hydrolysis of ATP, not only just on myosin but also on actin.

II. ATOMIC STRUCTURE OF THE ACTIN MONOMER

The actin molecule has two major domains, each of which is divided into two subdomains (Kabsch *et al.*, 1990). Subdomains 1 and 3 form a flexible base of the molecule. The purine of the ATP is sandwiched in a hydrophobic pocket between subdomains 3 and 4, and the polyphosphate tail is held by two loops originating from subdomains 1 (P1-loop) and 3 (P2-loop) (Fig. 1). The divalent cation, chelated by the ATP phosphates, makes contacts with residues around the base of the interdomain cleft. The actin monomer can

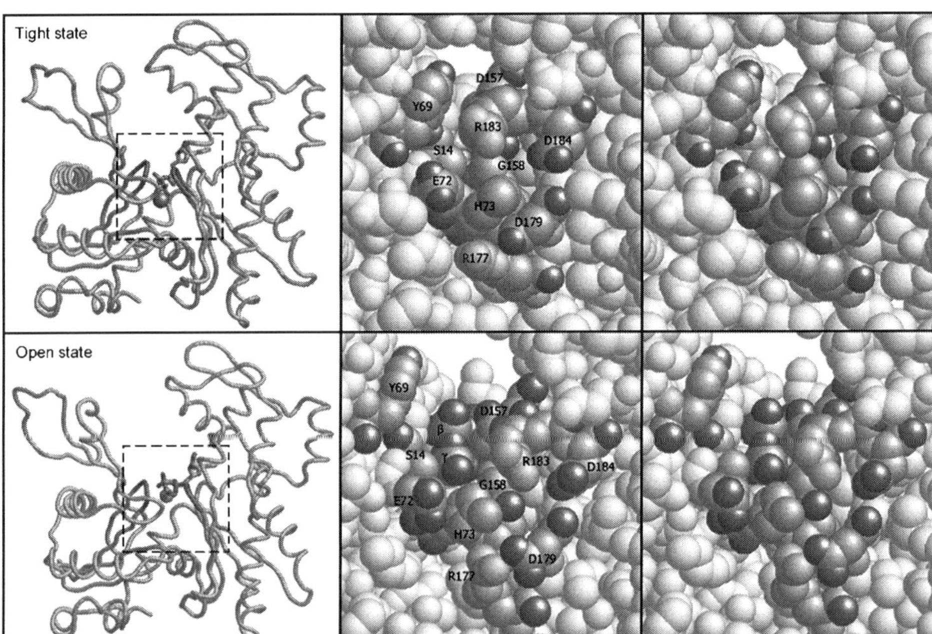

Figure 1. Comparison of the tight and open states of actin in profilin:β-actin crystals. The overview to the left outlines the area shown in stereo containing the barrier residues E72, H73, R177, D179, and R183, the phosphate-binding loop residues S14, D157, and G158, and finally the hydrogen bond-forming residues Y69 and R183. In the open state (bottom row) the β- and γ-phosphates (yellow) of the ATP are exposed. Pdb accession codes 2BTF and 1HLU, respectively. Figure published in *J. Mol. Biol.* (2002) 317, 577–589. (See Color Insert.)

exist in different conformations, depending on the status of the actin-bound ATP, the nature of the divalent cation (Ca^{2+} or Mg^{2+}) at the high-affinity site or at additional sites, and the degree of oligomerization (Moraczewska *et al.*, 1999; Schüler, 2001).

The structures of different actin orthologues cocrystallized with different actin-binding proteins all have a closed nucleotide-binding cleft (Vorobiev *et al.*, 2003), corresponding to the tight state found for β-actin in the profilin:β-actin crystals (Schutt *et al.*, 1993). The conformation of actin appears relatively unchanged regardless of whether the nucleotide is ATP or ADP, or the tightly bound cation is Ca^{2+} or Mg^{2+}. In the light of many observations indicating that the actin can attain different conformations depending on the nature of bound ligands, this may seem contradictory (Schüler, 2001; Strzelecka-Golaszewska, 2001). The relative invariability in most actin crystal structures, however, may be explained by clamping of the two domains by DNase I binding across the cleft between subdomains 2 and 4, or by packing interactions in the case of the gelsolin subfragment 1:α-actin crystals, as discussed previously (Nyman *et al.*, 2002; Sablin *et al.*, 2002). This may also be the case in the crystal structure of tetramethylrhodamine maleimide (TMR)-derivatized α-actin in the ADP state (Otterbein *et al.*, 2001). In this tight state structure, however, subdomain 2 has attained a different structure as compared with previously determined α-actin with a small helical stretch apparently formed through rotation around the same hinge region that is involved in the movement of subdomain 2 in the tight-to-open state transition of profilin:β-actin (Chik *et al.*, 1996). It should be noted that the rotation of subdomain 1 with respect to subdomains 3 and 4, intrinsic to the opening of the nucleotide-binding cleft in profilin:β-actin, is not seen in TMR–actin. It was reported that binding of TMR to actin does not significantly influence DNase I–actin interaction or the susceptibility of the actin to subtilisin cleavage in subdomain 2, which implies that the solution structure of TMR–actin is closely similar to the nonconjugated protein (Kudryashov and Reisler, 2003). Another possibility is that the binding of TMR between subdomains 1 and 3 locks the protein in the tight state. There is evidence for allosteric coupling between the C-terminus and subdomain 2. For instance, there is an increase in the K_{diss} for the profilin–actin interaction after introduction of a P38A mutation (subdomain 2), and reciprocally, replacing cysteine 374 with a serine lowered the affinity of the actin for DNase I binding (subdomain 2) (Aspenström *et al.*, 1993). Likewise, removal of C-terminal residues of actin affected the proteolytic sensitivity of subdomain 2 (Strzelecka-Golaszewska *et al.*, 1993). Figure 2 further illustrates the influence of the C-terminal C374S mutation on the thermal stability of actin in the Mg^{2+}- as well as the Ca^{2+}-bound state as assayed by DNase I inhibition (Schüler *et al.*, 2000a). Clearly, the mechanisms behind the allosteric coupling indicated by these modifications are still unclear.

III. PROFILIN:β-ACTIN CRYSTAL

In the profilin:β-actin crystal, the actin molecules are also bridged across subdomains 2 and 4. Here, a neighboring actin molecule is responsible for the bridging, and the cleft can open and close in response to changes in ionic conditions, despite the bridging (Fig. 3). This is possible through intramolecular hinge and shear movements

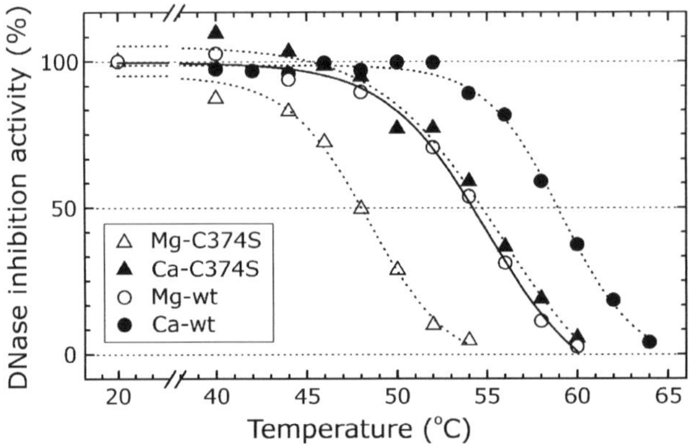

Figure 2. Thermal stability of β-actin carrying the C-terminal mutation C374S. Melting curves of monomeric actins determined with the DNase I inhibition assay as described earlier (Schüler *et al.*, 2000a).

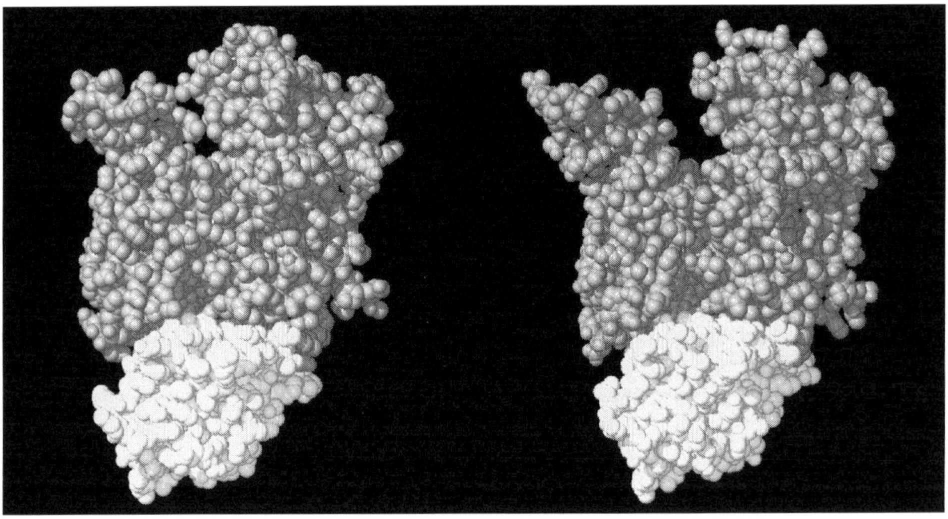

Figure 3. The profilin:β-actin closed and open states (pdb accession codes 2BTF and 1HLU, respectively). Overview of the crystal structures of profilin:β-actin solved in 3.5 M NH_4SO_4 (closed state, left) and 1.8 M potassium phosphate (open state, right) in the presence of Ca^{2+} and ATP, illustrating the magnitude of the conformational difference between the two states. The polyphosphate tail of the ATP is exposed in the open state (oxygen O1beta, red dot). See also Fig. 1. (See Color Insert.)

coordinated between the bound actin molecules (Schutt *et al.*, 1989; Chik *et al.*, 1996; Page *et al.*, 1998). Exchange of ADP and AMP for ATP in the profilin:β-actin crystals significantly influences their diffraction (Schutt *et al.*, 1989). In the tight state structure, the terminal phosphates of the nucleotide are buried (see also Fig. 1, upper panels), held by the β-hairpin loops, N12-C17 (P1) and D156-V159 (P2), which protrude into the interdomain cleft from subdomains 1 and 3, respectively. The β-phosphate is hydrogen bonded to the amide nitrogens of S14, G15, M16, and D157, and the γ-phosphate is bound to amide nitrogens of S14, D157, G158, and V159. The most dramatic difference between the two states, the opening of the interdomain cleft, results in an outward shift of the N12-C17 loop, exposing the phosphate tail of ATP to solution (Fig. 1, lower panels) (Chik *et al.*, 1996).

In the open state, the hydrogen bond with the amide nitrogen of G15 shifts from the O1 oxygen to the O2 oxygen of the β-phosphate, while the hydrogen bonds of G158 and V159 to the γ-phosphate are broken (Chik *et al.*, 1996). In the tight state, there are two hydrogen bonds that span the nucleotide-binding cleft. These two bridging hydrogen bonds, between MeH73 and the carbonyl oxygen of G158, and between the guanido group of R183 and π-electrons of the ring of Y69 (for bond type see Levitt and Perutz, 1988), stabilize the closed ATP form of the actin molecule. In the open state, these hydrogen bonds are broken.

IV. INTERDOMAIN CONNECTIVITY IN ACTIN

Exchange of ADP for ATP in actin results in a significant reduction in the stability of the protein, and removal of the nucleotide leads to rather rapid loss in polymerizability (Asakura and Oosawa, 1960), demonstrating the importance of the ATP γ-phosphate for holding the two major domains in position. In addition to the loop-phosphate-loop links and the bridging hydrogen bonds discussed earlier, the two major domains of actin are connected through a charge network involving mostly long-chained residues (Fig. 1): E72 and MeH73 from one side of the cleft and D157, R177, D179, and R183 from the other. The presence of a methyl group on the ε2-nitrogen of the imidazole ring of H73 increases the basicity of the δ1-nitrogen, thereby strengthening the hydrogen bond connecting this nitrogen with the carbonyl group of G158. These residues shield the ATP phosphates from the solvent on one side of the molecule. On the opposite side, there is another set of large residues (M16, K18, K336, and Y337) separating the polyphosphate tail from solvent in both the open and tight states. The transition from the tight to open state does not result in any major changes in this barrier region.

Investigations of the effect of mutations in the nucleotide-binding cleft on the spatial relationship between the major domains of actin have shown that the H73A mutation, as well as mutations in the loops binding the phosphates, causes a significant decrease in the affinity for DNase I (Chen and Rubenstein, 1995; Schüler *et al.*, 1999, 2000b; Nyman *et al.*, 2002). The mutations H73A, R177D, S14C, and the double mutation S14C/D157A all destabilized the molecule at increased temperatures, caused increased nucleotide exchange rates, and reduced polymerization rates. Replacing H73 with positively charged residues (arginine or lysine) made the actin more stable, whereas introduction of glutamic acid destabilized the protein (Yao *et al.*, 1999), further illustrating the coordinating position of H73 in the charge network

(Fig. 1) and its importance for the stability and polymerizability of actin (Nyman *et al.*, 2002). See also discussion of MeH73 later.

V. ACTIN ATPASE

A. Monomeric Actin Hydrolyzes ATP

Addition of salts (including Mg^{2+} ions) to a solution of monomeric actin, causing polymerization, increases the rate of ATP hydrolysis approximately by a factor of 100 (Pollard and Weeds, 1984), suggesting a tight coupling between actin–actin interactions and the hydrolysis of ATP. Thus, the slow hydrolysis of ATP in buffers stabilizing the monomeric form of actin has been seen as a consequence of the formation of unstable oligomers of actin in the solution and not as an expression of an *intrinsic* ATPase activity of the monomer (Mozo-Villarias and Ware, 1985; Newman *et al.*, 1985). However, as demonstrated in Fig. 4, the ATPase activity per actin monomer under nonpolymerizing conditions is independent of the total monomer concentration over a broad range of actin concentrations, strongly suggesting that the ATPase activity of actin is independent of oligomerization, that is, monomeric actin has an intrinsic ATPase activity. This activity ($0.6 \pm 0.11\ h^{-1}$) is in the same range as the activities of the heat shock protein Hsc70 or its isolated ATPase domain (Ha and McKay, 1994;

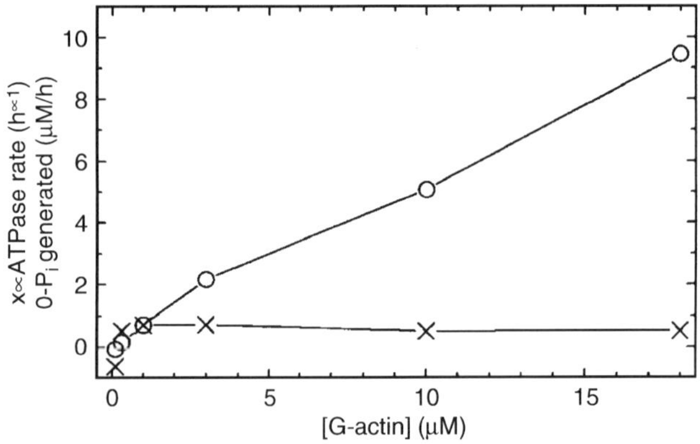

Figure 4. Actin in the monomeric state has ATPase activity. Bovine cytoplasmic β/γ-actin was prepared from calf thymus as described (Lindberg *et al.*, 1988). Monomeric actin in G-ATPase buffer (5 mM tris-HCl pH 7.6, 0.5 mM ATP, 0.67 μM [γ-^{32}P]-ATP (3000 Ci/mmol; Amersham Pharmacia), 0.1 mM $CaCl_2$, 0.5 mM DTT) was converted to Mg-actin by incubation with 0.2 mM EGTA + 50 μM $MgCl_2$ for 15 min at RT. The actin was diluted in the same buffer, including all components listed, to concentrations of 0.1, 0.3, 1.0, 3.0, 10.0, and 18.7 μM and incubated at 25 °C. Sample aliquots were removed at intervals over a 2-h period and spotted onto PEI-cellulose sheets (Merck). After drying under a light bulb, TLC was performed in 0.2 M ammonium bicarbonate, pH 8.0. Radioactivity in the ATP and the P_i spots was measured by phosphoimager analysis (Applied Biosystems). G-actin ATPase rates were determined by linear curve fitting (Schüler, 2000d).

Wilbanks *et al.*, 1994). The stimulation of the actin ATPase activity seen after addition of polymerizing salts is due to the changes in ionic conditions and conformational changes occurring during subsequent incorporation of the actin monomers into filaments.

B. Polymer Formation and ATP Hydrolysis

Filament formation *in vitro* is characterized by a distinct lag phase, the rate-limiting step being the formation of nuclei consisting of three to four actin monomers (Kasai *et al.*, 1962). Support for a linkage between ATPase activity and polymer formation has come from experiments in which filament formation was either stimulated by or interfered with different actin-binding proteins, mutations in actin, or actin-modifying reagents (Brenner and Korn, 1980; Tobacman and Korn, 1982; Tellam, 1986; Polzar *et al.*, 1989; Geipel *et al.*, 1990; Dancker *et al.*, 1991; Hayden *et al.*, 1993; Kasprzak, 1994; Schüler *et al.*, 2000b; Schüler, 2001).

Addition of salts to physiological concentrations is coupled to conformational changes, involving the nucleotide-binding site, making the actin assembly-competent (Higashi and Oosawa, 1965; Rich and Estes, 1976; Rouayrenc and Travers, 1981; Carlier *et al.*, 1986; Merkler *et al.*, 1987). Kinetic evidence suggest the formation of an intermediate referred to as G*-actin, whose formation by itself does not stimulate the actin ATPase activity. Apparently, the formation of this intermediate depends on the binding of monovalent and divalent cations to a polyanionic surface of the actin molecule (Barany *et al.*, 1962; Rouayrenc and Travers, 1981; Pardee and Spudich, 1982). There are also biochemical and physical evidence that the early phase of *de novo* polymerization involves the formation of a special actin dimer, which subsequently seems to be used for filament growth (Steinmetz *et al.*, 1997). In view of the results presented in Fig. 4, it would be interesting to use matrix-coupled actin monomers in an attempt to single out the effects of salt on the ATPase activity from the cooperative conformational changes occurring during the incorporation of actin monomers into filaments.

C. The Actin–ATP/ADP·P_i Cap

The actin filament is structurally, as well as functionally, asymmetric, which *in vitro* is reflected in a difference in rate of addition of actin monomers to the two ends. Elongation at the fast polymerizing end [the (+)-end (barbed end)], is 10- to 20-fold faster than at the slow polymerizing end [the (–)-end (pointed end)]. ATP–actin with bound Mg^{2+} has an on-rate that is faster than the on-rate of ADP–actin at both ends, and at the (–)-end ATP–actin dissociates faster than ADP–actin (Bonder and Mooseker, 1983; Lal *et al.*, 1984; Pollard, 1986; Selden *et al.*, 1986). Thus, during the initial phase of polymerization *in vitro* in the presence of excess ATP, ATP–actin is rapidly and preferentially incorporated into filaments at their (+)-end.

During fast filament elongation, ATP hydrolysis and subsequent P_i release is slower than addition of ATP monomers, resulting in the formation of a detectable ATP/ADP·P_i cap at the (+)-end of the growing filament (Carlier and Pantaloni, 1986; Korn *et al.*, 1987; Pinaev *et al.*, 1995; Melki *et al.*, 1996). It has been argued that hydrolysis occurs preferentially at the boundary between the ATP cap and the ADP·P_i-containing monomers inside the filament (Korn *et al.*, 1987). This would imply that the ADP·P_i–actin monomer has a different structure than the ATP monomer and that the ADP·P_i monomer has a propensity to accelerate ATP hydrolysis on the adjacent ATP monomer. Results reported by others suggest random hydrolysis of ATP within newly formed stretches of the ATP–actin polymer (Ohm and Wegner, 1994; Pieper and Wegner, 1996).

The polymerization reaction does not reach thermodynamic equilibrium. Instead, the different rates of monomer association and dissociation at the two ends eventually result in a steady state in which the net incorporation of ATP–actin at the (+)-end equals the loss of ADP–actin at the (–)-end. As long as there is ATP in the solution, the steady state is characterized by a constant flux of actin monomers through the filaments, a phenomenon referred to as treadmilling (Wegner, 1976; Neuhaus *et al.*, 1983). The rate of treadmilling is determined not only by the combined association and dissociation rate constants at the filament ends but also by the rate of nucleotide exchange on ADP monomers coming off the pointed end. In a solution of purified actin, the nucleotide exchange reaction is the rate-limiting step (Kinosian *et al.*, 1993), and the ATP-actin cap persists at steady state as long as there are ATP-actin monomers available for incorporation.

The atomic structure of the actin filament (F-actin) is not known. Consequently, structural transitions in actin that accelerate ATP hydrolysis also remain to be elucidated. Cryoelectron microscopy has demonstrated that there is a structural difference between ATP/ADP·P_i filaments and ADP-containing filaments, where the latter, that is, the ground state, has the most well-ordered structure (Lepault *et al.*, 1994). A difference between newly formed actin polymers, presumably consisting of ATP/ADP·P_i-actin and filaments consisting of ADP-actin, is further demonstrated by preferential binding of the filament-nucleating Arp2/3 complex to the former *in vitro* (Ichetovkin *et al.*, 2002). ATP hydrolysis and P_i release destabilizes monomer: monomer bonds at filament ends making the ADP polymer dynamic (Rickard and Sheterline, 1986, 1988; and the preceding references).

VI. MECHANISM OF ATP HYDROLYSIS ON ACTIN

A. Active Site Nucleophile

In the actin-related heat shock 70 proteins, ATP hydrolysis likely involves in-line attack on the ATP γ-phosphate by a hydroxyl ion coordinated by K71 (O'Brien *et al.*, 1996; Rajapandi *et al.*, 1998). In actin, the only basic side chains in the vicinity of the γ-phosphate, R177 and H73, have been shown to be nonessential for catalysis by directed mutagenesis (Schüler *et al.*, 2000b; Nyman *et al.*, 2002). High-affinity Mg^{2+} boosts the ATPase activity of monomeric actin 20- to 30-fold as compared with Ca^{2+} (Geipel *et al.*, 1990; Chen and Rubenstein, 1995; Schüler *et al.*, 1999). Thus, the catalytic activity is regulated via the coordination sphere of the metal cofactor at the base of the cleft. In immediate proximity of the divalent cation, Q137 or H161 may coordinate a hydroxyl ion or water molecule. Structures of nonvertebrate actins suggest that a Q137-bound water molecule may act as a catalytic nucleophile. For a detailed illustration of the catalytic mechanism see Vorobiev *et al.* (2003). A monovalent cation that may coordinate nucleophilic water in Hsc70 (Wilbanks and McKay, 1995) has not been observed in actin.

The location of the hydroxyl of S14 in actin is within hydrogen-bonding distance of the γ-phosphate of ATP, suggesting its involvement in the ATPase reaction. This residue is one of a number of ligands binding to the the γ-phosphate of ATP, thereby stabilizing the actin–ATP complex. In yeast actin, mutation of Ser-14 to Ala (S14A) causes a temperature-sensitive phenotype *in vivo* and temperature-sensitive polymerization defects *in vitro* (Chen and Rubenstein, 1995). It also decreases in the intrinsic ATPase activity of both Ca- and Mg-G-actin at 30 °C and alters the protease susceptibility of sites on subdomain 2. It was proposed that the Ser-14 hydroxyl forms a polar bridge between the ATP γ-phosphate and the amide nitrogen of Gly-74, thus conferring additional stability on the actin small domain.

The mutant S14C in yeast actin does not support growth (Chen and Rubenstein, 1995), but mutant S14C-β-actin can be coexpressed with endogenous yeast actin, and is isolated free of the endogenous protein allowing the investigation of its ATPase activity. The S14C-β-actin retains ATPase activity (Schüler *et al.*, 1999), and Cys-14 in S14C mutant actin reacts covalently with the sulfhydryl of ATPγS (Fig. 5 and Schüler *et al.*, 2000c). This leaves the possibility of a transient phosphoserine formation during the course of ATP hydrolysis. A phosphorylated actin has not been described as an intermediate in the ATPase reaction, but this might be due to instability of such a species. Heat shock 70 proteins are known to undergo autophosphorylation on a threonine residue (T199 in DnaK). However, the function of this reaction and its implications for the mechanism of ATP hydrolysis are still unclear, especially since they seem to vary between members of the protein class (McCarty and Walker, 1991; Gaut and Hendershot, 1993; O'Brian and McKay, 1993; Barthel *et al.*, 2001). Therefore, it is possible that actin with phosphoserine at position 14 is an intermediate in a switch mechanism partitioning the release of free energy after ATP hydrolysis.

B. Catalytic Base(s)

In many phosphoryl transferases, the active site nucleophile is activated by a nearby side chain. Mutational analyses of Hsp70 proteins have shown that not only the glutamate or aspartate in the position corresponding to actin Q137 but also other nearby carboxylic side chains are important for full catalytic activity (Gaut and Hendershot, 1993; McCarty and Walker, 1994; Wilbanks *et al.*, 1994; Kamath-Loeb

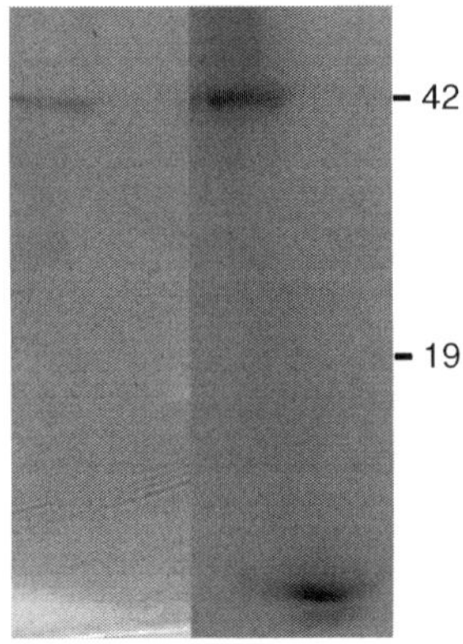

Figure 5. Covalent binding of ATPγS to S14C-actin. Monomeric Mg-actin (10 μM) carrying the S14C replacement was incubated with 0.1 mM of ATPγ-^{35}S for 1 h. Excess nucleotide was removed by gel filtration before the protein was denatured by addition of 2% SDS and subjected to SDS-PAGE under nonreducing conditions. An autoradiograph (right lane) of the Coomassie-stained gel (left lane) showed that approximately one-third of the total radioactivity resided in the protein band, illustrating that a transfer of the ^{35}S from the nucleotide to the protein had occurred, most likely to cysteine at position 14.

et al., 1995). Actin Q137 is unlikely to be deprotonated under physiological conditions and is therefore a poor base. Given that the ATPase reaction might proceed without conserved symmetry, residues D11 and D154 could be catalytic bases. Semiconservative replacements of D11 in yeast actin lead to mild defects, whereas charge reversions as well as the double replacement D154A, D157A are lethal (Cook *et al.*, 1992, 1993; Wertman *et al.*, 1992). The ATPase activities of these mutant proteins, however, have not been tested.

The replacement V159N in yeast actin causes an increased release of inorganic phosphate and a high rate of filament turnover, while ATP hydrolysis itself seems unaffected (Belmont *et al.*, 1999). These results were interpreted as an uncoupling of P_i release from a conformational change, which destabilizes the actin filament. Thus, V159 is necessary for harnessing the free energy change of P_i release for a discrete step. In actin V159 is conserved except for a few sequences that have isoleucine in this position, while in the 70-kDa heat shock proteins a threonine is highly conserved in the corresponding position. This small hydrophobic barrier shielding the γ-phosphate from solvent seems to have evolved as a special feature for actin.

VII. ACTIN METHYLHISTIDINE 73, ATPASE, PHOSPHATE RELEASE, AND POLYMERIZATION

Saccharomyces cerevisiae, which is used for the expression of wild-type and mutant β-actin, does not methylate histidine 73 (Yao *et al.*, 1999). This was utilized in setting up a series of experiments to elucidate the role of the histidine as well as its methylation in ATP hydrolysis and P_i release and polymerization of the actin. For this, wild-type β-actin (with MeHis73) isolated from calf thymus, β-actin expressed in yeast (nonmethylated), and mutant H73A β-actin also expressed in yeast were used (Nyman *et al.*, 2002). As shown in Fig. 6A, bovine β-actin hydrolyzed ATP only slightly ahead of polymerization and phosphate release, whereas β-actin expressed in yeast (nonmethylated) hydrolyzed ATP and released P_i well ahead of polymer formation (Fig. 6B). These results were at odds with earlier work reporting a sequence of events in which ATP hydrolysis and polymer formation went hand in hand, whereas P_i release was significantly delayed (Carlier *et al.*, 1986; Melki *et al.*, 1996). To clarify this, experiments were performed with rabbit skeletal muscle actin (α-actin). As shown in Fig. 6C, α-actin (methylated) released P_i only after polymers had formed, corroborating earlier results.

Comparison of β-actin expressed in yeast (nonmethylated) with the mutant H73A-β-actin also expressed in yeast showed that in both cases hydrolysis of ATP and P_i release preceded polymer formation, suggesting an uncoupling of the hydrolysis and product release from filament formation and that most likely in these cases polymers form from actin monomers with ADP on them. This is reasonable, since actin with bound ADP (Higashi and Oosawa, 1965; Kasai *et al.*, 1962; Pollard, 1984) and actin with nonhydrolyzable nucleotide analogues (Cooke and Murdoch, 1973) can polymerize. Thus, neither ATP hydrolysis nor bound ATP is needed for polymerization to occur, although there may be significant differences in the quality of the filament formed from the different actin states. The H73A mutant β-actin did not form filaments in the absence of Mg^{2+} ions and ATP hydrolysis was very slow (Nyman *et al.*, 2002). The nonmethylated β-actin did form filaments in the absence of Mg^{2+}, albeit at a slow speed, further emphasizing the importance of a histidine in position 73 in keeping the actin in a polymerizable state. The difference in the kinetics of polymer formation suggests that proper polymers form only with actin having a MeH73 in it. It should also be noted that there is a clear isoform difference in that α-actin holds on to the P_i much longer than native β-actin does, something which may be related to a difference in force generation in the highly organized myofibrillar actomyosin system as compared with the less stable MFS in nonmuscle cells, where instead rapid actin reorganization is crucial for the function (Nyman *et al.*, 2002).

As shown in Fig. 1, MeH73 and D184 have moved apart in the open state of the β-actin, allowing D184 to form a salt bridge with R183 rather than H73, and R177 of β-actin has moved from hydrogen bonding with the backbone atoms of MeH73 to a salt bridge interaction with the D179. In the presence of Mg^{2+} ions, the actin ATPase activity is greatly stimulated. Under these conditions, the region near Y69 (R62–K68) is protected from proteolysis (Strzelecka-Golaszewska *et al.*, 1993), suggesting that the interdomain cleft is closed. Thus, it is possible that the binding of Mg^{2+} ions stabilizes the tight state of actin allowing nucleotide hydrolysis to take place. Following ATP hydrolysis, breakage of the interdomain bridges might allow the opening of the cleft, facilitating release of the γ-phosphate directly into the solvent. Such a mechanism is supported by the fact that actin can attain a state in which the interdomain cleft is opened up (Chik *et al.*, 1996). It suggests that changes in the conformation of

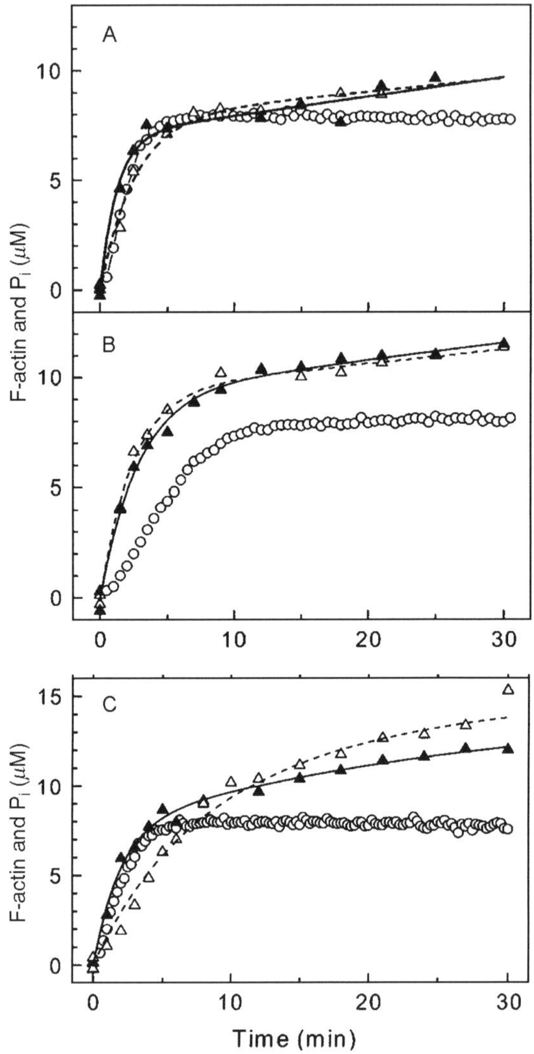

Figure 6. Comparison of native β-actin (MeHis73) with yeast-expressed β-actin (H73) and native α-actin with respect to ATP hydrolysis, phosphate release, and polymerization. Polymerization of Mg-actin (8 μM) was induced by 0.1 M KCl and 1 mM $MgCl_2$ and monitored using the pyrenyl assay (open circles). Samples were withdrawn at the time points indicated and analyzed for ATP hydrolysis (closed triangles) and P_i release (open triangles) as described earlier (Nyman *et al.*, 2002). (A) β-Actin from bovine thymus (methylated), (B) yeast-expressed β-actin (unmethylated), and (C) α-actin (methylated). (C) The measured amount of P_i released exceeds the total amount of ATP hydrolyzed. This anomaly depends on the experimental design and does not invalidate the conclusion. In fact, the delay in the P_i release from α-actin would be even more pronounced if the anomaly was corrected. Figure published in (2002) *J. Mol. Biol.* **317,** 577–589.

actin opens the interdomain cleft to allow the inorganic phosphate to escape directly into solution rather than through a narrow "backdoor" channel (Wriggers and Schulten, 1997).

VIII. IMPORTANCE OF THE STATUS OF THE ACTIN-BOUND NUCLEOTIDE

Only little information is available about the conformation of actin monomers in filaments formed under different conditions, but it is generally believed that hydrolysis of ATP on actin *in vivo* has a regulatory function in determining the structure and the dynamic turnover of actin monomers and their interactions with actin-binding proteins. In fact, actin depolymerizing factors (ADF/cofilins) (Carlier *et al.*, 1997; Blanchoin and Pollard, 1998), twinfilin (Palmgren *et al.*, 2001), adenylate cyclase-associated protein (CAP; Mattila *et al.*, 2004), and gelsolin (Laham *et al.*, 1993) bind preferentially to ADP–actin, while DNase I (Schüler *et al.*, 2000c), profilin (Vinson *et al.*, 1998), thymosin β4 (Carlier *et al.*, 1993), Arp2/3 (Ichetovkin *et al.*, 2000), and Wiskott–Aldrich syndrome protein (WASP) homology domain 2 (Chereau *et al.*, 2005) bind preferentially to ATP–actin. Binding of ADF–cofilin to actin filaments is strongly enhanced by P_i release (Blanchoin and Pollard, 1999). Thus, ATP hydrolysis is an integral part in the regulation of the function of actin and actin filaments *in vivo*.

In cells, the ultimate precursor in the formation of actin filaments is profilin:actin, and polymerization takes place either onto preexisting free filament ends or at specific sites formed by polymerization-promoting protein complexes (Hajkova *et al.*, 2000; Grenklo *et al.*, 2003; Higashida *et al.*, 2004; Romero *et al.*, 2004). Profilin effectively accelerates the exchange of the nucleotide on actin by opening up the nucleotide-binding cleft (Chik *et al.*, 1996). Also mutant profilins, which bind actin only weakly, stimulate nucleotide exchange (Korenbaum *et al.*, 1998). Profilin binds more strongly to actin–ADP than cofilin. Therefore, profilin efficiently forms profilin–ATP–actin from the cofilin–ADP–actin, which comes off the (–)-end of depolymerizing filaments. Profilin also inhibits the ATPase activity of the actin monomer, and thus ensures the delivery of ATP–actin for incorporation into growing filaments by actin-polymerizing machineries (Dickinson *et al.*, 2002). Thus, treadmilling occurs *in vivo* and is at the heart of myosin-independent, actin-based translocations, and the loss and addition of actin monomers is strictly regulated by auxiliary proteins.

REFERENCES

Asakura, S. and Oosawa, F. (1960). Dephosphorylation of adenosine triphosphate in actin solutions at low concentrations of magnesium. *Arch. Biochem. Biophys.* **87,** 273–280.

Aspenström, P., Schutt, C. E., Lindberg, U., and Karlsson, R. (1993). Mutations in beta-actin: Influence on polymer formation and on interactions with myosin and profilin. *FEBS Lett.* **329,** 163–170.

Barany, M., Finkelman, F., and Therattil-Antony, T. (1962). Studies on the bound calcium of actin. *Arch. Biochem. Biophys.* **98,** 28–45.

Barthel, T. K., Zhang, J., and Walker, G. C. (2001). ATPase-defective derivatives of Escherichia coli DnaK that behave differently with respect to ATP-induced conformational change and peptide release. *J. Bacteriol.* **183,** 5482–5490.

Belmont, L. D., Orlova, A., Drubin, D. G., and Egelman, E. H. (1999). A change in actin conformation associated with filament instability after Pi release. *Proc. Natl. Acad. Sci. USA* **96,** 29–34.

Blanchoin, L. and Pollard, T. D. (1998). Interaction of actin monomers with Acanthamoeba actophorin (ADF/cofilin) and profilin. *J. Biol. Chem.* **273,** 25106–25111.

Blanchoin, L. and Pollard, T. D. (1999). Mechanism of interaction of Acanthamoeba actophorin (ADF/cofilin) with actin filaments. *J. Biol. Chem.* **274,** 15538–15546.

Bonder, E. M. and Mooseker, M. S. (1983). Direct electron microscopic visualization of barbed end capping and filament cutting by intestinal microvillar 95-k dalton protein (villin): A new actin assembly assay using the Limulus acrosomal process. *J. Cell Biol.* **96,** 1097–1107.

Brenner, S. L. and Korn, E. D. (1980). The effects of cytochalasins on actin polymerization and actin ATPase provide insights into the mechanism of polymerization. *J. Biol. Chem.* **255,** 841–844.

Carlier, M. F. and Pantaloni, D. (1986). Direct evidence for ADP-Pi-F-actin as the major intermediate in ATP-actin polymerization. *Biochemistry* **25,** 7789–7792.

Carlier, M. F., Pantaloni, D., and Korn, E. D. (1986). Fluorescence measurements of the binding of cations to high-affinity and low-affinity sites on ATP-G-actin. *J. Biol. Chem.* **261,** 10778–10784.

Carlier, M. F., Jean, C., Rieger, K. J., Lenfant, M., and Pantaloni, D. (1993). Modulation of the interaction between G-actin and thymosin beta 4 by the ATP/ADP ratio: Possible implication in the regulation of actin dynamics. *Proc. Natl. Acad. Sci. USA* **90,** 5034–5038.

Carlier, M. F., Laurent, V., Santolini, J., Melki, R., Didry, D., Xia, G.-X., Hong, Y., Chua, N.-H., and Pantaloni, D. (1997). Actin depolymerizing factor (ADF/cofilin) enhances the rate of filament turnover: Implication in actin-based motility. *J. Cell Biol.* **136,** 1307–1322.

Chen, X. and Rubenstein, P. A. (1995). A mutation in an ATP-binding loop of *Saccharomyces cerevisiae* actin (S14A) causes a temperature-sensitive phenotype *in vivo* and *in vitro*. *J. Biol. Chem.* **270,** 11406–11414.

Chereau, D., Kerff, F., Graceffa, P., Grabarek, Z., Langsetmo, K., and Dominguez, R. (2005). Actin-bound structures of Wiskott–Aldrich syndrome protein (WASP)-homology domain 2 and the implications for filament assembly. *Proc. Natl. Acad. Sci. USA* **102,** 16644–16649.

Chik, J. K., Lindberg, U., and Schutt, C. E. (1996). The structure of an open state of beta-actin at 2. 65 A resolution. *J. Mol. Biol.* **263,** 607–623.

Cook, R. K., Blake, W. T., and Rubenstein, P. A. (1992). Removal of the amino-terminal acidic residues of yeast actin. Studies *in vitro* and *in vivo*. *J. Biol. Chem.* **267,** 9430–9496.

Cook, R. K., Root, D., Miller, C., Reisler, E., and Rubenstein, P. A. (1993). Enhanced stimulation of myosin subfragment 1 ATPase activity by addition of negatively charged residues to the yeast actin NH2 terminus. *J. Biol. Chem.* **268,** 2410–2415.

Cooke, R. and Murdoch, L. (1973). Interaction of actin with analogs of adenosine triphosphate. *Biochemistry* **12,** 3927–3932.

Dancker, P., Hess, L., and Ritter, K. (1991). Product release is not the rate-limiting step during cytochalasin B-induced ATPase activity of monomeric actin. *Z. Naturforsch. [C]* **46,** 139–144.

Dickinson, R. B., Southwick, F. S., and Purich, D. L. (2002). A direct-transfer polymerization model explains how the multiple profilin-binding sites in the actoclampin motor promote rapid actin-based motility. *Arch. Biochem. Biophys.* **406,** 296–301.

Gaut, J. R. and Hendershot, L. M. (1993). Mutations within the nucleotide binding site of immunoglobulin-binding protein inhibit ATPase activity and interfere with release of immunoglobulin heavy chain. *J. Biol. Chem.* **268,** 7248–7255.

Geipel, U., Just, I., and Aktories, K. (1990). Inhibition of cytochalasin D-stimulated G-actin ATPase by ADP-ribosylation with Clostridium perfringens iota toxin. *Biochem. J.* **266,** 335–339.

Grenklo, S., Geese, M., Lindberg, U., Wehland, J., Karlsson, R., and Sechi, A. S. (2003). A crucial role for profilin-actin in the intracellular motility of Listeria monocytogenes. *EMBO Rep.* **4,** 526–529.

Ha, J. H. and McKay, D. B. (1994). ATPase kinetics of recombinant bovine 70 kDa heat shock cognate protein and its amino-terminal ATPase domain. *Biochemistry* **33,** 14625–14635.

Hajkova, L., Nyman, T., Lindberg, U., and Karlsson, R. (2000). Effects of cross-linked profilin: β/γ-Actin on the dynamics of the microfilament system in cultured cells. *Exp. Cell Res.* **256,** 112–121.

Hayden, S. M., Miller, P. S., Brauweiler, A., and Bamburg, J. R. (1993). Analysis of the interactions of actin depolymerizing factor with G- and F-actin. *Biochemistry* **32,** 9994–10004.

Higashi, S. and Oosawa, F. (1965). Conformational changes associated with polymerization and nucleotide binding in actin molecules. *J. Mol. Biol.* **12,** 843–865.

Higashida, C., Miyoshi, T., Fujita, A., Oceguera-Yanez, F., Monypenny, J., Andou, Y., Narumiya, S., and Watanabe, N. (2004). Actin polymerization-driven molecular movement of mDia1 in living cells. *Science* **303,** 2007–2010.

Ichetovkin, I., Grant, W., and Condeelis, J. (2002). Cofilin produces newly polymerized actin filaments that are preferred for dendritic nucleation by the Arp2/3 complex. *Curr. Biol.* **12,** 79–84.

Kabsch, W., Mannherz, H. G., Suck, D., Pai, E. F., and Holmes, K. C. (1990). Atomic structure of the actin: DNase I complex. *Nature* **347,** 37–44.

Kamath-Loeb, A. S., Lu, C. Z., Suh, W. C., Lonetto, M. A., and Gross, C. A. (1995). Analysis of three DnaK mutant proteins suggests that progression through the ATPase cycle requires conformational changes. *J. Biol. Chem.* **270,** 30051–30059.

Kasai, M., Asakura, S., and Oosawa, F. (1962). The cooperative nature of G-F transformation of actin. *Biophys. Biochem. Acta* **57,** 22–31.

Kasprzak, A. A. (1994). Myosin subfragment 1 activates ATP hydrolysis on Mg^{2+}-G-actin. *Biochemistry* **33,** 12456–12462.

Kinosian, H. J., Selden, L. A., Estes, J. E., and Gershman, L. C. (1993). Nucleotide binding to actin. Cation dependence of nucleotide dissociation and exchange rates. *J. Biol. Chem.* **268,** 8683–8691.

Korenbaum, E., Nordberg, P., Bjorkegren-Sjögren, C., Schutt, C. E., Lindberg, U., and Karlsson, R. (1998). The role of profilin in actin polymerization and nucleotide exchange. *Biochemistry* **37,** 9274–9283.

Korn, E. D., Carlier, M. F., and Pantaloni, D. (1987). Actin polymerization and ATP hydrolysis. *Science* **238,** 638–644.

Kudryashov, D. S. and Reisler, E. (2003). Solution properties of tetramethylrhodamine-modified G-actin. *Biophys. J.* **85,** 2466–2475.

Laham, L. E., Lamb, J. A., Allen, P. G., and Janmey, P. A. (1993). Selective binding of gelsolin to actin monomers containing ADP. *J. Biol. Chem.* **268,** 14202–14207.

Lal, A. A., Brenner, S. L., and Korn, E. D. (1984). Preparation and polymerization of skeletal muscle ADP-actin. *J. Biol. Chem.* **259,** 13061–13065.

Lepault, J., Ranck, J. L., Erk, I., and Carlier, M. F. (1994). Small angle X-ray scattering and electron cryomicroscopy study of actin filaments: Role of the bound nucleotide in the structure of F-actin. *J. Struct. Biol.* **112,** 79–91.

Levitt, M. and Perutz, M. F. (1988). Aromatic rings act as hydrogen bond acceptors. *J. Mol. Biol.* **204,** 751–754.

Lindberg, U., Schutt, C. E., Hellsten, E., Tjader, A. C., and Hult, T. (1988). The use of poly(L-proline)-Sepharose in the isolation of profilin and profilactin complexes. *Biochem. Biophys. Acta* **967,** 391–400.

Mattila, P. K., Quintero-Monzon, O., Kugler, J., Moseley, J. B., Almo, S. C., Lappalainen, P., and Goode, B. L. (2004). A high-affinity interaction with ADP-actin monomers underlies the mechanism and *in vivo* function of Srv2/cyclase-associated protein. *Mol. Biol. Cell* **15,** 5158–5171.

McCarty, J. S. and Walker, G. C. (1991). DnaK as a thermometer: Threonine-199 is site of autophosphorylation and is critical for ATPase activity. *Proc. Natl. Acad. Sci. USA* **88,** 9513–9517.

McCarty, J. S. and Walker, G. C. (1994). DnaK mutants defective in ATPase activity are defective in negative regulation of the heat shock response: Expression of mutant DnaK proteins results in filamentation. *J. Bacteriol.* **176,** 764–780.

Melki, R., Fievez, S., and Carlier, M. F. (1996). Continuous monitoring of Pi release following nucleotide hydrolysis in actin or tubulin assembly using 2-amino-6-mercapto-7-methylpurine ribonucleoside and purine-nucleoside phosphorylase as an enzyme-linked assay. *Biochemistry* **35,** 12038–12045.

Merkler, I., Stournaras, C., and Faulstich, H. (1987). Changes in OD at 235 nm do not correspond to the polymerization step of actin. *Biochem. Biophys. Res. Commun.* **145,** 46–51.

Moraczewska, J., Wawro, B., Seguro, K., and Strzelecka-Golaszewska, H. (1999). Divalent cation-, nucleotide-, and polymerization-dependent changes in the conformation of subdomain 2 of actin. *Biophys. J.* **77,** 373–385.

Mozo-Villarias, A. and Ware, B. R. (1985). Actin oligomers below the critical concentration detected by fluorescence photobleaching recovery. *Biochemistry* **24,** 1544–1548.

Neuhaus, J. M., Wanger, M., Keiser, T., and Wegner, A. (1983). Treadmilling of actin. *J. Muscle Res. Cell Motil.* **4,** 507–527.

Newman, J., Estes, J. E., Selden, L. A., and Gershman, L. C. (1985). The presence of oligomers at subcritical actin concentrations. *Biochemistry* **24,** 1538–1544.

Nyman, T., Schüler, H., Korenbaum, E., Schutt, C. E., Karlsson, R., and Lindberg, U. (2002). The role of MeH73 in actin polymerization and ATP hydrolysis. *J. Mol. Biol.* **317,** 577–589.

O'Brien, M. C. and McKay, D. B. (1993). Threonine 204 of the chaperone protein Hsc70 influences the structure of the active site, but is not essential for ATP hydrolysis. *J. Biol. Chem.* **268,** 24323–24329.

O'Brien, M. C., Flaherty, K. M., and McKay, D. B. (1996). Lysine 71 of the chaperone protein Hsc70 is essential for ATP hydrolysis. *J. Biol. Chem.* **271,** 15874–15878.

Ohm, T. and Wegner, A. (1994). Mechanism of ATP hydrolysis by polymeric actin. *Biochim. Biophys. Acta* **1208,** 8–14.

Otterbein, L. R., Graceffa, P., and Dominguez, R. (2001). The crystal structure of uncomplexed actin in the ADP state. *Science* **293,** 708–711.

Page, R., Lindberg, U., and Schutt, C. E. (1998). Domain motions in actin. *J. Mol. Biol.* **280,** 463–474.

Palmgren, S., Ojala, P. J., Wear, M. A., Cooper, J. A., and Lappalainen, P. (2001). Interactions with PIP2, ADP-actin monomers, and capping protein regulate the activity and localization of yeast twinfilin. *J. Cell Biol.* **155,** 251–260.

Pardee, J. D. and Spudich, J. A. (1982). Mechanism of K^+-induced actin assembly. *J. Cell Biol.* **93,** 648–654.

Pieper, U. and Wegner, A. (1996). The end of a polymerizing actin filament contains numerous ATP-subunit segments that are disconnected by ADP-subunits resulting from ATP hydrolysis. *Biochemistry* **35,** 4396–4402.

Pinaev, G., Schutt, C. E., and Lindberg, U. (1995). The effect on actin ATPase of phalloidin and tetramethylrhodamine phalloidin. *FEBS Lett.* **369,** 144–148.

Pollard, T. D. (1984). Polymerization of ADP-actin. *J. Cell Biol.* **99,** 769–777.

Pollard, T. D. (1986). Rate constants for the reactions of ATP- and ADP-actin with the ends of actin filaments. *J. Cell Biol.* **103,** 2747–2754.

Pollard, T. D. and Weeds, A. G. (1984). The rate constant for ATP hydrolysis by polymerized actin. *FEBS Lett.* **170,** 94–98.

Polzar, B., Nowak, E., Goody, R. S., and Mannherz, H. G. (1989). The complex of actin and deoxyribonuclease I as a model system to study the interactions of nucleotides, cations and cytochalasin D with monomeric actin. *Eur. J. Biochem.* **182,** 267–275.

Rajapandi, T., Wu, C., Eisenberg, E., and Greene, L. (1998). Characterization of D10S and K71E mutants of human cytosolic hsp70. *Biochemistry* **37,** 7244–7250.

Rich, S. A. and Estes, J. E. (1976). Detection of conformational changes in actin by proteolytic digestion: Evidence for a new monomeric species. *J. Mol. Biol.* **104,** 777–792.

Rickard, J. E. and Sheterline, P. (1986). Cytoplasmic concentrations of inorganic phosphate affect the critical concentration for assembly of actin in the presence of cytochalasin D or ADP. *J. Mol. Biol.* **191,** 273–280.

Rickard, J. E. and Sheterline, P. (1988). Effect of ATP removal and inorganic phosphate on length redistribution of sheared actin filament populations. Evidence for a mechanism of end-to-end annealing. *J. Mol. Biol.* **201,** 675–681.

Romero, S., LeClainche, C., Didry, D., Egile, C., Pantaloni, D., and Carlier, M. F. (2004). Formin is a processive motor that requires profilin to accelerate actin assembly and associated ATP hydrolysis. *Cell* **119,** 419–429.

Rouayrenc, J. F. and Travers, F. (1981). The first step in the polymerisation of actin. *Eur. J. Biochem.* **116,** 73–77.

Sablin, E. P., Dawson, J. F., VanLoock, M. S., Spudich, J. A., Egelman, E. H., and Fletterick, R. J. (2002). How does ATP hydrolysis control actin's associations? *Proc. Natl. Acad. Sci. USA* **99,** 10945–10947.

Schutt, C. E., Lindberg, U., Myslik, J., and Strauss, N. (1989). Molecular packing in profilin: Actin crystals and its implications. *J. Mol. Biol.* **209,** 735–746.

Schutt, C. E., Myslik, J. C., Rozycki, M. D., Goonesekere, N. C., and Lindberg, U. (1993). Interaction of profilin with G-actin and poly(L-proline). *Nature* **365,** 810–816.

Schüler, H. (2000d). The molecular dynamics of actin. PhD Thesis, Stockholm University, ISBN 91-7265-100-8.

Schüler, H. (2001). ATPase activity and conformational changes in the regulation of actin. *Biochim. Biophys. Acta* **1549,** 137–147.

Schüler, H., Korenbaum, E., Schutt, C. E., Lindberg, U., and Karlsson, R. (1999). Mutational analysis of Ser14 and Asp157 in the nucleotide-binding site of beta-actin. *Eur. J. Biochem.* **265,** 210–220.

Schüler, H., Lindberg, U., Schutt, C. E., and Karlsson, R. (2000a). Thermal unfolding of G-actin monitored with the DNase I-inhibition assay stabilities of actin isoforms. *Eur. J. Biochem.* **267,** 476–486.

Schüler, H., Nyåkern, M., Schutt, C. E., Lindberg, U., and Karlsson, R. (2000b). Mutational analysis of arginine 177 in the nucleotide binding site of beta-actin. *Eur. J. Biochem.* **267,** 4054–4062.

Schüler, H., Schutt, C. E., Lindberg, U., and Karlsson, R. (2000c). Covalent binding of ATPgammaS to the nucleotide-binding site in S14C-actin. *FEBS Lett.* **476,** 155–159.

Selden, L. A., Gershman, L. C., and Estes, J. E. (1986). A kinetic comparison between Mg-actin and Ca-actin. *J. Muscle Res. Cell Motil.* **7,** 215–224.

Steinmetz, M. O., Goldie, K. N., and Aebi, U. (1997). A correlative analysis of actin filament assembly, structure, and dynamics. *J. Cell Biol.* **138,** 559–574.

Strzelecka-Golaszewska, H. (2001). Divalent cations, nucleotides, and actin structure. *Results Probl. Cell Differ.* **32,** 23–41.

Strzelecka-Golaszewska, H., Moraczewska, J., Khaitlina, S. Y., and Mossakowska, M. (1993). Localization of the tightly bound divalent-cation-dependent and nucleotide-dependent conformation changes in G-actin using limited proteolytic digestion. *Eur. J. Biochem.* **211,** 731–742.

Strzelecka-Golaszewska, H., Wozniak, A., Hult, T., and Lindberg, U. (1996). Effects of the type of divalent cation, Ca^{2+} or Mg^{2+}, bound at the high-affinity site and of the ionic composition of the solution on the structure of F-actin. *Biochem. J.* **316,** 713–721.

Tellam, R. L. (1986). Gelsolin inhibits nucleotide exchange from actin. *Biochemistry* **25,** 5799–5804.

Tobacman, L. S. and Korn, E. D. (1982). The regulation of actin polymerization and the inhibition of monomeric actin ATPase activity by Acanthamoeba profilin. *J. Biol. Chem.* **257,** 4166–4170.

Vinson, V. K., De La Cruz, E. M., Higgs, H. N., and Pollard, T. D. (1998). Interactions of Acanthamoeba profilin with actin and nucleotides bound to actin. *Biochemistry* **37,** 10871–10880.

Vorobiev, S., Strokopytov, B., Drubin, D. G., Frieden, C., Ono, S., Condeelis, J., Rubenstein, P. A., and Almo, S. C. (2003). The structure of nonvertebrate actin: Implications for the ATP hydrolytic mechanism. *Proc. Natl. Acad. Sci. USA* **100,** 5760–5765.

Wegner, A. (1976). Head to tail polymerization of actin. *J. Mol. Biol.* **108,** 139–150.

Wertman, K. F., Drubin, D. G., and Botstein, D. (1992). Systematic mutational analysis of the yeast ACT1 gene. *Genetics* **132,** 337–350.

Wilbanks, S. M. and McKay, D. B. (1995). How potassium affects the activity of the molecular chaperone Hsc70. II. Potassium binds specifically in the ATPase active site. *J. Biol. Chem.* **270,** 2251–2257.

Wilbanks, S. M., DeLuca-Flaherty, C., and McKay, D. B. (1994). Structural basis of the 70-kilodalton heat shock cognate protein ATP hydrolytic activity. I. Kinetic analyses of active site mutants. *J. Biol. Chem.* **269,** 12893–12898.

Wriggers, W. and Schulten, K. (1997). Stability and dynamics of G-actin: Back-door water diffusion and behavior of a subdomain 3/4 loop. *Biophys. J.* **73,** 624–639.

Yao, X., Grade, S., Wriggers, W., and Rubenstein, P. A. (1999). His(73), often methylated, is an important structural determinant for actin. *J. Biol. Chem.* **274,** 37443–37449.

Chapter 4

Spectrin Function: A Survey of Genetic Systems from *Drosophila* to Humans

Ronald R. Dubreuil

Department of Biological Sciences, University of Illinois, Chicago, Illinois 60607

Spectrin and ankyrin are major components of a plasma membrane-associated scaffold known as the spectrin cytoskeleton. Spectrin is thought to cross-link actin filaments in a two-dimensional submembrane network and ankyrin links the network to a number of integral membrane activities, including the sodium pump, voltage-dependent sodium channels, and L1-family cell adhesion molecules. Genetic studies have established that the spectrin cytoskeleton is required for the normal accumulation of these integral proteins at their normal sites of function. Defects in spectrin and ankyrin have been implicated in human disease, and studies in model genetic organisms are making it possible to elucidate the mechanisms behind their effects. Cell culture studies have also shed new light on the processes that govern formation of

ISBN: 0-444-52868-7
DOI: 10.1016/S1569-2558(06)37004-X

polarized membrane domains. This chapter surveys the results and their implications for the assembly and function of the spectrin cytoskeleton.

I. INTRODUCTION

From their study of human erythrocytes, Tilney and Detmers (1975) concluded that the spectrin cytoskeleton looks like a net woven by a myopic fisherman and speculated that the spectrin–actin complex was probably associated with the plasma membrane in many different cells. Since then a vast body of work has demonstrated that indeed spectrins are important structural elements of the plasma membrane that are present in virtually all animal tissues that have been examined thus far. Spectrin function is intimately associated with ankyrin, the adapter protein, that links spectrin to a number of integral membrane proteins. The molecular structures of spectrin and ankyrin have been conserved from invertebrates, such as the fruit fly to humans, and current models suggest that one of their conserved roles is in organizing specialized functional domains of the plasma membrane. Genetic approaches to understanding spectrin and ankyrin function have led to exciting progress. Spectrin and ankyrin defects have been implicated in human diseases affecting the heart and brain. Genetic studies in mice using spontaneous mutations and targeted knockouts have shown that molecular defects in spectrin and ankyrin alter the morphology, composition, and function of specialized plasma membrane domains. Genetic studies in *Drosophila* have shown that spectrin mutations have lethal phenotypes with drastic effects on cell morphology and tissue organization and function. Yet, despite significant progress in identifying problems associated with a loss of spectrin function, there has been little progress in understanding how and when these problems arise. One obstacle to answering these questions is the lack of information on cues that lead to polarized spectrin assembly. However, studies of epithelial cell differentiation *in vitro* have unexpectedly shed new light on the development of cell polarity and the possible contributions of spectrin and ankyrin to that process. Here I will survey these studies and discuss the implications and insights that have emerged from them. For a more comprehensive treatment of the spectrin literature, the reader is referred to two excellent reviews (Lux and Palek, 1995; Bennett and Baines, 2001).

II. SPECTRIN IN THE RED BLOOD CELL

The human erythrocyte is a terminally differentiated cell that lacks a nucleus and other cytoplasmic organelles (Lux and Palek, 1995). It consists of little more than a plasma membrane and a hemoglobin-rich cytoplasm. The normal biconcave disc shape of the cell relies on the spectrin-based network of peripheral membrane proteins that lines the cytoplasmic face of the membrane (Fig. 1). Erythrocytes from humans and mice that lack spectrin or ankyrin have abnormal cell shapes and their membranes are unusually fragile. As a result, these individuals exhibit inherited forms of anemia (Tse and Lux, 1999).

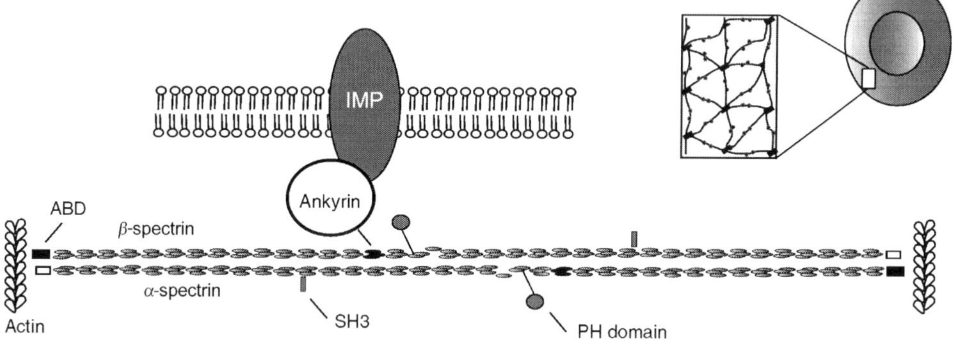

Figure 1. Protein interactions of the spectrin cytoskeleton. Human erythrocytes (upper right) are flattened biconcave disc-shaped cells. Cell shape depends on the integrity of a two-dimensional hexagonal network of spectrin and actin (inset) found immediately beneath the lipid bilayer. Each hub corresponds to a short actin filament attached to six spectrin tetramers. Each spectrin molecule has two binding sites for ankyrin near the center of the molecule. The native spectrin molecule (bottom) is a tetramer of antiparallel α- and β-subunits. Each subunit is divided into discrete structural domains that are referred to here as segments (repetitive and nonrepetitive). The β-subunit of spectrin consists of an N-terminal actin-binding domain (shaded rectangle), a string of 17 triple helical spectrin repeats, a partial repeat consisting of the first two helices of a structural repeat, and a C-terminal pleckstrin homology (PH) domain. An ankyrin-binding site is found in segment 16 of β-spectrin (shaded). The PH domain is absent from the major β-spectrin isoform expressed in erythrocytes due to alternative splicing. The α-subunit consists of a partial repeat consisting of the third helix of a structural repeat followed by a string of 20 spectrin repeats and a C-terminal EF hand calcium-binding domain (open rectangle). Segment 10 of α-spectrin is an SH3 domain that is inserted between helix 2 and 3 of repetitive segment 9. Ankyrin binds to repetitive segment 16 of β-spectrin and to the cytoplasmic domain of integral membrane proteins (IMP). The ankyrin-binding integral membrane protein in erythrocytes is the anion exchanger band 3. Other components of the erythroid spectrin cytoskeleton not shown here include protein 4.1, p55, dematin, and adducin at the spectrin–actin junction (Bennett and Baines, 2001). Spectrin tetramers have similar interactions with actin, ankyrin, and integral membrane proteins in nonerythroid cells. Membrane proteins that interact with ankyrin in nonerythroid cells include the Na^{+}- K^{+}-ATPase, voltage-dependent sodium channels, and L1 family cell adhesion molecules.

The relative simplicity of the erythrocyte membrane and its ease of purification made it possible to isolate all of the major membrane-associated proteins and to study their interactions *in vitro* (Bennett and Baines, 2001). The spectrin cytoskeleton is assembled as a two-dimensional network of spectrin molecules cross-linked together by short actin filaments. Each actin hub associates with about six spectrin molecules to form a geodesic structure beneath the membrane (Byers and Branton, 1985). Spectrin itself is an unusually large protein, composed of α- and β-subunits arranged as a tetramer (Fig. 1). The bulk of each subunit is made up of spectrin repeats that are each approximately 106 amino acids long (Speicher and Marchesi, 1984). Structural studies indicate that each repeat is composed of three α-helices folded into a triple barrel structure (Yan *et al.*, 1993). Interactions between α- and β-subunits have been mapped to repeats near the tail end of each spectrin dimer (Harper *et al.*, 2001) and at the tetramer formation site found at the head end of each dimer (Tse *et al.*, 1990).

Tetramers formed by the association of a partial structural repeat from the C-terminal region of β-spectrin (segment 18) with a complementing partial repeat (known as segment 0″) at the N-terminus of α-spectrin. As a result, the central helical domain of spectrin is essentially an uninterrupted string of repeats that connect the actin-binding domains at each tail end of the molecule.

There are also a number of nonrepetitive sequence segments in the spectrin tetramer. Segment 1 of β-spectrin includes two calponin homology (CH) sequence motifs that are responsible for actin-binding activity (Banuelos *et al.*, 1998). An adjacent nonrepetitive domain in α-spectrin is an EF hand Ca^{2+}-binding module (Wasenius *et al.*, 1989; Dubreuil *et al.*, 1991). Most isoforms of β-spectrin include the C-terminal pleckstrin homology (PH) domain (Tse *et al.*, 2001). PH domains are found in hundreds of different proteins and in many cases they exhibit lipid-binding activity (Lemmon and Ferguson, 2000). While the major β-spectrin isoform found in human erythrocytes is a splice variant that lacks the C-terminal PH domain, recent evidence indicates that there is also a population present that includes the PH domain (Pradhan *et al.*, 2004). When the PH domain is present, it is presumably positioned so as not to interfere with other neighboring activities (e.g., head-end interactions between α- and β-subunits, ankyrin binding to the β-subunit). Small globular extensions have been observed alongside the spectrin tetramer by electron microscopy of purified molecules (Dubreuil *et al.*, 1990). These globules may correspond to the PH domain. Finally, α-spectrin includes an src homology-3 (SH3) domain (Lehto *et al.*, 1988) inserted between two of the α-helices of repetitive segment 9.

The two best-characterized links between the spectrin-actin network and the lipid bilayer are through protein–protein interactions (Bennett and Baines, 2001). Ankyrin is an adapter that links the β-subunit of spectrin to the cytoplasmic domain of the major membrane protein of erythrocytes, the anion exchanger band 3. A second high-affinity interaction occurs via a ternary complex between proteins 4.1 and p55 (both components of the spectrin–actin junction, not shown here) and the integral protein glycophorin C (Marfatia *et al.*, 1994). However, defects in either band 3 (Peters *et al.*, 1996) or glycophorin C (Chasis and Mohandas, 1992) in mammalian erythrocytes produce surprisingly mild effects on assembly of the spectrin cytoskeleton. Thus, these two integral membrane proteins may make redundant contributions to the spectrin cytoskeleton assembly, or there may be a third as yet uncharacterized pathway. There are a number of proteins associated with the spectrin–actin junction in erythrocytes, including protein 4.1, and p55, as well as dematin and adducin (Bennett and Baines, 2001). Like spectrin and ankyrin, these proteins represent gene families that are expressed in diverse tissues. Protein 4.1 (Bretscher *et al.*, 2002) and p55 (Funke *et al.*, 2005) in particular have extensive family trees and are thought to have important roles at the plasma membrane of many cell types. The spatial relationship between these proteins and spectrin and ankyrin in nonerythroid cells is not yet known.

III. THE SPECTRIN PARADIGM IN NONERYTHROID CELLS

Spectrins and ankyrins are both members of multigene families. In humans there are five β-spectrin genes, two α-spectrin genes, and three ankyrin genes (Bennett and Baines, 2001). Family members are further diversified through extensive alternative splicing of transcripts. Invertebrates, such as *Drosophila* and *Caenorhabditis elegans*, have a smaller repertoire with one α-spectrin gene, two β-spectrins, and two ankyrin genes. Remarkably there is often greater sequence conservation between spectrins from *Drosophila* and human nonerythroid tissues than between the erythroid and nonerythroid forms within mammals (Table I). Sequence analyses suggest that the erythrocyte spectrin genes arose during vertebrate evolution (Dubreuil, 1991) and some of the sequence changes are thought to correspond to neofunctionalization of the erythroid spectrin genes (Salomao *et al.*, 2006).

Most models simply depict nonerythroid spectrin as an actin cross-linking protein that is linked to integral membrane proteins through the adapter ankyrin (as in Fig. 1). The two-dimensional spectrin–actin network found in erythrocytes (Fig. 1, inset) has been directly viewed by electron microscopy of artificially spread preparations (Byers and Branton, 1985). However, it has not been technically feasible to make that determination for nonerythroid spectrins because of the complexity of the plasma membrane in most other systems. One striking difference that has been observed is the complexity of different membrane proteins that can interact with the spectrin cytoskeleton through ankyrin (Bennett and Baines, 2001). Biochemical studies have detected interactions between ankyrin and membrane pumps, channels, exchangers, and cell adhesion molecules (described later).

Nonerythoid spectrins can be further subdivided into the conventional spectrins and those that possess an unusually large β_H-subunit. The β_H-subunit was first discovered in *Drosophila* as an unusually large variant of β-spectrin that forms unusually large spectrin tetramers (Dubreuil *et al.*, 1990). In both *Drosophila* and *C. elegans*, the β_H-subunit forms tetramers with the conventional α-spectrin subunit, but the $\alpha\beta_H$

Table I
Amino Acid Sequence Comparisons Between Human β-Spectrins and *Drosophila* β-Spectrin[a]

	hbspII	hbspIII	hbspIV	Fly
hbspI	60%	54	49	49
hbspII	–	64	53	57
hbspIII		–	50	50
hbspIV			–	42
Fly				–

[a]Comparisons based on sequences of *Drosophila* β-spectrin (Byers *et al.*, 1992), human β-spec IεII (P11277; Winkelmann *et al.*, 1990), human β-spec II (Q01082; Hu *et al.*, 1992), human β-spec III (NP008877; Stankewich *et al.*, 1998), and human β-spec IV (NP057726; Stabach and Morrow, 2000).

tetramers have a distinct subcellular distribution relative to $\alpha\beta$-spectrin, that is, they do not appear to associate with ankyrin, and their mutant phenotypes are markedly different from β-spectrin mutants. The human βV-spectrin gene appears to be a homologue of the invertebrate β_H-spectrins (Stabach and Morrow, 2000). Since these spectrins appear to be in a distinct functional class from other spectrins, they will not be considered further here. A review of the novel properties of β_H-spectrins and a more detailed comparison to conventional spectrins can be found in a review by Thomas (2001).

IV. GENETIC STUDIES OF SPECTRIN CYTOSKELETON FUNCTION IN NONERYTHROID CELLS

A. Human Genetic Studies

Loss of function mutations of nonerythroid spectrin and ankyrin (described later) often exhibits a recessive lethal phenotype. Dominant mutations affecting spectrin and ankyrin have also been discovered by sequencing of candidate genes from mapped disease loci. (1) Ankyrin-B is broadly expressed in human tissues including cardiac muscle. Defects in ankyrin-B were observed in patients exhibiting a rare cardiac arrhythmia (Mohler *et al.*, 2003, 2004a), and the defect led to reduced expression and altered localization of three proteins that normally interact with ankyrin: the Na, Ca exchanger, the Na^+-K^+-ATPase, and the InsP3 receptor. These proteins appear to exist in a specialized complex in cardiac muscle that normally localizes within the T tubules (Mohler *et al.*, 2005). (2) Sequencing of voltage-dependent sodium channel genes from a patient with Brugada syndrome, another rare cardiac arryhthmia, led to the identification of a point mutation in one of the cytoplasmic loops of the voltage-dependent sodium channel $Na_V1.5$ (Mohler *et al.*, 2004b). The mutation occurs precisely within the site that interacts with ankyrin-G. Loss of ankyrin-binding activity caused the mutant channel to mislocalize within the cytoplasm of cardiomyocytes, instead of at its normal sites of function within the T tubule, and intercalated disc (Mohler and Bennett, 2005). (3) A defect in human βIII-spectrin was identified in three families with spinocerebelar ataxia type V (Ikeda *et al.*, 2006). This dominant neurodegenerative disorder is associated with altered behavior of glutamate transporters and receptors. One lineage was traced to Abraham Lincoln's grandparents, leading the authors to speculate that this defect may explain ataxia-like symptoms in our former president. In each of the cases described earlier, a defect in spectrin or ankyrin led to abnormal behavior and function of physiologically important membrane activities that normally interact with the spectrin cytoskeleton.

B. Genetic Studies of Spectrin and Ankyrin in the Mouse

Loss of function mutations of ankyrin and spectrin in the mouse has also produced dramatic effects on the behavior of interacting membrane proteins. For example, sequencing of candidate genes identified defects in βIV-spectrin as the lesion in the quivering mouse, a spontaneous autosomal recessive mutation that affects the nervous system (Parkinson *et al.*, 2001). Further characterization of βIV-spectrin knockout mice (Komada and Soriano, 2002) demonstrated that this spectrin has important roles in the structure and stability of the node of Ranvier and the axon initial segment in the CNS. Ankyrin-G and voltage-gated sodium channels, which normally cocluster with βIV-spectrin, were mislocalized in the mutants. Interestingly, these authors also found that βIV-spectrin was mislocalized in ankyrin-G knockout mice, suggesting that the targeting of these two proteins is mutually dependent. Other studies revealed that voltage-dependent sodium channels, the L1 family cell adhesion molecule neurofascin, and βIV-spectrin (all of which interact with ankyrin *in vitro*) were mislocalized in ankyrin-G knockout mice (Zhou *et al.*, 1998; Jenkins and Bennett, 2001). A knockout of ankyrin-B in mouse produced a severe phenotype that partially overlapped the phenotype of L1 mutations, including a significant decrease in the amount of L1 (Scotland *et al.*, 1998). Thus, knockouts of spectrin and ankyrin have dire consequences for the development of the mammalian nervous system. They affect neuronal physiology by interfering with the normal targeting and stabilization of interacting sodium channels, adhesion molecules, and perhaps other physiologically important membrane activities.

V. STUDIES OF THE SPECTRIN CYTOSKELETON IN *DROSOPHILA*

A. Characterization of the Protein Spectrin in *Drosophila*

A conventional isoform of spectrin was originally discovered in *Drosophila* as an actin-binding protein from lysates of S3 tissue culture cells. Subsequent biochemical, structural, and molecular studies established that *Drosophila* spectrin is an $\alpha\beta$ tetramer that shares all of the major functional sites found in mammalian nonerythroid spectrins (Dubreuil *et al.*, 1987; Byers *et al.*, 1989; Dubreuil and Yu, 1994; Deng *et al.*, 1995). A comparison of overall amino acid identities between *Drosophila* β-spectrin and human βI–IV-spectrins is shown in Table I. The similarities between fly and human spectrins, which in many cases are greater than the similarities among human spectrins, suggest that the major functions conserved between human spectrin isoforms are likely to be conserved in the fly, as well.

Antibody-staining experiments suggest that *Drosophila* $\alpha\beta$-spectrin is primarily a plasma membrane-associated protein (as described later). In fact, spectrin antibodies are often used to monitor morphogenetic movements of cells during early embryonic development, because they brightly stain the plasma membrane of nearly all cells in the early embryo (Pesacreta *et al.*, 1989). Several observations in vertebrate systems suggest the presence of additional populations of spectrin associated with intracellular compartments, including the Golgi and ER (reviewed in DeMattheis and Morrow, 2000). One of the proposed functions for organellar spectrin is the attachment of microtubule motors to the organelle membrane. Mammalian spectrin interacts with components of the dynactin complex, which associates with dynein and is thought to link it to cargo (Holleran *et al.*, 2001). Binding of organellar spectrin to acidic phospholipids also appears to contribute to the attachment of dynein to its cargo (Muresan *et al.*, 2001). There is evidence from *Drosophila* that spectrin codistributes with a Golgi-like compartment in the early embryo (Sisson *et al.*, 2000; Papoulas *et al.*, 2005). And there is evidence from RNAi knockdown studies suggesting that spectrin has a role in axonal transport (Pielage *et al.*, 2005). However, a genetic test of dynein function in *C. elegans* failed to detect a requirement for spectrin in cargo binding (Koushika *et al.*, 2004). RNAi knockdowns of all of the spectrin subunits, alone or in combination, also failed to inhibit motor driven movements of peroxisomes in *Drosophila* S2 cells (V. Gelfand, personal communication). This assay has been shown to be a sensitive test of kinesin 1 and dynein heavy chain function (Kural *et al.*, 2005). Thus, further experiments will be necessary to determine if there are stage-specific and/or species-specific differences in the interactions of spectrin with microtubule motors and organelles.

B. *Drosophila* Ankyrins

Two isoforms of ankyrin have been described in *Drosophila*. The first, DAnk1, was cloned by its similarities to mammalian ankyrins using degenerate PCR (Dubreuil and Yu, 1994). DAnk1 appears to be broadly expressed throughout development. DAnk2 was cloned in a yeast 2-hybrid screen using the cytoplasmic domain of *Drosophila* neuroglian as bait (Bouley *et al.*, 2000). DAnk2 appears to be selectively expressed in the nervous system, and antibodies raised against long and short isoforms of DAnk2 demonstrated that they are segregated within different subdomains of neurons (Hortsch *et al.*, 2002). Both *Drosophila* ankyrins share the major conserved domains of mammalian ankyrins and are thought to function as adapters that link β-spectrin to integral membrane proteins. Mutations have not yet been identified for either ankyrin.

C. Genetic Studies of *Drosophila* Spectrin

Mutations in the α- and β-subunits of spectrin were identified using standard reverse genetic strategies (Lee *et al.*, 1993; Dubreuil *et al.*, 2000). In both cases, the mutations are lethal early in development: just before larval hatching in the case of β mutants and just after larval hatching in α mutants. Proof that spectrin gene defects were solely responsible for the observed phenotypes was obtained by mutant rescue with α- or β-spectrin transgenes. Survival through early stages of embryonic development is thought to depend on gene products loaded into the unfertilized egg by heterozygous mothers. The reason for the difference in survival between α and β mutants is not known. There may be differences in size or stability of the maternal protein pool, or there may be real differences in function between the two subunits (Dubreuil *et al.*, 2000). In either case, little or no residual maternal protein can be detected in mutant embryos or larvae as they approach the lethal phase (Lee *et al.*, 1993; Dubreuil *et al.*, 2000).

a. The Cell Shape Phenotype

Initial studies on cellular effects of the spectrin mutations were focused on the midgut epithelium, because it was amenable to staining with antibody probes and easy to detect changes in its morphology in the spectrin mutants. In both mutants, individual cell shapes and their order within the epithelial tube were conspicuously perturbed. As a result, the straight, smooth appearance of the gut lumen in wild type became crooked and irregular in the mutants. Electron microscopy of the α-spectrin mutants revealed gaps between interstitial cells and copper cells that were judged to be defects in cell–cell adhesion (Lee *et al.*, 1993). Our recent EM analysis of β-spectrin mutants also detected gaps between cells (unpublished observation). However, it was not clear that the gaps were due to effects on cell adhesion per se, as opposed to effects on the morphogenesis of interstitial cells.

b. Effects on the Na^+-K^+-ATPase

The Spectrin mutants made it possible to test the hypothesis that the basolateral distribution of the Na^+-K^+-ATPase depends on the integrity of the spectrin cytoskeleton (Hammerton *et al.*, 1991). An antibody raised against an α-subunit of vertebrate Na^+-K^+-ATPase (Lebovitz *et al.*, 1989) stained the basolateral domain of copper cells in the wild-type larval midgut. Unexpectedly the basolateral pattern of Na^+-K^+-ATPase staining appeared unaltered by a lethal mutation in the α-spectrin gene (Lee *et al.*, 1993). However, subsequent analysis of copper cells from β-spectrin mutants revealed a striking shift in the distribution of the Na^+-K^+-ATPase from the plasma membrane to an intracellular compartment (Dubreuil *et al.*, 2000). Thus it appears that the basolateral accumulation of the Na^+-K^+-ATPase is strictly dependent on β-spectrin function in copper cells. A working hypothesis is that the Na^+-K^+-ATPase is captured by a resident complex of β-spectrin and ankyrin in the basolateral domain of copper cells, and that α-spectrin is dispensable for their stabilizing activity.

c. Cell Polarity and Growth Control

One of the first developmental defects observed in *Drosophila* spectrin mutants was in cell polarity and regulation of cell growth in the somatic follicle epithelium. A P-element excision strategy was developed to produce somatic clones of homozygous mutant cells in the adult ovary (Lee *et al.*, 1997). Normally the follicle cells undergo a fixed number of cell divisions to produce a monolayer of approximately 1100 epithelial cells surrounding each oocyte. Loss of α-spectrin function in somatic clones of mutant cells resulted in overgrowth of follicle cells and consequently a loss of their monolayer organization, as well as a loss of markers of epithelial polarity.

Since these cells express both β- and β_H-spectrin (Dubreuil *et al.*, 1997), it was possible to envisage that the growth and polarity defects were due to an effect on the apical population of $\alpha\beta_H$-spectrin. However, subsequent studies failed to detect a hyperplastic or polarity phenotype in follicle tissue lacking β_H-spectrin (Zarnescu and Thomas, 1999). Thus it appears that the basolateral population of $\alpha\beta$-spectrin has an important role in the control of cell growth and polarity in the follicle epithelium.

d. Stomach Acid Secretion

A physiological effect of the α-spectrin mutation was detected in studies of stomach acid secretion in the larval middle midgut. This work was inspired by the observation that spectrin was dynamically associated with the proton pump in mammalian gastric parietal cells, which suggested a role in the regulation of stomach acid secretion (Mercier *et al.*, 1989). Wild-type larvae fed with the pH sensitive dye bromophenol blue exhibited a color change from brilliant blue in the anterior midgut to bright yellow in the copper cell region of the middle midgut (Dubreuil *et al.*, 1998). The change in dye color corresponds to a drop in pH to <2.3, thus identifying the stomach region of the larval gut. The color change was reversed in the posterior midgut where food absorption is thought to occur. No color change was observed in the middle midguts of *labial* mutant larvae in which copper cells failed to differentiate. Likewise, the acidification of the middle midgut was blocked by a mutation in α-spectrin. Based on these observations, copper cells appear to be responsible for the acid secretion activity of the middle midgut, and their function depends on the presence of α-spectrin. The copper cells bear a striking resemblance to the gastric parietal cells that are responsible for stomach acid secretion in mammals (Dubreuil, 2004). Loss of α-spectrin may have affected the activity of an apical proton pump ($\alpha\beta_H$-spectrin is present in the apical domain of copper cells) or it may have affected a basolateral activity that indirectly supports proton pumping through the apical domain. An apical effect seems more likely since there was only a small effect of β-spectrin mutations on acid secretion (Dubreuil *et al.*, 2000).

e. Synaptic Function

Both the α- and β-spectrin mutants are somewhat sluggish at the time of larval hatching, although they crawl, they eat, and they respond when poked with a needle. However, over a period of hours they slowly cease movement until they ultimately die, suggesting a likely effect of the mutations on nervous system function. The general morphology of neuromuscular junctions from mutant embryos appeared normal and the number of synaptic vesicles detected at the nerve terminal by electron microscopy appeared normal (Featherstone *et al.*, 2001). There was no detectable effect of spectrin mutations on the activity of postsynaptic receptors as monitored by electrophysiology. However, there was a presynaptic defect in the rates of synaptic vesicle fusion and several presynaptic markers including synaptotagmin, synapsin, syntaxin, and Dlg were mislocalized in the axons of spectrin mutants. These results suggest that spectrin has an effect on the function of the presynaptic vesicle release machinery at the nerve terminal. A further effect of spectrin mutations on synaptic structure was detected by RNAi knockdown of spectrin later in larval development (Pielage *et al.*, 2005). Loss of presynaptic spectrin, but not postsynaptic spectrin, led to a destabilization of cell adhesion molecules at the synapse and ultimately destabilization and retraction of the nerve terminal. Thus spectrin appears to be important for both the function and the long-term stability of synaptic connections.

f. Tetramer Formation

Another approach taken in *Drosophila* was to examine the effects of mutations that perturb tetramer formation in human erythoid spectrin on the structure and function of *Drosophila* (nonerythroid) spectrin (Deng *et al.*, 1995). The human mutations cause hereditary anemia by preventing spectrin dimers from interacting to form tetramers. Tetramers form through interactions between a partial repeat found near the C-terminus of β-spectrin and a complementary partial repeat from the N-terminus of α-spectrin (Fig. 1). A mutation at codon 28 of human erythroid αI-spectrin (a conserved arginine) was observed in multiple families with hereditary elliptocytosis (Coetzer *et al.*, 1991). Most of the spectrin in erythrocytes from these individuals was in the dimer form. An R22S mutation was introduced into the corresponding conserved position of *Drosophila* α-spectrin (Deng *et al.*, 1995). The mutation blocked the association of recombinant fragments of α- and β-spectrin in an *in vitro* assay of tetramer formation. When tested in mutant rescue experiments, an α-spectrin transgene carrying the R22S mutation was nearly as efficient as a wild-type transgene at rescuing the lethality of a null α-spectrin mutation when flies were reared at 19 °C. But rescue activity was greatly diminished at 29 °C indicating that the function of spectrin *in vivo* is dependent on its ability to form tetramers. Tetramer formation may be necessary for spectrin to produce short-range cross-linking effects (actin–actin, ankyrin–ankyrin, and so on) or it may reflect a requirement for long-range spectrin network formation, comparable to what is observed in the human erythrocyte.

D. Evaluation of the Genetic Model

The earlier examples illustrate some of the uses of *Drosophila* as a model experimental organism to address fundamental questions of protein function. In some cases the system was used to test the contribution of the entire spectrin molecule to a particular cellular process or formation of a particular structure. The elucidation of mutant phenotypes in synaptic function and gut acidification are examples of this approach. Another use of the system was to test the contribution of particular functional sites in the molecule to its function, as in the elucidation of a temperature-sensitive defect in spectrin tetramer formation. While these strategies are more streamlined in invertebrate model organisms, they have also been successfully employed in studies of spectrin and ankyrin in the mouse (Lacas-Gervais *et al.*, 2004; Mohler *et al.*, 2004c). These strategies are important because they make it possible to ask direct questions about the cellular roles of spectrin and ankyrin. But they do not take full advantage of the genetic approach because they are constrained by preconceptions about where to look for effects and which sites in the molecule are likely to produce effects.

The third genetic approach that is uniquely suited to simple model organisms is their use in genetic screens. The forward genetic approach makes it possible to address novel aspects of gene function independently of preconceptions from previous studies. From the phenotype analyses carried out so far, we can speculate that the effects of spectrin mutations on cell growth and polarity, on cell shape and tissue organization, and on synaptic vesicle release all arise through effects on molecules that normally interact with spectrin. Along the same lines, we know that the polarized assembly of spectrin and ankyrin in cells is likely to depend on interactions with as yet unidentified partners in the cell. Some of the relevant interactions between spectrin and other proteins may not be detectable by biochemical and cellular approaches. In these cases, genetic approaches are likely to make a unique and important contribution.

VI. SPECTRIN ASSEMBLY AND CELL POLARITY IN TISSUE CULTURE MODELS

Prior to the recent wave of studies in genetic model systems, tissue culture models were used to describe the distribution and behavior of spectrin and ankyrin in polarized cells. Madin–Darby canine kidney (MDCK) cells have been an especially useful model system that produced a number of valuable insights that are being tested now in genetic systems. One idea that emerged from MDCK cell studies was that cell adhesion provides an important spatial cue that translates into assembly of spectrin and ankyrin at sites of cell–cell contact (Fig. 2A; Yeaman *et al.*, 1999). However, it has not been possible to elucidate a chain of events or specific protein interactions to explain that observation. Two studies in other tissue culture systems have uncovered novel mechanisms that may operate alongside (or perhaps instead of) cell adhesion in the assembly and function of spectrin in polarized cells.

A. Cell Adhesion as a Cue for Spectrin Assembly

Antibody-staining experiments revealed that as MDCK cells grow to confluence in cell culture and form mature lateral cell–cell contacts, there is a concomitant redistribution of spectrin (fodrin) and the Na^+-K^+-ATPase to sites of contact (Nelson and Veshnock, 1986). Evidence that their redistribution was driven by cell–cell adhesion was obtained by expressing E-cadherin, which mediates Ca^{2+}-dependent adhesion in MDCK cells, in nonpolarized fibroblasts (McNeill *et al.*, 1990). E-cadherin induced polarized assembly of the spectrin cytoskeleton and the Na^+-K^+-ATPase at sites of cell–cell contact. The interaction between ankyrin and the Na^+-K^+-ATPase was found

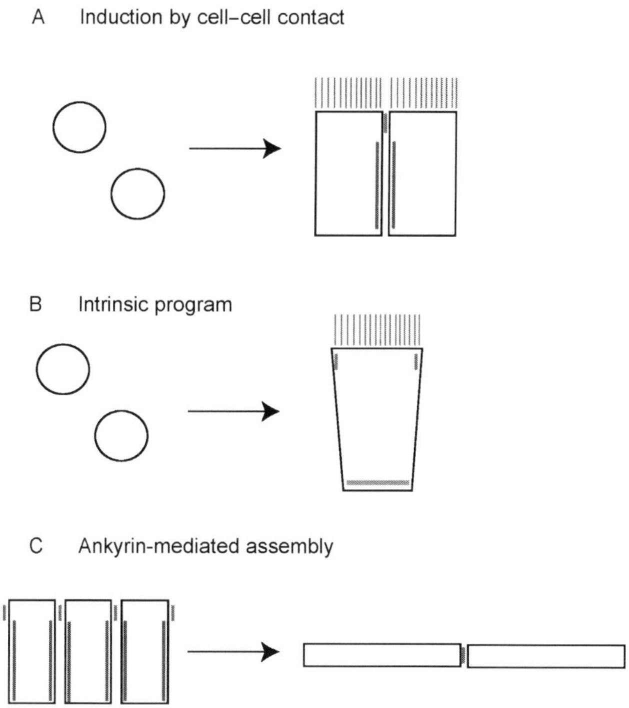

Figure 2. Three pathways to epithelial cell polarity *in vitro*. (A) Polarized assembly of the spectrin cytoskeleton (long bars) can occur in response to cell–cell adhesion (Yeaman *et al.*, 1999). (B) Some aspects of epithelial polarity can arise through activation of LKB1 including segregation of an apical-like microvillar domain, a subapical zone marked by the tight junction protein ZO-1 (short bars), and a basal domain containing the transferrin receptor (long bars; Baas *et al.*, 2004). (C) Human bronchial epithelial cells ordinarily differentiate as a columnar epithelium with apicolateral tight junctions (short bars) and a lateral domain associated with ankyrin-G and β2-spectrin (long bars). RNAi knockdown of ankyrin-G leads to a transformed cell shape in which the only remaining lateral membrane domain is the tight junction (Kizhatil and Bennett, 2004).

to be direct (Nelson and Veshnock, 1987). That interaction is thought to explain the selective stabilization of Na^+-K^+-ATPase at the lateral domain of MDCK cells, even in a cell line in which the pump is delivered equally to the apical and basolateral domains by the secretory pathway (Hammerton *et al.*, 1991). These observations support a general model in which (1) spectrin and ankyrin assemble in response to a positional cue such as cell adhesion, and (2) they form a multivalent protein scaffold that can capture and stabilize interacting membrane proteins to form a plasma membrane domain with unique composition and function.

Another connection between cell adhesion and the spectrin cytoskeleton was established using a biochemical approach. Davis *et al.* (1993) recovered an immunoglobulin superfamily cell adhesion molecule related to L1 from detergent extracts of rat brain membranes by affinity chromatography with immobilized ankyrin. L1 family members mediate homophilic cell–cell adhesion, and they also mediate heterophilic interactions with other ligands (Hortsch, 1996). The discovery that L1 family members directly interact with ankyrin provided a clear link between adhesion and the spectrin cytoskeleton, raising the question of whether L1 family members could trigger polarized assembly. Expression of the fly homologue neuroglian in *Drosophila* tissue culture cells induced a dramatic redistribution of ankyrin, spectrin, and the Na^{+}-K^{+}-ATPase to sites of cell–cell contact (Dubreuil *et al.*, 1996, 1997). It was especially intriguing that recruitment was limited to sites of cell contact even though neuroglian itself was abundant all over the surface of cells. Further studies revealed that ankyrin binding to the cytoplasmic domain of neuroglian could be activated by receptor clustering (Jefford and Dubreuil, 2000) and that interactions between mammalian L1 family members and ankyrin could be potently regulated by phosphorylation (Garver *et al.*, 1997). Studies of mammalian homologues in cell culture experiments also demonstrated recruitment of ankyrin to the plasma membrane by L1 family members, sometimes in a polarized fashion (Nishimura *et al.*, 2003) and sometimes not (Zhang *et al.*, 1998; Needham *et al.*, 2001).

Genetic studies have so far failed to detect a role for L1 family members in polarized ankyrin assembly *in vivo*. Instead, ankyrin mutations lead to decreased cell surface expression of L1 and neurofascin in mouse brain (Scotland *et al.*, 1998; Zhou *et al.*, 1998). Developmental studies indicate that in some cases the appearance of L1 molecules occurs later in time than the appearance of ankyrin at the plasma membrane (Jenkins and Bennett, 2001). Studies of neuroglian mutants in *Drosophila* also failed to detect effects on the axonal accumulation of DAnk2 (Bouley *et al.*, 2000) or the lateral accumulation of spectrin in the follicle epithelium (Wei *et al.*, 2004). Thus, despite the promising activities of L1 family members *in vitro,* these molecules do not have the expected properties of a cue that targets spectrin and ankyrin assembly *in vivo*.

B. Programed Formation of Specialized Regions of the Plasma Membrane

While an inductive program of cell polarization in response to cell–cell adhesion is appealing, it has been known for some time that neurons grown in culture form polarized plasma membrane domains (axons and dendrites) in the absence of cell–cell interactions. The mechanisms responsible for neuronal differentiation *in vitro* are poorly understood, but studies have demonstrated an important contribution of SAD kinases (Kishi *et al.*, 2005). The SAD kinase domain is related to PAR-1, one of several PAR proteins that contribute to embryonic polarity (Kemphues *et al.*, 1988; Wodarz, 2002). Baas *et al.* (2004) examined the effects of activating the PAR kinase LKB1 on the morphology of colon cell lines grown in culture. LKB1 (PAR-4) was discovered as a tumor suppressor that is mutated in Peutz–Jeghers syndrome patients (Hemminki *et al.*, 1998; Jenne *et al.*, 1998).

Activation of LKB1 produced striking effects on the distribution of a number of markers of epithelial polarity in cultured colon cell lines (Baas *et al.*, 2004). Prior to activation, cells exhibited an amoeboid-like morphology with no evidence of epithelial polarity. However, once activated in cells growing in suspension culture, LKB1 induced rearrangement of actin into a polar cap of microvillus-like structures (Fig. 2B). Remarkably, the tight junction marker ZO-1, which normally forms a belt at the top of the lateral domain of mammalian epithelial cells, formed a belt surrounding the microvillar cap. Other apical and basal plasma membrane markers were also sorted to their respective domains, all in the absence of cell–cell contact as a positional cue. The downstream targets of LKB1 activation that translate its activation into positional information and polarity are not yet known. Whether this pathway induces spectrin assembly in these tissue culture cells is also not known.

Preliminary evidence linking the PAR proteins to spectrin assembly in epithelial cells has been obtained in *Drosophila.* There is evidence that PAR-1 and LKB1 operate in a common pathway in *C. elegans* (Watts *et al.*, 2000) and in *Drosophila* where LKB1 is an *in vitro* substrate for PAR-1 and where LKB1 overexpression can suppress the PAR-1 phenotype (Martin and Johnston, 2003). Interestingly, loss of PAR-1 function in the somatic follicle epithelium leads to a striking increase in the level of lateral spectrin assembly (Doerflinger *et al.*, 2003). The staining pattern of α-spectrin in LKB1 mutants also appeared altered in somatic clones of mutant epithelial cells, but it was unclear whether or not there was an effect on spectrin targeting (Martin and Johnston, 2003). Whether there are effects of LKB1 on spectrin that are independent of cell–cell adhesion is not yet known.

C. Ankyrin-Dependent Formation of a Membrane Domain

Another intriguing new observation came from RNAi knockdown studies of ankyrin-G in human bronchial epithelial cells grown in culture (Kizhatil and Bennett, 2004). These cells normally differentiate into a columnar epithelium with a prominent lateral membrane domain at the site of cell–cell contact (Fig. 2C, left). The tight junction marker ZO-1 was distributed in a belt around the top of the lateral domain, as in other epithelia. An RNAi strategy effectively depleted ankyrin-G from these cells. Loss of ankyrin-G produced a striking transformation of these cells from their normal columnar morphology to a squamous-like morphology in which the diameter of each cell was greatly increased (Fig. 2C, right). The lateral domain, apart from the ZO-1-stained tight junction, was apparently absent and perhaps was absorbed into the larger apical and basal domains. Na^+-K^+-ATPase staining, which was normally concentrated in the lateral domain of the bronchial cell cultures, appeared reduced and redistributed to the cytoplasm, further indicating that the defining characteristics of the lateral domain were lost from the plasma membrane. The results of this study suggest that ankyrin (and presumably spectrin) has a role in the biogenesis of the lateral domain of epithelial cells, as opposed to a role in stabilizing a preexisting domain formed through cell–cell adhesion.

VII. ROLE OF THE PH DOMAIN IN SPECTRIN ASSEMBLY

Another tissue culture cell result that shed new light on the mechanism of spectrin assembly was the finding that a β-spectrin PH domain-GFP fusion protein was targeted to the plasma membrane in COS7 cells (Wang *et al.*, 1996). This result clearly established that the PH domain was sufficient for targeting spectrin to the plasma membrane. PH domains are found in hundreds of different proteins and in many cases they exhibit phosphoinositide-binding activity (Lemmon and Ferguson, 2000). Structures have been determined for spectrin PH domains (Macias *et al.*, 1994; Zhang *et al.*, 1995) and PIP2 binding has been demonstrated for mammalian spectrins (Hyvonen *et al.*, 1995; Wang and Shaw, 1995). The C-terminal region of β-spectrin was also identified as a membrane interaction site in studies of ankyrin-independent binding of spectrin to a stripped membrane fraction from mammalian brain (Steiner and Bennett, 1988; Lombardo *et al.*, 1994). The original biochemical studies that identified ankyrin-independent membrane-binding sites on spectrin were performed before the lipid-binding activity of the PH domain had been described, and, therefore, it seemed most likely that the membrane-binding sites were proteins. In fact, there is no evidence to date to exclude the possibility that the PH domain of spectrin interacts with both protein and lipid receptors. The interaction between the PH domain and lipids has not generally been thought of as a likely mechanism to explain spectrin targeting to particular domains of the plasma membrane (Wang *et al.*, 1996).

We obtained direct evidence for a role of the PH domain in spectrin targeting in *Drosophila* copper cells (Das *et al.*, 2006). The $\alpha\beta$-isoform of spectrin is normally targeted to the basal and lateral membrane domains of these cells (Dubreuil *et al.*, 1998). When the PH domain was removed from β-spectrin that targeting activity was lost and spectrin was found in the cytoplasm instead of at the plasma membrane. In contrast, removal of the ankyrin-binding site from β-spectrin had no detectable effect on its targeting in copper cells. Furthermore, the distribution of an ankyrin-1-EGFP reporter in these cells was found to be dependent on spectrin's ability to bind to the plasma membrane. Ankyrin was detected at the plasma membrane in wild-type copper cells, but it was shifted to the cytoplasm when β-spectrin was absent or when the β-spectrin bound to the plasma membrane lacked ankyrin-binding activity (Das *et al.*, 2006). These results suggest a sequence of events in spectrin cytoskeleton assembly that begins with the association of the PH domain of spectrin with either a protein or a lipid target at the plasma membrane (Fig. 3). Spectrin in turn recruits assembly of ankyrin and membrane proteins that interact with ankyrin to the original site of spectrin assembly. The schematic shown here illustrates the complexity of molecular interactions that occur near the center of the spectrin molecule. The spectrin molecule is bivalent with respect to all of its potential membrane interaction sites, even though only one interaction of each site is depicted here. It would be surprising if the activities of the various sites in this region of spectrin did not sterically affect one another.

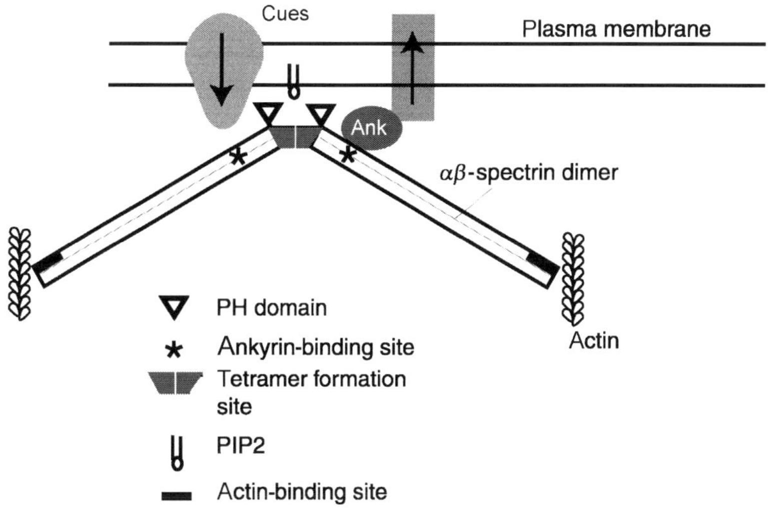

Figure 3. A model for assembly of the spectrin cytoskeleton in nonerythroid cells. In this model, the cue for spectrin assembly is a protein or lipid that interacts with the PH domain of β-spectrin. Ankyrin, in turn, conveys the positional information of spectrin assembly to interacting integral membrane proteins. The downward and upward arrows indicate the flow of information, from a membrane protein or lipid that recruits spectrin assembly or to a membrane protein that is captured and stabilized by the spectrin cytoskeleton, respectively. Depending on the system, cues may transmit information to the PH domain or other sites in β-spectrin (ankyrin-independent membrane-binding sites) (Lombardo *et al.*, 1994) or to ankyrin. Similarly proteins may be captured and retained in specialized plasma membrane domains by interactions with either spectrin or ankyrin.

VIII. FUTURE PROSPECTS

Genetic studies were invaluable in elucidating the role of the spectrin cytoskeleton in human erythrocytes, and now genetic studies are helping to delineate the functions of spectrin in other systems. The results of human genetic studies, and studies in the mouse and *Drosophila* models have now firmly established that one important role of spectrin is to organize specialized domains of plasma membrane composition and function. The same principle emerged in studies of Duchenne muscular dystrophy (DMD) (Campbell, 1995). Dystrophin, the defective molecule in DMD, is a close relative of β-spectrins. The loss of dystrophin in DMD patients is associated with the complete disappearance of a host of membrane proteins that normally associate with dystrophin at the sarcolemma. It seems reasonable to expect that the list of membrane activities that associate with and rely on spectrin and related proteins for their targeting and function will grow with ongoing studies in human and model systems.

While there has been great progress, there still are large gaps in our understanding of spectrin's activities in the cell. For example, the effects of spectrin and ankyrin mutations on cell growth and polarity cannot readily be explained in our current framework of ideas for their function. One area that I have focused on here is the mechanism responsible for spectrin assembly at the plasma membrane. There is clearly a link between spectrin assembly and sites of cell–cell adhesion in a number of cellular contexts. Yet, as pointed out in a number of studies, there is also clearly more to the story. The field has produced a fresh set of clues to the cellular activities of spectrins and ankyrins and the next wave of research promises to bring new insights to these fundamental questions in cell biology.

ACKNOWLEDGMENTS

This work was supported by NIH GM49301.

REFERENCES

Baas, A. F., Kuipers, J., Wel, N. N. V., Batlle, E., Koerten, H. K., Peters, P. J., and Clevers, H. C. (2004). Complete polarization of single intestinal epithelial cells upon activation of LKB1 by, STRAD. *Cell* **116,** 457–466.

Banuelos, S., Saraste, M., and Carugo, K. D. (1998). Structural comparisons of calponin homology domains: Implications for actin binding. *Structure* **6,** 1419–1431.

Bennett, V. and Baines, A. J. (2001). Spectrin and ankyrin-based pathways: Metazoan inventions for integrating cells into tissues. *Physiol. Rev.* **81,** 1353–1388.

Bouley, M., Tian, M.-Z., Paisley, K., Shen, Y.-C., Malhotra, J. D., and Hortsch, M. (2000). The, L1-type CAM neuroglian influences the stability of neural ankyrin in the *Drosophila* embryo but not its axonal localization. *J. Neurosci.* **20,** 4515–4523.

Bretscher, A., Edwards, K., and Fehon, R. G. (2002). ERM proteins and merlin: Integrators at the cell cortex. *Nat. Rev. Mol. Cell. Biol.* **3,** 586–599.

Byers, T. J. and Branton, D. (1985). Visualization of the protein associations in the erythrocyte membrane skeleton. *Proc. Natl. Acad. Sci. USA* **82,** 6153–6157.

Byers, T. J., Husain-chishti, A., Dubreuil, R. R., Branton, D., and Goldstein, L. S. B. (1989). *Drosophila* beta-spectrin: Sequence similarity to the amino-terminal domain of alpha-actinin and dystrophin. *J. Cell Biol.* **109,** 1633–1641.

Byers, T. J., Brandin, E., Winograd, E., Lue, R., and Branton, D. (1992). The complete sequence of *Drosophila* beta spectrin reveals supra-motifs comprising eight 106-residue segments. *Proc. Natl. Acad. Sci. USA* **89,** 6187–6191.

Campbell, K. P. (1995). Three muscular dystrophies: Loss of cytoskeleton-extracellular matrix linkage. *Cell* **80,** 675–679.

Chasis, J. A. and Mohandas, N. (1992). Red blood cell glycophorins. *Blood* **80,** 1869–1879.

Coetzer, T. L., Sahr, K., Prchal, J., Blacklock, H., Peterson, L., Koler, R., Doyle, J., Manaster, J., and Palek, J. (1991). Four different mutations in Codon 28 of alpha spectrin are associated with structurally and functionally abnormal spectrin alphaI-74 in hereditary elliptocytosis. *J. Clin. Invest.* **88,** 743–749.

Das, A., Base, C., Dhulipala, S., and Dubreuil, R. R. (2006). Spectrin functions upstream of ankyrin in a spectrin cytoskeleton assembly pathway. Submitted for publication.

Davis, J. Q., McLaughlin, T., and Bennett, V. (1993). Ankyrin-binding proteins related to nervous system cell adhesion molecules: Candidates to provide transmembrane and intercellular connections in adult brain. *J. Cell Biol.* **121,** 121–133.

DeMattheis, M. A. and Morrow, J. S. (2000). Spectrin tethers and mesh in the biosynthetic pathway. *J. Cell. Sci.* **113,** 2331–2343.

Deng, H., Lee, J. K., Goldstein, L. S. B., and Branton, D. (1995). *Drosophila* development requires spectrin network formation. *J. Cell Biol.* **128,** 71–79.

Doerflinger, H., Benton, R., Shulman, J. M., and Johnston, D. S. (2003). The role of, PAR-1 in regulating the polarised microtubule cytoskeleton in the *Drosophila* follicular epithelium. *Development* **130,** 3965–3975.

Dubreuil, R. R. (1991). Structure and evolution of the actin crosslinking proteins. *BioEssays* **13,** 219–226.

Dubreuil, R. R. (2004). Copper cells and stomach acid secretion in the *Drosophila* midgut. *Int. J. Biochem. Cell Biol.* **36,** 745–752.

Dubreuil, R., Byers, T. J., Branton, D., Goldstein, L. S. B., and Kiehart, D. P. (1987). *Drosophila* spectrin, I. Characterization of the purified protein. *J. Cell Biol.* **105,** 2095–2102.

Dubreuil, R. R. and Yu, J. (1994). Ankyrin and beta spectrin accumulate independently of alpha spectrin in *Drosophila. Proc. Natl. Acad. Sci. USA* **91,** 10285–10289.

Dubreuil, R. R., Byers, T. J., Stewart, C. T., and Kiehart, D. P. (1990). A beta spectrin isoform from *Drosophila* (betaH) is similar in size to vertebrate dystrophin. *J. Cell Biol.* **111,** 1849–1858.

Dubreuil, R. R., Brandin, E., Reisberg, J. H. S., Goldstein, L. S. B., and Branton, D. (1991). Structure, calmodulin-binding, and calcium-binding properties of recombinant alpha spectrin polypeptides. *J. Biol. Chem.* **266,** 7189–7193.

Dubreuil, R. R., MacVicar, G. R., Dissanayake, S., Liu, C., Homer, D., and Hortsch, M. (1996). Neuroglian-mediated adhesion induces assembly of the membrane skeleton at cell contact sites. *J. Cell Biol.* **133,** 647–655.

Dubreuil, R. R., Maddux, P. B., Grushko, T., and MacVicar, G. R. (1997). Segregation of two spectrin isoforms: Polarized membrane binding sites direct polarized membrane skeleton assembly. *Mol. Biol. Cell* **8,** 1933–1942.

Dubreuil, R. R., Frankel, J., Wang, P., Howrylak, J., Kappil, M., and Grushko, T. (1998). Mutations of alpha spectrin and labial block cuprophilic cell differentiation and acid secretion in the middle midgut of *Drosophila* larvae. *Dev. Biol.* **194,** 1–11.

Dubreuil, R. R., Wang, P., Dahl, S. C., Lee, J. K., and Goldstein, L. S. B. (2000). *Drosophila* beta spectrin functions independently of alpha spectrin to polarized the NaK ATPase in epithelial cells. *J. Cell Biol.* **149,** 647–656.

Featherstone, D. E., Davis, W. S., Dubreuil, R. R., and Broadie, K. (2001). *Drosophila* alpha and beta spectrin mutations disrupt presynaptic neurotransmitter release. *J. Neurosci.* **21,** 4215–4224.

Funke, L., Dakoji, S., and Bredt, D. S. (2005). Membrane-associated guanylate kinases regulate adhesion and plasticity at cell junctions. *Ann. Rev. Biochem.* **74,** 219–245.

Garver, T. D., Ren, Q., Tuvia, S., and Bennett, V. (1997). Tyrosine phosphorylation at a site highly conserved in the, L1 family of cell adhesion molecules abolishes ankyrin binding and increases lateral mobility of neurofascin. *J. Cell Biol.* **137,** 703–714.

Hammerton, R. W., Krzeminski, K. A., Mays, R. W., Ryan, T. A., Wollner, D. A., and Nelson, W. J. (1991). Mechanism for regulating cell surface distribution of Na^+, K^+-ATPase in polarized epithelial cells. *Science* **254,** 847–850.

Harper, S. L., Begg, G. E., and Speicher, D. W. (2001). Role of terminal nonhomologous domains in initiation of human red cell spectrin dimerization. *Biochemistry* **40,** 9935–9943.

Hemminki, A., Markie, M., Tomlinson, I., Avizienyte, E., Roth, S., Loukola, A., Bignell, G., Warren, W., Aminoff, M., Hoglund, P., Jarvinen, H., Kristo, P., *et al.* (1998). A serine/threonine kinase gene defective in Peutz–Jeghers syndrome. *Nature* **391,** 184–187.

Holleran, E. A., Ligon, L. A., Tokito, M., Stankewich, M. C., Morrow, J. S., and Holzbaur, E. L. F. (2001). Beta 3 spectrin binds to the Arp1 subunit of dynactin. *J. Biol. Chem.* **276,** 36598–36605.

Hortsch, M. (1996). The, L1 family of neural cell adhesion molecules–Old proteins performing new tricks. *Neuron* **17,** 587–593.

Hortsch, M., Paisley, K. L., Tian, M.-Z., Qian, M., Bouley, M., and Chandler, R. (2002). The axonal localization of large *Drosophila* ankyrin2 protein isoforms is essential for neuronal functionality. *Mol. Cell Neurosci.* **20,** 43–55.

Hu, R.-J., Watanabe, M., and Bennett, V. (1992). Characterization of human brain cDNA encoding the general isoform of beta-spectrin. *J. Biol. Chem.* **267,** 18715–18722.

Hyvonen, M., Macias, M. J., Nilges, M., Oschkinat, H., Saraste, M., and Wilmanns, M. (1995). Structure of the binding site for inositol phosphates in a PH domain. *EMBO J.* **14,** 4676–4685.

Ikeda, Y., Dick, K. A., Westherspoon, M. R., Gincel, D., Armbrust, K. R., Dalton, J. C., Stevanin, G., Durr, A., Zuhlke, C., Burk, K., Clark, H. B., Brice, A., *et al.* (2006). Spectrin mutations cause spinocerebellar ataxia type 5. *Nat. Genet.* **38,** 184–190.

Jefford, G. and Dubreuil, R. R. (2000). Receptor clustering drives polarized assembly of ankyrin. *J. Biol. Chem.* **275,** 27726–27732.

Jenkins, S. M. and Bennett, V. (2001). Ankyrin-G coordinates assembly of the spectrin-based membrane skeleton voltage-gated sodium channels and L1 CAMs at Purkinje neuron initial segments. *J. Cell Biol.* **155,** 739–745.

Jenne, D. E., Reimann, H., Nezu, J., Friedel, W., Loff, S., Jeschke, R., Muller, O., Back, W., and Zimmer, M. (1998). Peutz–Jeghers syndrome is caused by mutations in a novel serine threonine kinase. *Nat. Genet.* **18,** 38–43.

Kemphues, K. J., Priess, J. R., Morton, D. G., and Cheng, N. S. (1988). Identification of genes required for cytoplasmic localization in early *C. elegans* embryos. *Cell* **52,** 311–320.

Kishi, M., Pan, Y. A., Crump, J. G., and Sanes, J. R. (2005). Mammalian SAD kinases are required for neuronal polarization. *Science* **307,** 929–932.

Kizhatil, K. and Bennett, V. (2004). Lateral membrane biogenesis in human bronchial epithelial cells requires 190-kDa ankyrin-G. *J. Biol. Chem.* **279,** 16706–16714.

Komada, M. and Soriano, P. (2002). BetaIV-spectrin regulates sodium channel clustering through ankyrin-G at axon initial segments and nodes of Ranvier. *J. Cell Biol.* **156,** 337–348.

Koushika, S. P., Schaefer, A. M., Vincent, R., Willis, J. H., Bowerman, B., and Nonet, M. L. (2004). Mutations in *Caenorhabditis elegans* cytoplasmic dynein components reveal specificity of neuronal retrograde cargo. *J. Neurosci.* **24,** 3907–3916.

Kural, C., Kim, H., Syed, S., Goshima, G., Gelfand, V. I., and Selvin, P. R. (2005). Kinesin and dynein move a peroxisome *in vivo*: A tug-of-war or coordinated movement? *Science* **308,** 1469–1472.

Lacas-Gervais, S., Guo, J., Strenzke, N., Scarfone, E., Kolpe, M., Jahkel, M., DeCamilli, P., Moser, T., Rasband, M. N., and Solimena, M. (2004). BIVE1 spectrin stabilizes the nodes of Ranvier and axon initial segments. *J. Cell Biol.* **166,** 983–990.

Lebovitz, R. M., Takeyasu, K., and Fambrough, D. M. (1989). Molecular characterization and expression of the ($Na^+ + K^+$)-ATPase alpha-subunit in *Drosophila melanogaster*. *EMBO J.* **8,** 193–202.

Lee, J., Coyne, R., Dubreuil, R. R., Goldstein, L. S. B., and Branton, D. (1993). Cell shape and interaction defects in alpha-spectrin mutants of *Drosophila melanogaster*. *J. Cell Biol.* **123,** 1797–1809.

Lee, J. K., Brandin, E., Branton, D., and Goldstein, L. S. B. (1997). Alpha-spectrin is required for ovarian follicle monolayer integrity in *Drosophila melanogaster*. *Development* **124,** 353–362.

Lehto, V. P., Wasenius, V. M., Salven, P., and Saraste, M. (1988). Transforming and membrane proteins. *Nature* **334,** 388.

Lemmon, M. A. and Ferguson, K. M. (2000). Signal-dependent membrane targeting by pleckstrin homology (PH) domains. *Biochem. J.* **350,** 1–18.

Lombardo, Z. R., Weed, S. A., Kennedy, S. P., Forget, B. G., and Morrow, J. S. (1994). BetaII-spectrin (fodrin) and Beta, IE2-Spectrin (muscle) contain, NH2- and COOH-terminal membrane association domains (MAD-1 and MAD2). *J. Biol. Chem.* **269,** 29212–29219.

Lux, S. E. and Palek, J. (1995). Disorders of the red cell membrane. In: *Blood: Principles and Practice of Hematology* (Handin, R. I., Lux, S. E., and Stossel, T. P., Eds.), pp. 1701–1818. J. B. Lippincott Co, Philadelphia.

Macias, M. J., Musacchio, A., Ponstingl, H., Nilges, M., Saraste, M., and Oschkinat, H. (1994). Structure of the pleckstrin homology domain from beta spectrin. *Nature* **369,** 675–677.

Marfatia, S. M., Lue, R. A., Branton, D., and Chishti, A. H. (1994). *In vitro* binding studies suggest a membrane-associated complex between erythroid p55, protein 4.1 and glycophorin C. *J. Biol. Chem.* **269,** 8631–8634.

Martin, S. G. and Johnston, D. S. (2003). A role for *Drosophila* LKB1 in anterior-posterior axis formation and epithelial polarity. *Nature* **421,** 379–384.

McNeill, H., Ozawa, M., Kemler, R., and Nelson, W. J. (1990). Novel function of the cell adhesion molecule uvomorulin as an inducer of cell surface polarity. *Cell* **62,** 309–316.

Mercier, F., Teggio, H., Deviliers, G., Bataille, D., and Mangeat, P. (1989). Membrane-cytoskeleton dynamics in rat parietal cells: Mobilization of actin and spectrin upon stimulation of gastric acid secretion. *J. Cell Biol.* **108,** 441–453.

Mohler, P. J. and Bennett, V. (2005). Ankyrin-based cardiac arrhythmias: A new class of channelopathies due to loss of cellular targeting. *Curr. Opin. Cardiol.* **20,** 189–193.

Mohler, P. J., Schott, J.-J., Gramolini, A. O., Dilly, K. W., Guatimoisim, S., duBell, W. H., Song, L.-S., Haurogne, K., Kyndt, F., Ali, M. E., Rogers, T. B., Lederer, W. J., *et al.* (2003). Ankyrin-B mutations causes type 4 long-QT cardiac arrhythmia and sudden cardiac death. *Nature* 634–639.

Mohler, P. J., Splawski, I., Napolitano, C., Bottelli, G., Sharpe, L., Timothy, K., Priori, S. G., Keating, M. T., and Bennett, V. (2004a). A cardiac arrhythmia syndrome caused by loss of ankyrin-B function. *Proc. Natl. Acad. Sci. USA* **101,** 9137–9142.

Mohler, P. J., Rivolta, I., Napolitano, C., LeMaillet, G., Lambert, S., Priori, S. G., and Bennett, V. (2004b). Nav1. 5 E1053K mutation causing, Brugada syndrome blocks binding to ankyrin-G and expression of nav1. 5 on the surface of cardiomyocytes. *Proc. Natl. Acad. Sci. USA* **101,** 17533–17538.

Mohler, P. J., Yoon, W., and Bennett, V. (2004c). Ankyrin-B targets, B2-spectrin to an intracellular compartment in neonatal cardiomyocytes. *J. Biol. Chem.* **279,** 40185–40193.

Mohler, P. J., Davis, J. Q., and Bennett, V. (2005). Ankyrin-B coordinates the, Na/K ATPase, Na/Ca exchanger, and InsP3 receptor in a cardiac, T-tubule/SR microdomain. *PloS Biology* **3,** e423.

Muresan, V., Stankewich, M. C., Steffen, W., Morrow, J. S., Holzbaur, E. L. F., and Schnapp, B. J. (2001). Dynactin-dependent, dynein-driven vesicle transport in the absence of membrane proteins: A role for spectrin and acidic phospholipids. *Mol. Cell* **7,** 173–183.

Needham, L. K., Thelen, K., and Maness, P. F. (2001). Cytoplasmic domain mutations of the, L1 cell adhesion molecule reduce, L1-ankyrin interactions. *J. Neurosci.* **21,** 1490–1500.

Nelson, W. J. and Veshnock, P. J. (1986). Dynamics of membrane skeleton (fodrin) organization during development of polarity in Madin-Darby Canine Kidney epithelial cells. *J. Cell Biol.* **103,** 1751–1765.

Nelson, W. J. and Veshnock, P. J. (1987). Ankyrin binding to (Na^+ and K^+) ATPase and implications for the organization of membrane domains in polarized cells. *Nature* **328,** 533–536.

Nishimura, K., Yoshihara, F., Tojima, T., Ooashi, N., Yoon, W., Mikoshiba, K., Bennett, V., and Kamiguchi, H. (2003). L1-dependent neuritogenesis involves ankyrinB that mediates L1-CAM coupling with retrotrade actin flow. *J. Cell Biol.* **163,** 1077–1088.

Papoulas, O., Hays, T. S., and Sisson, J. C. (2005). The golgin Lava lamp mediates dynein-based Golgi movements during *Drosophila* cellularization. *Nat. Cell Biol.* **7,** 612–618.

Parkinson, N. J., Olsson, C. L., Hallows, J. L., McKee-Johnson, J., Keogh, B. P., Noben-Trauth, K., Kujawa, S. G., and Tempel, B. L. (2001). Mutant beta-spectrin 4 causes auditory and motor neuropathies in quivering mice. *Nat. Genet.* **29,** 61–65.

Pesacreta, T. C., Byers, T. J., Dubreuil, R. R., Keihart, D. P., and Branton, D. (1989). *Drosophila* spectrin: The membrane skeleton during embryogenesis. *J. Cell Biol.* **108,** 1697–1709.

Peters, L. L., Shivdasani, R. A., Liu, S.-C., Manspal, M., John, K. M., Gonzalez, J. M., Brugnara, C., Gwynn, B., Mohandas, N., Alper, S. L., Orkin, S. H., and Lux, S. E. (1996). Anion exchanger 1 (band 3) is required to prevent erythocyte membrane surface loss but not to form the membrane skeleton. *Cell* **86,** 917–927.

Pielage, J., Fetter, R. D., and Davis, G. W. (2005). Presynaptic spectrin is essential for synapse stabilization. *Curr. Biol.* **15,** 918–928.

Pradhan, D., Tseng, K., Cianci, C. D., and Morrow, J. S. (2004). Antibodies to beta I E2 spectrin identify in-homogeneities in the erythrocyte membrane skeleton. *Blood Cells Mol. Dis.* **32,** 408–410.

Salomao, M., An, X., Guo, X., Gratzer, W. B., Mohandas, N., and Baines, A. J. (2006). Mammalian alphaI-spectrin is a neofunctionalized polypeptide adapted to small highly deformable erythrocytes. *Proc. Natl. Acad. Sci. USA* **103,** 643–648.

Scotland, P., Zhou, D., Benveniste, H., and Bennett, V. (1998). Nervous system defects of ankyrin, B (–/–) mice suggest functional overlap between the cell adhesion molecule, L1 and 440 kD ankyrin, B in premyelinated axons. *J. Cell Biol.* **143,** 1305–1315.

Sisson, J. C., Field, C., Ventura, R., Royou, A., and Sullivan, W. (2000). Lava lamp, a novel peripheral golgi protein, is required for *Drosophila melanogaster* cellularization. *J. Cell Biol.* **151,** 905–917.

Speicher, D. W. and Marchesi, V. T. (1984). Erythrocyte spectrin is comprised of many homologous triple helical segments. *Nature* **311,** 177–180.

Stabach, P. R. and Morrow, J. S. (2000). Identification and characterization of betaV spectrin, a mammalian ortholog of *Drosophila* betaH spectrin. *J. Biol. Chem.* **275,** 21385–21395.

Stankewich, M. C., Tse, W. T., Peters, L. L., Ch'ng, Y., John, K. M., Stabach, P. R., Devarajan, P., Morrow, J. S., and Lux, S. E. (1998). A widely expressed betaIII spectrin associated with Golgi and cytoplasmic vesicles. *Proc. Natl. Acad. Sci. USA* **95,** 14158–14163.

Steiner, J. P. and Bennett, V. (1988). Ankyrin-independent membrane protein-binding sites for brain and erythrocyte spectrin. *J. Biol. Chem.* **263,** 14417–14425.

Thomas, G. H. (2001). Spectrin: The ghost in the machine. *BioEssays* **23,** 152–160.

Tilney, L. G. and Detmers, P. (1975). Actin in erythrocyte ghosts and its association with spectrin. *J. Cell Biol.* **66,** 508–520.

Tse, W. T. and Lux, S. E. (1999). Red blood cell membrane disorders. *Br. J. Hematol.* **104,** 2–13.

Tse, W. T., Lecomte, M. C., Costa, F. F., Garbarz, M., Feo, C., Boivin, P., Dhermy, D., and Forget, B. G. (1990). A point mutation in the beta-spectrin gene associated with alpha-I/74 hereditary elliptocytosis—implications for the mechanism of spectrin dimer self association. *J. Clin. Invest.* **86,** 909–916.

Tse, W. T., Tang, J., Jin, O., Korsgren, C., John, K. M., Kung, A. L., Gwynn, B., Peters, L. L., and Lux, S. E. (2001). A new spectrin beta4 has a major truncated isoform that associates with promyelocytic leukemia protein nuclear bodies and the nuclear matrix. *J. Biol. Chem.* **276,** 23974–23985.

Wang, D.-S. and Shaw, G. (1995). The association of the C-terminal region of beta1EII spectrin to brain membranes is mediated by a PH domain, does not require membrane proteins, and coincides with a inositol-1,4,5 trisphosphate binding site. *Biochem. & Biophys. Res. Comm.* **217,** 608–615.

Wang, D.-S., Miller, R., Shaw, R., and Shaw, G. (1996). The pleckstrin homology domain of human B1EII spectrin is targeted to the plasma membrane *in vivo*. *BBRC* **225,** 420–426.

Wasenius, V.-M., Saraste, M., Salven, P., Eramaa, M., Holm, L., and Lehto, V.-P. (1989). Primary structure of the brain alpha-spectrin. *J. Cell Biol.* **108,** 79–93.

Watts, J. L., Morton, D. G., Bestman, J., and Kemphues, K. J. (2000). The *C. elegans par-4* gene encodes a putative serine-threonine kinase required for establishing embryonic asymmetry. *Devlopment* **127,** 1467–1475.

Wei, J., Hortsch, M., and Goode, S. (2004). Neuroglian stabilizes epithelial structure during *Drosophila oogenesis*. *Dev. Dyn.* **230,** 800–808.

Winkelman, J. C., Costa, F. F., Linzie, B. L., and Forget, B. G. (1990). Beta spectrin in human skeletal muscle. *J. Biol. Chem.* **265,** 20449–20454.

Wodarz, A. (2002). Establishing cell polarity in development. *Nat. Cell. Biol.* **4,** E39–E44.

Yan, Y., Winograd, E., Viel, A., Cronin, T., Harrison, S. C., and Branton, D. (1993). Crystal structure of the repetitive segments of spectrin. *Science* **262,** 2027–2030.

Yeaman, C., Grindstaff, K. K., and Nelson, W. J. (1999). New perspectives on mechanisms involved in generating epithelial cell polarity. *Physiol. Rev.* **79,** 73–98.

Zarnescu, D. C. and Thomas, G. H. (1999). Apical spectrin is essential for epithelial morphogenesis but not apicobasal polarity in *Drosophila*. *J. Cell Biol.* **146,** 1075–1086.

Zhang, P., Talluri, S., Deng, H., Branton, D., and Wagner, G. (1995). Solution structure of the pleckstrin homology domain of *Drosophila* beta-spectrin. *Structure* **3,** 1185–1195.

Zhang, X., Davis, J. Q., Carpenter, S., and Bennett, V. (1998). Structural requirements for association of neurofascin with ankyrin. *J. Biol. Chem.* **273,** 30785–30794.

Zhou, D., Lambert, S., Malen, P. L., Carpenter, S., Boland, L. M., and Bennett, V. (1998). Ankyrin G is required for clustering of voltage-gated Na channels at axon initial segments and for normal action potential firing. *J. Cell Biol.* **143,** 1295–1304.

Chapter 5

Structure and Function of Villin

Seema Khurana

Department of Physiology, University of Tennessee Health Science Center, Memphis, Tennessee 38163

Villin is an epithelial cell-specific protein that belongs to a family of actin-binding proteins that display structural and functional homology and include among other proteins, gelsolin, severin, fragmin, adseverin/scinderin, supervillin, and protovillin. Villin is unique among these proteins in that it can sever, cap, nucleate, and crosslink actin filaments. The actin modifying functions of villin are regulated by calcium, phosphoinositides, and tyrosine phosphorylation of the villin protein. Villin also self-associates and this property of villin is likewise determined by its ligand-binding properties. Overexpression of villin regulates actin redistribution, cell morphology, cell migration, and cell death. Studies done with the villin knockout mice substantiate the significance of villin in actin reorganization and maintenance of epithelial cell plasticity. In this chapter I discuss how the ligand-binding properties of villin are mechanistically important to its functions in cell migration and cell death.

I. INTRODUCTION

Villin was first isolated from chicken intestinal microvilli and named by Bretscher and Weber (1979). Villin is a member of a conserved family of actin-binding proteins that is widely expressed from slime molds to humans. The villin gene is localized on

Advances in Molecular and Cell Biology, Vol. 37

ISBN: 0-444-52868-7

DOI: 10.1016/S1569-2558(06)37005-1

human chromosome 2q35-q36 and on mouse chromosome 1 and belongs to a cluster of genes that are conserved between the two species (Rousseau-Merck *et al.*, 1988). There is one copy of the villin gene per human haploid genome, about 25 kilobases (kb) and contains 19 exons (Pringault *et al.*, 1991). Villin is an acidic polypeptide with a mass of 95 kDa and is expressed in most significant amounts in renal and gastrointestinal epithelial cells, where it is localized to the microvillar core and the terminal web (Bretscher *et al.*, 1981; Robine *et al.*, 1985; Arpin *et al.*, 1988a). Villin has also been identified in exocrine glands of endodermic lineage, such as the thymus, as well as in those associated with the gastrointestinal tract, such as in pancreatic and biliary ductal cells. Villin in other epithelial cell types include brush cells that line the respiratory and gastrointestinal tract (Hofer and Drenckhahn, 1992), taste receptor cells (Hofer and Drenckhahn, 1999; Yoshie *et al.*, 2003), and osteocytes (Kamioka *et al.*, 2004). Villin expression has also been noted in cells that do not form microvilli-like structures such as undifferentiated crypt cells, embryonic intestinal cells, M-cells and in protists and plants (Robine *et al.*, 1985; Iovanna *et al.*, 1992; Tsukada *et al.*, 1995; Kerneis *et al.*, 1996; Klahre *et al.*, 2000). Most of these cells share either functional similarity or a common embryonic origin.

Villin is expressed in every cell of the intestinal epithelium along a concentration gradient (increases from the crypt to the villus tip and from colon to duodenum). Likewise, in cell models, differentiation of crypt-like cell lines including HT29 (cultured in glucose-free medium) and IEC-6 cells (transfected with the transcriptional factor Cdx1) the expression of villin increases significantly as the cells differentiate (Dudouet *et al.*, 1987; Soubeyran *et al.*, 1999). Since villin is detected in the immature digestive tract during the early stages of development and shows a transient elevation prior to the formation of microvilli, it has been suggested that villin may play a significant role in microvillar formation, maturation, and/or maintenance. It may be noted that while the villin knockout mice are viable and fertile, subtle changes in the microvillar structure have been noted at the ultrastructural level. The actin bundles in the microvillar core are not well organized or as tightly packed as seen in the wild-type littermates, likewise the terminal web appears thicker and less well organized in the villin-null mice (Pinson *et al.*, 1998). Thus, the uniform filament polarity and packing of actin filament bundles in the microvillar core may be determined by villin. Isolated brush borders from villin knockout mice also do not disassemble *in vitro* in the presence of elevated calcium suggesting that the actin-severing activity of villin is equally important (Mooseker *et al.*, 1980; Matsudaira and Burgess, 1982; Ferrary *et al.*, 1999). Since villin is expressed in the progenitor cells in the crypt, villin is regarded as an early marker of committed intestinal absorptive cells. Villin also appears as an early marker of the endodermal cell lineage and a marker of cells arising from mesenchymal/epithelial conversion in the developing kidney. Much of the published data suggest that villin responds to the morphogenetic cues associated with intestinal development including tube formation, regional differentiation and formation of epithelial boundaries along the gastrointestinal tract, as well as cell proliferation, differentiation and migration that occurs continually in the adult intestine. A 12.4-kb region in the mouse villin gene has been identified that regulates the reporter gene expression, and determines the expression of villin in most of the small and large intestinal epithelium (Madison *et al.*, 2002). However, this region does not determine villin expression in the stomach or the kidney. Thus, it appears that different regulatory mechanisms control the expression of villin in different organs. The expression of villin gene is also maintained in carcinomas derived from intestinal and renal epithelial cells, irrespective of the level of differentiation of the tumor cells, further villin is also expressed in other adenocarcinomas even though it is absent from normal tissue such as Barrett's metaplasia, esophageal adenocarcinomas, gastric cardia adenocarcinomas, and in chronic atrophic gastritis (Moll *et al.*, 1987; Regalado *et al.*, 1998; MacLennan *et al.*, 1999). Since villin expression is detected in all intestinal metaplasia, it has been speculated that induction of villin in these lesions may participate in the altered genetic program that gives rise to intestinal metaplasia. Rieder *et al.* (2005) demonstrated that *Helicobacter pylori* infection in a human adenocarconoma cell line or in a mouse model of infection results in the expression of villin. Further, the authors demonstrate that a 554-bp region of the villin promoter contains elements that respond to *H. pylori*. These studies suggest that villin is an important marker of the preneoplastic cell type that develops in the gut in response to chronic injury.

Like other proteins of its family including gelsolin, adseverin, and fragmin, villin contains repeats of an evolutionarily conserved domain. Villin contains six such domains (S1–S6). In addition, it contains a C-terminal (S7) domain called the headpiece (H), which is conserved among other proteins of the villin superfamily such as protovillin, supervillin, and advillin (Andre *et al.*, 1988) (Fig. 1). Villin domains S1–S6 are encoded by exons 1–8 and 10–16, and are separated by exon 9 which encodes a short

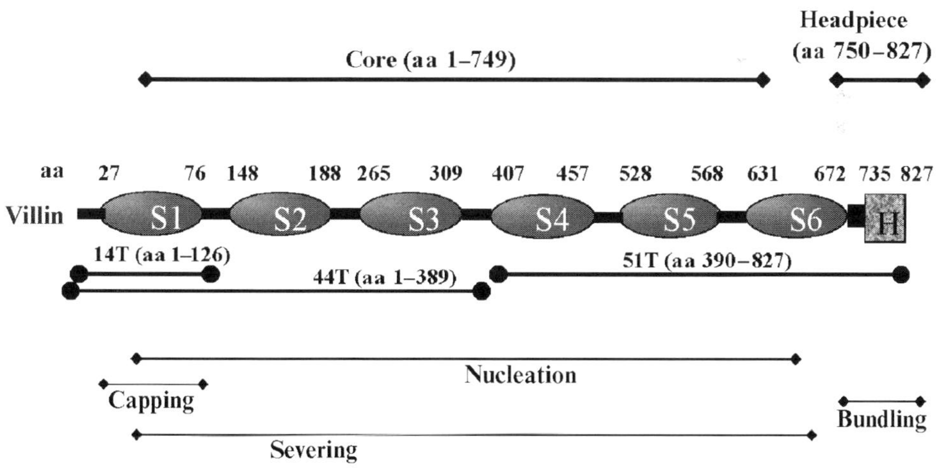

Figure 1. Domain structure of villin. The six homologous domains conserved between villin and other members of this superfamily are shown as S1–S6 and the C-terminal headpiece domain is indicated by the letter H. The villin core consists of domains S1–S6. The fragments of villin derived by proteolytic digestion are denoted as 14T, 44T, and 51T. Also shown in this figure is the relative placement of domains regulating the various actin-modifying functions of villin, including capping, nucleating, severing, and bundling of actin filaments.

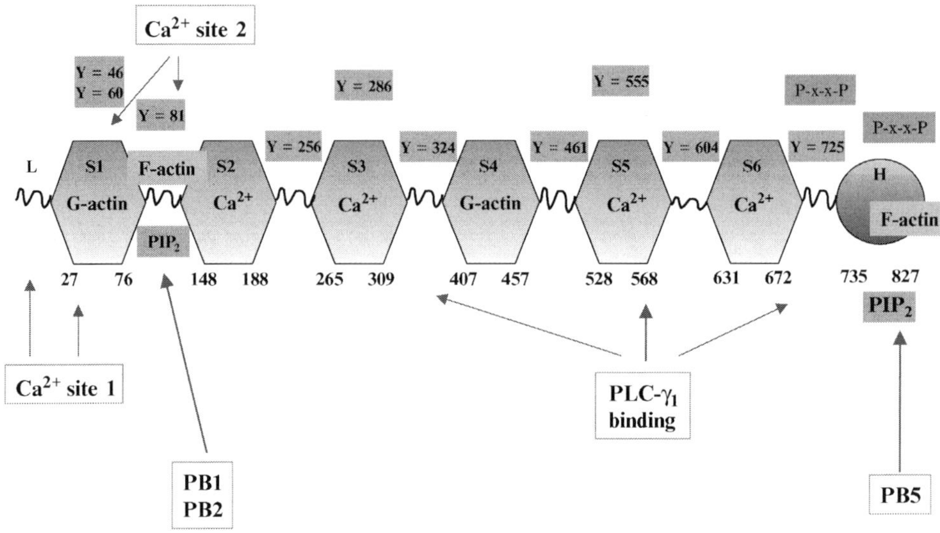

Figure 2. Schematic representation of the various ligand-binding and tyrosine phosphorylation sites identified in villin. Villin contains two F-actin–binding sites, one between S1–S2 and a second site in the headpiece. Villin also contains two monomer (G-actin)-binding sites, one each in S1 and S4 domains. Ten tyrosine (Y) phosphorylation sites have been identified in villin. These are identified by specific amino acids in human villin protein. Six of these C-terminal phosphorylation sites together (but not individually) determine the association of tyrosine-phosphorylated villin with PLC-γ_1 (Y-286, Y-324, Y-461, Y-555, Y-604, and Y-725). Three PIP_2-binding sites have been identified in villin, two in the core (PB1 and PB2) and the third in the headpiece (PB5). Two calcium-sensitive sites regulating calcium-induced conformational changes in villin have been identified in the N-terminal domains of villin. These include calcium site 1 which regulates actin capping and severing, while calcium site 2 regulates only actin depolymerization by villin. Four additional calcium-sensitive sites have been identified in intradomains S2, S3, S5, and S6 and two calcium-sensitive residues that function as latch residues (to release autoinhibited conformation), namely Asp-467 and Asp-715. Two putative SH3-binding domains are also shown in this figure.

hinge sequence. Exons 17 and 18 encode the headpiece domain of villin (Pringault *et al.*, 1991). This observation has led to the suggestion that the villin gene may have evolved by gene duplication and then recombination between a gelsolin-like gene and a gene encoding a different protein with a headpiece-like domain. This unique structure of villin makes it the most versatile member of its family since it can nucleate, cap, sever, and cross-link actin filaments. Several ligand-binding domains have been identified in villin and are discussed in greater detail later in this chapter (Fig. 2). All amino acid residues mentioned in this chapter refer to human villin protein unless indicated otherwise.

II. ACTIN-MODIFYING FUNCTIONS OF VILLIN

Villin is unique among the actin-binding proteins in that it can both polymerize as well as depolymerize actin filaments. Like gelsolin, adseverin, and CapG, villin can bind to the barbed end of actin filaments with high affinity (binding constant of

10^{11} M^{-1}) and inhibit the rate of filament elongation at a very low stoichiometry (villin:actin ratio of 1:1000), thus functioning as an actin-capping protein (Walsh *et al.*, 1984b). Like gelsolin and adseverin, villin can also sever actin filaments. Further, in the presence of calcium villin can form a complex with monomeric actin thus nucleating actin assembly from the pointed end (Glenney *et al.*, 1981b). In addition, the headpiece allows villin to cross-link actin filaments. Many but not all villins retain the ability to sever, nucleate, cap, and cross-link filaments. For instance, plant villin (isoform VILLIN 1) does not sever or cap actin filaments but bundles actin like the vertebrate villin protein (Huang *et al.*, 2005). However, other isoforms of villin that have been identified in plant pollen including VLN5 and VLN2 that can sever and cap actin filaments.

The protease V8 cleaves villin into the core (S1–S6) and the headpiece (Glenney and Weber, 1981; Glenney *et al.*, 1981b). Cleavage by trypsin yields two fragments 44T and 51T, and 44T can be cleaved further into a 14T fragment (Janmey and Matsudaira, 1988) (Fig. 1). The structural and functional domains in villin were first described in these proteolytic fragments of villin. Two F-actin–binding sites were identified, one retained in the core (between S1 and S2) and a second site in the headpiece of villin (Janmey and Matsudaira, 1988). The headpiece and the S1–S2 domains of villin bind actin with different affinities, K_d of 7 and 0.3 μM, respectively, at low ionic strength (Burgess *et al.*, 1987; Pope *et al.*, 1994; Zhai *et al.*, 2001). The F-actin–binding residues in the villin core were first characterized by de Arruda by footprinting/peptide mapping (de Arruda *et al.*, 1992; Friederich *et al.*, 1999). Several actin-binding residues have also been described in the headpiece and are discussed later in this section. In addition, two monomer-binding sites have been identified, one in villin 14T as well as the 44T fragment (speculated to be in domain S1), and a second monomer-binding site in 51T (by homology with gelsolin, speculated to be in domain S4 of villin). But the specific residues in these monomer-binding domains of villin have not yet been identified (Janmey and Matsudaira, 1988). Actin binding by villin is completely dependent on Ca^{2+} (Janmey and Matsudaira, 1988; Kumar and Khurana, 2004). Other factors that regulate villin's interaction with F-actin include tyrosine phosphorylation of villin, which inhibits the binding affinity of villin for F-actin (Zhai *et al.*, 2001). Tropomyosin, an actin-binding protein, also inhibits villin's association with F-actin (Burgess *et al.*, 1987; Weber *et al.*, 1987).

The actin-capping activity of villin is contained within the first domain of villin, that is within fragment 14T (Fig. 1) (Janmey and Matsudaira, 1988). In contrast, the entire villin core is required to duplicate the actin-severing and actin-nucleating activities of full-length villin protein (Glenney *et al.*, 1980; Matsudaira *et al.*, 1985; Janmey and Matsudaira, 1988). Some of these observations have since been confirmed by mutational analysis of human villin protein *in vitro* (Friederich *et al.*, 1999; Kumar *et al.*, 2004a; Tomar *et al.*, submitted for publication). While the actin capping and nucleating functions of villin are intimately linked, these functions of villin can be separated structurally (Glenney *et al.*, 1981b; Janmey and Matsudaira, 1988). It is speculated that actin capping by villin is regulated by the monomer-binding site in domain S1, while actin nucleation by villin may be regulated by monomer-binding

Table I
Regulation of Villin's Actin-Modifying Functions[a]

	Nucleation	Capping	Severing	Bundling
Ca^{2+}	Required	Increased	Increased	Decreased
PIP_2	No effect	Decreased	Decreased	Increased
Tyrosine phosphorylation	Decreased	No effect	Increased	Decreased

[a]Actin capping by villin is regulated by calcium and PIP_2 but not by tyrosine phosphorylation of villin. Actin nucleation by villin is regulated by tyrosine phosphorylation of villin and not by PIP_2 but requires calcium. Actin severing is enhanced by high calcium and tyrosine phosphorylation of villin and is inhibited by PIP_2. Actin bundling does not require calcium and is inhibited by tyrosine phosphorylation of villin and enhanced by PIP_2.

sites in domains S1 and S4 (Friederich *et al.*, 1999; Zhai *et al.*, 2002; Kumar *et al.*, 2004a; Tomar *et al.*, in press). Additional differences between the actin-capping and actin-nucleating functions have been identified. These include the regulation of actin capping by Ca^{2+} and phosphatidylinositol 4,5-bisphosphate (PIP_2) but not tyrosine phosphorylation of villin, and regulation of actin nucleation by Ca^{2+} and tyrosine phosphorylation but not by PIP_2 (Table I) (Northrop *et al.*, 1986; Tomar *et al.*, 2004; Kumar *et al.*, 2004a,b; Tomar *et al.*, in press). *In vitro* half-maximal actin capping by villin can be achieved at calcium concentrations between 10 and 30 nM while half-maximal severing by villin requires calcium concentrations close to 100 μM. This suggests that both capping and severing of villin can be separated by calcium-sensitive sites in villin regulating these functions (Kumar *et al.*, 2004b). Actin capping and severing by villin can also be separated by other regulatory mechanisms that modify these functions of villin. Actin capping is regulated by PIP_2 and not tyrosine phosphorylation of villin. In contrast, actin severing by villin is enhanced by tyrosine phosphorylation and not regulated by PIP_2 (Zhai *et al.*, 2001, 2002; Kumar *et al.*, 2004b; Tomar *et al.*, in press) (Table I).

The technique of mutagenesis has been used to examine the two F-actin–binding sites in villin. A cluster of basic residues between Arg-138 and Val-148 were identified as the F-actin–binding site in the villin core (de Arruda *et al.*, 1992). Substitution with alanine of three basic residues within this domain of Arg-138, Lys-143, and Lys-145 resulted in significant inhibition of the actin-severing activity of villin in the order of 83%, 55%, and 47%, respectively. Mutational analysis then identified Arg-138 as the major residue regulating actin severing, the implication being that each residue contributes to the total binding energy, hence the actin-severing activity of villin (de Arruda *et al.*, 1992; Way *et al.*, 1992). It is significant that neither of these mutants affected the actin-capping activity of villin, consistent with the idea that the actin-capping site is retained in domain S1 of villin, while actin-severing site is retained between domains S1 and S2. Since the actin-severing site in villin lies in the inter-domain between S1 and S2, a region of least homology between related proteins of this family, it may reflect differences in the actin-severing activities of these related proteins. Friederich *et al.* (1999) have demonstrated that mutation of these three residues also abolishes villin's ability to reorganize the actin cytoskeleton and regulate cell shape in transfected CV-1 cells.

The solution structure of the N-terminal domain of villin (14T) has been determined by Markus *et al.* (1994b). The 14T is a compact domain organized around a hydrophobic core of conserved residues and consists of a central β-sheet composed of four antiparallel strands and one parallel strand that is surrounded by one long α-helix and two shorter helices, and a short parallel β-sheet. The central β-sheet and long α-helix represent the conserved sequences between villin and other proteins of this family. The nonconserved sequences are present along the surface and form potential binding sites for actin monomers. All three helices are amphipathic which is consistent with their presence on the solvent surface of 14T. Using chemical shift analysis Markus *et al.* localized two calcium-binding sites in 14T, one at each end of the long α-helix (see discussion later in this chapter). The actin monomer-binding region has been mapped to residues 82–102 in 14T (Way *et al.*, 1992). Chemical cross-linking of cystiene replacement mutants of villin showed that Cys-91 can be cross-linked to actin Cys-374 (Doering and Matsudaira, 1996). In addition, residues corresponding to His-96, Arg-97, Glu-98, and Val-99 have been shown to regulate actin monomer binding by 14T (Way *et al.*, 1992). Some of these sites were estimated on the basis of the monomer-binding sites in gelsolin domain G1. However, the actin-binding properties of villin differ from those of gelsolin: the 14T fragment of villin binds actin monomers with less affinity than the first domain of gelsolin (Kwiatkowsk *et al.*, 1985; Janmey and Matsudaira, 1988). A comparison of the actin monomer-binding sites in villin 14T solution structure and gelsolin domain 1 crystal structure suggests differences in the hydrophobic residues. In gelsolin, residues Phe-49, Ile-103, Val-106, and Phe-149 contribute to the actin-binding surface. In contrast, in villin the Ile is conserved but the other residues include Met-26, Thr-82, and Met-125 which are less hydrophobic. Further, the charge distributions are more neutral in the monomer-binding domain in villin compared with gelsolin except for a more negative patch between strands β1 and β2 in villin 14T compared to gelsolin (Markus *et al.*, 1997). Some of these differences are speculated to be responsible for the differences in the actin-binding properties of villin and gelsolin N-terminal domains (Markus *et al.*, 1997).

A 76 amino acid C-terminal domain in villin constitutes the headpiece region (Glenney *et al.*, 1981a). The isolated villin headpiece does not bundle F-actin by itself but binds F-actin but not G-actin at a molar ratio of 1 headpiece:actin; and the binding is calcium independent, occurring equally in the absence or presence of EGTA (Glenney and Weber, 1981; Glenney *et al.*, 1981a; George *et al.*, submitted for publication). Actin bundles are formed at a 4:10 villin:actin ratio (Matsudaira and Burgess 1982) and bundling is regulated by Ca^{2+} (disrupted in high Ca^{2+}), pH (favored by a more acidic pH), tyrosine phosphorylation (inhibited), and PIP_2 (favored) (Matsudaira and Burgess, 1982; Zhai *et al.*, 2001; Kumar *et al.*, 2004b). Actin-bundling activity of villin is also inhibited by other actin-binding proteins such as tropomyosin (Burgess *et al.*, 1987). Using mutational analysis, a C-terminal cluster of charged amino acids (^{821}KKEK^{824}) was identified as a region corresponding to the F-actin–binding site in the villin headpiece (Friederich *et al.*, 1992). Transfection of CV-1 cells with point mutants of the KKEK motif were indistinguishable from wild-type villin and formed dorsal cell surface microvilli/microspikes similar to cells expressing wild-type protein. However, transfection with double-substitution mutants (EEEK vs KKEK) displayed significantly shorter microvilli. From such results it was concluded that the KKEK motif contributes to the F-actin binding and hence, bundling activity of villin. Mutation of the KKEK motif in recombinant villin protein was also shown to influence F-actin–bundling activity *in vitro* (Friederich *et al.*, 1999). In addition, mutation of the three F-actin–binding residues in the side-binding domain in the villin core ($^{138}R^{143}K^{145}K$), was also shown to affect actin-bundling by villin *in vitro* (Friederich *et al.*, 1999). While the authors noted no significant difference in the extent of bundle formation, the bundle morphology was affected, showing loosely packed filaments associated with electron dense material. From these studies it was deduced that two F-actin–binding sites which include the actin side-binding site in the core, as well as the F-actin–binding site in the headpiece, are required for actin bundling by villin. However, we have found that deletion of the entire F-actin–binding site in the villin core (ΔPB2, aa 138–146) is without effect on the actin-bundling functions of villin *in vitro* (George *et al.*, submitted for publication). To confirm these observations, we demonstrated that a recombinant protein expressing headpiece can only bundle actin filaments *in vitro*. However, to bundle actin with headpiece alone required the cross-linking of the headpiece domain. These data suggested that self-association of villin could generate a composite actin-binding surface, where each villin headpiece could bind one actin filament, thus effectively using a single F-actin–binding site to bundle actin filaments (Fig. 3). These data are also supported by our observation that villin self-associates in living cells in cell surface structures, including actin bundles in microvilli, microspikes, lamellipodia, and membrane ruffles (George *et al.*, submitted for publication).

Overexpression of villin in fibroblasts and other villin-null cells have shown a drastic change in cell morphology and actin reorganization. For example, the cells are more rounded and less flat, there is a loss on the ventral surface of stress fibers while on the dorsal surface the cells develop structures resembling microvilli and microspikes/filopodia (Friederich *et al.*, 1989; Franck *et al.*, 1990; Tomar *et al.*, 2004). It has been suggested that villin may associate only with actin filaments in a specific ordered way with a single polarity such as that found in intestinal microvilli. However, overexpression of villin in fibroblasts and the formation of cell surface microspikes and microvilli in these cells suggests that villin may inherently be able to organize the microfilament structure into ordered bundles of actin filaments. Overexpression of villin in fibroblasts and other villin-null cells has also clearly demonstrated the ability of villin to recruit actin from other microfilament structures, including stress fibers, thus suggesting that villin plays a key role in the regulation of actin assembly and actin dynamics in the cell.

It is estimated that there are at least 25 different headpiece domains in at least six different proteins so far identified and while most of them share 35–40% sequence

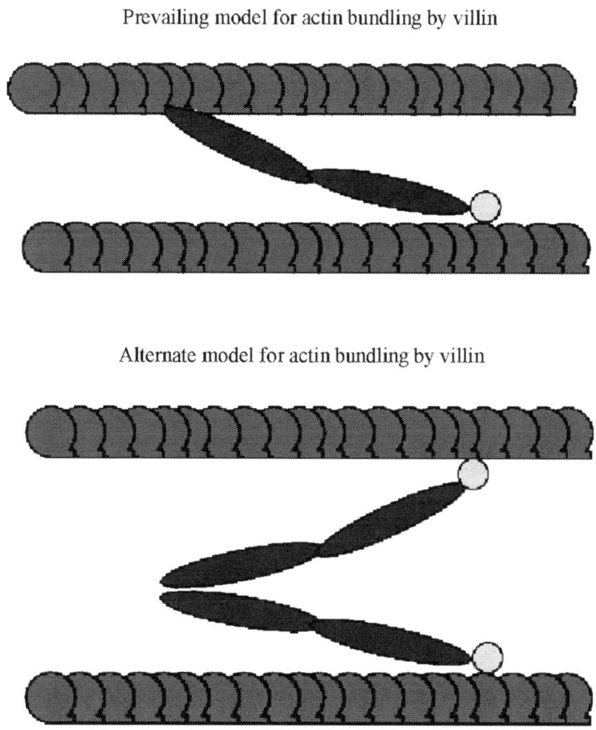

Figure 3. Model for actin cross-linking by villin oligomers. Upper image shows existing model for actin bundling by villin by the two F-actin–binding sites in villin, one in the core and the other in the headpiece domain. Lower panel shows our model for actin bundling by villin dimers, using the single F-actin–binding site in the headpiece domain.

identity, they are part of completely unrelated proteins. Further, not all these headpiece domains share the property to either bind F-actin and/or cross-link actin filaments. Two headpiece domains that have been examined in some detail include the headpiece of dematin and villin. The headpiece of villin is a 76 amino acid F-actin–binding C-terminal domain and is the smallest protein sequence that has been shown to fold autonomously and cooperatively. It does not require disulfide bonds, oligomerization, or ligand binding for stabilization of the folded structure, and has a high melting point (>70 °C) in aqueous solution. Because of these properties, it has been used by several groups of investigators to examine protein folding in both theoretical and computational studies aimed at understanding protein folding that will enable prediction of protein structure and synthesis of new molecules into the desired folded structures. This could then aid biologists to predict the structure of new proteins and ease in the design of drugs targeting these proteins. Notwithstanding the original intent, such studies have yielded some invaluable information on both the structure of the headpiece and the actin-binding properties of this domain in villin.

The headpiece of villin can be further divided into two subdomains namely, the N-terminal and C-terminal subdomains (McKnight *et al.*, 1997). All the residues that directly interact with F-actin have been localized to the C-terminal subdomain. The two subdomains are packed against each other into a compact structure having a hydrophobic core that consists of residues from both subdomains. These residues are well conserved among the various headpiece sequences. Charged residues also link the two subdomains (McKnight *et al.*, 1996, 1997; Vardar *et al.*, 1999). The C-terminal subdomain consists of three helices and is stabilized by a buried cluster of three phenylalanine, namely, Phe-798, Phe-802, and Phe-809 (Brewer *et al.*, 2005). The NMR structure also suggests a consensus sequence for the N-terminal subdomain starting at residue 16 of the headpiece (aa 767 in human villin), which is referred to as the *variable loop*, F/YxL/YxxLxx...L/PxG/DV/ID/NxxR/KxE/DxH/Y, where x denotes any residue. It is proposed that the length of the loop is not important and the conserved residues determine the correct N-terminal subdomain folding of the headpiece (Vardar *et al.*, 1999).

Analysis of headpiece-folding simulations has suggested that there are no intermediates in the folding of the headpiece: the folding process is driven by the collapse of the hydrophobic core and a secondary structure forms before the collapse. The helices are unstable; they form and break. The secondary structures do not appear all at once. And the two hydrophobic cores made up of the N- and C-terminal halves of the headpiece form an extended hydrophobic core for the entire headpiece. The two residues that bridge these two domains include His-792 and Leu-793. Thus, the N-terminal subdomain can or may influence the C-terminal subdomain. Tang *et al.* were able to demonstrate that above pH 6.0 the headpiece is completely folded and compactly arranged around the hydrophobic core. But not at pH below 6.0 (which would result in protonation of His-792) since unfolding of the N-terminal subdomain occurs without altering the folding of the C-terminal subdomain. On the other hand, a conservative mutation of His-792 to tyrosine stabilized the headpiece in that it remained folded over a much wider pH range and had a higher melting point temperature (an increase by more than 10 °C) (Tang *et al.*, 2005). Further, stabilization of the N-terminal subdomain in this mutant resulted in stabilization of the C-terminal subdomain as well, thus demonstrating that the unfolding of the two domains is interdependent (Tang *et al.*, 2005). These data suggest that unfolding of the N-terminal subdomain is important for actin-bundling function of villin even though it contains no F-actin–binding residues perhaps by orienting the C-terminal subdomain for effective actin binding. Thus, localized unfolding of the N-terminal subdomain with a concomitant change in the actin-binding site(s) in the C-terminal subdomain of the headpiece may be the mechanism for regulating actin bundling by villin (Grey *et al.*, 2005).

Using cystiene-scanning mapping and site-directed mutagenesis the residues involved in F-actin binding have been mapped to the third helix of the C-terminal subdomain (Friederich *et al.*, 1992; Doering and Matsudaira, 1996). Based on NMR solution structures, Vardar *et al.* put forward a model for F-actin binding by the headpiece in which there are prominent structural features: first, a surface exposed *hydrophobic patch* with a conserved Trp-815, second, a band of altering charges termed *crown* which surrounds this patch (Arg-788, Lys-816, Lys-822, Glu-823, Phe-827), and third, a *positive patch* below the charged crown (Lys-789, Lys-824). These investigators suggested that the hydrophobic patch docks the headpiece to F-actin which is then locked in place by electrostatic interaction with the crown. Rossenu *et al.* (2003), on the other hand, had shown that some of these residues are in fact dispensable for F-actin binding, namely Glu-823, Lys-824, and Phe-827. The only residue found necessary for F-actin binding by the headpiece was Lys-822. Although one additional residue implicated in F-actin binding is Leu-826 (Friederich *et al.*, 1992; Doering and Matsudaira, 1996), Vermeulen *et al.* point out that Leu-826 may be involved in maintaining the structure of the headpiece but may not directly participate in F-actin binding (Friederich *et al.*, 1992; Doering and Matsudaira, 1996; Vardar *et al.*, 2002; Vermeulen *et al.*, 2004). Likewise the reasoning is that loss of actin binding by Lys-816 may not be due to direct interaction of this residue with F-actin but more likely the result of a change in the local stacking of the adjacent residue, Trp-815 (Vermeulen *et al.*, 2004). Similarly, Lys-789 in the positive patch was suggested to regulate the structural stability of the headpiece rather than provide a positive charge for interaction with F-actin (Vermeulen *et al.*, 2004). Based on these data, Vermeulen *et al.* revised their model of F-actin binding by villin headpiece that takes into account the following: a hydrophobic patch (Trp-815), two positive charges from the cap (Lys-816, Lys-822), and a second hydrophobic patch from the side chain of Phe-827. In other words, Trp-815, Phe-827 binding to the complementary hydrophobic patches on actin, and Lys-816 and Lys-822 interacting with the negative charges on the actin determines actin binding by the villin headpiece domain. Since the residues interacting directly with F-actin lie within the C-terminal subdomain, it is possible that the N-terminal subdomain only serves to create a favorable conformation in the actin-binding residues.

More recently, Meng *et al.* (2005) determined the high-resolution crystal structure of chicken villin headpiece (HP67) and two mutants Arg-788 to Ala and Trp-815 to Tyr. The crystal structure was determined to be more compact than the NMR structure largely because the turn between the two C-terminal helices were closer to the hydrophobic core, and the position of Trp-815 was shifted closer to the hydrophobic core in the crystal structure relative to that seen in the NMR structures. In the crystal structure of HP67, a limited secondary structure was found in the N-terminal subdomain consisting of a short α-helix lying between Leu-18 and Val-22 (in human villin aa 769–784) and two 3–10 helices between Ala-26 and Asp-28 (in human villin aa 776–779), and Lys-38 and His-41 (in human villin aa 789–792). The most significant difference lay in the variable loop which was seen as a type-I β-turn in the NMR structure. In the crystal structure, however, it was seen as a 3–10 helix resulting from the intermolecular salt bridges from the side chain of Asp-28 (which is Glu-779 in human villin). Both the crystal and NMR structures of the C-terminal subdomain are very similar. The residues in the hydrophobic core were essentially the same as identified in the NMR structure. This includes the three phenylalanine residues (Phe-47, Phe-51, and Phe-58 equivalent to Phe-798, Phe-802, and Phe-809 in human villin) that form an aromatic cluster in the core (Frank *et al.*, 2002). Vermeulen *et al.* (2004) have previously mutated Trp-815 to alanine in HP35, while Meng *et al.* mutated this residue in HP67 and demonstrated a significant change in F-actin–binding affinity of the mutant (K_d for W815Y of 150 μM compared to 4.1 μM for wild type). They also found a fivefold decrease in F-actin binding affinity by the headpiece with Arg-788 mutated to Ala (K_d of 4.1 μM for wild type vs 24 μM for R788A) (Meng *et al.*, 2005). Since the crystal structures of the two mutants were similar to that of HP67, and since they did not show any significant change in the thermal unfolding midpoint, the authors inferred that the loss of actin binding with R788A is due to the loss of positively charged side chain of arginine. The loss also of actin binding by Trp-815Tyr suggested that the smaller aromatic ring of tyrosine and the extra hydroxyl group on tyrosine were less favorable for the interaction of the hydrophobic domain with actin. Based on comparisons of the crystal structures with the NMR structures, Meng and his coworkers proposed yet another revised model: a *hydrophobic cap* consisting of mostly Arg-815, a *crown of positive and negative charges,* including Arg-787, Lys-816, and Lys-822, and a *positive patch below the crown* consisting of Lys-789. They also suggested that Glu-823 and Lys-824 do not contribute to actin binding by the headpiece (Meng *et al.*, 2005). Yet none of these residues have been examined thus far in the context of the full-length villin protein and several of these residues have not been confirmed in functional assays examining actin-bundling by these mutants. Proteins that do not share the ability of the villin headpiece to either bind F-actin with high affinity or to cross-link F-actin share many of these conserved sequences which suggests that other residues in the villin headpiece, including those in the N-terminal subdomain of the headpiece, may determine the affinity of these headpiece domains for F-actin. Alternatively, they may be modified by the rest of the protein conformation and/or by association of villin with other ligands *in vivo*.

III. LIGAND-BINDING PROPERTIES OF VILLIN

There are several references in the literature to the possible interaction of villin with the plasma membrane. For instance, early in embryonic development, villin is the first microvillar actin-binding protein that appears when the density of the microvilli on the apical surface is quite low (Chambers and Grey, 1979). Villin is localized within the apical cytoplasm at a time when there is little actin-based cytoskeleton in the brush border region. Clearly, the apical distribution of villin is not determined entirely by the microfilament density. In fact, Shibayama *et al.* (1987) hypothesized that villin may interact with the cytoplasmic surface of the apical membrane where it may function to assemble the microvillar core filaments. These and other studies have suggested that villin can interact with the plasma membrane. Using peptides encompassing PIP_2-binding homologous domains in villin and gelsolin, two phospholipid-binding domains were identified in villin (Janmey *et al.*, 1992). However, the direct interaction of villin protein with PIP_2 was first characterized by us using human recombinant villin protein (Panebra *et al.*, 2001; Kumar *et al.*, 2004b). Villin does not associate with unilamellar vesicles containing pure phosphatidylcholine, phosphatidylethanolamine, or phosphatidylserine, but binds phosphatidylinositol. Villin binds phosphoinositides with the following binding affinity: $PIP_2 > PIP > PI$ (Kumar *et al.*, 2004b). Based on these results, we suggest that electrostatic interactions determine the association of villin with PIP_2. The dissociation constant for villin binding to PIP_2 is

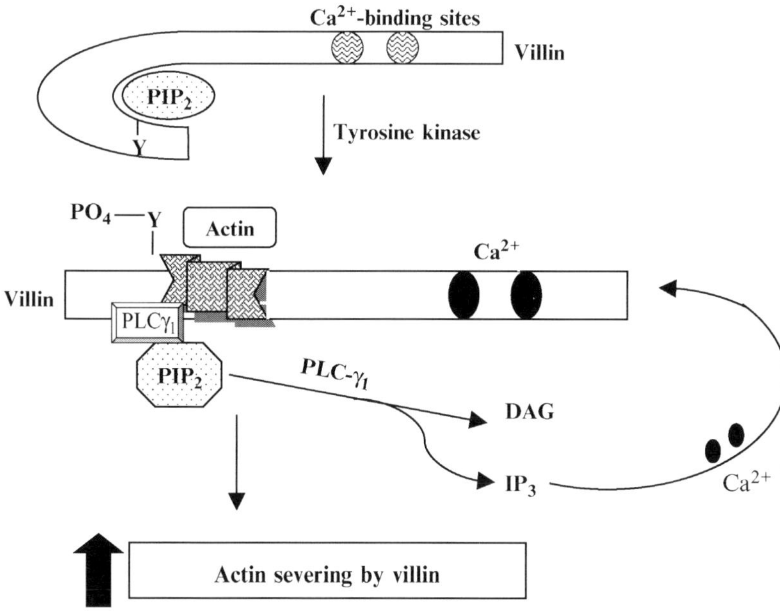

Figure 4. Model for the molecular mechanism of actin remodeling and catalytic activation of PLC-γ_1 by tyrosine-phophorylated villin. Unphosphorylated villin binds PIP_2. Tyrosine phosphorylation of villin by c-src kinase recruits PLC-γ_1 to the brush border membrane of intestinal epithelial cells and alters the ability of villin to sequester PIP_2. Removal of PIP_2 from villin vacates this site (PB2) for F-actin binding before severing. Activation of PLC-γ_1 in proximity of villin generates D-*myo*-inositol 1,4,5-trisphosphate (IP_3) which releases calcium from intracellular stores. Removal of PIP_2 from villin, as well as binding of calcium to villin would both enhance the actin-severing activity of villin. Reprinted with permission from Panebra *et al.* (2001) *Am. J. Physiol. Cell Physiol.*

39.4 μM, which is comparable to that reported for gelsolin and CapG (Lin *et al.*, 1997; Kumar *et al.*, 2004b). This binding affinity constant also suggests that villin has a much higher affinity for PIP_2 than phospholipase C-γ_1 (PLC-γ_1), another ligand of villin's, which may provide a molecular basis for the role of villin in phosphoinositide-mediated signal cascades (Khurana *et al.*, 1997; Panebra *et al.*, 2001; Wang *et al.*, submitted for publication) (Fig. 4). More recently, we have demonstrated the direct interaction of villin with PIP_2 in living cells by coexpressing DsRed2-tagged villin and enhanced green fluorescent protein (EGFP)-tagged pleckstrin homology domain of PLC-δ_1 using fluorescence resonance energy transfer (FRET) (George *et al.*, submitted for publication). These studies confirm that PIP_2 is a biologically relevant ligand of villin and that villin associates with the plasma membrane in living cells.

Using site-directed mutagenesis, we have identified three PIP_2-binding sites in human villin, two in the core (PB1, aa 112–119; PB2, aa 138–146) and a third PIP_2-binding site in the headpiece domain (PB5, aa 816–824) (Kumar *et al.*, 2004b) (Fig. 2). All three sites are conserved among related actin-binding proteins including gelsolin, scinderin, adseverin, CapG, supervillin, advillin, dematin, and pervin (Kumar *et al.*, 2004b). The PIP_2-binding sites in villin consist of a cluster of basic amino acid residues with the motif x_4R/KxR/KR/K, where x is any residue. While the association of villin with PIP_2 does not result in global changes in the conformation of the villin protein, there are localized changes in the secondary structure in domains PB1 and PB5, but not PB2 (Kumar *et al.*, 2004b). The PIP_2-binding domain, PB2 also corresponds to the F-actin side-binding site in villin which regulates the actin-severing activity of villin. Arg-138 in domain PB2 determines 80% of the PIP_2 binding and 83% of the actin-severing activity of villin (de Arruda *et al.*, 1992). Such results are suggestive of competitive inhibition of actin severing by villin in the presence of PIP_2. Note that the binding affinity of villin for F-actin is higher than for PIP_2. The phosphoinositide-binding domain PB5 includes residues that have been identified as the F-actin–binding site in the villin headpiece. PB1 on the other hand, lies in the vicinity of the actin monomer-binding site (which has been identified between residues 72 and 99 in human villin), as well as the F-actin side-binding site. Thus, the three PIP_2-binding sites in villin are either in proximity or overlapping with actin-binding sites. Consistent with this the association of villin with PIP_2 regulates the actin-modifying activities of villin (Janmey and Matsudaira, 1988; Kumar *et al.*, 2004b). Association of villin with PIP_2 inhibits its actin-severing function which may be explained by the binding of PIP_2 to PB1 and/or PB2, which are in proximity of the monomer-binding site, and in the case of Arg-138 overlapping with the F-actin side-binding site in villin core. The conformational changes induced in PB1 following its association with PIP_2 suggest that actin severing and actin capping could also be affected by structural changes in the monomer-binding domain. PIP_2 also inhibits the capping activity of villin (a rate constant of 1.75 s^{-1} in the absence and 2.38 s^{-1} in the presence of PIP_2) and enhances actin bundling by villin *in vitro* (Kumar *et al.*, 2004b). The only actin-modifying function of villin that is not regulated by PIP_2 is actin nucleation (Table I). By sequestering PIP_2, villin also inhibits the catalytic activation of PLC-γ_1 (Panebra *et al.*, 2001). Taken together, these studies imply that villin has the ability to modify the phosphoinositide-signaling pathways, as well as the actin cytoskeleton in response to changes in lipid signaling (Fig. 4).

Our work showed for the first time the self-association of villin both *in vitro* and in living cells. Using FRET, self-associating villin molecules were located in microspikes and microvilli, suggesting a role for villin oligomers in actin bundling. Villin oligomers were also found in lamellipodia and membrane ruffles in migrating cells, suggesting a role for villin self-association in cell motility (George *et al.*, submitted for publication). Increasing concentrations of PIP_2 *in vitro,* as well as overexpression of PIP_5-kinase-γ (which increases intracellular PIP_2 levels) enhanced villin oligomerization in the intestinal cell line, Caco-2. Likewise F-actin *in vitro* enhanced villin self-association while tyrosine phosphorylation inhibited villin self-association (George *et al.*, submitted for publication). A significant observation is that the microvillus core in the intestine contains about 20 hexagonally packed microfilaments (Mukherjee and Staehelin, 1971) of which approximately two-thirds are on the outside of the core and could be linked to the plasma membrane by villin in addition to myosin 1. It has been estimated that the protein molar ratio expected for the protein to cross-link the actin to the plasma membrane is approximately 1:15 if the protein is a monomer and 1:7.5 if it is a dimer. The molar ratio of villin to actin in the microvillus core is known to be around 1:10 (Bretscher and Weber, 1979). It is thus reasonable to presume that villin dimers, regulated by association of villin with PIP_2, could bundle actin filaments and anchor these filaments to the plasma membrane.

Villin sediments at 5.1 S and behaves almost like a globular protein (Bretscher and Weber, 1980). Like gelsolin, villin exists in an autoinhibited conformation and undergoes a calcium-induced conformational change referred to as the "hinge mechanism" (Hesterberg and Weber, 1983; Kumar and Khurana, 2004). We have identified two residues Asp-467 and Asp-715 in human villin that function as hinge residues and regulate this major calcium-induced conformational change in villin (Kumar and Khurana, 2004) (Fig. 2). Neither Asp-86 (predicted to open the S1–S3 latch in gelsolin) nor Asp-648 (predicted to release the S4–S6 latch in gelsolin) induce significant calcium-induced conformational changes in villin. These studies suggest that while gelsolin and villin both exist in an autoinhibited conformation that is regulated by high calcium, unlike gelsolin, villin may undergo a single calcium-induced transition (Kumar *et al.*, 2004a). It is also conjectured that calcium-induced conformational changes that release the autoinhibited conformation of villin additionally expose the F-actin side-binding domain in S2, thus regulating the actin-severing function of villin. Since actin-severing activity by villin requires high calcium concentrations, it is speculated that the hinge residues that release the autoinhibited conformation of villin are low-affinity calcium-binding sites. Some very elegant biochemical analysis done over a period of two decades has demonstrated that the capping and cutting activities of villin are regulated by different calcium-binding sites and that calcium-binding sites, regulating cutting have a much lower affinity for calcium than the sites regulating capping of actin filaments by villin (Walsh *et al.*, 1984a,b). Northrop *et al.* (1986) determined that 10 μM Ca^{2+} was saturating for actin capping, while 100–200 μM Ca^{2+} was required for actin severing by villin. Glenney *et al.* (1980) reported for the first time the presence of a single tight calcium-binding site with a K_d of 2.5 μM in villin. However, Hesterberg and Weber (1983) reported three calcium-binding sites in villin, including two rapidly exchanging (not tightly bound) sites, one each in the core and the headpiece, and a third nonexchangeable (tightly bound) high-affinity calcium-binding site in the villin core. The binding constants reported were 4.6 μM for full-length villin, 3.5 μM for the villin core, and 7.4 μM for the headpiece (Hesterberg and Weber, 1983). The fact that actin severing by villin requires much higher calcium concentrations suggests that these values are an underestimate of the calcium-binding sites in villin and more likely reflect the binding constants of the calcium-binding sites regulating actin capping by villin.

The specific residues that may be involved in calcium binding were first predicted by Markus *et al.* (1994a) who employed the solution structure of 14T. Using calcium titration, these authors identified two calcium-binding sites in 14T, one as a strong calcium-binding site (Glu-25, Asp-44, Glu-74) with a K_d of 1.8 μM, and a second as a weak calcium-binding site (Asp-86 or Glu-87) with a K_d of 11 μM. Based on what we know now about villin, these values appear to be overestimates and suggest that the calcium affinity of villin increases in the context of the full-length protein. Although the study identified some of the calcium-binding residues in villin, calcium-induced conformational changes were not related to these sites, nor were the sites functionally characterized. These calcium-binding sites appear to have been best described in gelsolin. Even so, the calcium-binding sites regulating capping and severing have not been functionally distinguished in gelsolin. Using mutational analysis, we functionally dissected the calcium-dependent actin-capping and actin-depolymerizing sites in villin for the first time. In our studies villin binds calcium with a K_d of 80.5 μM, closer to the half-maximal binding constant for actin severing (100 μM); that is a stoichiometry of approximately 6, suggesting that six Ca^{2+} ions are bound per molecule of villin. A Hill coefficient of 1.2 indicates positive cooperativity among the calcium-binding sites in villin, with the binding of one Ca^{2+} ion facilitating the binding of other calcium ions to villin (Kumar *et al.*, 2004a). Using mutagenesis, we identified two major calcium-sensitive regions that induce calcium-induced conformational changes in the N-terminal domain of villin (Fig. 2). The first, calcium site 1 (Glu-25, Asp-44, and Glu-74) regulates actin capping and actin severing, while the second site, calcium site 2 (Asp-86, Ala-93, Asp-61) regulates only actin depolymerization by villin. In addition, four sites were identified within the villin homologous repeats (S2–S6) which also regulated calcium-induced conformational changes in villin, and the actin-depolymerizing activity of villin. While capping was exclusively regulated by calcium site 1, practically most of the villin core and the calcium-sensitive sites contained within the core regulated the actin-severing activity of villin. Such observations are in line with the observations of Janmey and Matsudaira (1988) that 44T was less effective than intact villin in filament severing *in vitro* and that the actin-capping activity was entirely contained in the villin 14T fragment. On the basis of the comparison of the crystal structures of gelsolin and the villin 14T refined structure, we tend to think that calcium site 1 is most likely a calcium-binding site shared between villin and actin, thus regulating the interaction between actin monomers and the villin S1 domain. The second calcium-binding site, site 2 as well as the intradomain calcium-sensitive sites could be entirely in the villin.

An analysis of the calcium-sensitive sites in villin, and those found or identified in gelsolin, suggest that while villin and gelsolin may share some structural similarities, they are not functionally identical. Whereas both proteins exist in an autoinhibited conformation, the folding of the two proteins may not be entirely identical, particularly since there are significant differences between the N- and C-terminal domains of these proteins, more specifically, the presence of the C-terminal headpiece in villin. This would suggest that while these proteins share some sequence homology in the ligand-binding domains, the functional sites in these two proteins may not lie in the same region of the molecule (Arpin *et al.*, 1988b). Some of our studies with the latch residues point to such structural differences in addition to showing differences in the calcium requirements and binding profiles of villin and gelsolin. The calcium requirement for severing is higher for villin than gelsolin; the severing domain of villin is calcium dependent, while the analogous domain in gelsolin is calcium independent. Villin requires calcium for its association with actin, while gelsolin can bind actin even in the absence of calcium. It seems certain that structural studies with villin, such as the crystal structures of gelsolin in the presence of actin and calcium, would provide the needed confirmatory data, and help explain some of these functional differences.

IV. POSTTRANSLATIONAL MODIFICATION OF VILLIN

Previous studies with chicken and rat villin have demonstrated the presence of two isoelectric variants of villin with different p*I* values of 6.08 and 6.11 in the rat versus 6.26 and 6.34 in the chicken (Alicea and Mooseker, 1988). It is also known that the differences in the isoelectric points originate from the villin core (Glenney and Weber, 1981). Posttranslational modifications, for example, phosphorylation could explain the presence of these isoelectric variants of villin. However, phosphorylation of villin was not explored in these studies. We reported for the first time the results obtained with tyrosine phosphorylation of villin both *in vitro* and *in vivo* (Khurana *et al.*, 1997; Panebra *et al.*, 2001). Villin is tyrosine phosphorylated both in the intestine and in epithelial cell lines in response to receptor activation (lysophosphatidic acid, epidermal growth factor: EGF, hepatocyte growth factor: HGF, and carbachol) as well as in response to wounding and induction of apoptosis (Khurana *et al.*, 1997; Tomar *et al.*, 2004; Tomar *et al.*, in press; Wang *et al.*, submitted for publication). We have previously demonstrated that in addition to high calcium, tyrosine phosphorylation of villin can also release its autoinhibited conformation thus allowing villin to sever actin at physiologically relevant calcium concentrations (Kumar and Khurana, 2004). The reduced calcium affinity of villin might possibly be physiologically relevant, since the intestine is the major site of calcium absorption. Hence, the relative calcium insensitivity of villin may prevent unregulated actin severing by villin, thus maintaining the brush border morphology and the absorptive functions of the gastrointestinal epithelium. Since the calcium concentrations required for actin severing by villin or any other member of its family are not physiologically relevant, it is very likely that tyrosine phosphorylation rather than high calcium is the mechanism by which villin and perhaps related proteins sever actin *in vivo*.

Ten tyrosine phosphorylation sites have been identified in human villin, all of which are contained within the villin core, six in the interdomains (Tyr-81, Tyr-256, Tyr-324, Tyr-461, Tyr-604, and Tyr-725) and four in the conserved domains S1, S3, and S5 (Tyr-46, Tyr-60, Tyr-286, and Tyr-555) (Fig. 2; Table II) (Zhai *et al.*, 2002; Tomar *et al.*, 2004; Tomar *et al.*, in press). The nonreceptor tyrosine kinase c-src regulates the phosphorylation of villin both *in vitro* and in cell lines (Panebra *et al.*, 2001; Zhai *et al.*, 2002; Tomar *et al.*, 2004; Wang *et al.*, submitted for publication). Tyrosine phosphorylation regulates the actin-modifying activities of villin (Table II) (Zhai *et al.*, 2001). Actin nucleation by villin is inhibited by multiple phosphorylation sites in villin. Actin severing is enhanced by phosphorylation at residues, 46, 60, and 286 (Zhai *et al.*, 2001, 2002; Tomar *et al.*, in press). Actin bundling by villin is also inhibited by tyrosine phosphorylation. However, the specific tyrosine residues regulating this function of villin remain to be identified (Zhai *et al.*, 2001). The only actin-modifying activity of villin that is not regulated by tyrosine phosphorylation is actin capping (Tables I and II). Since phosphorylation sites regulating the actin-severing and actin-nucleating functions of villin

Table II
Tyrosine Phosphorylation Sites in Villin and Regulation of Villin Function by These Sites

Tyrosine phosphorylation sites	Nucleation[a]	Severing	Capping	Bundling	PLC-γ_1 association[b]	Cell shape	F-actin redistribution
46	+	+	−	N.D.[c]	−	−	−
60	+	+	−	N.D.	−	+	+
81	+	−	−	N.D.	−	+	+
256	+	−	−	N.D.	−	+	+
286	+	+	−	N.D.	+	−	−
324	+	−	−	N.D.	+	−	−
461	+	−	−	N.D.	+	−	−
555	+	−	−	N.D.	+	−	−
604	+	−	−	N.D.	+	−	−
725	+	−	−	N.D.	+	−	−

[a]Two or more of these phosphorylation sites regulate the actin-nucleating function of villin.

[b]The six C-terminal phosphorylation sites Y-286, Y-324, Y-461, Y-555, Y-604, and Y-725 collectively but not individually regulate villin's association with PLC-γ_1 and villin-induced cell migration.

[c]N.D., not done.

lie in proximity of the actin-binding domains in villin, we have suggested that phosphorylation of these residues may change the surface charge distribution, which may influence the villin–actin interactions, thereby regulating the severing and nucleating functions of villin. This is consistent with previous studies that have demonstrated that actin severing by villin is sensitive to monovalent salt concentrations, suggesting that electrostatic interactions between villin and actin determine villin–actin interactions and villin's actin-severing function (Janmey and Matsudaira, 1988).

Tyrosine phosphorylation also regulates villin's ligand-binding properties including its association with PIP_2, PLC-γ_1, and F-actin. Tyrosine phosphorylation decreases the binding affinity of villin for F-actin (Zhai *et al.*, 2001). Tyrosine phosphorylated villin associates with the C-terminal SH2 domain of PLC-γ_1 and does not bind PIP_2, thus regulating the catalytic activity of this enzyme (Panebra *et al.*, 2001; Wang *et al.*, submitted for publication). From these studies, we learnt that villin regulates PLC-γ_1 activity by modifying its own ability to bind PIP_2 (Fig. 4). This study provided biochemical proof of the functional relevance of tyrosine phosphorylation of villin and identified the molecular mechanism involved in the activation of PLC-γ_1 by villin. Moreover, these studies showed that there was significant cross talk between components of the transmembrane signal machinery, and the actin cytoskeleton at multiple levels, including the generation of second messengers. Based on these studies, we proposed a working model, which suggested that villin's ligand-binding properties were mechanistically important to its role in actin cytoskeletal remodeling (Fig. 4). According to our model, the regulation of villin's actin-severing function may be intimately linked to its tyrosine phosphorylation (Panebra *et al.*, 2001). Since then, several groups have used this model to demonstrate a role for tyrosine-phosphorylated villin in the regulation of Na^+/P_i cotransport, in calcium storage and release properties of F-actin and in HGF-induced actin reorganization and cell migration (Lange and Brandt, 1996; Papakonstanti *et al.*, 2000; Athman *et al.*, 2003). These studies reinforced the importance of tyrosine-phosphorylated villin in actin remodeling and the regulation of epithelial functions. Other proteins of the villin superfamily including gelsolin, fragmin, and CapG have likewise been shown to be tyrosine phosphorylated *in vitro* by c-src (De Corte *et al.*, 1997). Proteomic analysis has also been used to identify tyrosine-phosphorylated gelsolin from human lumbar cerebrospinal fluid (Yuan and Desiderio, 2003). These reports combined with our studies point to a more general mechanism and add another level of regulation involving the tyrosine phosphorylation of these proteins, which will be recognized in future studies following the identification of the phosphorylated tyrosine residues in these homologous proteins.

Tyrosine phosphorylation of villin is accompanied by the redistribution of phosphorylated villin and a concomitant decrease in the F-actin content of intestinal epithelial cells (Khurana *et al.*, 1997; Khurana, 2000). The temporal and spatial correlation of these two events suggested that tyrosine phosphorylation of villin may be involved in the rearrangement of the microvillar cytoskeleton. Since phosphorylation also changes the actin-modifying activities of villin, we thought that tyrosine-phosphorylated villin could prevent actin assembly *in vivo* by distinct mechanisms, including lower binding affinity for F-actin, inhibiting nucleation of new filaments, and cutting of preexisting filaments.

Thus, phosphorylation of villin and a local decrease in the affinity of villin for actin could generate a dynamic state and an increase in the fluidity of the cytoskeleton that could also affect the mechanical properties of the actin cytoskeleton (Panebra *et al.*, 2001; Zhai *et al.*, 2001). We have demonstrated that such spatial and temporal changes in the physical properties of the cytoskeleton enhance cell motility (this is discussed later in greater detail). Furthermore, these studies suggested that filament turnover in cells might be defined by the regulated action of actin-binding proteins interacting with signaling molecules.

V. VILLIN'S ROLE IN CELL MIGRATION

Like other proteins of its family, overexpression of villin increases cell motility (Tomar *et al.*, 2004). Villin-induced cell migration is enhanced by treatment of cells with EGF, HGF, or LPA (Athman *et al.*, 2003; Tomar *et al.*, 2004; Tomar *et al.*, in press; Wang *et al.*, submitted for publication). Peptide YY (PYY) has also been shown to induce villin expression in nontumorigenic intestinal cell line, hBRIE 380i, and reduce adhesion and enhance cell migration (Lee *et al.*, 2005). This study also demonstrated that increased expression of matrix metalloproteinase-3 (MMP-3) enhanced villin expression and cell migration, while an MMP inhibitor eliminated the increase in villin mRNA levels and PYY induced increase in cell migration. This study suggests that MMP-3 or MMP-3–induced increase in cell migration modulate villin expression in these cells. These studies also propose that in epithelial cells, the actin cytoskeleton can be directed by external signals to coordinate directional movement by using villin to translate cell surface receptor activation into cell locomotion. Both basal as well as growth factor induced increase in epithelial cell migration is associated with increased tyrosine phosphorylation of villin and its association with PLC-γ_1 (Athman *et al.*, 2003; Tomar *et al.*, 2004; Tomar *et al.*, in press; Wang *et al.*, submitted for publication). This is consistent with our *in vitro* experiments suggesting that the ligand-binding properties of villin are mechanistically important to villin's role in cell migration (Fig. 4) (Panebra *et al.*, 2001). Athman *et al.* (2003) likewise demonstrated the role of villin in cell migration in response to HGF. Given that their study showed a temporal relationship between tyrosine phosphorylation of villin and association of phosphovillin with PLC-γ_1, no causal relationship was demonstrated. The most compelling data demonstrating a causal relationship between tyrosine phosphorylation of villin and increased cell migration have come from studies in which we have demonstrated that overexpression of a villin mutant lacking all phosphoryable tyrosine residues had a dominant-negative effect in the intestinal cell line Caco-2, inhibiting both basal as well as growth factor-induced Caco-2 cell migration (Wang *et al.*, submitted for publication). Caco-2 cells express villin endogenously, however, ectopic expression of this mutant in villin-null cell lines including HeLa, MDCK, or IEC-6 demonstrated a similar decrease in cell migration compared to cells expressing wild-type villin (Tomar *et al.*, in press; Wang *et al.*, submitted for publication). These studies suggested that the effects of villin on cell migration were not cell type specific and were the same in transformed versus nontransformed cell lines and in differentiated versus undifferentiated intestinal cell lines. Tyrosine phosphorylation of villin, as well as villin-induced cell migration were also significantly inhibited by overexpression of a dominant-negative mutant of c-src kinase, confirming our *in vitro* observation that tyrosine phosphorylation of villin is regulated by c-src kinase (Panebra *et al.*, 2001; Zhai *et al.*, 2002; Tomar *et al.*, 2004). Using PLC-$\gamma_1^{-/-}$ cells as well as by downregulation of endogenous PLC-γ_1, an absolute requirement for PLC-γ_1 in villin-induced cell migration has also been established (Wang *et al.*, submitted for publication).

VI. STUDIES WITH POINT MUTANTS

Point mutants of the 10 phosphorylation sites in villin (tyrosine to phenylalanine and tyrosine to glutamic acid) have allowed us to characterize the cellular and molecular mechanisms involved in phosphovillin-induced regulation of cell shape and cell migration. Tyrosine phosphorylation at residues Tyr-60, Tyr-81, and Tyr-256 play an essential role in the reorganization of the actin cytoskeleton and villin-induced change in cell morphology and cell migration (Tomar *et al.*, 2004). In contrast, all six tyrosine residues in the C-terminal domain of villin core collectively inhibited villin-induced cell migration without any change in either actin reorganization or cell shape (Tomar *et al.*, in press). These studies indicated that whereas the three residues Y-60, Y-81, and Y-256 regulated villin-induced changes in cell morphology, all but one (no effect of Y-46) tyrosine phosphorylation site regulated villin-induced cell migration. The C-terminal phosphorylation sites (Y-286, Y-324, Y-461, Y-555, Y-604, and Y-725) were found to be critical for villin's interaction with its ligand PLC-γ_1 as well as for villin's redistribution to the plasma membrane and the lamellipodia in migrating cells.

A very important step in cell locomotion is the development of a polarized phenotype with the formation of a pseudopod in the direction of cell movement. In addition to PI3-kinase, PLC-γ_1 has been shown to regulate cell motility by its recruitment to the leading edge and its role in the formation of a protrusion (Chang *et al.*, 1997; Chou *et al.*, 2002; Piccolo *et al.*, 2002). We have suggested that villin-induced cell migration is regulated by a rapid increase in actin polymerization at the leading edge which is enhanced by tyrosine phosphorylation of villin and by the catalytic activation of PLC-γ_1 by phosphovillin. The spatial and temporal correlation of these events was further confirmed by overexpressing villin mutants lacking the C-terminal phosphorylation sites, which inhibited the association of phosphovillin with PLC-γ_1, as well as villin-induced increase in cell migration (Tomar *et al.*, in press; Wang *et al.*, submitted for publication).

VII. EPITHELIAL-TO-MESENCHYMAL TRANSITION

Epithelial-to-mesenchymal transition (EMT) is an important process during development by which epithelial cells acquire mesenchymal properties and show reduced intercellular adhesion and increased motility. The mechanisms that govern EMT are only now being unraveled and many parallels have been found between the processes during embryonic development and in adult tissue maintenance, such as wound healing, organ remodeling during fibrogenesis, as well as between mechanisms that propel invasive growth and metastases. Most studies point to a critical role of EMT during tumor progression and malignant transformation, bestowing the developing cancer cell with invasive and metastatic properties. EMT is always associated with cell scattering, defined by the loss of intercellular junctions, massive actin reorganization, and acquisition of cell motility. Overexpression of villin has been shown to enhance HGF-induced tubulogenesis in an *in vitro* model of EMT (Athman *et al.*, 2003). Using MDCK Tet-Off cells expressing wild-type villin or various phosphorylation site mutants of villin, we have examined the role of villin in EMT during HGF-induced tubulogenesis. In three-dimensional culture, full-length villin was localized to the apical surface of MDCK cysts and enhanced HGF-induced tubulogenesis (Chatman and Khurana, unpublished data). In contrast, ectopic expression of the phosphorylation site mutants of villin known to downregulate PLC-γ_1 signaling, as well as regulate actin reorganization, cell morphology, and cell migration, all abolished the effects of villin on tubulogenesis. These studies suggest that villin's function in actin reorganization and cell migration may play a role in enhancing the induction of EMT and that this function of villin is regulated by its tyrosine phosphorylation and ligand-binding properties, particularly its interaction with PLC-γ_1 (Chatman and Khurana, unpublished data).

VIII. VILLIN AS A MARKER IN ONCOLOGY

Villin has been used as a sensitive and a relatively specific marker of gastrointestinal adenocarcinomas including colonic, gastric, and pancreatic carcinomas. The expression levels and the pattern of villin distribution in the gastrointestinal adenocarcinomas are comparable to those seen in normal tissue, with apical distribution of the protein. No correlation has been found between the expression of villin and the differentiation of the tumor cells (Bacchi and Gown, 1991). No single gene has been implicated as a metastasis-specific gene, which has led to the speculation that the same mechanisms that drive tissue morphogenesis and repair also propel metastases. The transition from an epithelial to a mesenchymal phenotype and the epithelial plasticity are essential mechanisms for the dissemination of metastatic tumor cells. Since the molecular mechanisms that propel invasive growth and metastases are also found in embryonic development and in adult tissue injury repair, a role for villin in metastases is possible. HGF stimulates invasive growth in almost every tissue (Trusolino and Comoglio, 2002; Zhang *et al.*, 2003). Villin's role in HGF-induced cell migration and tubulogenesis suggests that villin is a downstream target of HGF and involved in the regulation of epithelial cell plasticity. Adhesion/de-adhesion and cytoskeletal remodeling are prerequisite for cell motility but are also linked to proliferation and survival (antiapoptotic) pathways. In addition, proinvasive and prosurvival messages by villin *vide infra* would also make it a strong candidate in enhancing the metastatic potential of an epithelial cell.

IX. OTHER FUNCTIONS OF VILLIN

We have already shown that the actin-depolymerizing functions of villin are important to its role in cell migration (Tomar *et al.*, 2004; Tomar *et al.*, in press). Given the close structural and functional similarities between villin and gelsolin, we hypothesized that the *in vivo* functions of the two proteins might overlap and that double-null villin$^{-/-}$gelsolin$^{-/-}$ mice could have a more severe phenotype than either villin$^{-/-}$ or gelsolin$^{-/-}$ mice. For these studies, we crossed the villin$^{-/-}$ and gelsolin$^{-/-}$ mice to generate a double knockout mice (Wang *et al.*, submitted for publication). The double knockout mice demonstrates no obvious phenotype, similar to the observations made with the villin$^{-/-}$ and gelsolin$^{-/-}$ single-null mice. However, electron micrographs of the intestinal and renal epithelial cells of double knockout mice showed the presence of a large number of autophagic vacuoles, including autophagosomes containing morphologically abnormal mitochondria. There was also a significant increase in the number of lysosomes in the renal proximal tubule cells from these mice (Khurana and Wang, unpublished data). Acute apoptosis induced within 4 h of gamma radiation demonstrated that the double knockout mice were also significantly more susceptible to apoptosis. Thus, the absence of villin and gelsolin together made the intestinal and renal epithelial cells more susceptible to both apoptosis and autophagy. Consistent with these observations, overexpression of villin in MDCK Tet-Off cells inhibited camptothecin-induced apoptosis as well as camptothecin-induced activation of caspase-9 and caspase-3 (Wang and Khurana, unpublished data). These studies led to evidence that villin delays apoptosis mediated by the intrinsic or mitochondrial pathway. Overexpression of villin truncation mutants expressing villin core (domains S1–S6 or S1–S3) have likewise been determined to induce cell death, suggesting that like gelsolin, the unfolded villin fragment may result in unregulated actin-severing and apoptotic cell death (Tomar *et al.*, in press). A detailed analysis of villin mutants suggested that the antiapoptotic and/or antiautophagic effects of villin were determined exclusively by its ability to sever actin filaments and not by its ligand-binding functions including its association with PIP_2 (Wang and Khurana, unpublished data). Villin mutants that failed to sever actin filaments also failed to delay campothecin-induced apoptosis. It is therefore tempting to speculate that the antiapoptotic functions of villin may be important in disease states, such as, in chronic pancreatitis, a disease believed to be related to abnormal apoptosis/necrosis, villin staining is irregular and even absent (Elsasser *et al.*, 1991). These studies suggest a role for villin in apoptosis, an event that occurs frequently in the terminally differentiated villus cells as part of normal physiology and also in other disorders such as ischemia or nephrotoxicity. The disruption of the actin system during cell death has been related to specific changes in actin-binding proteins and a direct link between actin depolymerization and DNA degradation has been suggested. Actin is also a prominent substrate for caspase which is consistent with evidence of overexpression of several actin regulatory proteins that either enhance or delay apoptosis (Ohtsu *et al.*, 1997).

Numerous attempts have been made to relate PLC-γ_1–mediated signaling pathways and cytoskeletal rearrangements in intestinal and renal epithelial cells. That this involves a role for villin is suggested by a few prominent observations notably: (1) regulation of intestinal and renal transport protein, thus regulation of fluid absorption; (2) nutrient absorption; (3) calcium absorption in intestinal cells; (4) endocytosis and transcytosis; (5) invasion of intestinal cells by bacterial and viral pathogens; (6) cytoskeletal remodeling following renal ischemia or perfusion. Some more studies have also provided either direct or indirect evidence for a role of villin in these functions. For instance, Athman *et al.* (2005) showed that primary cultures of intestinal epithelial cells derived from villin-null mice were resistant to *Shigella flexneri* infection and dissemination. Since invasive enteropathogens are known to require both actin depolymerization and extensive rearrangement of host actin filaments, the suggestion put forward is that villin expression, and thus its ability to remodel the actin cytoskeleton, is crucial for *S. flexneri* infection. Other enteropathogens that have been shown to elevate host cell PLC-γ_1, increase intracellular calcium, remodel host actin cytoskeleton and/or induce villin redistribution include *Escherichia coli* (Finlay and Cossart, 1997), *Salmonella typhimurium* (Finlay *et al.*, 1991), *Listeria monocytogenes* (Finlay and Cossart, 1997), and rotavirus (Dong *et al.*, 1997). Similarly, other intracellular mediators that stimulate secretion or inhibit fluid absorption in the intestine *via* activation of PLC-γ_1 and/or calcium include: bradykinin, acetylcholine, prostaglandins, gastrin-releasing peptide, substance P, serotonin, and neurotensin. PLC-γ_1 activation and concomitant changes in the apical cytoskeleton have likewise been demonstrated to regulate fluid, nutrient, and calcium absorption in renal and intestinal epithelial cells, respectively. Thus, both PLC-γ_1 activation and the cytoskeletal remodeling are critical to the basic functions of the intestine and kidney. Further, nutritional stress, anoxia, metabolic inhibitors, hydrostatic pressure, hormones, antibodies, and glutens have been reported to affect disassembly of the microvillar core bundle, vesiculation of the membrane, and shortening of the microvillus. A role for villin in the disassembly of microvillar core is well established (Mooseker *et al.*, 1980; Matsudaira and Burgess, 1982; Ferrary *et al.*, 1999). During ischemia and reperfusion of the proximal tubule, the initial disruption of the actin cytoskeleton during reperfusion injury as well has been related to the redistribution of villin, and perhaps, increased actin-severing activity of villin (Brown *et al.*, 1997). Changes in pH may also regulate microvillar cytoskeleton, and since intestinal and renal epithelial cells may undergo such changes, it is likely that villin participates in these functions. Thus, villin's ability to modulate phosphoinositide turnover, second messenger generation, and its ability to modify the microfilaments structure suggest that understanding the molecular and cellular mechanisms regulated by villin are crucial to our understanding of intestinal and renal physiology and pathophysiology.

The only clinical defect that has been directly associated with defects in the villin gene is in biliary atresia-like disorder in children (Phillips *et al.*, 2003). Three pediatric patients with progressive cholestasis and liver failure resembling biliary atresia also exhibited structural abnormalities within the microvilli of bile duct canaliculi, which were determined to be due to a lack of villin mRNA as well as villin protein in these children.

ACKNOWLEDGMENTS

The work described in this chapter has been supported by the National Institute of Diabetes and Digestive and Kidney Diseases. Grants DK-65006 and DK-54755 (to S.K.).

REFERENCES

Alicea, H. A. and Mooseker, M. S. (1988). Characterization of villin from the intestinal brush border of the rat, and comparative analysis with avian villin. *Cell Motil. Cytoskeleton* **9,** 60–72.

Andre, E., Lottspeich, F., Schleicher, M., and Noegel, A. (1988). Severin, gelsolin, and villin share a homologous sequence in regions presumed to contain, F-actin severing domains. *J. Biol. Chem.* **263,** 722–727.

Arpin, M., Blair, L., Coudrier, E., Dudouet, B., Finidori, J., Carcia, A., Huet, C., Pringault, E., Robine, S., Sahuguillo-Merino, C., *et al.* (1988a). Villin, a specific marker for some epithelia specialized in transport, to study the differentiation of intestinal and kidney cells *in vivo* and in a human colon adenocarcinoma line, HT29 in culture. *Mol. Aspects Med.* **10,** 257–272.

Arpin, M., Pringault, E., Finidori, J., Garcia, A., Jeltsch, J. M., Vandekerckhove, J., and Louvard, D. (1988b). Sequence of human villin: A large duplicated domain homologous with other actin-severing proteins and a unique small carboxy-terminal domain related to villin specificity. *J. Cell Biol.* **107,** 1759–1766.

Athman, R., Louvard, D., and Robine, S. (2003). Villin enhances hepatocyte growth factor-induced actin cytoskeleton remodeling in epithelial cells. *Mol. Biol. Cell* **14,** 4641–4653.

Athman, R., Fernandez, M. I., Gounon, P., Sansonetti, P., Louvard, D., Philpott, D., and Robine, S. (2005). Shigella flexneri infection is dependent on villin in the mouse intestine and in primary cultures of intestinal epithelial cells. *Cell Microbiol.* **7,** 1109–1116.

Bacchi, C. E. and Gown, A. M. (1991). Distribution and pattern of expression of villin, a gastrointestinal-associated cytoskeletal protein, in human carcinomas: A study employing paraffin-embedded tissue. *Lab. Invest.* **64,** 418–424.

Bretscher, A. and Weber, K. (1979). Villin: The major microfilament-associated protein of the intestinal microvillus. *Proc. Natl. Acad. Sci. USA* **76,** 2321–2325.

Bretscher, A., Osborn, M., Wehland, J., and Weber, K. (1981). Villin associates with specific microfilamentous structures as seen by immunofluorescence microscopy on tissue sections and cells microinjected with villin. *Exp. Cell Res.* **135,** 213–219.

Bretscher, A. and Weber, K. (1980). Villin is a major protein of the microvillus cytoskeleton which binds both, G and F actin in a calcium-dependent manner. *Cell* **20,** 839–847.

Brewer, S. H., Vu, D. M., Tang, Y., Li, Y., Franzen, S., Raleigh, D. P., and Dyer, R. B. (2005). Effect of modulating unfolded state structure on the folding kinetics of the villin headpiece subdomain. *Proc. Natl. Acad. Sci. USA* **102,** 16662–16667.

Brown, D., Lee, R., and Bonventre, J. V. (1997). Redistribution of villin to proximal tubule basolateral membranes after ischemia and reperfusion. *Am. J. Physiol.* **273,** F1003–F1012.

Burgess, D. R., Broschat, K. O., and Hayden, J. M. (1987). Tropomyosin distinguishes between the two actin-binding sites of villin and affects actin-binding properties of other brush border proteins. *J. Cell Biol.* **104,** 29–40.

Chambers, C. and Grey, R. D. (1979). Development of the structural components of the brush border in absorptive cells of the chick intestine. *Cell Tissue Res.* **204,** 387–405.

Chang, J. S., Noh, D. Y., Park, I. A., Kim, M. J., Song, H., Ryu, S. H., and Suh, P. G. (1997). Overexpression of phospholipase, C-gamma1 in rat 3Y1 fibroblast cells leads to malignant transformation. *Cancer Res.* **57,** 5465–5468.

Chou, J., Stolz, D. B., Burke, N. A., Watkins, S. C., and Wells, A. (2002). Distribution of gelsolin and phosphoinositol 4,5-bisphosphate in lamellipodia during, EGF-induced motility. *Int. J. Biochem. Cell Biol.* **34,** 776–790.

de Arruda, M. V., Bazari, H., Wallek, M., and Matsudaira, P. (1992). An actin footprint on villin. Single site substitutions in a cluster of basic residues inhibit the actin severing but not capping activity of villin. *J. Biol. Chem.* **267,** 13079–13085.

De, Corte, V., Gettemans, J., and Vandekerckhove, J. (1997). Phosphatidylinositol 4,5-bisphosphate specifically stimulates, PP60(c-src) catalyzed phosphorylation of gelsolin and related actin-binding proteins. *FEBS Lett.* **401,** 191–196.

Doering, D. S. and Matsudaira, P. (1996). Cysteine scanning mutagenesis at 40 of 76 positions in villin headpiece maps the F-actin binding site and structural features of the domain. *Biochemistry* **35,** 12677–12685.

Dong, Y., Zeng, C. Q., Ball, J. M., Estes, M. K., and Morris, A. P. (1997). The rotavirus enterotoxin, NSP4 mobilizes intracellular calcium in human intestinal cells by stimulating phospholipase, C-mediated inositol 1,4,5-trisphosphate production. *Proc. Natl. Acad. Sci. USA* **94,** 3960–3965.
Dudouet, B., Robine, S., Huet, C., Sahuquillo-Merino, C., Blair, L., Coudrier, E., and Louvard, D. (1987). Changes in villin synthesis and subcellular distribution during intestinal differentiation of, HT29–18 clones. *J. Cell Biol.* **105,** 359–369.
Elsasser, H. P., Kloppel, G., Mannherz, H. G., Flocke, K., and Kern, H. F. (1991). Immunohistochemical demonstration of villin in the normal human pancreas and in chronic pancreatitis. *Histochemistry* **95,** 383–390.
Ferrary, E., Cohen-Tannoudji, M., Pehau-Arnaudet, G., Lapillonne, A., Athman, R., Ruiz, T., Boulouha, L., El Marjou, F., Doye, A., Fontaine, J. J., Antony, C., Babinet, C., *et al.* (1999). *In vivo*, villin is required for, $Ca^{(2+)}$-dependent, F-actin disruption in intestinal brush borders. *J. Cell Biol.* **146,** 819–830.
Finlay, B. B. and Cossart, P. (1997). Exploitation of mammalian host cell functions by bacterial pathogens. *Science* **276,** 718–725.
Finlay, B. B., Ruschkowski, S., and Dedhar, S. (1991). Cytoskeletal rearrangements accompanying salmonella entry into epithelial cells. *J. Cell Sci.* **99,** 283–296.
Frank, B. S., Vardar, D., Buckley, D. A., and McKnight, C. J. (2002). The role of aromatic residues in the hydrophobic core of the villin headpiece subdomain. *Protein Sci.* **11,** 680–687.
Franck, Z., Footer, M., and Bretscher, A. (1990). Microinjection of villin into cultured cells induces rapid and long-lasting changes in cell morphology but does not inhibit cytokinesis, cell motility, or membrane ruffling. *J. Cell Biol.* **111,** 2475–2485.
Friederich, E., Huet, C., Arpin, M., and Louvard, D. (1989). Villin induces microvilli growth and actin redistribution in transfected fibroblasts. *Cell* **59,** 461–475.
Friederich, E., Vancompernolle, K., Huet, C., Goethals, M., Finidori, J., Vandekerckhove, J., and Louvard, D. (1992). An actin-binding site containing a conserved motif of charged amino acid residues is essential for the morphogenic effect of villin. *Cell* **70,** 81–92.
Friederich, E., Vancompernolle, K., Louvard, D., and Vandekerckhove, J. (1999). Villin function in the organization of the actin cytoskeleton. Correlation of *in vivo* effects to its biochemical activities *in vitro*. *J. Biol. Chem.* **274,** 26751–26760.
George, S., Ramagiri, S., Chatman, L., Tomar, A., and Khurana, S. (submitted for publication). Localization and function of villin oligomers in living cells.
Glenney, J. R., Jr. and Weber, K. (1981). Calcium control of microfilaments: Uncoupling of the F-actin-severing and -bundling activity of villin by limited proteolysis *in vitro*. *Proc. Natl. Acad. Sci. USA* **78,** 2810–2814.
Glenney, J. R., Jr., Bretscher, A., and Weber, K. (1980). Calcium control of the intestinal microvillus cytoskeleton: Its implications for the regulation of microfilament organizations. *Proc. Natl. Acad. Sci. USA* **77,** 6458–6462.
Glenney, J. R., Jr., Geisler, N., Kaulfus, P., and Weber, K. (1981a). Demonstration of at least two different actin-binding sites in villin, a calcium-regulated modulator of, F-actin organization. *J. Biol. Chem.* **256,** 8156–8161.
Glenney, J. R., Jr., Kaulfus, P., and Weber, K. (1981b). F actin assembly modulated by villin: Ca ++ -dependent nucleation and capping of the barbed end. *Cell* **24,** 471–480.
Grey, M. J., Tang, Y., Alexov, E., McKnight, C. J., Raleigh, D. P., and Palmer, A. G., III. (2005). Characterizing a partially folded intermediate of the villin headpiece domain under non-denaturing conditions: Contribution of His41 to the pH-dependent stability of the N-terminal subdomain. *J. Mol. Biol.* **335**(5), 1066–1077.
Hesterberg, L. K. and Weber, K. (1983). Demonstration of three distinct calcium-binding sites in villin, a modulator of actin assembly. *J. Biol. Chem.* **258,** 365–369.
Hofer, D. and Drenckhahn, D. (1992). Identification of brush cells in the alimentary and respiratory system by antibodies to villin and fimbrin. *Histochemistry* **98,** 237–242.
Hofer, D. and Drenckhahn, D. (1999). Localisation of actin, villin, fimbrin, ezrin and ankyrin in rat taste receptor cells. *Histochem. Cell Biol.* **112,** 79–86.
Huang, S., Robinson, R. C., Gao, L. Y., Matsumoto, T., Brunet, A., Blanchoin, L., and Staiger, C. J. (2005). Arabidopsis, VILLIN1 generates actin filament cables that are resistant to depolymerization. *Plant Cell* **17,** 486–501.

Iovanna, J. L., Lechene de la, Porte, P., and Dagorn, J. C. (1992). Expression of genes associated with dedifferentiation and cell proliferation during pancreatic regeneration following acute pancreatitis. *Pancreas* **7,** 712–718.

Janmey, P. A. and Matsudaira, P. T. (1988). Functional comparison of villin and gelsolin. Effects of, Ca2+, KCl, and polyphosphoinositides. *J. Biol. Chem.* **263,** 16738–16743.

Janmey, P. A., Lamb, J., Allen, P. G., and Matsudaira, P. T. (1992). Phosphoinositide-binding peptides derived from the sequences of gelsolin and villin. *J. Biol. Chem.* **267,** 11818–11823.

Kamioka, H., Sugawara, Y., Honjo, T., Yamashiro, T., and Takano-Yamamoto, T. (2004). Terminal differentiation of osteoblasts to osteocytes is accompanied by dramatic changes in the distribution of actin-binding proteins. *J. Bone Miner. Res.* **19,** 471–478.

Kerneis, S., Bogdanova, A., Colucci-Guyon, E., Kraehenbuhl, J. P., and Pringault, E. (1996). Cytosolic distribution of villin in M cells from mouse Peyer's patches correlates with the absence of a brush border. *Gastroenterology* **110,** 515–521.

Khurana, S. (2000). Role of actin cytoskeleton in regulation of ion transport: Examples from epithelial cells. *J. Membr. Biol.* **178,** 73–87.

Khurana, S., Arpin, M., Patterson, R., and Donowitz, M. (1997). Ileal microvillar protein villin is tyrosine-phosphorylated and associates with PLC-gamma1. Role of cytoskeletal rearrangement in the carbachol-induced inhibition of ileal, NaCl absorption. *J. Biol. Chem.* **272,** 30115–30121.

Klahre, U., Friederich, E., Kost, B., Louvard, D., and Chua, N. H. (2000). Villin-like actin-binding proteins are expressed ubiquitously in Arabidopsis. *Plant Physiol.* **122,** 35–48.

Kumar, N. and Khurana, S. (2004). Identification of a functional switch for actin severing by cytoskeletal proteins. *J. Biol. Chem.* **279,** 24915–24918.

Kumar, N., Tomar, A., Parrill, A. L., and Khurana, S. (2004a). Functional dissection and molecular characterization of calcium-sensitive actin-capping and actin-depolymerizing sites in villin. *J. Biol. Chem.* **279,** 45036–45046.

Kumar, N., Zhao, P., Tomar, A., Galea, C. A., and Khurana, S. (2004b). Association of villin with phosphatidylinositol 4,5-bisphosphate regulates the actin cytoskeleton. *J. Biol. Chem.* **279,** 3096–3110.

Kwiatkowski, D. J., Janmey, P. A., Mole, J. E., and Yin, H. L. (1985). Isolation and properties of two actin-binding domains in gelsolin. *J. Biol. Chem.* **260,** 15232–15238.

Lange, K. and Brandt, U. (1996). Calcium storage and release properties of F-actin: Evidence for the involvement of F-actin in cellular calcium signaling. *FEBS Lett.* **395,** 137–142.

Lee, M., Hadi, M., Hallden, G., and Aponte, G. W. (2005). Peptide, YY and neuropeptide Y induce villin expression, reduce adhesion, and enhance migration in small intestinal cells through the regulation of CD63, matrix metalloproteinase-3, and Cdc42 activity. *J. Biol. Chem.* **280,** 125–136.

Lin, K. M., Wenegieme, E., Lu, P. J., Chen, C. S., and Yin, H. L. (1997). Gelsolin binding to phosphatidylinositol 4,5-bisphosphate is modulated by calcium and pH. *J. Biol. Chem.* **272,** 20443–20450.

MacLennan, A. J., Orringer, M. B., and Beer, D. G. (1999). Identification of intestinal-type, Barrett's metaplasia by using the intestine-specific protein villin and esophageal brush cytology. *Mol. Carcinog.* **24,** 137–143.

Madison, B. B., Dunbar, L., Qiao, X. T., Braunstein, K., Braunstein, E., and Gumucio, D. L. (2002). Cis elements of the villin gene control expression in restricted domains of the vertical (crypt) and horizontal (duodenum, cecum) axes of the intestine. *J. Biol. Chem.* **277,** 33275–33283.

Markus, M. A., Dayie, K. T., Matsudaira, P., and Wagner, G. (1994a). Effect of deuteration on the amide proton relaxation rates in proteins. Heteronuclear, NMR experiments on villin 14T. *J. Magn. Reson. B.* **105,** 192–195.

Markus, M. A., Nakayama, T., Matsudaira, P., and Wagner, G. (1994b). Solution structure of villin 14T, a domain conserved among actin-severing proteins. *Protein Sci.* **3,** 70–81.

Markus, M. A., Matsudaira, P., and Wagner, G. (1997). Refined structure of villin 14T and a detailed comparison with other actin-severing domains. *Protein Sci.* **6,** 1197–1209.

Matsudaira, P. T. and Burgess, D. R. (1982). Partial reconstruction of the microvillus core bundle: Characterization of villin as a Ca++-dependent, actin-bundling/depolymerizing protein. *J. Cell Biol.* **92,** 648–656.

Matsudaira, P., Jakes, R., and Walker, J. E. (1985). A gelsolin-like, Ca2+-dependent actin-binding domain in villin. *Nature* **315,** 248–250.

McKnight, C. J., Doering, D. S., Matsudaira, P. T., and Kim, P. S. (1996). A thermostable 35-residue subdomain within villin headpiece. *J. Mol. Biol.* **260,** 126–134.

McKnight, C. J., Matsudaira, P. T., and Kim, P. S. (1997). NMR structure of the 35-residue villin headpiece subdomain. *Nat. Struct. Biol.* **4,** 180–184.

Meng, J., Vardar, D., Wang, Y., Guo, H. C., Head, J. F., and McKnight, C. J. (2005). High-resolution crystal structures of villin headpiece and mutants with reduced F-actin binding activity. *Biochemistry* **44,** 11963–11973.

Moll, R., Robine, S., Dudouet, B., and Louvard, D. (1987). Villin: A cytoskeletal protein and a differentiation marker expressed in some human adenocarcinomas. *Virchows Arch. B. Cell Pathol. Incl. Mol. Pathol.* **54,** 155–169.

Mooseker, M. S., Graves, T. A., Wharton, K. A., Falco, N., and Howe, C. L. (1980). Regulation of microvillus structure: Calcium-dependent solation and cross-linking of actin filaments in the microvilli of intestinal epithelial cells. *J. Cell Biol.* **87,** 809–822.

Mukherjee, T. M. and Staehelin, L. A. (1971). The fine-structural organization of the brush border of intestinal epithelial cells. *J. Cell Sci.* **8,** 573–599.

Northrop, J., Weber, A., Mooseker, M. S., Franzini-Armstrong, C., Bishop, M. F., Dubyak, G. R., Tucker, M., and Walsh, T. P. (1986). Different calcium dependence of the capping and cutting activities of villin. *J. Biol. Chem.* **261,** 9274–9281.

Ohtsu, M., Sakai, N., Fujita, H., Kashiwagi, M., Gasa, S., Shimizu, S., Eguchi, Y., Tsujimoto, Y., Sakiyama, Y., Kobayashi, K., and Kuzumaki, N. (1997). Inhibition of apoptosis by the actin-regulatory protein gelsolin. *EMBO J.* **16,** 4650–4656.

Panebra, A., Ma, S. X., Zhai, L. W., Wang, X. T., Rhee, S. G., and Khurana, S. (2001). Regulation of phospholipase, C-gamma(1) by the actin-regulatory protein villin. *Am. J. Physiol. Cell Physiol.* **281,** C1046–C1058.

Papakonstanti, E. A., Emmanouel, D. S., Gravanis, A., and Stournaras, C. (2000). PLC-gamma1 signaling pathway and villin activation are involved in actin cytoskeleton reorganization induced by Na+/Pi cotransport upregulation. *Mol. Med.* **6,** 303–318.

Phillips, M. J., Azuma, T., Meredith, S. L., Squire, J. A., Ackerley, C. A., Pluthero, F. G., Roberts, E. A., Superina, R. A., Levy, G. A., and Marsden, P. A. (2003). Abnormalities in villin gene expression and canalicular microvillus structure in progressive cholestatic liver disease of childhood. *Lancet* **362,** 1112–1119.

Piccolo, E., Innominato, P. F., Mariggio, M. A., Maffucci, T., Iacobelli, S., and Falasca, M. (2002). The mechanism involved in the regulation of phospholipase, Cgamma1 activity in cell migration. *Oncogene* **21,** 6520–6529.

Pinson, K. I., Dunbar, L., Samuelson, L., and Gumucio, D. L. (1998). Targeted disruption of the mouse villin gene does not impair the morphogenesis of microvilli. *Dev. Dyn.* **211,** 109–121.

Pope, B., Way, M., Matsudaira, P. T., and Weeds, A. (1994). Characterisation of the F-actin binding domains of villin: Classification of F-actin binding proteins into two groups according to their binding sites on actin. *FEBS Lett.* **338,** 58–62.

Pringault, E., Robine, S., and Louvard, D. (1991). Structure of the human villin gene. *Proc. Natl. Acad. Sci. USA* **88,** 10811–10815.

Regalado, S. P., Nambu, Y., Iannettoni, M. D., Orringer, M. B., and Beer, D. G. (1998). Abundant expression of the intestinal protein villin in Barrett's metaplasia and esophageal adenocarcinomas. *Mol. Carcinog.* **22,** 182–189.

Rieder, G., Tessier, A. J., Qiao, X. T., Madison, B., Gumucio, D. L., and Merchant, J. L. (2005). Helicobacter-induced intestinal metaplasia in the stomach correlates with, Elk-1 and serum response factor induction of villin. *J. Biol. Chem.* **280,** 4906–4912.

Robine, S., Huet, C., Moll, R., Sahuquillo-Merino, C., Coudrier, E., Zweibaum, A., and Louvard, D. (1985). Can villin be used to identify malignant and undifferentiated normal digestive epithelial cells? *Proc. Natl. Acad. Sci. USA* **82,** 8488–8492.

Rossenu, S., Leyman, S., Dewitte, D., Peelaers, D., Jonckheere, V., Van Troys, M., Vandekerckhove, J., and Ampe, C. (2003). A phage display-based method for determination of relative affinities of mutants.

Application of the actin-binding motifs in thymosin beta 4 and the villin headpiece. *J. Biol. Chem.* **278,** 16642–16650.

Rousseau-Merck, M. F., Simon-Chazottes, D., Arpin, M., Pringault, E., Louvard, D., Guenet, J. L., and Berger, R. (1988). Localization of the villin gene on human chromosome 2q35-q36 and on mouse chromosome 1. *Hum. Genet.* **78,** 130–133.

Shibayama, T., Carboni, J. M., and Mooseker, M. S. (1987). Assembly of the intestinal brush border: Appearance and redistribution of microvillar core proteins in developing chick enterocytes. *J. Cell Biol.* **105,** 335–344.

Soubeyran, P., Andre, F., Lissitzky, J. C., Mallo, G. V., Moucadel, V., Roccabianca, M., Rechreche, H., Marvaldi, J., Dikic, I., Dagorn, J. C., and Iovanna, J. L. (1999). Cdx1 promotes differentiation in a rat intestinal epithelial cell line. *Gastroenterology* **117,** 1326–1338.

Tang, Y., Grey, M. J., McKnight, J., Palmer, A. G., III, and Raleigh, D. P. (2005). Multistate folding of the villin headpiece domain. *J. Mol. Biol.* **355,** 1066–1072.

Tomar, A., George, S., Kansal, P., Wang, Y., and Khurana, S. (2006). Interaction of phospholipase C-g1 with villin regulates epithelial cell migration. *J. Biol. Chem.* (epublished ahead of print).

Tomar, A., Wang, Y., Kumar, N., George, S., Ceacareanu, B., Hassid, A., Chapman, K. E., Aryal, A. M., Waters, C. M., and Khurana, S. (2004). Regulation of cell motility by tyrosine phosphorylated villin. *Mol. Biol. Cell* **15,** 4807–4817.

Trusolino, L. and Comoglio, P. M. (2002). Scatter-factor and semaphorin receptors: Cell signalling for invasive growth. *Nat. Rev. Cancer* **2,** 289–300.

Tsukada, N., Ackerley, C. A., and Phillips, M. J. (1995). The structure and organization of the bile canalicular cytoskeleton with special reference to actin and actin-binding proteins. *Hepatology* **21,** 1106–1113.

Vardar, D., Buckley, D. A., Frank, B. S., and McKnight, C. J. (1999). NMR structure of an F-actin-binding "headpiece" motif from villin. *J. Mol. Biol.* **294,** 1299–1310.

Vardar, D., Chishti, A. H., Frank, B. S., Luna, E. J., Noegel, A. A., Oh, S. W., Schleicher, M., and McKnight, C. J. (2002). Villin-type headpiece domains show a wide range of F-actin-binding affinities. *Cell Motil. Cytoskeleton* **52,** 9–21.

Vermeulen, W., Vanhaesebrouck, P., Van Troys, M., Verschueren, M., Fant, F., Goethals, M., Ampe, C., Martins, J. C., and Borremans, F. A. (2004). Solution structures of the, C-terminal headpiece subdomains of human villin and advillin, evaluation of headpiece, F-actin-binding requirements. *Protein Sci.* **13,** 1276–1287.

Walsh, T. P., Weber, A., Davis, K., Bonder, E., and Mooseker, M. (1984a). Calcium dependence of villin-induced actin depolymerization. *Biochemistry* **23,** 6099–6102.

Walsh, T. P., Weber, A., Higgins, J., Bonder, E. M., and Mooseker, M. S. (1984b). Effect of villin on the kinetics of actin polymerization. *Biochemistry* **23,** 2613–2621.

Wang, Y., Tomar, A., Chatman, L., George, S., and Khurana, S. (submitted for publication). Villin mediated activation of phosphoinositide signaling pathway is required for intestinal cell migration.

Way, M., Pope, B., and Weeds, A. G. (1992). Are the conserved sequences in segment 1 of gelsolin important for binding actin? *J. Cell Biol.* **116,** 1135–1143.

Weber, A., Northrop, J., Bishop, M. F., Ferrone, F. A., and Mooseker, M. S. (1987). Kinetics of actin elongation and depolymerization at the pointed end. *Biochemistry* **26,** 2537–2544.

Yoshie, S., Kumakura, M., and Toyoshima, K. (2003). Villin is a possible marker of receptor cells in frog taste organs. *Histochem. Cell Biol.* **119,** 447–450.

Yuan, X. and Desiderio, D. M. (2003). Proteomics analysis of phosphotyrosyl-proteins in human lumbar cerebrospinal fluid. *J. Proteome Res.* **2,** 476–487.

Zhai, L., Zhao, P., Panebra, A., Guerrerio, A. L., and Khurana, S. (2001). Tyrosine phosphorylation of villin regulates the organization of the actin cytoskeleton. *J. Biol. Chem.* **276,** 36163–36167.

Zhai, L., Kumar, N., Panebra, A., Zhao, P., Parrill, A. L., and Khurana, S. (2002). Regulation of actin dynamics by tyrosine phosphorylation: Identification of tyrosine phosphorylation sites within the actin-severing domain of villin. *Biochemistry* **41,** 11750–11760.

Zhang, Y. W., Su, Y., Volpert, O. V., and Vande Woude, G. F. (2003). Hepatocyte growth factor/scatter factor mediates angiogenesis through positive VEGF and negative thrombospondin 1 regulation. *Proc. Natl. Acad. Sci. USA* **100,** 12718–12723.

Chapter 6

Roles of the Actin Cytoskeleton and Myosins in the Endomembrane System

Karl R. Fath

Biology Department, Queens College and Graduate Center of the City University of New York, 65-30 Kissena Boulevard, Flushing, New York 11367

The positioning of the endomembrane system (biosynthetic/secretory and endocytic pathways) and the constant movement of constituents between these compartments requires the involvement of the cytoskeleton and cytoskeletal-based motors. Whereas much is known about the roles of the microtubule (MT) cytoskeleton in these events, considerably less is known about the roles of the actin cytoskeleton. Recent work has shown that the Golgi complex is linked to an actin-based network that is important for Golgi morphology and the trafficking of Golgi-derived membranes. Moreover, many of the cellular players used to transport cargo near the actin-rich network subjacent to the plasma membrane (PM) may function near the Golgi.

Advances in Molecular and Cell Biology, Vol. 37

ISBN: 0-444-52868-7
DOI: 10.1016/S1569-2558(06)37006-3

I. INTRODUCTION

The structure and cellular positioning of the major membrane compartments of the endomembrane system are determined by the cytoskeleton and its associated motors. In addition, the constant movement of materials between these compartments engaged in both synthesis and degradation also depends on the cytoskeleton and cytoskeletal-based motors. It has been proposed that they are involved in cargo selection, membranes budding from source compartments, transport to the target compartment and, finally, fusion with the target. Many times these materials must move relatively great distances through the cytoplasm between compartments. The highly cross-linked cytoplasm obviates a major role of diffusion in moving membranes and other cellular organelles over significant distances in the cytoplasm (Luby-Phelps, 1994). Thus, a facilitated transport is required to deliver membranes in a timely fashion from their sites of synthesis to their places of utilization and to help components reach their proper destination.

The forces that transport membranes inside cells can be classified into four main categories (Pollard and Earnshaw, 2002): (1) Cargo moves along microtubule (MT) tracks using kinesin or dynein motors and the hydrolysis of ATP. This method is used generally by higher eukaryotes to maintain Golgi and endoplasmic reticulum (ER) morphology and positioning and longer movements of transport carriers in the deep cytoplasm. (2) The cargo that is tethered to the plus ends of MTs moves as the attached MT lengthens or shortens. This method is used to move chromosomes in karyokinesis and may have a role in positioning the ER and loading membranes onto MTs for subsequent transport (Vaughan, 2005). (3) Cargo moves along actin filaments using myosin motors and the hydrolysis of ATP. This process is used by yeast and plant cells for the majority of long-distance membranous movements. Movement along actin filaments in metazoans is most prevalent in the actin-rich cytoplasm adjacent to the plasma membrane (PM). (4) The force generated by the polymerization of actin filaments can move cellular constituents. Actin polymerization extends the PM at the leading edge of migrating cells and has been proposed to propel membranes over short distances close to the PM and in the deep cytoplasm (Pantaloni *et al.*, 2001; Ridley, 2001).

II. ACTIN CYTOSKELETON IN MOVEMENTS ALONG THE BIOSYNTHETIC PATHWAY

The role of the actin cytoskeleton has been less well studied than those of the MT cytoskeleton and MT-based motors during the last two decades. This is due in a large part because the longer distance movements driven by MT-based motors in metazoan cells are easier to discern both biochemically because of relatively greater distances between source and target, and morphologically because individual MTs are more easily visualized in the electron and light microscopes than the actin cytoskeleton. Furthermore, MTs in most regions of the cell form regular and long tracks whose polarity (and thus motor directionality) can be inferred from position relative to the centrosome. By contrast, actin filaments in nonmuscle cells do not generally form long tracks but are generally short filaments, the organization of which is more difficult to discern (Fath and Lasek, 1988; Mallik and Gross, 2004; Snider *et al.*, 2004).

The focus of this chapter will be on the less known role the actin cytoskeleton and myosin motors play in the movement of cargo in the biosynthetic pathways of metazoan cells. Other systems will be briefly discussed, where necessary. Lately there has been an increased realization that the actin cytoskeleton, once thought to function primarily at the PM of nonmuscle cells, has significant roles in the deep cytoplasm, especially in the movement of materials to and from the Golgi complex. The roles of MT-based movements will only be cursorily discussed and the reader should consult the following reviews on the MT cytoskeleton (Allan *et al.*, 2002; Müsch, 2004; Vaughan, 2005). The third major cytoskeletal system, the intermediate filaments, is thought to have a major role in intracellular membrane motility. It has been shown that they may have limited roles in intracellular trafficking of endocytic organelles (Styers *et al.*, 2005).

A. Golgi Actin Cytoskeleton

Work from many laboratories has characterized an actin-based matrix associated with the Golgi as shown by biochemical and morphological methods (Lorra and Huttner, 1999; De Matteis and Morrow, 2000). This matrix comprises in part of actin, specific spectrin, tropomyosin, and ankyrin isoforms (Percival *et al.*, 2004) and a growing list of other actin, MT, and intermediate filament-binding proteins. Also present are the small GTPases ADP ribosylation factor 1 (ARF1) and Cdc42, which are considered to regulate the binding of components of this matrix. The matrix also includes the force-generating molecules myosins 1, 2, 5, and 6 as well as cytoplasmic dynein and dynamin (Cao *et al.*, 2005). The binding of several components of this matrix is regulated by components of the coatomer protein complex 1(COPI) transport vesicles. This will be discussed later.

One role of the MT cytoskeleton in the form and positioning of the Golgi complex has been elucidated and appreciated (Allan *et al.*, 2002; Barr and Egerer, 2005). Evidence is now accruing that the actin cytoskeleton may assist a MT-based system in maintaining the structural organization of the Golgi. Experiments in which the actin cytoskeleton is disrupted with cytochalasin D or botulinum toxin C2 suggest that actin filaments are essential in the positioning and morphology of the Golgi complex in normal rat kidney (NRK) cells (Valderrama *et al.*, 1998; di Campli *et al.*, 1999; Valderrama *et al.*, 2000, 2001). The disruption of the actin cytoskeleton in NRK cells induces a concentration of the Golgi complex to a tight spot near the centrosome (Valderrama *et al.*, 1998). In sharp contrast, when MTs are disrupted, or the motor dynein is inhibited, Golgi fragments into vesicles that accumulate at ER exit sites (Murshid and Presley, 2004). It is thus deduced that the actin cytoskeleton in conjunction with MTs is crucial to the spatial organization of the Golgi complex.

Furthermore, the work of Cobbold *et al.* (2004) in which he used drugs to depolymerize actin filaments and/or MTs in HeLa cells suggests that the *trans*-Golgi network (TGN) may have distinct subdomains that are regulated by the actin and MT cytoskeleton and that both cytoskeletal systems are required to concentrate the TGN adjacent to the nucleus (Cobbold *et al.*, 2004).

B. Actin Inhibition Affects Movements to and from Golgi

In addition to maintaining Golgi morphology, experiments in which the actin cytoskeleton is disrupted with drugs that lead to filament depolymerization suggest that actin filaments play important roles in the movements of some transport carriers near the Golgi complex. A role for the actin cytoskeleton has been shown for membranes leaving the Golgi in the anterograde pathway to the basolateral membrane in polarized cells (Müsch *et al.*, 1997, 2001; Duran *et al.*, 2003), in the detachment of vesicular stomatitis virus glycoprotein (VSV-G)–containing tubules from the Golgi in COS cells (Hirschberg *et al.*, 1998), and the budding of some classes of coated vesicles from isolated Golgi (Fucini *et al.*, 2000, 2002). Percival *et al.* (2004) using ultrastructural immunolocalization methods in Madin Darby canine kidney cells have identified small Golgi-associated vesicles (lacking identifiable coats) closely associated with short actin filaments. Actin depolymerization also markedly decreased the trafficking of the enzyme cathepsin D from the TGN to lysosomes (Carreno *et al.*, 2004). Furthermore, actin is important in materials leaving the Golgi in retrograde trafficking to the ER (Valderrama *et al.*, 2001; Duran *et al.*, 2003).

In the trafficking studies reported thus far, the depolymerization of actin decreases the efficiency but does not block proteins arriving at or leaving the Golgi complex. This may be related in part to the requirement that actin depolymerizing agents are effective only on filaments that are in equilibrium with monomer. Relatively stable filaments, such as the short filaments associated with Swiss 3T3 Golgi, may survive treatment (Percival *et al.*, 2004). Additionally, disruption and redistribution of the endomembrane compartments may obviate requirements for actin filaments occurring in untreated cells. Similar decreases in efficiency but not complete inhibition in membrane trafficking are seen in disrupted MT (Müsch, 2004). Alternatively, inhibition of one pathway may allow a nonphysiological pathway to function albeit at an less efficient level.

C. Role of Cdc42 in Recruiting Actin to the Golgi

The recruitment of Golgi actin may be driven by the same machinery that is required for cargo selection and vesicle formation at the Golgi (Stamnes, 2002). Golgi-bound active ARF1 recruits coat protein COPI to the membrane. It is well established that components of COPI induce conformational changes in the membrane to form a vesicle and also work directly or indirectly through associated receptors to accumulate select cargo for export. This COPI complex of proteins is used primarily for intra-Golgi and retrograde Golgi-to-ER membrane carrier movements (Murshid and Presley, 2004). Actin filaments may also be an important component of this forming vesicle coat.

How is actin polymerization induced and regulated at the sites of Golgi export? Dubois and his colleagues have proposed that coat assembly and actin polymerization may be coordinated by an interaction between ARF1 and Cdc42 (Dubois and Chavrier, 2005). Cdc42 is a small GTPase in the Rho family that regulates cortical actin dynamics at the PM. Cdc42 is also a component of the Golgi, which has been shown to function in the movement of Golgi vesicles. Cdc42 may be targeted to Golgi membranes by binding to the gamma subunit of COPI (Erickson *et al.*, 1996; Wu *et al.*, 2000; Chen *et al.*, 2004). Once bound to the membrane and activated by an unidentified guanine nucleotide exchange factor, GTP-Cdc42 can activate the actin-related protein 2/3 (Arp2/3) complex through the neural Wiskott–Aldrich syndrome protein (N-WASP) pathway (Luna *et al.*, 2002; Chen *et al.*, 2004). The activated Arp2/3 may then serve as nuclei for actin filament polymerization near the Golgi. An alternative model suggests that Arp2/3 may require Golgi actin for binding (Chen *et al.*, 2004). The induction of actin polymerization may be switched off when Golgi-bound Cdc42 is inactivated by a Rho GTPase-activating protein (Rho-GAP) that is also recruited to Golgi membranes through binding GTP-ARF1 (Dubois and Chavrier, 2005).

What is the evidence for the role of Cdc42 in the early steps in the biosynthetic pathway? The overexpression of wild-type Cdc42 or a constitutively active form (Cdc42V12; GTPase deficient) in HeLa cells inhibited the retrograde transport of Shiga toxin from the Golgi complex to the ER and the low temperature-induced redistribution of the Lys-Asp-Glu-Leu KDEL receptor from the Golgi (Luna *et al.*, 2002). These data were interpreted to mean that the nonphysiological overactivation of the Arp2/3 pathway might induce a dense Golgi actin network that may restrict membranes from leaving the Golgi area.

Other studies have provided evidence that Cdc42 plays a role in the anterograde movement of components from ER-to-Golgi. Fast cycling mutants of Cdc42 that have a normal GTPase (Cdc42F28L; GTPase does not require a guanine nucleotide exchange factor for GDP/GTP exchange) stimulated the movement of VSV-G from the ER to the Golgi, while two different mutants that are GTPase defective (constitutively active) either blocked (Wu *et al.*, 2000; Chen *et al.*, 2005) or had no effect on transport (Luna *et al.*, 2002). In polarized epithelia, both constitutively active and inactive mutants of Cdc42 inhibited the exit of basolateral membrane proteins from the TGN, while stimulating the exit of an apical membrane protein (Cohen *et al.*, 2001; Müsch *et al.*, 2001). In the same studies, the trafficking of soluble proteins to either the basolateral or apical domain was unaffected.

The variable effects obtained with Cdc42 mutants in different laboratories as mentioned earlier highlight the view that some of the functions of Cdc42 may require GTP hydrolysis and are not controlled simply by the species of bound nucleotide (Symons and Settleman, 2000; Cohen *et al.*, 2001). Further, different amino acid substitutions that apparently affect Cdc42 enzymatic properties in a comparable manner may function differently *in vivo* (i.e., different mutants behave differently). Also as studies in polarized epithelia show, there are differences in the requirements for Cdc42 in the same cell, depending on the PM domain targeted and the identity of the cargo.

In addition to a role in actin recruitment to membranes, Cdc42 may also function in intracellular movements along the MT pathway. Chen *et al.* (2005) have shown that Cdc42 regulates the binding of the MT-dependent motor protein dynein to COPI-coated membranes. They proposed that Cdc42 is targeted to the membrane by binding to coatomer and when bound stimulates actin assembly via Arp2/3. Although bound Cdc42 inhibits the MT-based motor dynein from binding to the same vesicle, when Cdc42 is released from the membrane by inactivation by Rho-GAP, dynein can now bind. Such results are consistent with our finding that dynein binding is a late event in the budding process, occurring when vesicles are budding from the TGN (Fath *et al.*, 1997). This order of events would allow actin to function in vesicle release for transfer to the MT cytoskeleton for long-distance transport.

Cao *et al.* (2005) provided evidence that in addition to Cdc42, ARF1 recruits the actin-binding protein cortactin and dynamin 2 to the Golgi. These proteins appear to function in the release of both clathrin-dependent and clathrin-independent cargos. In another study, clathrin-coated vesicles leaving the TGN in HeLa cells were shown to express both Arp2/3 and Huntingtin-interacting protein 1 related (Hip1R) an actin and clathrin-binding protein (Kaksonen *et al.*, 2005). Using RNA interference (RNAi) knockdowns, these workers showed that Hip1R is necessary for linking actin dynamics and the movement of enzyme cathepsin D from the TGN to the lysosomes.

The Cdc42 regulated recruitment of actin and associated proteins, as has been described, implies whether it is the formation of an endocytic vesicle at the PM or a budded transport carrier at the Golgi, there could be significant overlap or conservation of the main proteins used. These results extend the idea that the assembled coat not only alters the membrane to form a vesicle but also selects cargo and cytoskeletal components necessary for membrane locomotion, small N-ethylmaleimide-sensitive factor attachment protein receptors (SNAREs) and Rabs that are necessary for regulated fusion with the correct target.

D. Actin Polymerization Driven Transport in the Endomembrane System

It has been proposed that the force generated by actin polymerization moves membranous vesicles through the cytoplasm in a manner reminiscent of *Listeria* and other pathogens (Ridley, 2001). The focalized polymerization of actin on one side of the membrane has been likened to a comet's tail that provides force to move the vesicle through the cytoplasm. There is an extensive literature regarding potential roles of actin comet tails in endocytosis (Schafer *et al.*, 2000; Taunton, 2001; Kjeken *et al.*, 2004; Kaksonen *et al.*, 2005), with suggestion that they may also be involved in movements near the Golgi (Carreno *et al.*, 2004). Although actin comet tails have been seen on endomembranes in the cell, it has been suggested that the use of constitutively active mutants may stimulate movements that are amplifications of normal processes in which actin is polymerized around transport vesicles (Rozelle *et al.*, 2000; Benesch *et al.*, 2002;

Dubois and Chavrier, 2005). In one set of experiments, it was determined that the movement of membranes from the Golgi to the ER was myosin driven (see later) and not by actin comets (Duran *et al.*, 2003). Clearly, much more work needs to be done to explore this intriguing possibility that actin polymerization provides the force for cytoplasmic membrane movements over short or long distances.

III. ROLES OF MYOSINS IN MEMBRANE MOVEMENTS ALONG ENDOMEMBRANE PATHWAYS

The requirement for the actin cytoskeleton in endomembrane compartment morphology, positioning, and membrane trafficking may also partly require the action of members of the myosin family, which use shape changes driven by ATP hydrolysis and actin binding to transport cargo along actin tracks. There are at least 20 members in the growing myosin superfamily (Krendel and Mooseker, 2005). To describe the various classes of vertebrate myosin heavy chains, we will adopt the nomenclature "Myo" + Arabic numerals corresponding to the myosin class (Krendel and Mooseker, 2005). Among members of the myosin family identified to date, myosins 1, 2, 5, and 6 have been shown to participate in the dynamics of the endomembrane system (Allan *et al.*, 2002). These motors will be discussed individually.

Here the concept of processivity of molecular motors is important. A processive motor is one that moves relatively long distances before detaching from the cytoskeletal track (De La Cruz and Ostap, 2004). High processivity may result from motor heavy chain dimerization in which the two heads work in coordination so that one does not get released from the track unless the other is bound. Motors with a high duty ratio, those that spend a large portion of the ATPase cycle in association with the track, are also processive motors. Processive motors are thought to be better suited for cargo motility than a nonprocessive motor that may quickly detach and allow the bound cargo to drift away from the track and stop moving. It is thought that a high density of a nonprocessive motor must be attached to cargo for such a motor to function in intracellular transport (De La Cruz and Ostap, 2004).

There are at least three ways in which myosins may function in actin-dependent transport carrier movements (Krendel and Mooseker, 2005). The myosin may provide the force to transport cargo along actin filaments. Alternatively, the myosin may help reorganize a dense actin network near the membrane that may otherwise inhibit the passage of a vesicle to or from the membrane. Last, myosin may assist in the formation of a vesicle carrier by providing force driving the constriction/scission of the neck of the forming vesicle.

A. Myo1

The class 1 myosins are nonprocessive, monomeric myosins that move toward the plus (barbed) ends of actin filaments. Myosins 1 have been found associated with membranes in many cell types. Among the best-characterized myosins 1 is Myo1a, which is found in highest concentrations in the microvilli of polarized epithelial cells. This myosin links the core bundle of actin filaments with the apical PM, but its function there is still unknown (Tyska *et al.*, 2005). Although a microvillar protein, Myo1a was assigned roles in transporting membranes and proteins to the apical PM of intestinal epithelial cells (Fath and Burgess, 1993, 1994; Montes de Oca *et al.*, 1997). It was proposed that vesicles enriched with glycosphingolipids (lipid rafts) move from the Golgi to the apical cytoplasm using the MT cytoskeleton and then Myo1a to move through the actin-rich, MT-poor cortical cytoplasm. Jacob *et al.* (2003) thought that Myo1a is the motor that delivers sucrase-isomaltase along peripheral actin filaments to the PM. However, lactase-phlorizin hydrolase (which is not a component of lipid rafts) is not delivered using actin. Studies in the Mooseker laboratory with expressed dominant-negative Myo1a tail fragments or involving complete removal in knockout mice also suggest that Myo1a functions in the apical locale of microvillus raft proteins sucrase-isomaltase and galectin (Tyska *et al.*, 2005). However, it is proposed that Myo1a functions in retaining these proteins in the PM once delivered but not in the delivery process itself. Intuitively, such a function is reflected by the low duty ratio properties of Myo1a, which makes it better suited for roles at higher concentrations in the microvillus than on carrier vesicles that would require many copies of the motor to function. However, there is agreement that Myo1 in polarized epithelia may be targeted to glycosphingolipid-rich membrane domains that are thought to serve as an apical PM-targeting mechanism in such cells.

In addition to apical delivery in polarized epithelia, Myo1 has also been suggested to function in moving vesicles away from the membrane during endocytosis (Raposo *et al.*, 1999; Salas-Cortes *et al.*, 2005) in the insulin-stimulated movement of recycling endosomes containing GLUT4 receptors to the PM (Bose *et al.*, 2002) and movements of lysosomes (Raposo *et al.*, 1999; Cordonnier *et al.*, 2001).

B. Nonmuscle Myo2

Members of the nonmuscle Myo2 class are two-headed myosins that are directed toward the plus ends of actin filaments. Three nonmuscle Myo2 genes have been identified as Myosin2a,-b, and -c (also called MYH9,-10, and -14, respectively, in humans). Of these isoforms, Myo2a is known to be a low duty ratio, nonprocessive motor, while the 2b isoform has a high duty ratio with unknown processivity (Wang *et al.*, 2003). Although these myosins are best known to function in whole cell motility and cytokinesis, they have also been implicated in moving transport carriers to and from the Golgi complex in several cell types.

Nonmuscle Myo2 has been immunolocalized to the Golgi complex by biochemical and morphological means in a number of polarized and unpolarized cell types (Ikonen *et al.*, 1997; Müsch *et al.*, 1997; Pepperkok *et al.*, 1998; Heimann *et al.*, 1999; Charron *et al.*, 2000; Mills *et al.*, 2000; Bustos *et al.*, 2001; Togo and Steinhardt, 2003; Fath, 2005). It is found not only in Golgi but also as a physiological component of the Golgi. In NRK and HeLa cells, the retrograde movement of the Shiga toxin fragment B and the brefeldin A (an inhibitor of ARF1)-induced movement of materials from the Golgi-to-ER require an actin-based Myo2 step (Duran *et al.*, 2003). A role for Myo2 was shown using the inhibitors BDM (2,3-butanedione monoxime) and the specific Myo2 inhibitor ML-7. It is important to note that research from several laboratories has shown that BDM does not inhibit nonmuscle Myo2, although used extensively for this purpose (Cheung *et al.*, 2002; Ostap, 2002). Because BDM causes a loss of the actin polymerization machinery from the PM (Yarrow *et al.*, 2003), it is likely that the observed inhibition in trafficking is the result of a block in the assembly of actin filaments on Golgi membranes as described earlier. The methods used by Duran *et al.* (2003) could not explain whether Myo2 was required for budding from the Golgi and transport between the membrane compartments or fusion with the ER. Another group of investigators using Swiss 3T3 cells, localized Myo2a to the Golgi, and antisense knockdown experiments revealed that Myo2a (but not Myo2b) is involved in brefeldin A-sensitive facilitated PM repair requiring membrane trafficking at the Golgi (Togo and Steinhardt, 2003). However, they failed to explore directly at what step Myo2 is required (e.g., budding, transport, or fusion). In a third type of study using permeabilized cells and isolated Golgi, the results obtained suggested that Myo2 is required for the budding of basolateral transport vesicles containing VSV-G from the TGN of polarized cells (Ikonen *et al.*, 1997; Müsch *et al.*, 1997; Stow *et al.*, 1998). Because the myosin was released from the vesicles soon after budding, it was thought to play a role at the Golgi and not for long-distance trafficking throughout the cell.

Although there is little evidence that post-Golgi transport carriers utilize Myo2 to move long distances in metazoan cells (unlike plants or yeast), several studies have suggested that Myo2 may function in moving membranes through an actin-rich cell cortex to fuse with the PM. Antibody inhibition in detergent-permeabilized cells have implicated Myo2 in the transport of vesicles in the sub-plasmalemmal area during chromaffin cell secretory activity (Ñeco *et al.*, 2002). In neurons, the microinjection of C-terminus tail fragments of Myo2b into superior cervical ganglion cells suggested that Myo2b participates in vesicle trafficking in presynaptic nerve terminals (Takagishi *et al.*, 2005). In another study, Myo2a was found to coisolate with a population of vesicles that may be recycling endosomes containing chloride channels in epithelial cells (Ecay *et al.*, 1997).

C. Myo5

Members of the Myo5 class are two-headed myosins that move toward the plus ends of actin filaments. Members of this family that are expressed in vertebrates are processive motors, while those from yeast are nonprocessive (Reck-Peterson *et al.*, 2001). Of all the myosin motors implicated in intracellular motility, Myo5 is the most known since it is a versatile motor, which has been implicated in the movement of many types of subcellular components such as ER, melanosomes, secretory granules, endosomes, and neurofilaments (Krendel and Mooseker, 2005). This multitude of functions is generated in part by tissue-specific isoforms with different tail domains resulting from differential RNA splicing that may alter cargo choice (Seperack *et al.*, 1995; Au and Huang, 2002). Further, the GTP-binding Rab proteins that control vesicular transport by regulating the timing of vesicle fusion with a specific target may also regulate the binding of Myo5 with specific cargo (Seabra and Coudrier, 2004).

Myo5 has roles in many compartments of the endomembrane system. Myo5 may be important in the morphology and positioning of the ER (Wagner and Hammer, 2003). Although the MT cytoskeleton is required for the extension of the ER throughout animal cells, Myo5 can drive the formation of ER networks in *Xenopus* extracts (Wöllert *et al.*, 2002). Genetic evidence shows that it is vital for ER inheritance in budding yeast (Estrada *et al.*, 2003). Thus, it is likely that Myo5 operates in concert with MT-based motors in the positioning of this organelle. Myo5 has also been identified to have roles in the movement of transport carriers in the endomembrane system both in the deep cytoplasm and near the cell surface. In budding yeast, Myo2p (a yeast Myo5) is involved in targeting membranes and proteins to the developing daughter bud during cell division (Schott *et al.*, 2002). In melanophores, Myo5 is involved in the distribution of melanosomes working in coordination with the MT cytoskeleton. In neuronal cells, Myo5 is thought to be involved in the transport of smooth ER-derived vesicles in the squid giant axon (Tabb *et al.*, 1998). With biochemical techniques, it has been shown to bind to synaptic vesicles (Prekeris and Terrian, 1997) and to unknown high-density vesicles isolated from chicken brain (Miller and Sheetz, 2000). When nearer the cell surface, Myo5 may be important for the local delivery of chromaffin vesicles in chromaffin cells (Rose *et al.*, 2003; Watanabe *et al.*, 2005; although see Ñeco *et al.*, 2002) and insulin-containing vesicles in pancreatic β-cells (Varadi *et al.*, 2005). Myo5c has also been shown to regulate an unknown step in transferrin receptor uptake in epithelial cells (Rodriguez and Cheney, 2002) and some movements of mouse macrophage phagosomes near the PM (Al-Haddad *et al.*, 2001).

D. Myo6

Myosins of class 6 are multifunctional motors that are implicated in cellular processes such as vesicle membrane traffic both near the PM and at the Golgi, cell migration, and mitosis. Although Myo6 can exist as a processive dimer and nonprocessive monomer *in vitro*, its form *in vivo* is not known (Buss *et al.*, 2004). This class of myosin is unique in that it translocates toward the minus (pointed) ends of actin filaments (Nishikawa *et al.*, 2002). Because the majority of membrane-associated actin filaments are regarded as being oriented with the plus ends toward the membranes (Cramer, 1999), this motor appears important in actin-based movements away from membranes such as during endocytosis. Myo6 is known to be required for moving uncoated endocytic vesicles through the dense cortical actin network of cultured epithelial cells (Aschenbrenner *et al.*, 2004). The reader is referred to the excellent review by Buss for more details concerning this remarkable motor (Buss *et al.*, 2004).

Myo6 has been shown to have important roles in Golgi membrane dynamics. It has been localized to the Golgi complex and Golgi-associated vesicles (Buss *et al.*, 1998;

Warner *et al.*, 2003). It is targeted to the Golgi via its tail domain, thus leaving the head domain free to interact with the Golgi actin cytoskeleton. When the Golgi is disrupted by the inhibition of ARF1 function with brefeldin A, Myo6 redistributes with the TGN near the centrosome. The protein optineurin has been shown to be essential in linking the tail of Myo6 to the Golgi membrane (Sahlender *et al.*, 2005). The knockdown of optineurin protein levels by RNAi induces Golgi complex disruption, but not vesiculation, and decreases the constitutive secretion of VSV-G in cultured cells (Sahlender *et al.*, 2005). Thus, Myo6 has roles both in Golgi morphology and in transport of components to the PM. These results are in agreement with those of Warner *et al.* (2003) in which fibroblasts from Snell's waltzer mice, which lack Myo6, have a disrupted Golgi, as well as reduction in the constitutive secretion of alkaline phosphatase. In these studies, it is not known whether Myo6 plays a role in cargo selection, vesicle formation, or the transport of vesicles from the Golgi to MTs for long-range transport (Warner *et al.*, 2003).

IV. CONCLUSIONS

We have now come close to full circle in our thinking about intracellular transport in metazoan cells. Earlier studies focused almost exclusively on the roles of the actin cytoskeleton in intracellular membrane trafficking. With the discovery of kinesin and cytoplasmic dynein in the 1980s, the focus moved to the MT cytoskeleton and the actin-based movements were relegated to movements near the PM. We now know that both the actin cytoskeleton and MTs are of central importance in terms of shape and positioning in the major endomembrane compartments throughout the cell. Although the MT cytoskeleton is considered as the major track for long-distance transport, local movements near intracellular and plasma membranes also require actin-based motility. In the coming years, those researching into the fields that work on MT- and actin-based motors will find that their once independent pathways have merged. Future work will of necessity focus on how motors are targeted to selected membranes and how multiple motors on the same cargo are regulated in different regions of the cell.

ACKNOWLEDGMENTS

This work was supported by grants from The City University of New York PSC-CUNY Research Award Program.

REFERENCES

Al-Haddad, A., Shonn, M. A., Redlich, B., Blocker, A., Burkhardt, J. K., Yu, H., Hammer, J. A., III, Weiss, D. G., Steffen, W., Griffiths, G., and Kuznetsov, S. A. (2001). Myosin Va bound to phagosomes binds to F-actin and delays microtubule-dependent motility. *Mol. Biol. Cell* **12**(9), 2742–2755.

Allan, V. J., Thompson, H. M., and McNiven, M. A. (2002). Motoring around the Golgi. *Nat. Cell Biol.* **4**(10), E236–E242.

Aschenbrenner, L., Naccache, S. N., and Hasson, T. (2004). Uncoated endocytic vesicles require the unconventional myosin, Myo6, for rapid transport through actin barriers. *Mol. Biol. Cell* **15**(5), 2253–2263.

Au, J. S. and Huang, J. D. (2002). A tissue-specific exon of myosin Va is responsible for selective cargo binding in melanocytes. *Cell Motil. Cytoskeleton* **53**(2), 89–102.

Barr, F. A. and Egerer, J. (2005). Golgi positioning: Are we looking at the right MAP? *J. Cell Biol.* **168**(7), 993–998.

Benesch, S., Lommel, S., Steffen, A., Stradal, T. E., Scaplehorn, N., Way, M., Wehland, J., and Rottner, K. (2002). Phosphatidylinositol 4,5-biphosphate (PIP2)-induced vesicle movement depends on N-WASP and involves Nck, WIP, and Grb2. *J. Biol. Chem.* **277**(40), 37771–37776.

Bose, A., Guilherme, A., Robida, S. I., Nicoloro, S. M., Zhou, Q. L., Jiang, Z. Y., Pomerleau, D. P., and Czech, M. P. (2002). Glucose transporter recycling in response to insulin is facilitated by myosin Myo1c. *Nature* **420**(6917), 821–824.

Buss, F., Kendrick-Jones, J., Lionne, C., Knight, A. E., Cote, G. P., and Paul Luzio, J. (1998). The localization of myosin VI at the Golgi complex and leading edge of fibroblasts and its phosphorylation and recruitment into membrane ruffles of A431 cells after growth factor stimulation. *J. Cell Biol.* **143**(6), 1535–1545.

Buss, F., Spudich, G., and Kendrick-Jones, J. (2004). Myosin VI: Cellular functions and motor properties. *Annu. Rev. Cell Dev. Biol.* **20,** 649–676.

Bustos, R., Kolen, E. R., Braiterman, L., Baines, A. J., Gorelick, F. S., and Hubbard, A. L. (2001). Synapsin I is expressed in epithelial cells: Localization to a unique trans-Golgi compartment. *J. Cell Sci.* **114**(Pt. 20), 3695–3704.

Cao, H., Weller, S., Orth, J. D., Chen, J., Huang, B., Chen, J. L., Stamnes, M., and McNiven, M. A. (2005). Actin and Arf1-dependent recruitment of a cortactin-dynamin complex to the Golgi regulates post-Golgi transport. *Nat. Cell Biol.* **7**(5), 483–492.

Carreno, S., Engqvist-Goldstein, A. E., Zhang, C. X., McDonald, K. L., and Drubin, D. G. (2004). Actin dynamics coupled to clathrin-coated vesicle formation at the trans-Golgi network. *J. Cell Biol.* **165**(6), 781–788.

Charron, A. J., Bacallao, R. L., and Wandinger-Ness, A. (2000). ADPKD: A human disease altering Golgi function and basolateral exocytosis in renal epithelia. *Traffic* **1**(8), 675–686.

Chen, J. L., Lacomis, L., Erdjument-Bromage, H., Tempst, P., and Stamnes, M. (2004). Cytosol-derived proteins are sufficient for Arp2/3 recruitment and ARF/coatomer-dependent actin polymerization on Golgi membranes. *FEBS Lett.* **566**(1–3), 281–286.

Chen, J. L., Fucini, R. V., Lacomis, L., Erdjument-Bromage, H., Tempst, P., and Stamnes, M. (2005). Coatomer-bound Cdc42 regulates dynein recruitment to COPI vesicles. *J. Cell Biol.* **169**(3), 383–389.

Cheung, A., Dantzig, J. A., Hollingworth, S., Baylor, S. M., Goldman, Y. E., Mitchison, T. J., and Straight, A. F. (2002). A small-molecule inhibitor of skeletal muscle myosin II. *Nat. Cell Biol.* **4**(1), 83–88.

Cobbold, C., Coventry, J., Ponnambalam, S., and Monaco, A. P. (2004). Actin and microtubule regulation of trans-Golgi network architecture, and copper-dependent protein transport to the cell surface. *Mol. Membr. Biol.* **21**(1), 59–66.

Cohen, D., Müsch, A., and Rodriguez-Boulan, E. (2001). Selective control of basolateral membrane protein polarity by cdc42. *Traffic* **2**(8), 556–564.

Cordonnier, M. N., Dauzonne, D., Louvard, D., and Coudrier, E. (2001). Actin filaments and myosin I alpha cooperate with microtubules for the movement of lysosomes. *Mol. Biol. Cell* **12**(12), 4013–4029.

Cramer, L. P. (1999). Organization and polarity of actin filament networks in cells: Implications for the mechanism of myosin-based cell motility. *Biochem. Soc. Symp.* **65,** 173–205.

De La Cruz, E. M. and Ostap, E. M. (2004). Relating biochemistry and function in the myosin superfamily. *Curr. Opin. Cell. Biol.* **16**(1), 61–67.

De Matteis, M. A. and Morrow, J. S. (2000). Spectrin tethers and mesh in the biosynthetic pathway. *J. Cell Sci.* **113**(Pt. 13), 2331–2343.

di Campli, A., Valderrama, F., Babià, T., De Matteis, M. A., Luini, A., and Egea, G. (1999). Morphological changes in the Golgi complex correlate with actin cytoskeleton rearrangements. *Cell Motil. Cytoskeleton* **43,** 334–348.

Dubois, T. and Chavrier, P. (2005). ARHGAP10, a novel RhoGAP at the cross-road between ARF1 and Cdc42 pathways, regulates Arp2/3 complex and actin dynamics on Golgi membranes. *Med. Sci. (Paris)* **21**(8–9), 692–694.

Duran, J. M., Valderrama, F., Castel, S., Magdalena, J., Tomas, M., Hosoya, H., Renau-Piqueras, J., Malhotra, V., and Egea, G. (2003). Myosin motors and not actin comets are mediators of the actin-based Golgi-to-endoplasmic reticulum protein transport. *Mol. Biol. Cell* **14**(2), 445–459.

Ecay, T., Conner, T., and Decker, E. (1997). Nonmuscle myosin IIA copurifies with chloride channel-enriched membranes from epithelia. *Biochem. Biophys. Res. Commun.* **231**(2), 369–372.

Erickson, J. W., Zhang, C., Kahn, R. A., Evans, T., and Cerione, R. A. (1996). Mammalian Cdc42 is a brefeldin A-sensitive component of the Golgi apparatus. *J. Biol. Chem.* **271**(43), 26850–26854.

Estrada, P., Kim, J., Coleman, J., Walker, L., Dunn, B., Takizawa, P., Novick, P., and Ferro-Novick, S. (2003). Myo4p and She3p are required for cortical ER inheritance in *Saccharomyces cerevisiae*. *J. Cell Biol.* **163**(6), 1255–1266.

Fath, K. R. (2005). Characterization of myosin-II binding to Golgi stacks *in vitro*. *Cell Motil. Cytoskeleton* **60**(4), 222–235.

Fath, K. R. and Burgess, D. R. (1993). Golgi-derived vesicles from developing epithelial cells bind actin filaments and possess myosin-I as a cytoplasmically oriented peripheral membrane protein. *J. Cell Biol.* **120**(1), 117–127.

Fath, K. R. and Burgess, D. R. (1994). Membrane motility mediated by unconventional myosin. *Curr. Opin. Cell Biol.* **6,** 131–135.

Fath, K. R. and Lasek, R. (1988). Two classes of actin microfilaments are associated with the inner cytoskeleton of axons. *J. Cell Biol.* **107,** 613–621.

Fath, K. R., Trimbur, G. M., and Burgess, D. R. (1997). Molecular motors and a spectrin matrix associate with Golgi membranes *in vitro*. *J. Cell Biol.* **139,** 1169–1181.

Fucini, R. V., Navarrete, A., Vadakkan, C., Lacomis, L., Erdjument-Bromage, H., Tempst, P., and Stamnes, M. (2000). Activated ADP-ribosylation factor assembles distinct pools of actin on golgi membranes. *J. Biol. Chem.* **275**(25), 18824–18829.

Fucini, R. V., Chen, J. L., Sharma, C., Kessels, M. M., and Stamnes, M. (2002). Golgi vesicle proteins are linked to the assembly of an actin complex defined by mAbp1. *Mol. Biol. Cell* **13**(2), 621–631.

Heimann, K., Percival, J. M., Weinberger, R., Gunning, P., and Stow, J. L. (1999). Specific isoforms of actin-binding proteins on distinct populations of Golgi-derived vesicles. *J. Biol. Chem.* **274**(16), 10743–10750.

Hirschberg, K., Miller, C. M., Ellenberg, J., Presley, J. F., Siggia, E. D., Phair, R. D., and Lippincott-Schwartz, J. (1998). Kinetic analysis of secretory protein traffic and characterization of Golgi to plasma membrane transport intermediates in living cells. *J. Cell Biol.* **143**(6), 1485–1503.

Ikonen, E., de Almeid, J. B., Fath, K. R., Burgess, D. R., Ashman, K., Simons, K., and Stow, J. L. (1997). Myosin II is associated with Golgi membranes: Identification of p200 as nonmuscle myosin II on Golgi-derived vesicles. *J. Cell Sci.* **110**(Pt. 18), 2155–2164.

Jacob, R., Heine, M., Alfalah, M., and Naim, H. Y. (2003). Distinct cytoskeletal tracks direct individual vesicle populations to the apical membrane of epithelial cells. *Curr. Biol.* **13**(7), 607–612.

Kaksonen, M., Toret, C. P., and Drubin, D. G. (2005). A modular design for the clathrin- and actin-mediated endocytosis machinery. *Cell* **123**(2), 305–320.

Kjeken, R., Egeberg, M., Habermann, A., Kuehnel, M., Peyron, P., Floetenmeyer, M., Walther, P., Jahraus, A., Defacque, H., Kuznetsov, S. A., and Griffiths, G. (2004). Fusion between phagosomes, early and late endosomes: A role for actin in fusion between late, but not early endocytic organelles. *Mol. Biol. Cell* **15**(1), 345–358.

Krendel, M. and Mooseker, M. S. (2005). Myosins: Tails (and heads) of functional diversity. *Physiology (Bethesda)* **20,** 239–251.

Lorra, C. and Huttner, W. B. (1999). The mesh hypothesis of Golgi dynamics. *Nat. Cell Biol.* **1,** E113–E115.

Luby-Phelps, K. (1994). Physical properties of cytoplasm. *Curr. Biol.* **6,** 3–9.

Luna, A., Matas, O. B., Martinez-Menarguez, J. A., Mato, E., Duran, J. M., Ballesta, J., Way, M., and Egea, G. (2002). Regulation of protein transport from the Golgi complex to the endoplasmic reticulum by CDC42 and N-WASP. *Mol. Biol. Cell* **13**(3), 866–879.

Mallik, R. and Gross, S. P. (2004). Molecular motors: Strategies to get along. *Curr. Biol.* **14**(22), R971–R982.

Miller, K. E. and Sheetz, M. P. (2000). Characterization of myosin V binding to brain vesicles. *J. Biol. Chem.* **275**(4), 2598–2606.

Mills, J. W., Falsig Pedersen, S., Walmod, P. S., and Hoffmann, E. K. (2000). Effect of cytochalasins on F-actin and morphology of Ehrlich ascites tumor cells. *Exp. Cell Res.* **261**(1), 209–219.

Montes de Oca, G., Lezama, R., Mondragon, R., Castillo, A., and Meza, I. (1997). Myosin I interactions with actin filaments and trans-Golgi-derived vesicles in MDCK cell monolayers. *Arch. Med. Res.* **28**(3), 321–328.

Murshid, A. and Presley, J. F. (2004). ER-to-Golgi transport and cytoskeletal interactions in animal cells. *Cell. Mol. Life Sci.* **61**(2), 133–145.

Müsch, A. (2004). Microtubule organization and function in epithelial cells. *Traffic* **5**(1), 1–9.

Müsch, A., Cohen, D., and Rodriguez-Boulan, E. (1997). Myosin II is involved in the production of constitutive transport vesicles from the TGN. *J. Cell Biol.* **138**(2), 291–306.

Müsch, A., Cohen, D., Kreitzer, G., and Rodriguez-Boulan, E. (2001). Cdc42 regulates the exit of apical and basolateral proteins from the trans-Golgi network. *EMBO J.* **20**(9), 2171–2179.

Ñeco, P., Gil, A., Del Mar Frances, M., Viniegra, S., and Gutierrez, L. M. (2002). The role of myosin in vesicle transport during bovine chromaffin cell secretion. *Biochem. J.* **368**(Pt. 2), 405–413.

Nishikawa, S., Homma, K., Komori, Y., Iwaki, M., Wazawa, T., Hikikoshi, I. A., Saito, J., Ikebe, R., Katayama, E., Yanagida, T., and Ikebe, M. (2002). Class VI myosin moves processively along actin filaments backward with large steps. *Biochem. Biophys. Res. Commun.* **290**(1), 311–317.

Ostap, E. M. (2002). 2,3-Butanedione monoxime (BDM) as a myosin inhibitor. *J. Muscle Res. Cell Motil.* **23**(4), 305–308.

Pantaloni, D., Le Clainche, C., and Carlier, M. F. (2001). Mechanism of actin-based motility. *Science* **292**(5521), 1502–1506.

Pepperkok, R., Lowe, M., Burke, B., and Kreis, T. E. (1998). Three distinct steps in transport of vesicular stomatitis virus glycoprotein from the ER to the cell surface *in vivo* with differential sensitivities to GTP gamma S. *J. Cell Sci.* **111**(Pt. 13), 1877–1888.

Percival, J. M., Hughes, J. A., Brown, D. L., Schevzov, G., Heimann, K., Vrhovski, B., Bryce, N., Stow, J. L., and Gunning, P. W. (2004). Targeting of a tropomyosin isoform to short microfilaments associated with the Golgi complex. *Mol. Biol. Cell* **15**(1), 268–280.

Pollard, T. D. and Earnshaw, W. E. (2002). *Cell Biology*. W. B. Saunders, New York, NY.

Prekeris, R. and Terrian, D. M. (1997). Brain myosin V is a synaptic vesicle-associated motor protein: Evidence for a $Ca2^{+}$-dependent interaction with the synaptobrevin-synaptophysin complex. *J. Cell Biol.* **137**(7), 1589–1601.

Raposo, G., Cordonnier, M. N., Tenza, D., Menichi, B., Dürrbach, A., Louvard, D., and Coudrier, E. (1999). Association of myosin I alpha with endosomes and lysosomes in mammalian cells. *Mol. Biol. Cell* **10**(5), 1477–1494

Reck-Peterson, S. L., Tyska, M. J., Novick, P. J., and Mooseker, M. S. (2001). The yeast class V myosins, Myo2p and Myo4p, are nonprocessive actin-based motors. *J. Cell Biol.* **153**(5), 1121–1126.

Ridley, A. J. (2001). Rho proteins: Linking signaling with membrane trafficking. *Traffic* **2**(5), 303–310.

Rodriguez, O. C. and Cheney, R. E. (2002). Human myosin-Vc is a novel class V myosin expressed in epithelial cells. *J. Cell Sci.* **115**(Pt. 5), 991–1004.

Rose, S. D., Lejen, T., Casaletti, L., Larson, R. E., Pene, T. D., and Trifaro, J. M. (2003). Myosins II and V in chromaffin cells: Myosin V is a chromaffin vesicle molecular motor involved in secretion. *J. Neurochem.* **85**(2), 287–298.

Rozelle, A. L., Machesky, L. M., Yamamoto, M., Driessens, M. H., Insall, R. H., Roth, M. G., Luby-Phelps, K., Marriott, G., Hall, A., and Yin, H. L. (2000). Phosphatidylinositol 4,5-bisphosphate induces actin-based movement of raft-enriched vesicles through WASP-Arp2/3. *Curr. Biol.* **10**(6), 311–320.

Sahlender, D. A., Roberts, R. C., Arden, S. D., Spudich, G., Taylor, M. J., Luzio, J. P., Kendrick-Jones, J., and Buss, F. (2005). Optineurin links myosin VI to the Golgi complex and is involved in Golgi organization and exocytosis. *J. Cell Biol.* **169**(2), 285–295.

Salas-Cortes, L., Ye, F., Tenza, D., Wilhelm, C., Theos, A., Louvard, D., Raposo, G., and Coudrier, E. (2005). Myosin Ib modulates the morphology and the protein transport within multi-vesicular sorting endosomes. *J. Cell Sci.* **118**(Pt. 20), 4823–4832.

Schafer, D. A., D'Souza-Schorey, C., and Cooper, J. A. (2000). Actin assembly at membranes controlled by ARF6. *Traffic* **1**(11), 896–907.

Schott, D. H., Collins, R. N., and Bretscher, A. (2002). Secretory vesicle transport velocity in living cells depends on the myosin-V lever arm length. *J. Cell Biol.* **156**(1), 35–40.

Seabra, M. C. and Coudrier, E. (2004). Rab GTPases and myosin motors in organelle motility. *Traffic* **5**(6), 393–399.

Seperack, P. K., Mercer, J. A., Strobel, M. C., Copeland, N. G., and Jenkins, N. A. (1995). Retroviral sequences located within an intron of the dilute gene alter dilute expression in a tissue-specific manner. *EMBO J.* **14**(10), 2326–2332.

Snider, J., Lin, F., Zahedi, N., Rodionov, V., Yu, C. C., and Gross, S. P. (2004). Intracellular actin-based transport: How far you go depends on how often you switch. *Proc. Natl. Acad. Sci. USA* **101**(36), 13204–13209.

Stamnes, M. (2002). Regulating the actin cytoskeleton during vesicular transport. *Curr. Opin. Cell. Biol.* **14**(4), 428–433.

Stow, J. L., Fath, K. R., and Burgess, D. R. (1998). Budding roles for myosin II on the Golgi. *Trends Cell Biol.* **8**(4), 138–141.

Styers, M. L., Kowalczyk, A. P., and Faundez, V. (2005). Intermediate filaments and vesicular membrane traffic: The odd couple's first dance? *Traffic* **6**(5), 359–365.

Symons, M. and Settleman, J. (2000). Rho family GTPases: More than simple switches. *Trends Cell Biol.* **10**(10), 415–419.

Tabb, J. S., Molyneaux, B. J., Cohen, D. L., Kuznetsov, S. A., and Langford, G. M. (1998). Transport of ER vesicles on actin filaments in neurons by myosin V. *J. Cell Sci.* **111,** 3221–3234.

Takagishi, Y., Futaki, S., Itoh, K., Espreafico, E. M., Murakami, N., Murata, Y., and Mochida, S. (2005). Localization of myosin II and V isoforms in cultured rat sympathetic neurones and their potential involvement in presynaptic function. *J. Physiol.* **569**(Pt. 1), 195–208.

Taunton, J. (2001). Actin filament nucleation by endosomes, lysosomes and secretory vesicles. *Curr. Opin. Cell. Biol.* **13**(1), 85–91.

Togo, T. and Steinhardt, R. A. (2003). Nonmuscle myosin IIA and IIB have distinct functions in the exocytosis-dependent process of cell membrane repair. *Mol. Biol. Cell* **15,** 688–695.

Tyska, M. J., Mackey, A. T., Huang, J. D., Copeland, N. G., Jenkins, N. A., and Mooseker, M. S. (2005). Myosin-1a is critical for normal brush border structure and composition. *Mol. Biol. Cell* **16**(5), 2443–2457.

Valderrama, F., Babià, T., Ayala, I., Kok, J. W., Renau-Piqueras, J., and Egea, G. (1998). Actin microfilaments are essential for the cytological positioning and morphology of the Golgi complex. *Eur. J. Cell Biol.* **76**(1), 9–17.

Valderrama, F., Luna, A., Babià, T., Martinez-Menárguez, J. A., Ballesta, J., Barth, H., Chaponnier, C., Renau-Piqueras, C., and Egea, G. (2000). The Golgi-associated COP1-coated buds and vesicles contain b/g-actin. *Proc. Natl Acad. Sci. USA* **97,** 1560–1565.

Valderrama, F., Duran, J. M., Babià, T., Barth, H., Renau-Piqueras, J., and Egea, G. (2001). Actin microfilaments facilitate the retrograde transport from the Golgi complex to the endoplasmic reticulum in mammalian cells. *Traffic* **2**(10), 717–726.

Varadi, A., Tsuboi, T., and Rutter, G. A. (2005). Myosin Va transports dense core secretory vesicles in pancreatic MIN6 β-cells. *Mol. Biol. Cell* **16**(6), 2670–2680.

Vaughan, K. T. (2005). Microtubule plus ends, motors, and traffic of Golgi membranes. *Biochim. Biophys. Acta* **1744**(3), 316–324.

Wagner, W. and Hammer, J. A., III (2003). Myosin V and the endoplasmic reticulum: The connection grows. *J. Cell Biol.* **163**(6), 1193–1196.

Wang, F., Kovács, M., Hu, A., Limouze, J., Harvey, E. V., and Sellers, J. R. (2003). Kinetic mechanism of non-muscle myosin IIB: Functional adaptations for tension generation and maintenance. *J. Biol. Chem.* **278**(30), 27439–27448.

Warner, C. L., Stewart, A., Luzio, J. P., Steel, K. P., Libby, R. T., Kendrick-Jones, J., and Buss, F. (2003). Loss of myosin VI reduces secretion and the size of the Golgi in fibroblasts from Snell's waltzer mice. *EMBO J.* **22**(3), 569–579.

Watanabe, M., Nomura, K., Ohyama, A., Ishikawa, R., Komiya, Y., Hosaka, K., Yamauchi, E., Taniguchi, H., Sasakawa, N., Kumakura, K., Ushiki, T., Sato, O., *et al.* (2005). Myosin-Va regulates exocytosis through the submicromolar $Ca2^{+}$-dependent binding of syntaxin-1A. *Mol. Biol. Cell* **16**(10), 4519–4530.

Wöllert, T., Weiss, D. G., Gerdes, H. H., and Kuznetsov, S. A. (2002). Activation of myosin V-based motility and F-actin-dependent network formation of endoplasmic reticulum during mitosis. *J. Cell Biol.* **159**(4), 571–577.

Wu, W. J., Erickson, J. W., Lin, R., and Cerione, R. A. (2000). The gamma-subunit of the coatomer complex binds Cdc42 to mediate transformation. *Nature* **405**(6788), 800–804.

Yarrow, J. C., Lechler, T., Li, R., and Mitchison, T. J. (2003). Rapid de-localization of actin leading edge components with BDM treatment. *BMC Cell Biol.* **4**(1), 5.

Chapter 7

The BRG1 and the Actin Filament System

Ann-Kristin Östlund Farrants, Erica Cavellán, and Patrik Asp

Department of Cell Biology, The Wenner-Gren Institute, Stockholm University, SE-106 91 Stockholm, Sweden

The DNA in eukaryotic cells is organized with proteins into chromatin. The chromatin structure is dynamic, altered when factors that use DNA as a template need access or to silence regions of the chromosome. The alteration of the chromatin structure is primarily the responsibility of ATP-dependent chromatin-remodeling complexes and histone-modifying complexes. ATP-dependent chromatin-remodeling factors of the SWI/SNF family affect the actin filament orgainzation in the cytoplasm. The SWI/SNF family of chromatin-remodeling complexes is required in transcription of many genes, and the effect on the actin filament organization depends on cell type context, in particular, the various signaling pathways activated in each cell type. In many cell types, SWI/SNF complexes affect the protein level of adhesion molecules and Rho pathway effector molecules. Actin, together with actin-related proteins (ARPs) are also present in the nucleus and are implicated in mRNA export, transcription, DNA repair, and chromatin remodeling. Actin associate to all three types of eukaryotic RNA polymerases (pol). In addition, actin and ARPs are part of several chromatin-remodeling complexes purified from yeast, *Drosophila*, and mammalian cells. The function of actin and ARPs in the nucleus is still unclear, and it is proposed that they work as motor proteins or as structural proteins in different complexes.

Advances in Molecular and Cell Biology, Vol. 37

ISBN: 0-444-52868-7
DOI: 10.1016/S1569-2558(06)37007-5

I. CHROMATIN REMODELING

The DNA in eukaryotic cells is organized into chromatin, a dynamic structure that must be altered in order for DNA-binding factors to gain or loose access to their sites on DNA (Berger, 2002; Tsukiyama, 2002; Lusser and Kadonaga, 2003; Längst and Becker, 2004; Sif, 2004). Alterations in the chromatin structure that occur over regions are the result of two different kinds of activity, histone modification and ATP-dependent chromatin remodeling. The histone proteins constitute the protein core around which DNA is wrapped, together forming the nucleosome. The histone proteins can be covalently modified by acetylation, methylation, phosphorylation, ubiquitinylation, and ribosylation, and these modifications both alter the nucleosome structure and constitute an epigenetic code, the histone code, in which each histone modification recruits proteins specific for certain processes (Strahl and Allis, 2000; Jenuwein and Allis, 2001; Berger, 2002). ATP-dependent chromatin remodeling requires ATP to change the histone–DNA contacts. The exact mechanism is not clear, but several models have been proposed in which ATP is used to introduce a twisting of the DNA or sliding of the histone core (Flaus and Owen-Hughes, 2004; Längst and Becker, 2004). Chromatin remodeling and histone modification activities are required for gene expression and repression, in replication, DNA repair and DNA recombination (Berger, 2002; Sif, 2004; Cairns, 2005; Gibbons, 2005), and often both types of activities are needed in these processes (Narlikar *et al.,* 2002).

Each cell has a number of different ATP-dependent chromatin remodeling complexes, which can be grouped into at least four different groups depending on the central ATPase: the SWI/SNF complexes, ISWI complexes, Mi-2 complexes, and INO80 complexes (Sif, 2004; Cairns, 2005). These complexes are evolutionarily conserved: the sequences of the proteins, the composition of the complexes, and the processes in which they take part. SWI/SNF complexes are mainly associated with transcriptional regulation in yeast, *Drosophila* and mammalian cells (Wang, 2003; Mohrmann and Verrijzer, 2005). The SWI/SNF complexes contain 9–15 subunits, and in yeast and mammalian cells, two homologous ATPases have been found (Wang, 2003; Mohrmann and Verrijzer, 2005). The two yeast homologues SWI2/SNF2 and Sth1 are components of two distinct complexes: the SWI/SNF complex that is required for transcription of a subset of genes (Cote *et al.,* 1994) and the RSC that is involved in transcription, chromatid segregation, and DNA repair (Cairns *et al.,* 1996). Two different SWI/SNF ATPases have been identified in mammalian cells, *BRG1* (*brahma related gene 1*) (Khavari *et al.,* 1993) and *hbrm* (*human brahma*) (Muchard and Yaniv, 1993). These proteins are not present in the same complex, but many of the other 9–12 subunits are shared (Kwon *et al.,* 1994; Wang *et al.,* 1996; Xue *et al.,* 2000; Lemon *et al.,* 2001; Sif *et al.,* 2001; Mohrmann *et al.,* 2004; Yan *et al.,* 2005). Several forms of SWI/SNF complex have been purified from mammalian cells, and BRG1 is present in at least two different complexes, BAF and PBAF (Xue *et al.,* 2000; Lemon *et al.,* 2001; Mohrmann *et al.,* 2004; Yan *et al.,* 2005). These complexes are separated by specific subunits, which are only present in one form of complex (Wang *et al.,* 1996; Xue *et al.,* 2000; Lemon *et al.,* 2001; Mohrmann *et al.,* 2004; Yan *et al.,* 2005). BAF and PBAF have been ascribed slightly different functions, BAF regulates genes in proliferation and differentiation while PBAF is involved in transcription mediated by nuclear receptors and in sister chromatid segregation (Xue *et al.*, 2000; Lemon *et al.*, 2001; Mohrmann *et al.*, 2004; Yan *et al.*, 2005). BRG1-containing complexes can thus be regarded as global transcription factors, affecting the transcription of genes involved in a wide variety of processes in cells.

II. EFFECT OF CHROMATIN REMODELING ON THE ACTIN FILAMENT SYSTEM

Changes in cell morphology are usually caused by a change in the actin filament organization (reviewed in Pawlak and Helfman, 2001; Jaffe and Hall, 2005). This phenomenon often occurs when cells are transformed into tumor cells, where changes in morphology are accompanied by dramatic alterations of the cytoplasmic actin filament structures (Pawlak and Helfman, 2001; Rao and Li, 2004). Transformation of cells is also often coupled to a changed expression of adhesion molecules, such as integrins and extracellular matrix proteins, and this leads to changes in the adhesion to different substrata (in tissue cultures) or stroma (for tissues) (reviewed in Hirohashi and Kanai, 2003). Many tumor cell lines and tumors have nonfunctional ATP-dependent chromatin-remodeling activities, mainly having mutations or deletion of components in the SWI/SNF complexes (reviewed in Neely and Workman, 2002; Gibbons, 2005). The most common SWI/SNF subunits that are mutated or missing in tumor cells are the ATPases, BRG1 and BRM (Wong *et al.*, 2000; Decristofaro *et al.*, 2001; Reisman *et al.*, 2002; Reisman *et al.*, 2003; Fukuoka *et al.*, 2004; Rosson *et al.*, 2005), the 47-kDa protein SNF5/INI1 (Versteege *et al.*, 1998; Biegel *et al.*, 1999; Grand *et al.*, 1999; Sévenet *et al.*, 1999a,b; Biegel *et al.*, 2002), and the BAF57 protein (Decristofaro *et al.*, 2001; Belandia *et al.*, 2002; Wang *et al.*, 2005). *BRG1/BRM* are lost or truncated in several tumors, and the loss is closely linked to a poor prognosis in lung cancers (Reisman *et al.*, 2003; Fukuoka *et al.*, 2004). The *SNF5/INI1* was first found to be mutated in malignant rhabdoid tumors and in rhabdoid predisposition syndrome (Versteege *et al.*, 1998; Sévenet *et al.*, 1999b), but mutations in *INI1/SNF5* have subsequently been found in other types of tumor such as in chorion plexus carcinomas (Roberts and Orkin, 2004). BAF57 is deregulated or missing in certain breast cancer cells and liver cancers (Belandia *et al.*, 2002; Wang *et al.*, 2005). Ectopic expression of the normal proteins in cells deficient for the BRG1, BRM, or SNF5/INI1 results in cell cycle arrest accompanied by morphological changes (Dunaief *et al.*, 1994; Stober *et al.*, 1996; Shanahan *et al.*, 1999; Stobeck *et al.*, 2000; Ae *et al.*, 2002; Versteege *et al.*, 2002; Reincke *et al.*, 2003), suggesting that SWI/SNF components act as tumor suppressors (Dunaief *et al.*, 1994; Stober *et al.*, 1996; Versteege *et al.*, 1998; Biegel *et al.*, 1999; Shanahan *et al.*, 1999; Stobeck *et al.*, 2000).

This is further supported by the fact that heterozygous mice, with one allele deleted of *BRG1* and *SNF5/INI1*, are more prone to develop tumors (Bultman *et al.*, 2000; Roberts *et al.*, 2000). In the case of SNF5/INI1, rhabdoid tumors progress in cells where the expression from the second SNF5/INI1 allele also is lost (Rousseau-Merckm et al., 1999; Roberts et al., 2000). It has not been possible to generate homozygous null mice of *SNF5/INI1* and *BRG1*, since blastocysts die at the preimplantation stage (Bultman et al., 2000; Klochendler-Yeivin et al., 2000; Roberts et al., 2000; Guidi et al., 2001). *BRM*-null mice, in contrast, are relatively healthy, but larger than wild-type mice (Reyes et al., 1998).

A. SWI/SNF Chromatin-Remodeling Complexes Are Involved in the Actin Filament Organization

The first two cell lines identified with deficiencies in BRG1 and BRM expression were the epithelial cell lines SW13, an adrenal adenocarcinoma cell line, and C33A, a cervix carcinoma cell line (Muchard and Yaniv, 1993; Dunaief et al., 1994; Stober et al., 1996). Introducing *BRG1* into SW13 cells by transfection induces cell cycle arrest and a change in morphology (Dunaief et al., 1994; Stober et al., 1996; Shanahan et al., 1999; Asp et al., 2002). The underlying molecular mechanism of the cell cycle arrest is not fully clear, but the activation of the retinoblastoma protein (Rb) pathway and the concomitant repression of E2F targets are involved in the cell cycle arrest (Dunaief et al., 1994; Stober et al., 1996; Stobeck et al., 2000). SW13 cells expressing BRG1 transiently have been used to identify BRG1-dependent genes that change their expression within 24 h of expression (Liu et al., 2001). BRG1 expression affects 80 genes, two of which are repressed, in a genome-wide microarray screen. Only a few of the genes affected have a direct known link to the actin filament organization, one such gene is transgelin. The actin filament is organized differently in different cell types (Presland et al., 2001; Frames, 2004; Fujita and Braga, 2005; Jaffe and Hall, 2005): fibroblasts have actin filaments at the cell edges and thick myosin-actin filament bundles (stress fibers) transversing the cell body, whereas epithelial cells in culture have cortical actin filament structures at the edge and very thin filaments transversing the cell body. The thickness of the actin filament bundles also varies in tissue culture cells of an epithelial origin, and HeLa cells have bundles transversing the cell body. SW13 cells resemble epithelial cells with almost no thick actin bundles in the cell body (Asp et al., 2002). The actin filament system in SW13 is rearranged on BRG1 expression, resulting in stress fiber-like actin filaments in the cell body 24 h after expression (Fig. 1) (Asp et al., 2002) and a subsequent formation of large, flat cells (Dunaief et al., 1994; Shanahan et al., 1999; Asp et al., 2002). However, the change in actin filament system is not caused by an upregulation of transgelin, since overexpression of the protein in SW13 cells does not result in the formation of actin filament bundles (Asp and Östlund Farrants, unpublished data). The organization of actin filament structures is regulated by several signaling pathways, and the small GTPases RhoA, Rac, and Cdc42 control directly the actin filament system and adhesion complexes (Pawlak and Helfman, 2001; Hirohashi and Kanai, 2003; Frames, 2004; Rao and Li, 2004; Birgersdotter et al., 2005; Fujita and Braga, 2005; Jaffe and Hall, 2005; Larue and Ballacosa, 2005). In addition, Ras and phosphoinositol signaling affects the actin

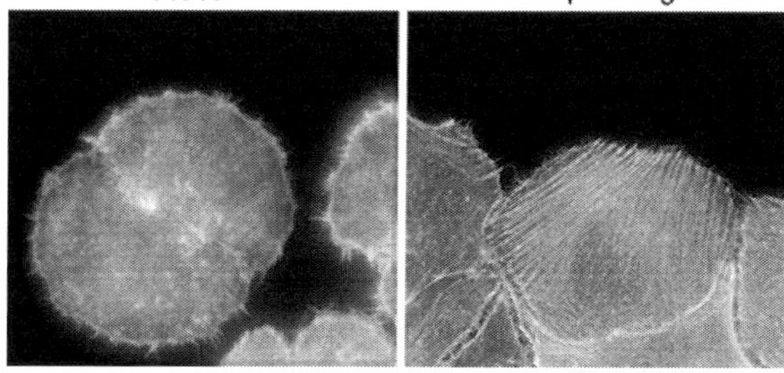

Figure 1. SW13 cells were grown on cover slips overnight (right panel) and compared to a clone expressing low levels of BRG1 (left panel) grown on cover slips for 48 h, subsequently fixed with 4% formaldehyde, and stained with phalloidin.

filament organization (Frames, 2004; Rao and Li, 2004; Jaffe and Hall, 2005; Larue and Ballacosa, 2005). Epithelial cells can be converted into mesenchyme cells, such as fibroblasts, by certain signaling pathways. This transition includes a rearrangement of the actin filament system, and occurs in development and cancer transformation (Larue and Ballacosa, 2005). Since many different pathways, such as the Notch pathway, Ras/Rho pathways, and signaling through adhesion molecules, are involved, the complete picture of the order of events is still not clear. The activation of the different GTPases is regulated by specific GTP-GDP exchange factors (GEFs) and GTPase-activating proteins (GAPs) (Frames, 2004; Jaffe and Hall, 2005; Larue and Ballacosa, 2005). BRG1 expression in SW13 cells does not cause a change in the RhoA activation, but it results in a changed balance between the downstream RhoA effectors ROCK (p160 Rho-associated kinase) and mDia (Watanabe *et al.*, 1999; Riento and Ridley, 2003; Larue and Ballacosa, 2005). The balance between these effectors is one factor that determines the thickness of the actin filaments formed in cells (Watanabe *et al.*, 1999; Larue and Ballacosa, 2005). A high level of ROCK gives thick myosin-actin stress fibers, whereas a high level of mDia gives thin filaments. It is possible that the low level of ROCK in SW13 cells decouples the RhoA activation from downstream effectors, a coupling that is restored by BRG1 expression. Defects in the Rho/Ras pathways, such as decoupling of the GTPases with downstream effectors, are observed in Ras-transformed cells and could be contributing to the change in actin filament organization (Riento and Ridley, 2003; Lee and Helfman, 2004). The increased level of the ROCK1 protein in *BRG1*-expressing SW13 cells is not caused by an increased ROCK1 mRNA level (Asp *et al.*, 2002), a fact that may explain why this gene was not found as a *BRG1*-dependent gene by Liu *et al.*, (2001). Instead, the ROCK1 level is increased indirectly, one possible mechanism being that *BRG1* expression affects the degradation rate of the ROCK proteins. This is supported by the fact that proteasome inhibitors induce stress fiber formation in SW13 cells (Asp and Östlund Farrants, unpublished data).

The response of the actin filament organization observed in SW13 cells to BRG1 expression is not valid for all cell types, however. MiaPaCa2 cells, pancreatic carcinoma cells that express BRG1 but lack BRM, gain actin filament bundles and become larger when the BRG1 protein is knocked down (Rosson *et al.*, 2005). This is in contrast to the effect of BRG1 on the actin filament organization in SW13 cells, but may be a consequence of cell type differences. Rosso *et al.* point out that whereas the SW13 cells have a defect in the Rb pathway (Stober *et al.*, 1996; Strobeck *et al.*, 2002; Hill *et al.*, 2004), MiaPaCa2 cells have defects both in the Rb and the p53 pathways, and the cells express an oncogenic K-ras (Aoki *et al.*, 1997; Furuwatati *et al.*, 1998). The Ras-signaling pathway affects the actin filament organization (Pawlak and Helfman, 2001; Birgersdotter *et al.*, 2005), where an activated Ras pathway influences the Rho, Rac, and Cdc42 responses. The differences in cellular signaling context between the SW13 and MiaPaCa2 cells may contribute to the different response of the actin filament organization to BRG1 expression. The importance of the cell type in response to BRG1 expression is shown in the study performed on another *BRG1*-deficient and *BRM*-deficient cell line, the breast tumor cell line ALAB (Hendricks *et al.*, 2004). Expression of BRG1 in these cells causes cell cycle arrest, and this is linked to an induction of the cdk inhibitor p21. These cells have a defect in the p53 pathway, and the induction of p21 may be a result of activation of the TFG-β pathway (Hendricks *et al.*, 2004). We have used ALAB cells to study the effect of BRG1 expression on the actin filament organization (Asp and Östlund Farrants, unpublished results). The actin filament organization in ALAB cells resembled that found in SW13 cells, with no stress fiber-like structures in the cytoplasm. BRG1 expression in these cells resulted in the formation of thicker filaments in the cell body, similar to the effect seen in SW13 cells. Hendricks *et al.* (2004) performed a microarray screen on BRG1-expressing ALAB cells and detected several novel target genes that had not been found in BRG1-expressing SW13 cells (Liu *et al.*, 2001). The genes were grouped into seven groups (cell cycle, transcriptional regulation, signal transduction and regulation, Ca^{2+} regulation and signaling, cell–cell or cell–matrix interactions, cytoskeleton, and cell architecture), and several actin regulatory proteins, such as tropomyosin and actinin, were upregulated in BRG1-expressing ALAB cells. Interestingly, several cell adhesion molecules, CD44, integrins, and moesin, were among the target genes found in this cell line. Of these adhesion molecules, CD44 is a BRG1-regulated gene in SW13 cells (Strobeck *et al.*, 2001). ALAB cells, as SW13 cells, do not express oncogenic Ras, as MiaPaCa2 cells do, which could explain the similar response on the actin filament organization to BRG1 expression in ALAB cells and SW13 cells.

Most research into the regulation of the actin filament organization in the cytoplasm has been conducted on fibroblasts. Fibroblasts display both cortical actin filaments and thick stress fibers transversing the cell body (Presland *et al.*, 2001; Fujita and Braga, 2005; Jaffe and Hall, 2005). All fibroblasts investigated express BRG1 and BRM proteins together with the other SWI/SNF subunits, and fibroblasts thus have functional SWI/SNF complexes. Expression of an ATP-deficient BRG1 in fibroblasts results in an increase in cell size without major changes in the actin filament organization (Hill *et al.*, 2004). Cells expressing the dominant-negative BRG1 still harbor thick stress fibers similar to those found in the wild-type cell, the difference displayed is that the number of focal adhesions increases. The protein level of the focal adhesion protein paxillin is not increased in these cells, nor is the actin level. Instead, these cells have higher levels of cell adhesion molecules, such as urokinase-type plasminogen activator receptor (uPAR), αV-integrins and α5-integrins. However, the higher amounts in cells are not caused by transcriptional upregulation of the genes, since the mRNA level for uPAR stays the same regardless whether the cells express the dominant negative BRG1 ATPase. The formation of stress fibers is not necessary for the change in cell size and shape. Reintroducing an ATP-deficient BRG1 into SW13 cells results in larger cells, without the formation of stress fiber-like structures (Asp *et al.*, 2002).

Mutations or deletions of the *SNF5/INI1* result in an altered actin filament organization. Since *SNF5/INI1* is defect in rhabdoid tumors, many cell lines deficient in the protein have been cloned. Introducing the SNF5/INI1 protein, restoring functional SWI/SNF complexes (Phelan *et al.*, 1999), into one of the rhabdoid tumor cell lines induces cell cycle arrest in G1 (Ae *et al.*, 2002; Versteege *et al.*, 2002; Reincke *et al.*, 2003). The arrest is ascribed to the activation of the Rb pathway, giving a downregulation of E2F targets, such as cyclin A, induction of ink4a and ink6 expression (Ae *et al.*, 2002; Versteege *et al.*, 2002). In contrast to BRG1 expression in *BRG1*-deficient cell line, SNF5/INI1 affects steps prior to the activation of the Rb protein, such as regulation of cyclin D (Ae *et al.*, 2002; Versteege *et al.*, 2002; Zhang *et al.*, 2002). Expression of SNF5/INI1 also leads to dramatic changes in the cell shape (Ae *et al.*, 2002; Betz *et al.*, 2002) and to actin filament rearrangement (Medjkane *et al.*, 2004). However, different effects have been reported. In many malignant rhabdoid tumor cell lines (MTRs), including MON cells, introduction of SNF5/INI1 results in large flat cells, indicating a similar response to SNF1/INI1 as that of SW13 to BRG1 expression (Ae *et al.*, 2002; Betz *et al.*, 2002; Zhang *et al.*, 2002). In a later study, Medjkane *et al.* (2004) examined the actin filament rearrangements in MON cells on SNF5/INI1 expression. MON cells are adherent cells that attach to the substratum, and, in this study, the cells became smaller and rounder on SNF5/INI1 expression. The actin filaments are rearranged and the stress fiber-like structures present in parental cells are lost 8 days after SNF5/INI1 induction. The global gene expression pattern of MON cells expressing SNF5/INI1 was examined using microarrays of 22,000 human genes. 482 genes were affected by the expression of SNF5/INI1: 332 activated and 150 repressed. These genes were grouped into five groups: cell cycle regulation, cytoskeletal regulation, cell matrix adhesion, signal transduction, and nuclear factors. The genes activated by SNF5/INI1 include *RhoE*, *Rho GTPase-activating protein 4*, *N-WASP*, *WAVE 3*, and *cortactin*. In SNF5/INI1-expressing MON cells, in contrast to the BRG1-induced SW13 cells, the RhoA GTPase is affected, and inactivation of the GTPase occurs, causing the stress fibers to disassemble. The activation of cytoskeleton regulator genes occurs rather late, indicating that these genes are not direct targets. In contrast, signal transduction molecules are induced early after SNF5/INI1 expression, and the signal molecules probably induce genes that control directly the actin filament organization. The discrepancies in response between cells and studies may, also in the case of SNF5/INI1, depend on cell context. The actin filament rearrangement seems to be an indirect response to the restoration of functional SWI/SNF complexes, and the determining factor must then be what specific direct targets are regulated in each cell. This, in turn, depends on cell type, cell origin, cancer development, and substrata.

Another SWI/SNF component, BAF57, is deleted in several breast tumor cell lines (Decristofaro *et al.*, 2001; Wang *et al.*, 2005), but no report of actin filament rearrangements has been presented on reintroducing *BAF57* into deficient cell lines.

Since the SWI/SNF complexes are evolutionarily conserved and are present not only in mammals but also in yeast, insects, and amphibians, it is not surprising that these complexes are involved in actin filament rearrangement in other species. Microarray studies performed on yeast strains in which the SWI/SNF ATPase SNF2/SWI2 has been deleted (*swi2*$^-$ strains) show that approximately 400 genes depend on a functional SWI/SNF complex (Holstege *et al.*, 1998; Sudarsanam *et al.*, 2000). Many of the genes affected in *swi2*$^-$ cells are actin-regulatory proteins, such as tropomyosin, but no alteration in the actin filament organization has been reported in *swi2*$^-$ cells. In contrast, mutations or deletions in the yeast SWI2/SNF2 homologue Sth1p give actin filament dislocations (Chai *et al.*, 2002). Sth1 is part of the 15-subunit ATP-dependent chromatin-remodeling complex RSC, which is essential for yeast cells (Cairns *et al.*, 1996). A number of functional differences exist between the SWI/SNF complexes, regulating different target genes, but more importantly, RSC is involved in sister chromatin segregation and DNA repair (reviewed in Wang, 2003). Sth1 mutations perturb the actin organization required for polarized growth during mitosis and for forming projections during mating (Chai *et al.*, 2002). Cells expressing a mutated sth1 at the permissive temperature do not form actin filament cables but display large actin lumps in the cytoplasm. These effects can be suppressed by stress sensors and by protein kinase C 1 (PKC1), indicating that sth1 functions through a PKC1-mediated signaling pathway and affects an upstream effector of the PKC.

ATP-dependent chromatin remodeling complexes of the SWI/SNF type exhibit global effects on many different cellular processes, such as actin filament reorganizations both in yeast and mammalian cells. In particular, the levels of adhesion molecules and signal transduction proteins, which affect the actin filament organization, are changed in cells with nonfunctional complexes. However, the effect of ATP-dependent chromatin remodeling complexes on the actin filament organization in mammalian cells depends on cell type, and the result of altering the expression of the same component in different cell lines can give opposite results. The regulation of the actin filament organization is complex and no SW1/SNF target gene has been identified as the direct cause for actin rearrangements. The fact that the SWI/SNF complexes are global transcription factors, activating or repressing many genes, makes it harder to link one particular gene product to the actin filament reorganization. It is more likely that many different genes, including signal transduction pathways, contribute to the end result, and that the characteristics of each cell line are the determining factors. Nevertheless, all cell lines investigated have responded by displaying altered levels of cell adhesion molecules and proteins involved in Rho signaling. It is likely that these proteins in turn influence the actin filament organization in cells, although in different ways that depend on what other pathways are activated in particular cell line. The alterations caused by the SWI/SNF complexes may be of great importance in development since they resemble those occurring in epithelial–mesenchymal transitions, a process which is vital during embryonic morphogenesis (Larue and Ballacosa, 2005).

III. ACTIN IN THE NUCLEUS

A. Actin in Nuclear Processes

Interest in actin in the nucleus has risen in recent years and it has become clear that actin is involved in several nuclear processes (reviewed in Olave *et al.*, 2002; Pedersen and Aebi, 2002; Blessing *et al.*, 2004; Delanerol *et al.*, 2005; Gettemans *et al.*, 2005; Pedersen and Aebi, 2005; Visa, 2005). However, it is not clear whether phalloidin-reactive actin filaments (F-actin) can be detected in the nucleus of eukaryotic cells under normal conditions. Filaments, or "nuclear rods or bundles," are present in the nuclei of stressed yeast cells and DMSO or heat-treated mammalian cells, as well as in cells overexpressing the actin-binding protein cofilin (Gettemans *et al.*, 2005). These structures may have a different conformation than that of F-actin in the cytoplasm. Indication that actin filaments assemble in a different structure than the phalloidin-reactive form is the "rod" structures found in *Dictyostelium discoideum* (Sameshima *et al.*, 2001) and that monoclonal antibodies react differently to nuclear and cytoplasmic actin filaments (Gonsior *et al.*, 1999). Thin filament structures composed of actin and protein-4.1 were found associated with the nuclear pore complex in *Xenopus* oocytes, stretching into the nucleus (Kiseleva *et al.*, 2004). Nuclear actin filaments also appear in *Xenopus* oocytes during meiosis after nuclear envelope breakdown (Lénárt *et al.*, 2005). It is unclear, however, whether these filaments are built from thin filaments that are already present in the nucleus or if actin molecules from the cytoplasm move into the "nucleus" after nuclear envelope breakdown. F-actin has also been reported to be present in mammalian nuclei: F-actin surrounds the nucleolus in a transcription-dependent manner (Zhang *et al.*, 2004) and nuclear preparations treated with the F-actin (or actin filament)-depolymerizing drug latrunculin release chromatin-modifying complexes (Andrin and Hendzel, 2004).

The presence of actin in the nucleus has been observed by immunoflourescence of fixed cells by several antibodies. Figure 2 shows a staining of HeLa cells with the monoclonal β-actin antibody after staining with 4% formaldehyde. Indications of a role for actin in the nucleus were reported as early as the 1980s, when a link between actin and transcription was demonstrated (Egly *et al.*, 1984; Sheer *et al.*, 1984). The abundance of actin in the cytoplasm, however, raised concerns of contamination from the cytoplasm. Nevertheless, studies have reported that actin is involved in several nuclear processes and that it is part of nuclear complexes. Actin has been implicated in nuclear export of mRNAs (Kimura *et al.*, 2000; Hofmann *et al.*, 2001; Percipalle *et al.*, 2001), and several lines of evidence for a role of actin in transcription have emerged. Actin is present at transcriptional sites in *Chironomus tentans* (Percipalle *et al.*, 2003; Kukalev *et al.*, 2005), and it associates with RNP particles, such as hRNP36 (Percipalle *et al.*, 2001) and hRNP65 in *C. tentans* (Percipalle *et al.*, 2003), and hnRNP U and hnRNP A/B in human cells (Percipalle *et al.*, 2002; Kukalev *et al.*, 2005). Sjölinder *et al.* (2005) have shown that actin binds to the growing mRNA during RNA pol II transcription and helps recruit a histone acetyl transferase (HAT), the p2D10 protein, that is required for transcriptional elongation in *C. tentans*.

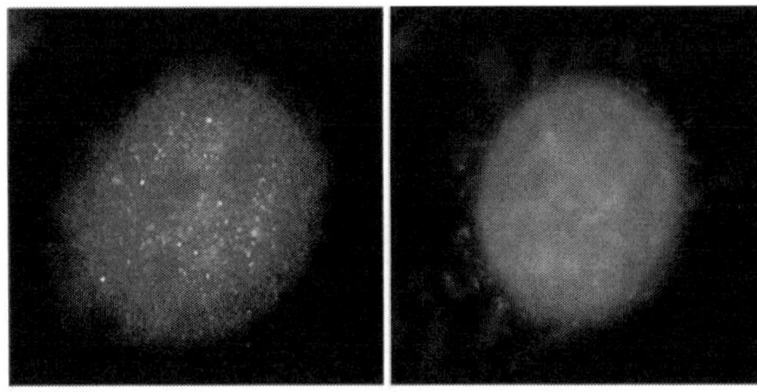

Figure 2. Confocal images of HeLa cell nucleus, stained with a β-actin monoclonal antibody (Sigma ac-15) (left panel), after having reduced a fuzzy cytoplasmic staining. The nucleoli are marked as lighter areas in the dotted actin pattern in the nucleus. The nucleus stained with DAPI.

Evidence for a role of actin in transcription is now emerging. Actin colocalizes, at least partially, with actin transcription sites in immuflorescence images of actin and BrU incorporation (Fig. 3A). Actin associates with all three types of eukaryotic RNA polymerases: RNA pol I (Fomproix and Percipalle, 2004; Philimonenko *et al.*, 2004), pol II (Hofmann *et al.*, 2004; Philimonenko *et al.*, 2004), and pol III (Hu *et al.*, 2004; Philimonenko *et al.*, 2004), and it is necessary for the activity or enhances the activity of the RNA polymerases *in vitro*. These studies show that actin has a role in transcriptional initiation and elongation, a function independent of chromatin (Hofmann *et al.*, 2004; Hu *et al.*, 2004; Philimonenko *et al.*, 2004). The three eukaryotic RNA polymerases are specific for certain types of genes; pol I transcribes the nucleolar rRNA gene, pol II transcribes mainly protein coding genes, and pol III transcribes genes for small RNAs, such as 5S rRNA and tRNAs. The RNA polymerases have distinct structures, but they share some subunits: RPABC2 and RPABC3 (Hu *et al.*, 2004). It is tempting to suggest that actin binds to the RNA polymerases through one of these shared subunits, and in this way be recruited to transcriptional sites. Actin also binds to the phosphorylated CTD of RNA pol II (Kukalev *et al.*, 2005). What structure actin takes in transcription is not known: as monomers or as actin filaments. There is no experimental evidence that actin in a filamentous form is associated with the RNA polymerases, but it cannot be excluded that very short oligomers of actin or atypic actin filament structures bind to the RNA polymerases. Actin-binding proteins, such as cofilin, profilin, and gelsolin are also present in the nucleus (Gettemans *et al.*, 2005), but it is still unknown whether these protein bind to actin as part of a complex with RNA polymerases. Myosin, which is a component of stress fibers in the cytoplasm, is also present in the nucleus in a specific isoform, nuclear myosin 1 (NM1) (Pestic-Dragovich *et al.*, 2000). NM1 is implicated in pol I (Fomproix and Percipalle, 2004; Philimonenko *et al.*, 2004; Percipalle *et al.*, 2006), pol II (Pestic-Dragovich *et al.*, 2000), and pol III transcription (Cavellán *et al.*, 2006). NM1 is found at the pol I gene by chromatin immunoprecipitation (ChIP) (Philimonenko *et al.*, 2004; Percipalle

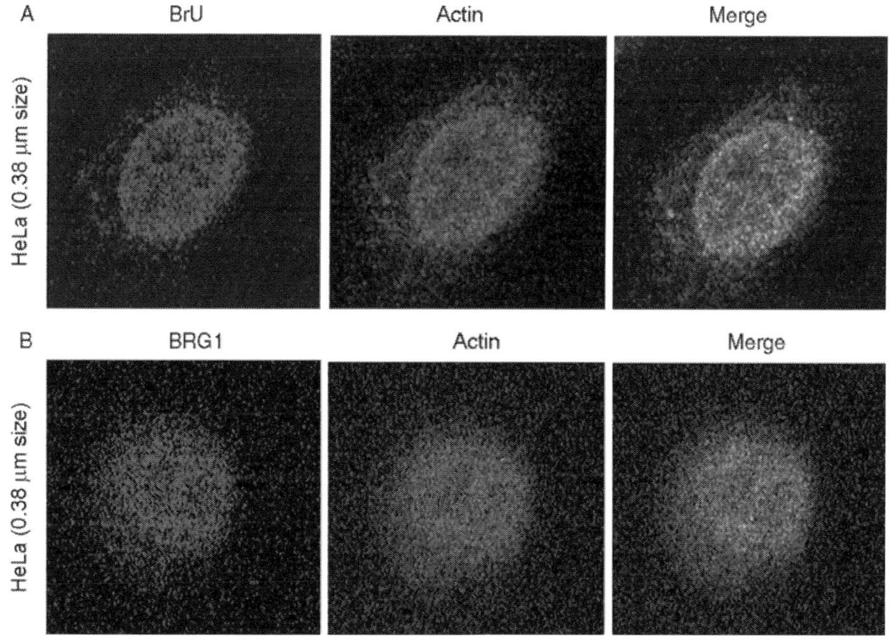

Figure 3. (A) HeLa cells grown on cover slips were treated with 10 μM BrU for 10 minutes after osmotic shock, fixed, and antibodies against BrU and actin were used to visualize transcription sites and the location of actin in the nucleus. In the confocal images, BrU is red and actin green, and colocalization is shown by yellow dots in the panel labeled merge. (B) HeLa cells grown on cover slips, fixed, and stained with antibodies against BRG1 (red) and actin (red), colocalization is shown by yellow spots in the panel labeled merge, which is a merge of the two confocal images to the left. (See Color Insert.)

et al., 2006), and interacts with actin in pol I transcription, although in a weak and transient manner (Fomproix and Percipalle, 2004; Philimonenko *et al.*, 2004; Percipalle *et al.*, 2006). NM1 also interacts with an ISWI-containing ATP-dependent chromatin remodeling complex, the WSTF–ISWI (Cavellán *et al.*, 2006; Percipalle *et al.*, 2006), indicating that NM1 is involved in recruiting chromatin remodeling complexes to the pol I gene (Percipalle and Östlund Farrant, 2006). NM1 together with the WSTF–ISWI is also found at certain pol III genes, 5S rRNA and 7SL RNA genes, but not on snU6 gene (Cavellán *et al.*, 2006). The transient interaction between NM1 and actin may be a prerequisite for the function of these proteins in transcription. Rather than being involved in transcription as motor proteins in actomyosin filaments, it has been suggested that actin and NM1 act as molecular switches, interacting when transcription is activated (Fomproix and Percipalle, 2004; Visa, 2005). Actin changes conformation on ATP binding and hydrolysis (Muller *et al.*, 2005; Pedersen and Aebi, 2005) and this feature could allow actin to bind to different proteins in the different states. Actin could, therefore, act as a molecular switch in the transition between transcriptional initiation and elongation.

Several ARPs are located to the nucleus, and these are often associated with actin or other ARPs (Pedersen and Aebi, 2005). The ARPs share a common ancestor with actin, having the same structure, and an ATP-binding pocket but with an ATPase activity weaker than actin *in vitro*, if any (Muller *et al.*, 2005; Pedersen and Aebi, 2005). The ARPs evolutionarily furthest away from actin have lost the ATPase activity, but it is still possible that the ATP binding is retained (Muller *et al.*, 2005). In one of the nuclear ARPs, Arp4, the binding of ATP to the pocket changes the conformation and the binding properties of Arp4 to other proteins (Sundana *et al.*, 2005). ARPs are not associated with the RNA polymerases, but are found together with actin in many other complexes in the nucleus. Actin and ARPs are components of chromatin-modifying complexes and chromatin-remodeling complexes (Blessing *et al.*, 2004; Pedersen and Aebi, 2005).

B. Actin and ARPs in Chromatin-Remodeling Complexes

Actin is a component of chromatin-remodeling complexes from several species (Table I). Two yeast complexes have been purified with actin and ARPs, INO80.com and SWR1.com present. INO80.com has the SWI2/SNF2-related ATPase INO80 and it contains actin and three ARPs: Arp4, Arp5, and Arp8 (Shen *et al.*, 2000). INO80.com is involved in transcription and DNA repair, where it requires actin and ARPs for

Table I
Actin and ARPs in Chromatin-Remodeling Complexes

Yeast		*Drosophila*		Mammalian cells	
SWI/SNF complexes					
SWI/SNF	RSC	BAP	PBAP	BAF	PBAF
Arp7, Arp9	Arp7, Arp9	Actin, BAP55	Actin, BAP55	Actin(?), BAF53	Actin(?), BAF53
INO80 complexes					
INO80.com	SWR1.com	–	–	INO80.com	SRCAP
Actin, Arp4, Arp5, Arp8	Actin, Arp4, Arp6			BAF53, Arp5, Arp8	BAF53, Arp6
HAT complexes					
NuA4		TIP60		TIP60/ TRRAP	
Actin, Arp4		Actin, BAP55		Actin, BAF53	

BAF53/BAP55/Arp4 are orthologues.

proper activity. The deletion of Arp5 or Arp8 compromises the ATPase activity, chromatin remodeling, DNA binding, and the conversion by INO80.com of dDNA in double strand breaks (DSB) into ssDNA (Shen *et al.*, 2003). It has been shown that this complex is recruited to the site of DSB by the phophorylation of histone H2A, possibly through Arp4 (Morrison *et al.*, 2004; van Attikum *et al.*, 2004). Both Arp4 and Arp8 bind histones (Harata *et al.*, 1999; Galarneau *et al.*, 2000; Shen *et al.*, 2003), Arp8 with a specificity that preferentially binds histone H3 and H4, and Arp4 for H3, H2A, and H2B (Harata *et al.*, 1999; Shen *et al.*, 2003). However, another protein in INO80.com is also responsible for the interaction with histones, Nhp1 (Morrison *et al.*, 2004), and it is possible that Nhp1 and the ARPs together are required for the binding to chromatin. The ARPs also have structural functions in the INO80.com. Arp8 is important for the integrity of the INO80 complex, since the presence of Arp4 and actin are heavily reduced in purified INO80.com from Arp8Δ yeast cells (Jónsson *et al.*, 2004). Furthermore, the association of Arp5 with the complex depends on the presence of the Ruv-B proteins, INO80p, and ATP (Jónsson *et al.*, 2004). SWR1.com contains the ATPase SWR1 and loads H2AX into chromatin, thereby preventing heterochromatin regions from spreading along chromosomes (Mizuguchi *et al.*, 2004). It is also involved in the DNA repair mechanism, most probably by exchanging histones close to the damaged sites (Kobor *et al.*, 2004; Mizuguchi *et al.*, 2004; Zhang *et al.*, 2004). SWR1.com contains actin and two ARPs: Arp4 and Arp6, and shares the two Ruv-B proteins with INO80.com (Mizuguchi *et al.*, 2004). A further yeast complex that shares actin, ARPs, and Ruv-B proteins with SWR1.com and INO80.com is NuA4, a histone acetyl transferase. NuA4 contains 13 subunits, including actin and Arp4 (Allard *et al.*, 1999; Downs *et al.*, 2004), but it can disassemble to the small, trimeric piccolloNuA4 complex (Boudreault *et al.*, 2003). ATP bound to Arp4 promotes the disassembly of NuA4 into piccolloNuA4 (Sundana *et al.*, 2005). This further supports the idea that actin and ARPs, with their ability to change conformation on ATP binding, provide a dynamic structural function in complexes. PiccolloNuA4 has a slightly different HAT activity from that of NuA4 and is involved in global acetylation of nucleosomes (Boudreault *et al.*, 2003). NuA4 is involved in more restricted transcription and DNA repair and is recruited to phosphorylated H2A sites by Arp4 (Downs *et al.*, 2004). All three complexes, INO80.com, SWR1.com, and NuA4, are believed to act in concert at DSBs, possibly by actin and ARPs recruiting the complexes to histones at the site of DNA damage, and, since actin and ARPs are shared, tether the three different complexes to these sites (Morrison and Shen, 2005; van Attikum and Gasser, 2005).

The yeast SWI/SNF complex and the related RSC complex contain two Arps: the yeast specific Arp 7 and Arp 9 (Cairns *et al.*, 1998; Peterson *et al.*, 1998), both of which are essential for the chromatin remodeling activity. The ATPase activity of these ARPs, on the other hand, is not required for activity, suggesting that the Arps provide a structural role (Cairns *et al.*, 1998). Arp7 and Arp9 form heterodimers in the SWI/SNF and RSC complexes, and it has been suggested that this dimer forms a connection between complexes, and between complexes and chromatin (Szerlong *et al.*, 2003). Since the SWI/SNF complex and the RSC complex perform different roles in the yeast cell, the ARPs do not provide the only docking proteins to chromatin. Other subunits in the complexes harbor protein-interacting domains and domains that interact with different histone modifications (Wang, 2003; Mohrmann and Verrijzer, 2005). The RSC plays roles in diverse processes, most likely recruited to the sites of action by different subunits. The complex is essential in such diverse processes as transcription, DNA repair (Peterson and Côté, 2004), and chromosome segregation (Hsu *et al.*, 2003; Baetz *et al.*, 2004; Huang *et al.*, 2004), in which it is important for the association of cohesin. In DNA repair process of DSB, the RSC complex is recruited at an early step and close to the site of the DSB (Shim *et al.*, 2005), and it is suggested to act before the INO80.com, SWR1.com, and NuA4 (van Attikum and Gasser, 2005).

A human INO80.com (Jin *et al.*, 2005) complex and a human SWR1.com (Cai *et al.*, 2005) complex, SRCAP, have been isolated. They contain approximately 12–15 subunits, most of which are conserved between yeast and mammals. The two complexes contain orthologues to the yeast ARPs: the mammalian INO80.com contains BAF53 (Arp4), Arp5, and Arp8; and SRCAP contains BAF53 and Arp6. Actin, however, is not mentioned as a component of these mammalian complexes (Cai *et al.*, 2005; Jin *et al.*, 2005). BAF53 is also part of a third complex, with which INO.80.com and SRCAP sometimes are copurified, the human orthologue to the yeast histone 4-HAT NuA4, TIP60/TRRAP (McMahon *et al.*, 1998; Ikura *et al.*, 2000; Cai *et al.*, 2003; Doyon *et al.*, 2004). The Ruv-B proteins TIP48 and TIP49 (pontin and reptin, respectively) are also shared between INO80.com, SRCAP, and TIP60/TRRAP. In addition to the shared subunits, TIP60/TRRAP has unique components and β-actin is present in some of the preparation of the human TIP60/TRRAP complex (McMahon *et al.*, 1998; Cai *et al.*, 2003). The yeast *SWR1* gene has two close orthologues in mammalian cells: the *SRCAP* gene and the *p400* gene display a high degree of homology to the *SWR1* gene (Doyon and Côté, 2004; Eissenberg *et al.*, 2005). p400 is an SWI2/SNF2 ATPase/helicase and one of the unique TIP60/TRRAP components. p400/DOM is the only SWR1 orthologue in *Drosophila* and it is part of a large TIP60 complex, which also contains both actin and Arp4 (Kusch *et al.*, 2004). In mammalian cells TIP60/TRRAP has a wide variety of functions and is involved in oncogenic transformation (McMahon *et al.*, 1998; Deleu *et al.*, 2001; Fuchs *et al.*, 2001), transcription (McMahon *et al.*, 2000; Frank *et al.*, 2001), and DNA repair (Murr *et al.*, 2005). Mammalian INO80.com and SRCAP have been linked only to DNA repair (Cai *et al.*, 2005; Jin *et al.*, 2005). The large TIP60 isolated from *Drosophila* combines the TIP60/NuA4 HAT activity with the histone exchange activity (exchanging phospho-H2Av for unphosphorylated H2Av) of the yeast SWR1.com (Kusch *et al.*, 2004), a combination of activities that has also been suggested for the human TIP60/TRRAP (Doyon and Côté, 2004). In contrast to the yeast and the *Drosophila* complexes, purified human SRCAP can only bind to the phosphorylated histone H2AX, but not exchange it for H2A at sites of DSB (Cai *et al.*, 2005). Further studies must be conducted to elucidate this difference in activity between complexes. The functions of actin and ARPs in human INO80.com, SRCAP, and TIP60/TRRAP complexes are still unclear, but it has been suggested that they recruit other subunits to the complexes by providing binding surfaces (Visa, 2005).

The mammalian SWI/SNF complexes, both BAF complexes and the PBAF (Mohrmann and Verrijzer, 2005), have been purified with actin and the mammalian

Arp4 orthologue BAF53 (Zhao *et al.*, 1998; O'Neill *et al.*, 1999; Nie *et al.*, 2000; O'Neill *et al.*, 2000; Xue *et al.*, 2000; Lemon *et al.*, 2001; Olave *et al.*, 2002; Nie *et al.*, 2003; Xu *et al.*, 2004; Yan *et al.*, 2005). The *Drosophila* SWI/SNF complexes, BAP and PBAP, also contain actin and the *Drosophila* Arp4 orthologue BAP55 (Mohrmann *et al.*, 2004). The function of actin in the mammalian BRG1 complexes is to enhance the ATPase activity of the SWI/SNF complex and attach BRG1-containing complexes to the nuclear matrix (Zhao *et al.*, 1998). In addition, BRG1 binds actin filaments and it has been proposed that SWI/SNF complexes are in this way recruited to an actin filament structure in nuclear matrix or at transcription sites (Rando *et al.*, 2002). However, actin is not part of all SWI/SNFs complexes isolated. Actin was not present in the first purifications of SWI/SNF complexes, while BAF53 was present in all preparations (Kwon *et al.*, 1994; Wang *et al.*, 1996a,b; Wang *et al.*, 1998). When actin had been reported to be present in the BRG1-containing BAF complex (Zhao *et al.*, 1998), others still could not find actin in purified complexes. BRG1-containing complexes and BRM-containing complexes associated with Sin3 or Sin3-PRMT5 (Sif *et al.*, 2001; Pal *et al.*, 2003), in the BAF-complex (Sif *et al.*, 1998), associated with EKLF in E-RC1 (Armstrong *et al.*, 1998), associated with IKAROS (Kim *et al.*, 1999), with N-Cor (Underhill *et al.*, 2000), or in WINAC (Kitagawa *et al.*, 2003) does not contain actin. BAF53, on the other hand, is present in nearly all reported SWI/SNF complexes (see Table II). The differences in the complex compositions suggest that actin is not a *bona fide* component of mammalian SWI/SNF. We have addressed this by first carefully separating the cytoplasm from nuclei of HeLa cells and then extract the nuclei at different KCl concentration (Fig. 4A). Tubulin and vinculin were used as cytoplasmic markers to assess the purity of the nuclear fraction. Actin and BRG1 are clearly present in the nuclear fraction, and both proteins require salt to be released from the chromatin fraction, requiring similar KCl concentrations, approximately 0.3 M KCl. To further examine whether they are released in a complex, we performed size exclusion fractionations of nuclear extracts, which were prepared in the presence and absence of ATP. Actin and BRG1 did not coelute with BRG1: actin was found in low Mr fractions without ATP, and in the presence of ATP, it eluted at a higher Mr than BRG1 and at a low Mr (Fig. 4B). We have also isolated BRG1-containing SWI/SNF complexes devoid of actin, both by biochemical chromatography and by immunoprecipitations (Fig. 4C). However, ChIP analysis of BRG1-target genes showed that actin, BRG1, and other SWI/SNF components were present at the promoters in HeLa cells. This suggests that SWI/SNF and actin operate in the vicinity of transcriptional initiation sites, and actin may, therefore, copurify with SWI/SNF complexes under certain preparatory conditions. It is likely that SWI/SNF and actin associate transiently, possibly by interactions with the BAF53 subunit. BRG1 and actin also partially colocalize in immunofluoresence of fixed HeLa cells (Fig. 3B). In a similar manner, SWI/SNF complexes locate in the proximity to RNA pol II at genes and RNA pol II has also been purified with SWI/SNF components from human cells (Wilson *et al.*, 1996; Neish *et al.*, 1998). Since actin associates with RNA polymerases, it may provide a bridge between SWI/SNF complexes and RNA polymerases, and since they act so in close proximity they may copurify.

Table II
Subunit Composition of Mammalian and *Drosophila* SWI/SNF Complex Preparation

Kwon *et al.*, 1994	Wang *et al.*, 1996 (Cloning of BAF47/INI1)				Wang *et al.*, 1996 (Cloning of BAF55, BAF70 & BAF60)	Wang *et al.*, 1998 (Cloning of BAF57)		Zhao 1998 (Cloning of BAF53. β-Actin as a subunit)	Sif			
SWI/SNF-A SWI/SNF-B	SWI/SNF-A		SWI/SNF-B		SWI/SNF-A	IP BAF57	IP BRG1	SWI/SNF				
	BAF250	BAF250			BAF250	BAF250	BAF250	BAF250	BAF250	BAF250		
			BAF200	BAF200								
BRG1	BRG1	BRM	BRG1	BRM	BRG1/BRM	BRG1	BRG1	BRG1	BRG1/BRM	BRG1	BRG1	BRG1
			BAF180	BAF180								
BAF170	BAF170	BAF170	BAF170	BAF170	BAF170	BAF170	BAF170	BAF170	BAF170	BAF170		BAF170
BAF155	BAF155	BAF155	BAF155	BAF155	BAF155	BAF155	BAF155	BAF155	BAF155	BAF155	BAF155	BAF155
	BAF110		BAF110		BAF110							
BAF60	BAF60		BAF60	BAF60	BAF60a,b,c	BAF60	BAF60	BAF60	BAF60	BAF60	BAF60	BAF60a
BAF50	BAF57	BAF57	BAF57	BAF57	BAF57	BAF57	BAF57	BAF57	BAF57	BAF57		BAF57
	BAF53	BAF53	BAF53	BAF53	BAF53	BAF53	BAF53	BAF53	BAF53	BAF53		BAF53 like
								Actin				Actin
	BAF47	BAF47	BAF47	BAF47	BAF47	BAF47	BAF47	BA47	BAF47	BAF47		BAF47
BAF45												
BAF43												

Underhill *et al.*, 2000 (SWI/SNF with N-Cor, HDACs and splicing)		Xue *et al.*, 2000 (BAF/PBAF division)					Nie *et al.*, 2000 (Characterization of BAF250)	Sif						
N-CoR 1	N-CoR 2	BAF	PBAF	BAF	PBAF	PBAF	BAF	BRG1 Complex 1	BRG1 Complex 2	BRM	FLAG			
SRCAP														
N-Cor	N-Cor													
		BAF250	BAF240	BAF250		BAF240	BAF250	BAF250	BAF250	BAF250	BAF250	BAF250	BAF250	
			BAF200	Unknown	BAF200	BAF200		BAF220			BAF220		BAF220	
BRG1	BRG1	BRG1	BRG1	BRG1	BRG1	BRG1	BRG1	BRG1	BRG1	BRM	BRG1/BRm	BRG1/BRm	BRG1/BRM	BRG1
			BAF180ab		BAF180a	BAF180a								
BAF170	BAF170	BAF170	BAF170	BAF170	BAF170	BAF170	BAF170	BAF170	BAF170	BAF170	BAF170	BAF170	BAF170	BAF170
BAF155	BAF155	BAF155	BAF155	BAF155	BAF155	BAF155	BAF155	BAF155	BAF155	BAF155	BAF155	BAF155	BAF155	BAF155
SAP130														
SF3a120														
KAP-1														
		BAF110	BAF110											
								HDAC2	HDAC2	HDAC12	HDAC1/2	HDAC1/2		
		BAF60a	BAF60a	BAF60a	BAF60a	BAF60a	BAF60	BAF60					BAF60a	BAF60ab
HDAC 3	HDAC 1–3													
	Sin3A							Sin3A		Sin3A	Sin3A	Sin3A		
		BAF57	BAF57	BAF57	BAF57	BAF57	BAF57	BAF57		BAF57	BAF57	BAF57	BAF57	BAF57
		BAF53	BAF53	BAF53	BAF53	BAF53	BAF53	BAF53	BAF53	BAF53	BAF53	BAF53	BAF53	BAF53
BAF47		Actin	Actin	Actin	Actin	Actin								
	–	BAF47	BAF47	BAF47	BAF47	BAF47	BAF47	BAF47	BAF47	BAF47	BAF47	BAF47	BAF47	BAF47
SAP30	SAP30							RbAp48	RbAp48	RbAp48	RbAp48	RbAp4		

Table II (*continued*)

Olave *et al.*, 2002 (Neuro.spec Baf53b)	Pal *et al.*, 2003 (SWI/SNF with methylation)	Nie *et al.*, 2003 (SWI/SNF and leukemia)	Kitagawa *et al.*, 2003 (SWI/SNF with replication)	Xu *et al.*, 2004 (SWI/SNF with methylation)			
bBAF	BRG1/ PRMT1	EBAF	WINAC	NUMAC	BAP	PBAP	PBAF
BAF250	BAF250	BAF250a/b	BAF250	BAF250		OSA	
BAF200							
BRG1/BRM	BRG1/BRM	BRG1	BRG1/BRM	BRG1	BRM	BRM	BRG1
p180			TOPOIIβ				
BAF170	BAF170	BAF170	BAF170	BAF170	BAP170		BAF170
BAF160	BAF155	BAF155	BAF155	BAF155	MOR	MOR	BAF155
(BAF155)		EBAF140	FACT	P105	BAP111	BAP111	
		EBAF100	CAF-1p				
		ENL					
	PRMT5	EBAF70					
	HDAC2		VDR	CARM1			
BAF60	BAF60a/b	BAF60a/b	BAF60a	BAF60	BAP60	BAP60	BAF60a
	Sin3A						
BAF57	BAF57	BAF57	BAF57	BAF57	BAP55	BAP55	BAF57
BAF53b	BAF53	BAF53a	BAF53				
Actin		Actin		β-Actin	Actin	Actin	Actin
BAF47	BAF47	BAF47	BAF47	BAF47	SNR1	SNR1	BAF47/SNF5

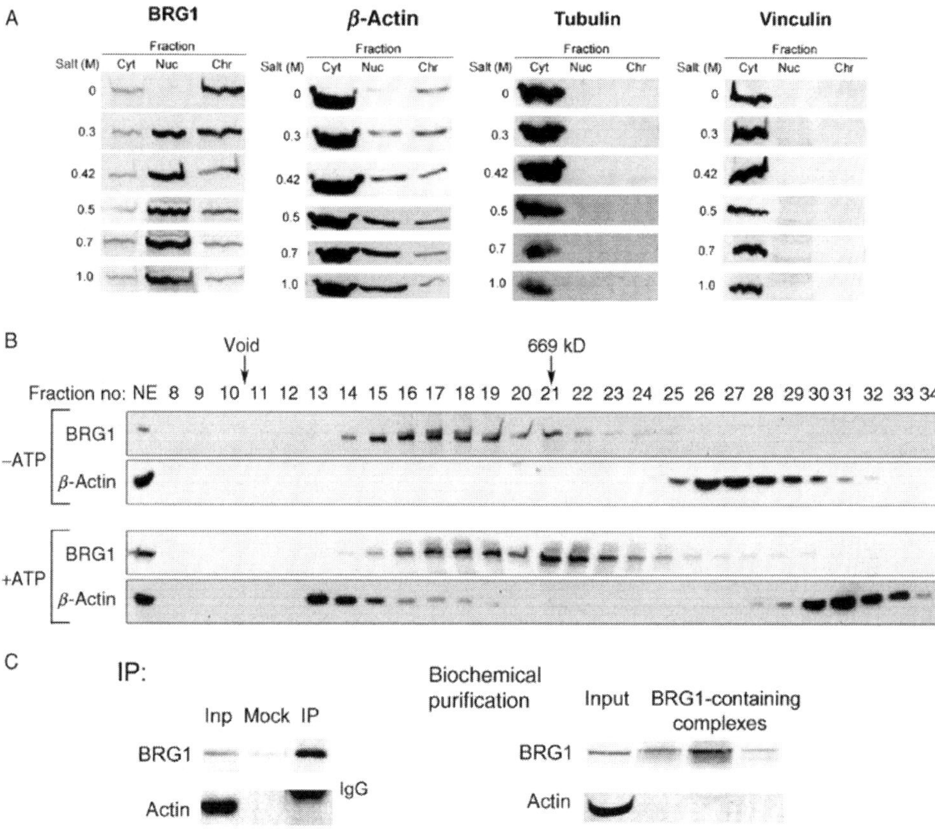

Figure 4. (A) Cellular fractionation of HeLa cells where nuclei and cytoplasm were first separated and the nuclear fraction subsequently extracted with different KCl concentrations as depicted to the left of each panel. Immunoblots show the level of each protein present in each fraction. (B) 0.5 mg nuclear extract prepared at 0.42 M KCl, in the absence (top) and presence (bottom) of ATP, was fractionated on a Superose 6 HR column. Immunoblots of the distribution of the BRG1 protein and actin, as marked to the left, is shown. (C) Immunoprecipitation (left panel) of the BRG1-containing complexes was performed from nuclear extract prepared at 0.42 M KCl, the precipitated protein separated by SDS-PAGE and transferred to an Immobilon membrane. The membranes were probed with antibodies against BRG1 and actin. Biochemical purification (left panel) of BRG1-containing SWI/SNF complexes was performed according to Lemon *et al.* (2001), and immunoblots of fractions from the final Superose 6 HR step is shown. Crude nuclear extract is input, followed by two complexes corresponding to BAF complexes, eluting at slightly different KCl concentrations from MonoQ column, and one complex corresponding to the PBAF complex. The immunoblots were probed with antibodies against BRG1 and actin, as indicated to the left.

REFERENCES

Ae, K., Kobayashi, N., Sakuma, R., Ogata, T., Kuroda, H., Kawaguchi, N., Shinomiya, K., and Kitamura, Y. (2002). Chromatin remodeling factor encoded by ini1 induces G1 arrest and apoptosis in ini1-deficient cells. *Oncogene* **21,** 3112–3120.

Allard, S., Utley, R. T., Savard, J., Clarke, A., Grant, P., Brandl, C. J., Pillus, L., Workman, J. L., and Côté, J. (1999). NuA4, and essential transcription adaptor/histone H4 acetyltransferase complex containing Esa1p and the ATM-related cofactor Tra1p. *EMBO J.* **18,** 5108–5119.

Andrin, C. and Hendzel, M. J. (2004). F-actin-dependent insolubility of chromatin-modifying components. *J. Biol. Chem.* **279,** 25017–25023.

Aoki, K., Yoshida, T., Matsomoto, N., Ide, H., Sugimura, T., and Terada, M. (1997). Suppression of Ki-ras p21 levels leading to growth inhibition of pancreatic cancer cell lines with Ki-ras mutation but not those without Ki-ras mutation. *Mol. Carcinog.* **20,** 251–258.

Armstrong, J. A., Bieker, J. J., and Emerson, B. M. (1998). A SWI/SNF-related chromatin remodeling complex, E-RC1, is required for tissue-specific transcriptional regulation by EKLF *in vitro*. *Cell* **95,** 93–104.

Asp, P., Wihlborg, M., Karlén, M., and Östlund Farrants, A.-K. (2002). Expression of BRG1, a human SWI/SNF component, affects the organisation of actin filaments through the rho pathway. *J. Cell Sci.* **115,** 2735–2746.

Baetz, K. K., Krogan, N. J., Emili, A., Greenblatt, J., and Hieter, P. (2004). The ctf13/CTF13 genomic haploidsufficiency modifier screen identifies the yeast chromatin remodeling complex RSC, which is required for the establishment of sister chromatid cohesion. *Mol. Cell. Biol.* **24,** 1232–1244.

Belandia, B., Orford, R. L., Hurst, H. C., and Parker, M. G. (2002). Targeting of SWI/SNF chromatin remodelling complexes to estrogen-responsive genes. *EMBO J.* **21,** 4094–4103.

Berger, S. L. (2002). Histone modifications in transcriptional regulation. *Curr. Opin. Genet. Dev.* **12,** 142–148.

Betz, B. L., Strobeck, M. W., Reisman, D. N., Knudsen, E. S., and Weissman, B. E. (2002). Re-expression of hSNF5/INI1/BAF47 in pediatric tumor cells leads to G1 arrest associated with induction of p16ink4a and activation of RB. *Oncogene* **21,** 5193–5203.

Biegel, J. A., Zhou, J. Y., Rorke, L. B., Stenström, C., Wainwright, L. M., and Fogelgren, B. (1999). Germ-line and acquired mutations of INI1 in atypical teratoid and rhabdoid tumors. *Cancer Res.* **59,** 74–79.

Biegel, J. A., Tan, L., Zhang, F., Wainwright, L., Russo, P., and Rorke, L. B. (2002). Alterations of the hSNF5/INI1 gene in central nervous system atypical teratoid/rhabdoid tumors and renal and extra-renal rhabdoid tumors. *Clin. Cancer Res.* **8,** 3461–3467.

Birgersdotter, A., Sandberg, R., and Ernberg, I. (2005). Gene expression perturbation *in vitro*: A growing case for three-dimensional (3D) culture systems. *Semin. Cancer Biol.* **15,** 405–412.

Blessing, C. A., Ugrinova, G. T., and Goodson, H. V. (2004). Actin and ARPs: Action in the nucleus. *Trends Cell Biol.* **14,** 435–442.

Boudreault, A. A., Cronier, D., Selleck, W., Lacoste, N., Utley, R. T., Allard, S., Savard, J., Lane, W. S., Tan, S., and Côté, J. (2003). ESA1 is a histone acetyltransferase that is essential for growth in yeast. *Genes Dev.* **17,** 1415–1428.

Bultman, S., Gebuhr, T., Yee, D., La Mantia, C., Nicholson, J., Gilliam, A., Randazzo, F., Metzger, D., Chambon, P., Crabtree, G., and Magnuson, T. A. (2000). Brg1 null mutation in the mouse reveals functional differences among mammalian SWI/SNF complexes. *Mol. Cell* **6,** 1287–1295.

Cai, Y., Jin, J., Florens, L., Swanson, S. K., Kusch, T., Li, B., Workman, J. L., Washburn, M. P., Conaway, R. C., and Conaway, J. W. (2005) The mammalian YL1 protein is a shared subunit of the TRRAP/TIP60 histone acetyltransferase and SRCAP complexes. *J. Biol. Chem.* **280,** 13665–13670.

Cai, Y., Jin, J., Tomomori-Sato, C., Sato, S., Sorokina, I., Parmely, T. J., Conaway, R. C., and Conaway, J. W. (2003). Identification of new subunits of the multiprotein mammalian TRRAP/TIP60-containing histone acetyltransferase complex. *J. Biol. Chem.* **278,** 42733–42736.

Cairns, B. (2005). Chromatin remodeling complexes: Strength in diversity, precision through specialization. *Curr. Opin. Genet. Dev.* **15,** 185–190.

Cairns, B. R., Lorch, Y., Li, Y., Zhang, M., Lacomis, L., Erdjument Bromage, H., Tempst, P., Du, J., Laurent, B., and Kornberg, R. D. (1996). RSC, an essential, abundant chromatin-remodeling complex. *Cell* **87,** 1249–1260.

Cairns, B. R., Erdjument-Bromage, H., Tempst, P., Winston, F., and Kornberg, R. D. (1998). Two actin-related proteins are shared functional components of the chromatin-remodeling complexes RSC and SWI/SNF. *Mol. Cell* **2,** 639–651.

Cavellán, E., Asp, P., Percipalle, P., and Östlund Farrants, A.-K. (2006). The WSTF-SNF 2h chromatin remodelling complex interacts with several nuclear proteins in transcription. *J. Biol. Chem.* **281,** 16264–16271.

Chai, B., Hsu, J.-M., Du, J., and Laurent, B. C. (2002). Yeast RSC function is required for organization of the cellular cytoskeleton via an alternative PKCI pathway. *Genetics* **161,** 575–584.

Cote, J., Quinn, J., Workman, J. L., and Peterson, C. L. (1994). Stimulation of GAL4 derivative binding to nucleosomal DNA by the yeast SWI/SNF complex. *Science* **265,** 53–60.

de Lanerolle, P., Johnson, T., and Hofmann, W. A. (2005). Actin and myosin I in the nucleus: What next? *Nat. Struct. Mol. Biol.* **12,** 742–746.

Decristofaro, M. F., Betz, B. L., Rorie, C. J., Reisman, D. N., Wang, W., and Weissman, B. E. (2001). Characterization of SWI/SNF protein expression in human breast cancer cell lines and other malignancies. *J. Cell Physiol.* **186,** 136–145.

Deleu, L., Shellard, S., Alevizopoulos, K., Amati, B., and Land, H. (2001). Recruitment of TRRAP required for oncogenic transformation by E1A. *Oncogene* **20,** 8270–8275.

Downs, J. A., Allard, S., Jobin-Robitaille, O., Javaheri, A., Auger, A., Bouchard, N., Kron, S. J., Jackson, S. P., and Côté, J. (2004). Binding of chromatin-modifying activities to phosphorylated histone H2A at DNA damage sites. *Mol. Cell* **16,** 979–990.

Doyon, J. and Côté, J. (2004). The highly conserved and multifunctional NuA4 HAT complex. *Curr. Opin. Gen. Dev.* **14,** 147–154.

Doyon, Y., Selleck, W., Lane, W. S., and Côté, J. (2004). Structural and functional conservation of the NuA4 histone acetyltransferase complex from yeast to humans. *Mol. Cell. Biol.* **24,** 1884–1896.

Dunaief, J. I., Strober, B. E., Guha, S., Khavari, P. A., Alin, K., Luban, J., Begemann, M., Crabtree, G. R., and Goff, S. P. (1994). The retinoblastoma protein and BRG1 form a complex and cooperate to induce cell cycle arrest. *Cell* **79,** 119–130.

Egly, J. M., Miyamoto, N. G., Moncollin, V., and Chambon, P. (1984). Is actin a transcription initiation factor for RNA polymerase B? *EMBO J.* **3,** 2363–2371.

Eissenberg, J. C., Wong, M., and Chrivia, J. C. (2005). Human SRCAP and *Drosophila melanogaster* DOM are homologs that function in the notch signaling pathway. *Mol. Cell. Biol.* **25,** 6559–6569.

Flaus, A. and Owen-Hughes, T. (2004). Mechanisms for ATP-dependent chromatin remodelling: Farewell to the tuna-can octamer? *Curr. Opin. Genet. Dev.* **14,** 165–173.

Fomproix, N. and Percipalle, P. (2004). An actin-myosin complex on actively transcribing genes. *Exp. Cell Res.* **294,** 140–148.

Frames, M. C. (2004). Newest finding on the oldest oncogene: How activated scr does it review. *J. Cell Sci.* **117,** 989–998.

Frank, S. R., Schroeder, M., Fernadez, P., Taubert, S., and Amati, B. (2001). Binding of c-Myc to chromatin mediates mitogen-induced acetylation of histone 4 and gene activation. *Genes Dev.* **15,** 2069–2082.

Fuchs, M., Gerber, J., Drapkin, R., Sif, S., Ikura, T., Ogryzko, V., Lane, W. S., Nakatani, Y., and Livingston, D. M. (2001). The p400 complex is and essential E1A transformation targets. *Cell* **106,** 297–307.

Fujita, Y. and Braga, V. (2005). Epithelial cell shape and Rho small GTPases. *Novskis Foundation Symposium* **269,** 144–145.

Fukuoka, J., Fujii, T., Shih, J. H., Dracheva, T., Meerzaman, D., Player, A., Hong, K., Settnek, S., Gupta, A., Buetow, K., Hewitt, S., Travis, W. D., *et al.* (2004). Chromatin remodeling factors and BRM/BRG1 expression as prognostic indicators in non-small cell lung cancer. *Clin. Cancer Res.* **10,** 4314–4324.

Furuwatati, C., Yagi, A., Yamagami, O., Ishikawa, M., Hidaka, E., Ueno, I., Furihata, K., Ogiso, Y., and Katsuyama, T. (1998). A comparative system to explore p53 mutations. *Am. J. Clin. Pathol.* **110,** 368–373.

Galarneau, L., Nourani, A., Boudreault, A. A., Zhang, Y., Heliot, L., Allard, S., Savard, J., Lane, W. S., Stillman, D. J., and Côté, J. (2000). Multiple links between the NuA4 histone acetyltransferase complex and epigenetic control of transcription. *Mol. Cell* **5,** 927–937.

Gettemans, J., Van Impe, K., Delanote, V., Hubert, T., Vandekerckhove, J., and De Corte, V. (2005). Nuclear actin-binding proteins as modulators of gene transcription. *Traffic* **6,** 847–857.

Gibbons, R. J. (2005). Histone modifying and chromatin remodelling enzymes in cancer and dysplastic syndrome. *Hum. Mol. Genet.* **14,** R85–R95.

Gonsior, S. M., Platz, S., Buchmeier, S., Scheer, U., Jockusch, B. M., and Hinssen, H. (1999). Conformation differences between nuclear and cytoplasmic actin as detected by a monoclonal antibody. *J. Cell Sci.* **112,** 797–809.

Grand, F., Kulkarni, S., Chase, A. S., Goldman, J. M., Gordon, M., and Cross, N. C. (1999). Frequent deletion of hSNF5/INI1, a component of the SWI/SNF complex, in chronic myeloid leukemia. *Cancer Res.* **59,** 3870–3874.

Guidi, C., Sands, A., Zambrowicz, B., Turner, T., Demers, D., Webster, W., Smith, T., Imbalzano, A., and Jones, S. (2001). Disruption of Ini1 leads to periimplantation lethality and tumorogeneisis in mice. *Mol. Cell. Biol.* **21,** 3598–3603.

Harata, M., Oma, Y., Mizuno, S., Jiang, Y. W., Stillman, D. J., and Wintersberger, U. (1999). The nuclear actin-related protein of *S. cerevisiae*, Act3p/Arp4, interacts with core histones. *Mol. Biol. Cell* **10,** 2595–2605.

Hendricks, K. B., Shanahan, F., and Lees, E. (2004). Role for BRG1 in cell cycle control and tumor suppression. *Mol. Cell. Biol.* **24,** 362–371.

Hill, D. A., Chiosea, S., Jamaluddin, S., Roy, K., Fischer, A. H., Boyd, D. D., Nickerson, J. A., and Imbalzano, A. N. (2004). Inducible changes in cell size and attachment area due to expression of a mutant SWI/SNF chromatin remodeling enzyme. *J. Cell Sci.* **117,** 5847–5854.

Hirohashi, S. and Kanai, Y. (2003). Cell adhesion systems and human cancer morphogenesis. *Cancer Sci.* **94,** 575–581.

Hofmann, W., Reichart, B., Ewald, A., Müller, E., Schmitt, I., Stauber, R. H., Lottspeich, F., Jockusch, B. M., Scheer, U., Hauber, J., and Dabauvalle, M. C. (2001). Cofactor requirements for nuclear export of Rev response element (RRE)- and constitutive transport element (CTE)-containing retroviral RNAs. An unexpected role for actin. *J. Cell Biol.* **152,** 895–910.

Hofmann, W. A., Stojiljkovic, L., Fuchsova, B., Vargas, G. M., Mavrommatis, E., Philimonenko, V., Kysela, K., Goodrich, J. A., Lessard, J. L., Hope, T. J., Hozak, P., and de Lanerolle, P. (2004). Actin is part of pre-initiation complex and is necessary for transcription by RNA polymerase II. *Nat. Cell Biol.* **6,** 1094–1101.

Holstege, F. C., Jennings, E. G., Wyrick, J. J., Lee, T. I., Hengartner, C. J., Green, M. R., Golub, T. R., Landers, E. S., and Young, R. A. (1998). Dissecting the regulatory circuitry of a eukaryotic genome. *Cell* **95,** 717–728.

Hsu, J. M., Huang, J., Meluh, P. B., and Laurent, B. C. (2003). The RSC chromatin-remodeling complex is required for kinetochore function in chromosome segregation. *Mol. Cell. Biol.* **23,** 3202–3215.

Hu, P., Wu, S., and Hernandez, N. (2004). A role for β-actin in RNA polymerase III transcription. *Genes Dev.* **18,** 3010–3015.

Huang, J., Hsu, B. C., and Laurent, B. C. (2004). The RSC nucleosome-remodeling complex is required for cohesin's association with chromosome arms. *Mol. Cell* **13,** 739–750.

Ikura, T., Ogryzko, V., Gigoriev, M., Groisman, R., Wang, J., Horikoshi, M., Scully, R., Qin, J., and Nakatani, Y. (2000). Involvement of the TIP60 histone acetylase complex in DNA repair and apoptosis. *Cell* **102,** 463–473.

Jaffe, A. B. and Hall, A. (2005). Rho GTPases: Biochemistry and biology. *Ann. Rev. Cell Dev. Biol.* **21,** 247–269.

Jenuwein, T. and Allis, C. D. (2001). Translating the histone code. *Science* **293,** 1074–1080.

Jin, J., Cai, Y., Yao, T., Gottshalk, A. J., Florens, L., Swanson, S. K., Guitérrez, J. L., Coleman, M. K., Workman, J. L., Mushegian, A., Washburn, M. P., Conaway, R. C., *et al.* (2005). A mammalian chromatin remodeling complex with similarities to the yeast INO80 complex. *J. Biol. Chem.* **280,** 41207–41212.

Jónsson, Z. O., Jha, S., Wohlschlegel, J. A., and Dutta, A. (2004). Rvb1p/Rvb2p recruits Arp5p and assemble a functional Ino80 chromatin remodeling complex. *Mol. Cell* **16,** 465–477.

Khavari, P. A., Peterson, C. L., Tamkun, J. W., Mendel, D. B., and Crabtree, G. R. (1993). BRG1 contains a conserved domain of the SWI2/SNF2 family necessary for normal mitotic growth and transcription. *Nature* **366,** 170–174.

Kim, J., Sif, S., Jones, B., Jackson, A., Koipally, J., Heller, E., Winandy, S., Viel, A., Sawyer, A., Ikeda, T., Kingston, E. R., and Georgopoulos, K. (1999). Ikaros DNA-binding proteins direct formation of chromatin remodeling complexes in lymphocytes. *Immunity* **10,** 345–355.

Kimura, T., Hashimoto, I., Yamamoto, A., Nishikawa, M., and Fujisawa, J. I. (2000). Rev-dependent association of the intron-containing HIV-1 gag mRNA with the nuclear actin bundles and the inhibition of its nucleocytoplasmic transport by latrunculin-B. *Genes Cells* **5,** 289–307.

Kiseleva, E., Drummond, S. P., Goldberg, M. W., Rutherford, S. A., Allen, T. D., and Wilson, K. L. (2004). Actin and protein-4.1 containing filaments link nucelar pore complexes to subnuclear organelles in *Xenopus* oocytes nuclei. *J. Cell Sci.* **117,** 2481–2490.

Kitagawa, H., Fujiki, R. R., Yoshimura, K., Mezaki, Y., Uematsu, Y., Matsui, D., Ogawa, S., Unno, K., Okubo, M., Tokita, A., Nakagawa, T., Ito, T., *et al.* (2003). The chromatin-remodeling complex WINAC targets a nuclear receptor to promoters and is impaired in Williams syndrome. *Cell* **113**(7), 905–917.

Klochendler-Yeivin, A., Fiette, L., Barra, K., Muchardt, C., Babinet, C., and Yaniv, M. (2000). The murin SNF5/INI1 chromatin remodeling factor is essential for embryonic development and tumor suppression. *EMBO Rep.* **1,** 500–506.

Kobor, M. S., Venkatasubrahmanyam, S., Meneghini, M. D., Gin, J. W., Jennings, J. L., Link, A. J., Madhani, H. D., and Rine, J. (2004). A protein complex containing the conserved Swi2/Snf2-related ATPase Swr1p deposits histone variant H2A.Z into euchromatin. *PLoS Biol.* **2,** E131.

Kukalev, A., Nord, Y., Palmberg, C., Bergman, T., and Percipalle, P. (2005). Actin and hnRNP U cooperate for productive transcription by RNA polymerase II. *Nat. Struct. Mol. Biol.* **12,** 238–244.

Kusch, T., Florens, L., Macdonald, W. H., Swanson, S. K., Glaser, R. L., Yates, J. R., Abmayr, S. M., Washburn, M. P., and Workman, J. L. (2004). Acetylation by Tip60 is required for selective histone variant exchange at DNA lesions. *Science* **306,** 2084–2087.

Kwon, H., Imbalzano, A. N., Khavari, P. A., Kingston, R. E., and Green, M. R. (1994). Nucleosome disruption and enhancement of activator binding by a human SW1/SNF complex. *Nature* **370,** 477–481.

Längst, G. and Becker, P. B. (2004). Nucleosome remodelling: One mechanism, many phenomena? *Biochim. Biophys. Acta* **1677,** 58–63.

Larue, L. and Ballacosa, A. (2005). Epithelial-mesenchymal transition in development and cancer: Role of phosphatidylinositol 3′ kinase/AKT pathways. *Oncogene* **24,** 7443–7454.

Lee, S. and Helfman, D. M. (2004). Cytoplasmic p21cip1 is involved in ras-induced inhibition of the ROCK/LIMK/cofilin pathway. *J. Biol. Chem.* **279,** 1885–1891.

Lemon, B., Inouye, C., King, D. S., and Tjian, R. (2001). Selectivity of chromatin-remodelling cofactors for ligand-activated transcription. *Biochim. Biophys. Acta* **414,** 924–928.

Lénárt, P., Bacher, C. P., Daigle, N., Hand, A. R., Eils, R., Terasaki, M., and Ellenberg, J. (2005). A contractile nuclear actin network drives chromosome congression in oocytes. *Nature* **436,** 812–818.

Li Wang, L., Baiocchi, R. A., Pal, S., Mosialos, G., Caligiuri, M., and Sif, S. (2005). The BRG1- and hBRM-associated factor BAF57 induces apoptosis by stimulating expression of the cylindromatosis tumor suppressor gene. *Mol. Cell. Biol.* **25,** 7953–7965.

Liu, R., Lui, H., Chen, X., Kirby, M., Brown, P. O., and Zhao, K. (2001). Regulation of CSF1 promoter by the SWI/SNF-like BAF complex. *Cell* **106,** 309–318.

Lusser, A. and Kadonaga, J. T. (2003). Chromatin remodeling by ATP-dependent molecular machines. *Bioessays* **25,** 1192–1200.

McMahon, S. B., Van Buskirk, H. A., Dugan, K. A., Copeland, T. D., and Cole, M. D. (1998). The novel ATM-related protein TRRAP is an essential cofactor for the c-Myc and E2F oncoproteins. *Cell* **94,** 363–374.

McMahon, S. B., Wood, M. A., and Cole, M. D. (2000). The essential cofactor TRRAP recruits the histone acetyltransferase hGCN5 to c-Myc. *Mol. Cell. Biol.* **20,** 556–562.

Medjkane, S., Novikov, E., Versteege, I., and Delattre, O. (2004). The tumour suppressor hSNF5/INI1 modulates cell growth and actin cytoskeleton organisation. *Cancer Res.* **64,** 3406–3413.

Mizuguchi, G., Shen, X., Landry, J., Wu, W. H., Sen, S., and Wu, C. (2004). ATP-driven exchange of histone H2AZ variant catalysed by SWR1 chromatin remodeling complex. *Science* **303,** 343–348.

Mohrmann, L. and Verrijzer, C. P. (2005). Composition and functional specificity of SWI2/SNF2 class chromatin remodeling complexes. *Biochim. Biophys. Acta* **16,** 8159–8173.

Mohrmann, L., Langenberg, K., Krijgsveld, J., Kal, A. J., Heck, A. J. R., and Verrijzer, C. P. (2004). Differential targeting of two distinct SWI/SNF-related Drosophila chromtin-remodeling complexes. *Mol. Cell. Biol.* **24,** 3077–3088.

Morrison, A. J. and Shen, X. (2005). DNA repair in the context of chromatin. *Cell Cycle* **4,** 568–571.

Morrison, A. J., Highland, J., Krogan, N. J., Arbel-Eden, A., Greenblatt, J. F., Haber, J. E., and Shen, X. (2004). INO80 and γ-H2aX interaction links ATP-dependent chromatin remodeling to DNA damage repair. *Cell* **119,** 767–775.

Muchard, C. and Yaniv, M. (1993). A human homologue of *Saccharomyces cerevisiae* SNF2/SWI2 and Drosophila brm genes potentiates transcriptional activation by the glucocorticoid receptor. *EMBO J.* **12,** 4279–4290.

Muller, J., Oma, Y., Vallar, L., Friederich, E., Poch, O., and Winsor, B. (2005). Sequence and comparative genomic analysis of actin-related proteins. *Mol. Biol. Cell* **16,** 5736–5748.

Murr, R., Loizou, J. I., Yang, Y.-G., Cuenin, C., Li, H., Wang, Z.-Q., and Herceg, Z. (2005). Histone acetylation by Trrap-Tip60 modulates loading of repair proteins and repair of DNA double-strand breaks. *Nat. Cell Biol.* Online publication.

Narlikar, G. J., Fan, H. Y., and Kingston, R. E. (2002). Cooperation between complexes that regulate chromatin structure and transcription. *Cell* **108,** 475–487.

Neely, K. E. and Workman, J. L. (2002). The complexity of chromatin remodeling and its link to cancer. *Biochim. Biophys. Acta* **1603,** 19–29.

Neish, A. S., Anderson, S. F., Schlegel, B. P., Wei, W., and Parvin, J. D. (1998). Factors associated with the mammalian RNA polymerase II holoenzyme. *Nucl. Acids Res.* **26,** 847–853.

Nie, Z., Xue, Y., Yang, D., Zhou, S., Deroo, B. J., Archer, T. K., and Wang, W. (2000). A specificity and targeting subunit of a human SWI/SNF family-related chromatin-remodeling complex. *Mol. Cell. Biol.* **20,** 8879–8888.

Nie, Z., Yan, Z., Chen, E. H., Sechi, S., Ling, C., Zhou, S., Xue, Y., Yang, D., Murray, D., Kanakubo, E., Cleary, M. L., and Wang, W. (2003). Novel SWI/SNF chromatin-remodeling complexes contain a mixed-lineage leukemia chromosomal translocation partner. *Mol. Cell. Biol.* **23,** 2942–2952.

O'Neill, D., Yang, J., Erdjument-Bromage, H., Bornschlegel, K., Tempst, P., and Bank, A. (1999). Tissue-specific and developmental stage-specific DNA binding by a mammalian SWI/SNF complex associated with human fetal-to-adult globin gene switching. *Proc. Natl. Acad. Sci. USA* **96,** 349–354.

O'Neill, D. W., Schoetz, S. S., Lopez, R. A., Castle, M., Rabinowitz, I., Shor, E., Krawchuk, D., Goll, M. G., Renz, M., Seelig, H. P., Han, S., Seong, R. H., *et al.* (2000). An ikaros-containing chromatin-remodeling complex in adult-type erythroid cells. *Mol. Cell. Biol.* **20,** 7572–7582.

Olave, I. A., Reck-Petersson, S. L., and Crabtree, G. R. (2002). Nuclear actin and actin related proteins in chromatin remodeling. *Annu. Rev. Biochem.* **71,** 755–781.

Olave, I., Wang, W., Xue, Y., Kuo, A., and Crabtree, G. R. (2002). Identification of a polymorphic, neuron-specific chromatin remodeling complex. *Genes Dev.* **16,** 2509–2517.

Pal, S., Yun, R., Datta, A., Lacomis, L., Erdjument-Bromage, H., Kumar, J., Tempst, P., and Sif, S. (2003). mSin3A/histone deacetylase 2- and PRMT5-containing Brg1 complex is involved in transcriptional repression of the Myc target gene cad. *Mol. Cell. Biol.* **23,** 7475–7487.

Pawlak, G. and Helfman, D. M. (2001). Cytoskeletal changes in cell transformation and tumorigenesis. *Curr. Opin. Genet Dev.* **11,** 41–47.

Pedersen, T. and Aebi, U. (2002). Actin in the nucleus: What form and what for? *J Struct. Biol.* **140,** 3–9.

Pedersen, T. and Aebi, U. (2005). Nuclear actin extends, with no contraction in sight. *Mol. Biol. Cell* **16,** 5055–5060.

Percipalle, P. and Östlund Farrants, A.-K. (2006). Chromatin remodelling and transcription: Be-WICHed by nuclear myosin 1. *Curr. Opin. Cell Biol.* **18,** 267–274.

Percipalle, P., Zhao, J., Pope, B., Weeds, A., Lindberg, U., and Daneholt, B. (2001). Actin bound to the heterogeneous nuclear ribonucleoprotein hrp36 is associated with Balbiani ring mRNA from the gene to polysomes. *J. Cell Biol.* **153,** 229–236.

Percipalle, P., Jonsson, A., Nashchekin, D., Karlsson, C., Bergman, T., Guialis, A., and Daneholt, B. (2002). Nuclear actin associated with a specific subset of hnRNP A/B-type proteins. *Nucl. Acids Res.* **30,** 1725–1734.

Percipalle, P., Fomproix, N., Kylberg, K., Miralles, F., Björkrot, B., Daneholt, B., and Visa, N. (2003). An actin-ribonucleoprotein interaction is involved in transcription by RNA polymerase II. *Proc. Natl. Acad. Sci. USA* **100,** 6475–6480.

Percipalle, P., Fomproix, N., Voigt, R., Cavellán, E., Reimer, G., Krüger, T., Thyberg, J., Grummt, I., and Östlund Farrants, A.-K. (2006). Actin and nuclear myosin 1 facilitate association of the chromatin remodeling complex WSTF-ISWI with RNA polymerase I. *EMBO Rep.* **7,** 525–530.

Pestic-Dragovich, L., Stojiljkovic, L., Philimonenko, A. A., Nowak, G., Ke, Y., Settlage, R. E., Shabanowitz, J., Hunt, D. F., Hozak, P., and de Lanerolle, P. (2000). A myosin I isoform in the nucleus. *Science* **290,** 337–341.

Peterson, C. L. and Côté, J. (2004). Cellular machineries for chromosomal DNA repair. *Genes Dev.* **18,** 602–616.

Peterson, C. L., Zhao, Y., and Chait, B. T. (1998). Subunits of the yeast SWI/SNF complex are members of the actin-related protein (ARP) family. *J. Biol. Chem.* **273,** 23641–23644.

Phelan, M. L., Sif, S., Narlikar, G. J., and Kingston, R. E. (1999). Reconstitution of a core chromatin remodeling complex from SWI/SNF subunits. *Mol. Cell* **3,** 247–253.

Philimonenko, V. V., Zhao, J., Ibn, S., Dingova, H., Kysela, K., Kahle, M., Zentgraf, H., Hofmann, W. A., de Lanerolle, P., Hozak, P., and Grummt, I. (2004). Nuclear actin and myosin I are required for RNA polymerase I transcription. *Nat. Cell Biol.* **6,** 1165–1172.

Presland, R. B., Kuechle, M. K., Lewis, S. P., Fleckman, P., and Dale, B. A. (2001). Regulated expression of human filaggrin in keratinocytes results in cytoskeletal disruption, loss of cell-cell adhesion, and cell cycle arrest. *Exp. Cell Res.* **270,** 199–213.

Rando, O. J., Zhao, K., Janmey, P., and Crabtree, G. R. (2002). Phosphatidylinositol-dependent actin filament binding by the SWI/SNF-like BAF chromatin remodeling complex. *Proc. Natl. Acad. Sci. USA* **99,** 2824–2829.

Rao, J. and Li, N. (2004). Microfilament actin remodeling as a potential target for cancer drug development. *Curr. Cancer Drug Targets* **4,** 345–354.

Reincke, B. S., Rosson, G. B., Oswald, B. W., and Wright, C. F. (2003). INI1 expression induces cell cycle arrest and markers of senescence in malignant rhabdoid tunor cells. *J. Cell Physiol.* **294,** 303–313.

Reisman, D. N., Strobeck, M. W., Betz, B. L., Sciariotta, J., Funkhouser, W., Jr., Muchardt, C., Yaniv, M., Sherman, L. S., Knudsen, E. S., and Weissman, B. E. (2002). Concomitant down-regulation of BRM and BRG1 in human tumor cell lines: Differential effects on RB-mediated growth arrest vs CD44 expression. *Oncogene* **21,** 1196–1207.

Reisman, D. N., Sciariotta, J., Wang, W., Funkhouser, W. K., and Weissman, B. E. (2003). Loss of GRG1/BRM in human lung cancer cell lines and primary lung cancers: Correlation with poor prognosis. *Cancer Res.* **63,** 560–566.

Reyes, J.-C., Barra, J., Muchard, C., Camus, A., Babinet, C., and Yaniv, M. (1998). Altered control of cellular proliferation in the absence of mammalian brahma (SNF2a). *EMBO J.* **17,** 6979–6991.

Riento, K. and Ridley, A. (2003). Rocks: Multifunctional kinases in cell behaviour. *Nat. Rev. Mol. Cell. Biol.* **4,** 446–456.

Roberts, C. W. and Orkin, S. H. (2004). The SWI/SNF complex—chromatin and cancer. *Nat. Rev. Cancer* **4,** 133–142.

Roberts, C. W., Galusha, S. A., McMenamin, M. E., Fletcher, C. D., and Orkin, S. H. (2000). Haploinsufficiency of Snf5 (integrase interactor 1) predeposes to malignant rhabdoid tumors in mice. *Proc. Natl. Acad. Sci. USA* **97,** 13796–13800.

Rosson, G. B., Bartlett, C., Reed, W., and Weissman, B. E. (2005). BRG1 loss in MiaPaCa2 cells induces and altered cellular morphology after a few generations simultaneous with a disruption in the organisation of the actin cytoskeleton. *J. Cell. Physiol.* **205,** 286–294.

Rousseau-Merckm, M.-F., Versteege, I., Legrand, I., Coururier, J., Mairal, A., Delattre, O., and Aurias, A. (1999). hSNF5/INI1 inactivation is mainly associated with homozygous deletions and mitotic recombinations in rhabdoid tumors. *Cancer Res.* **59,** 3152–3156.

Sameshima, M., Kishi, Y., Osumi, M., Minamikawa-Tashino, R., Mahadeo, D., and Cotter, D. A. (2001). The formation of actin rods composed of actin tubules in Dictyostelium discoideum spores. *J. Struct. Biol.* **136,** 7–19.

Sévenet, N., Lelloch-Tubiana, A., Schofield, D., Hoang-Xuan, K., Gessler, M., Birnbaum, D., Jeanpierre, C., Jouvet, A., and Delattre, O. (1999a). Spectrum of hSNF5/INI1 somatic mutations in human cancer and genotype-phenotype correlations. *Hum. Mol. Genet.* **8,** 2359–2368.

Sévenet, N., Sheridan, E., Amram, D., Schneider, P., Handgreitinger, R., and Delattre, O. (1999b). Constitutional mutations of the hSNF5/INI1 gene predispose to a variety of cancers. *Am. J. Hum. Genet.* **65,** 1342–1348.

Shanahan, F., Seghezzi, W., Parry, D., Mahony, D., and Lees, E. (1999). Cyclin E associates with BAF155 and BRG1, components of the mammalian SWI/SNF complex, and alters the ability of BRG1 to induce growth arrest. *Mol. Cell. Biol.* **19,** 1460–1469.

Sheer, U., Hinssen, H., Franke, W. W., and Jockusch, B. M. (1984). Microinjection of actin binding proteins and actin antibodies demonstrates involvement of nuclear actin in transcription of lampbrush chromosomes. *Cell* **39,** 111–122.

Shen, X., Mizuguchi, G., Hamiche, A., and Wu, C. (2000). A chromatin remodeling complex involved in transcription and DNA processing. *Nature* **406,** 541–544.

Shen, X., Ranallo, R., Choi, E., and Wu, C. (2003). Involvement of actin-related proteins in ATP-dependent chromatin remodeling. *Mol. Cell* **12,** 147–155.

Shim, E. Y., Ma, J. L., Oum, J. H., Yaner, Y., and Lee, S. E. (2005). The yeast chromatin remodeler RSC complex facilitates end joining repair of DNA double strand breaks. *Mol. Cell. Biol.* **25,** 3934–3944.

Sif, S. (2004). ATP-dependent nucleosome remodeling complexes: Enzymes tailored to deal with chromatin. *J. Cell. Biochem.* **91,** 1087–1098.

Sif, S., Stukenberg, P. T., Kischner, M. W., and Kingston, R. E. (1998). Mitotic inactivation of a human SWI/SNF chromatin remodeling complex. *Genes Dev.* **12,** 2842–2851.

Sif, S., Saurin, A. J., Imbalzano, A. N., and Kingston, R. E. (2001). Purification and characterization of mSin3A-containing Brg1 and hBrm chromatin remodeling complexes. *Genes Dev.* **15,** 603–618.

Sjölinder, M., Björk, P., Söderberg, E., Sabri, N., Östlund Farrants, A.-K., and Visa, N. (2005). The growing pre-mRNA recruits actin and chromatin modifying factors to transcriptionally active genes. *Genes Dev.* **19,** 1871–1884.

Stobeck, M. W., Knudsen, K. E., Fribourg, A. F., DeCristofaro, M. F., Weissman, B. E., Imbalzano, A. N., and Knudsen, E. S. (2000). BRG-1 is required for RB-mediated cell cycle arrest. *Proc. Natl. Acad. Sci. USA* **97,** 7748–7753.

Stober, B. E., Dunaief, J. L., Guha, S., and Goff, S. P. (1996). Functional interactions between the hBRM/hBRG1 transcriptional activators and the pRB family of proteins. *Mol. Cell. Biol.* **16,** 1576–1583.

Strahl, B. D. and Allis, C. D. (2000). The language of covalent histone modifications. *Nature* **403,** 41–45.

Strobeck, M. W., DeCristofaro, M. F., Banine, F., Weissmann, B. E., Sherman, L. S., and Knudsen, E. S. (2001). The BRG1 subunit of the SWI/SNF complex regulates the CD44 expression. *J. Biol. Chem.* **276,** 9273–9278.

Strobeck, M. W., Reisman, D. N., Gunawardena, R. W., Betz, B. L., Angus, S. P., Knudsen, K. E., Kowalik, T. F., Weismann, B. E., and Knudsen, E. S. (2002). Compensation of BRG1 function by BRM: Insight into the role of the core SWI-SNF subunits in retinoblastoma protein signaling. *J. Biol. Chem.* **277,** 4782–4789.

Sudarsanam, P., Iyer, V. R., Brown, P. O., and Winston, F. (2000). Whole-genome expression analysis of snf/swi mutants of *Saccharomyces cerevisiae*. *Proc. Natl. Acad. Sci. USA* **97,** 3364–3369.

Sundana, R., Görzer, I., Oma, Y., Yoshida, T., Suka, N., Wintersberger, U., and Harata, M. (2005). The nuclear actin-related protein Act3p/Arp4p is involved in the dynamics of chromatin-modulating complexes. *Yeast* **22,** 753–768.

Szerlong, H., Saha, A., and Cairns, B. R. (2003). The nuclear actin-related proteins Arp7 and Arp9: A dimeric module that cooperates with architectural proteins for chromatin remodeling. *EMBO J.* **22,** 3175–3187.

Tsukiyama, T. (2002). The in vivo functions of ATP-dependent chromatin-remodelling factors. *Nat. Rev. Mol. Cell. Biol.* **3,** 422–429.

Underhill, C., Qutob, M. S., Yee, S. P., and Torchia, J. (2000). A novel nuclear receptor corepressor, N-Cor, containing components of the mammalian SWI/SNF complex and the corepressor KAP-1. *J. Biol. Chem.* **275,** 40463–40470.

van Attikum, H. and Gasser, S. M. (2005). ATP-dependent chromatin remodeling and DNA double-strand break repair. *Cell Cycle* **4,** 1011–1014.

van Attikum, H., Fritsch, O., Hohn, B., and Gasser, S. M. (2004). Recruitment of the INO80 complex by H2A phosphorylation links ATP-dependent chromatin remodeling with DNA double-strand break repair. *Cell* **119,** 777–788.

Versteege, I., Sevenet, N., Lange, J., Rousseau-Merck, M. F., Ambros, P., Handgretinger, R., Aurias, A., and Delattre, O. (1998). Truncating mutations of hSNF5/INI1 in aggressive paediatric cancer. *Nature* **394,** 203–206.

Versteege, I., Medjkane, S., Rouillard, D., and Delattre, O. (2002). A key role of the hSNF5/INI1 tumour suppressor in the control of the G1-S transition of the cell cycle. *Oncogene* **21,** 6403–6412.

Visa, N. (2005). Actin in transcription. Actin is required for transcription by all three RNA polymerases in the eukaryotic cell nucleus. *EMBO Rep.* **6,** 218–219.

Wang, L., Baiocchi, R. A., Pal, S., Mosialos, G., Caligiuri, M., and Sif, S. (2005). The BRG1- and HBRM-associated factor BAF57 induces apoptosis by stimulating expression of the cylindromatosis tumor suppressor gene. *Mol. Cell. Biol.* **25,** 7953–7965.

Wang, W. (2003). The SWI/SNF family of ATP-dependent chromatin remodelers: Similar mechanism for diverse functions. *Curr. Top. Microbiol. Immunol.* **274,** 143–169.

Wang, W., Cote, J., Xue, Y., Zhou, S., Khavari, P. A., Biggar, S. R., Muchardt, C., Kalpana, G. V., Goff, S. P., Yaniv, M., Workman, J. L., and Crabtree, G. R. (1996a). Purification and biochemical heterogeneity of the mammalian SWI/SNF complex. *EMBO J.* **15,** 5370–5382.

Wang, W., Xue, Y., Zhou, S., Kuo, A., Cairns, B. R., and Crabtree, G. R. (1996b). Diversity and specialization of mammalian SWI/SNF complexes. *Genes Dev.* **10,** 2117–2130.

Wang, W., Chi, T., Xue, S., Kuo, S., and Crabtree, G. R. (1998). Architectural DNA binding by a high-mobility-group/kinesin-like subunit in mammalian SWI/SNF-related complexes. *Proc. Natl. Acad. Sci. USA* **95,** 492–498.

Watanabe, N., Kato, T., Fujita, A., Ishizaki, T., and Narumiya, S. (1999). Cooperation between mDia1 and ROCK in Rho-induced actin reorganization. *Nat. Cell Biol.* **1,** 136–143.

Wilson, C. J., Chao, D. M., Imbalzano, A. N., Schnitzler, G. R., Kinsgston, R. E., and Young, R. A. (1996). RNA polymerase II holoenzyme contains SWI/SNF regulators involved in chromatin remodeling. *Cell* **84,** 235–244.

Wong, A. K. C., Shanahan, F., Chen, Y., Lian, L., Ha, P., Hendricks, K., Ghaffari, S., Iliev, D., Penn, B., Woodland, A.-M., Smith, R., Salada, G., *et al.* (2000). BRG1, a component of the SWI-SNF complex, is mutated in multiple human tumor cell lines. *Cancer Res.* **60,** 6171–6177.

Xu, W., Cho, H., Kadam, S., Banayo, E. M., Anderson, A., Yates, J. R., III, Emerson, B. M., and Evans, R. M. (2004). A methylation-mediator complex in hormone signaling. *Genes Dev.* **18,** 144–156.

Xue, Y., Canman, J. C., Lee, C. S., Nie, Z., Yang, D., Moreno, G. T., Young, M. K., Salmon, E. D., and Wang, W. (2000). The human SWI/SNF-B chromatin-remodelling complex is related to yeast Rsc and localizes at kinetochores of mitotic chromosomes. *Proc. Natl. Acad. Sci. USA* **97,** 13015–13020.

Yan, Z., Cui, K., Murray, D. M., Ling, C., Xue, Y., Gerstein, A., Parsons, R., Zhao, K., and Wang, W. (2005). PBAF chromatin-remodeling complex requires a novel specificity subunit, BAF200, to regulate expression of selective interferon-responsive genes. *Genes Dev.* **19,** 1662–1667.

Zhang, H., Richardson, D. O., Roberts, D. N., Utley, R., Erdjument-Bromage, H., Tempst, P., Côté, J., and Cairns, B. R. (2004). The Yaf9 component of the SWR1 and NuA4 complexes is required for proper gene expression, histone H4 acetylation, and Htz1 replacement near telomeres. *Mol. Cell. Biol.* **24,** 9424–9436.

Zhang, S., Kohler, C., Hemmerich, P., and Grosse, F. (2004). Nuclear DNA helicase II (RNA helicase A) binds to an F-actin containing shell that surrounds the nucleolus. *Exp. Cell Res.* **293,** 248–258.

Zhang, Z. K., Davies, K. P., Allen, J., Zhu, L., Pestell, R. G., Zagzag, D., and Kalpana, G. V. (2002). Cell cycle arrest and repression of cyclin D1 transcription by INI1/hSNF5. *Mol. Cell. Biol.* **22,** 5975–5988.

Zhao, K., Wang, W., Rando, O. J., Xue, Y., Swuderek, A., Kuo, A., and Crabtree, G. R. (1998). Rapid and phosphoinositol-dependent binding of the SWI/SNF-like BAT complex to chromatin after T lymphocytes receptor signaling. *Cell* **95,** 625–636.

Chapter 8

Ionic Waves Propagation Along the Dendritic Cytoskeleton as a Signaling Mechanism

Avner Priel,[1] Jack A. Tuszynski,[2] and Horacio F. Cantiello[3]

[1]*Department of Physics, University of Alberta Edmonton, AB, T6G 2J1, Canada*
[2]*Division of Experimental Oncology, Cross Cancer Institute, Edmonton, Alberta, T6G 1Z2, Canada*
[3]*Massachusetts General Hospital East and Harvard Medical School, Charlestown, Massachusetts*

We outline the physical conditions that enable cytoskeletal polymers, such as actin filaments (AFs) and microtubules, to act as electrical transmission lines for ion flows along their lengths. For AFs we propose a model in which each protein subunit is an electric element with a capacitive, inductive, and resistive property due to the molecular structure of the filament and viscosity of the solution. Based on the conductivity rules that apply to electrical circuits, we discuss the properties of ionic waves that propagate along actin filaments. We then discuss the dynamics of C-termini states in microtubules and their networks, including the interactions with ions and signal transmission via microtubule-associated proteins. Experiments on ionic conductivity along AFs and microtubules validate the basic assumptions postulated in our models. As a consequence of these results we propose a new signaling mechanism in the cell, especially in neurons, that involves clouds of ions surrounding protein filaments which may travel without significant decay along the axon or the dendritic tree. These signals may be utilized to control various membrane properties, for example, the transition rate of ion channel opening and closing, local membrane conductivity, and vesicle trafficking.

Advances in Molecular and Cell Biology, Vol. 37

ISBN: 0-444-52868-7

DOI: 10.1016/S1569-2558(06)37008-7

I. INTRODUCTION

The prevailing view of neural information processing continues to be based on passive properties of the membrane derived from the application of linear cable theory to dendrites (Rall, 1959). Studies, however, have suggested nonlinear models that accommodate new experimental evidence. Apparently these models are inconsistent with the classical view in explaining the following observations: (1) significant fluctuations of synaptic efficacy over short periods of time in response to a recent burst of activation (Hempel *et al.*, 2000), (2) a variety of active ion channels capable of affecting the local membrane electric properties, and (3) local nonlinear responses, pointing to highly specialized mechanisms correlated with specific inputs. On the other hand, nonlinearities inherent in the new models (Koch and Segev, 2000; Segev and London, 2000) give rise to a wider repertoire of computational capabilities such as multiplication and fast correlation.

In this chapter we propose that functional electrodynamic interactions between cytoskeletal structures and ion channels are central to the neural information-processing mechanism. These interactions are supported by long-range ionic wave propagation along microtubule networks (MTNs) and actin filaments (AFs) and exhibit subcellular control of ionic channel activity, hence impact the computational capabilities of the whole neural function. Cytoskeletal biopolymers, including AFs and microtubules (MTs), constitute the backbone for wave propagation, and in turn interact with membrane components to modulate synaptic connections and membrane ion channels. Indeed, only recently clear functional interactions between these cytoskeletal structures have become apparent. Association of MTs with AFs in neuronal filopodia appears to guide MT growth and plays a key role in neurite initiation (Dehmelt and Halpain, 2004). This is further evidenced in neurons by the presence of proteins that interact with both MTs and F-actin, and proteins that can mediate signaling between both types of filaments. This is likely used to control microtubular invasion. The microtubule-associated proteins (MAPs) MAP1B and MAP2, for example, are known to interact with actin *in vitro* (Pedrotti *et al*, 1994; Togel *et al.*, 1998). Cross-linking, MAP2 and/or MAP1B are very likely associated with both types of filaments contributing to the guidance of MTs along AF bundles. Extensive evidence confirms a direct interaction between AFs and ion channels and a regulatory functional role associated with actin. Thus, it is clear that the cytoskeleton has a direct connection to membrane components, in particular ion channels and synapses.

The possibility of an evolvable, dynamic, and responsive electrical circuitry within the cell provided by actin and microtubule filaments could be of enormous consequences to our understanding of the way cells operate internally and interact with their environment. In particular, it would cast an entirely new light on cell differentiation, cell division, and cell–cell communication. While an integrated theory of this type of behavior is far from being constructed, its individual elements are gradually taking shape.

Later, we demonstrate how each individual cytoskeletal component is capable of supporting ionic wave propagation and how they may interact to manipulate incoming signals and control their effect on the neuron's dynamic behavior. In Section II we present a description of how the neural cytoskeleton interacts with the cell membrane. In Section III we outline the physical mechanism of ion wave propagation along AFs. The following section discusses the mechanism of long-range spatiotemporal ion waves along MTs. Section V integrates the mechanisms of signaling in the two types of filaments into a cytoskeletal network. The chapter closes with a discussion and conclusions.

II. THE INTERRELATION BETWEEN THE NEURAL CYTOSKELETON AND THE MEMBRANE

Usually, F-actin and microtubular cytoskeletal networks are thought to fulfill separate, independent cellular roles. Highly dynamic actin networks are known for their role in cell motility, in particular the spreading of the leading edge and contraction. The more stable microtubule cytoskeleton is best known for its importance in cell division and organelle trafficking. However, studies provide a more unified role, ascribing important roles to the actin cytoskeleton in cell division and trafficking and important roles for MTs in the generation and plasticity of cellular morphology. Coordination between the actin- and microtubule-based cytoskeletons has been observed during cellular migration and morphogenesis, processes that share some similarities with neurite initiation (Dehmelt and Halpain, 2004). A direct physical association between both cytoskeletons has been suggested because MTs often preferentially grow along actin bundles and transiently target actin-rich adhesion complexes. In neurons certain plakins and neuron-specific MAPs, like MAP1B and MAP2, may play a role in linking MTs and AFs, helping in the transition from an undifferentiated state to neurite-bearing morphology.

MAP1B and MAP2 are both known to interact with actin *in vitro* (Selden and Pollard, 1983; Sattilaro, 1986; Pedrotti and Islam, 1996; Togel *et al.*, 1998). It is likely that by cross-linking, MAP2 and/or MAP1B associated with both cytoskeletons could be involved in guidance of MTs along AF bundles. Alternatively, MAPs could shuttle from MTs to actin and alter F-actin behavior by actively cross-linking AFs.

Little is known, however, about interactions between neuronal ion channels and the cytoskeleton. Whole-cell and single channel recordings showed that acute disruption of endogenous AFs with cytochalasin D activated voltage-gated K^+ currents in these cells, which was largely prevented by intracellular perfusion with the AF-stabilizer agent, phalloidin. Direct addition of actin to excised, inside-out patches activated and/or increased single K^+ channels. Thus, acute changes in actin-based cytoskeleton dynamics regulate voltage-gated ion channel activity in bipolar neurons. This may be indicative of a more general and quite appealing mechanism by which cytoskeletal structures control feedback mechanisms in neuronal channels.

Theoretical and experimental studies of the electrical properties of AFs and MTs in solution revealed their capability to act as biopolymer wires (Lin and Cantiello, 1993; Tuszynski *et al.*, 2004). This means that these protein polymer filaments are

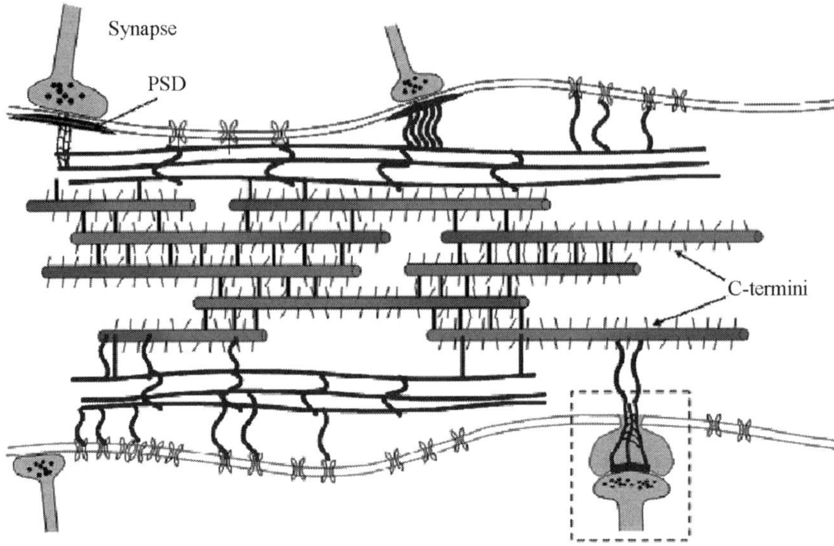

Figure 1. A scheme of the dendritic shaft with MTs arranged in networks of mixed polarity which receive signals in the form of electric perturbations from synapses via AFs connected to MTs by MAP2, or via direct MT connections to postsynaptic density proteins by molecules such as CRIPT. A spiny synapse is depicted in the bottom right where an actin bundle enters the spine neck. (See Color Insert.)

capable of conducting nonlinear ionic waves and even amplifying the signal with respect to the conducting solution (Priel *et al.*, 2005). We conjecture a mechanism in which a direct regulation of ion channels and thus synaptic strength by AFs and associated cytoskeletal structures controls and modifies the electrical response of the neuron. According to this scenario (Fig. 1), MTs arranged in networks of mixed polarity receive signals in the form of electric perturbations from synapses via AFs connected to MTs by MAP2 (Rodriguez *et al.*, 2003) or via direct MT connections to postsynaptic density proteins by molecules such as CRIPT (Passafaro *et al.*, 1999). These signals propagate in the form of ionic waves. Specific physical properties of the propagating ionic waves will be discussed in sections that follow.

III. ACTIN FILAMENTS SUPPORT NONLINEAR IONIC WAVES

AFs are one-dimensional polymers with an uneven distribution of electric charges along the polymer's length giving rise to a spatially dependent electric field arranged in peaks and troughs as originally postulated by Oosawa (1971). This implies large changes in the density of small ions around the polymer with a large dielectric discontinuity in the ionic distribution (Anderson and Record, 1990). The electroconductive medium is a condensed cloud of ions surrounding the polymer and separated from it due to the thermal fluctuations in the solution. The distance beyond which thermal fluctuations are stronger than the electrostatic attractions or repulsions between charges in solution, is defined as the so-called Bjerrum length, λ_B. With the dielectric constant of the medium denoted by ε, the Bjerrum length is given by

$$\frac{\varepsilon^2}{4\pi\varepsilon\varepsilon_0\lambda_B} = k_B T \tag{1}$$

for a given temperature T in Kelvin. Here e is the electronic charge, ε_0 the permittivity of the vacuum and k_B is Boltzmann's constant. For a temperature of 293 K we find that $\lambda_B = 7.13 \times 10^{-10}$ m. Counterion condensation occurs when the mean distance between charges, b, is such that $\lambda_B/b = S > 1$. Each actin monomer carries an excess of 14 negative charges in vacuum, and accounting for events, such as protonation of histidines, and assuming there to be three histidines per actin monomer, there exist 11 fundamental charges per actin subunit (Tang and Janmey, 1996). Assuming an average of 370 monomers per μm we find that there are approximately 4 e/nm in agreement with an earlier statement. Thus we expect a linear charge spacing of $b = 2.5 \times 10^{-10}$ m so $S = 2.85$. As the effective charge, q_{eff}, or renormalized rod charge is the bare value divided by S we find $q_{\text{eff}} = 3.93$ e/monomer. Consequently, it can be shown that approximately 99% of the counterion population is predominantly constrained within a radius of 8 nm (Pollard and Cooper, 1986) round the polymer's radial axis (Zimm, 1986). Significant ionic movements within this "tightly bound" ionic cloud are therefore allowed along the length of the actin, provided that it is shielded from the bulk solution (Oosawa, 1970; Parodi *et al.*, 1985).

Several intriguing experiments (Cantiello *et al.*, 1991; Lin and Cantiello, 1993) indicate the possibility of ionic wave generation along AFs. As the condensed cloud of counterions separates the filament core from the rest of the ions in the bulk solution, we expect this cloud to act as a dielectric medium between the two. It has both resistive and capacitive components associated with each monomer that makes up the AF. Ion flow is expected to occur at a radial distance from the surface of the filament approximately equal to the Bjerrum length. An inductive component is proposed to emerge due to the actin's double-stranded helical structure that induces the ionic flow in a solenoidal manner. Due to the presence of the sheath of counterions around the AF, these polymers act as biological "electrical wires" (Lin and Cantiello, 1993) and have been modeled as nonlinear inhomogeneous transmission lines propagating nonlinear dispersive solitary waves.

Motivated by this picture, physical properties of the ionic distribution along a short stretch of the polymer (the average pitch ~35–40 nm) have been modeled by Tuszynski *et al.* (2004) as an electrical circuit with nonlinear components (Fig. 2). The main elements of the circuit are: (1) a nonlinear capacitor associated with the spatial charge distribution between the ions located in the outer and inner regions of the polymer, (2) an inductance due to the helical geometry of the filament, and (3) a resistor due to the viscosity imposed by the solution.

The characteristic values of the electric components of the circuit are derived using basic laws of electrodynamics. Starting with the conductive part we note that for actin

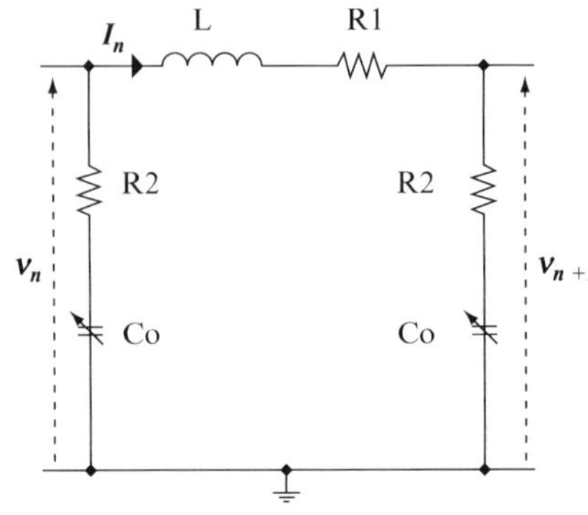

Figure 2. An equivalent electrical circuit for a segment of an actin filament.

in solution, a key feature is that the positively charged end assembles more quickly than the negatively charged end (Sept *et al.*, 1999). This results in an asymmetry in the charges at the ends of the filaments and F-actin's electric polarization. Actin monomers arrange themselves head-to-head to form actin dimers resulting in an alternating distribution of electric dipole moments along the length of the filament (Kobayasi, 1964; Kobayasi *et al.*, 1964). We assume, therefore, that there is a helical distribution of ions winding around the filament at approximately one Bjerrum length (~1 nm). This corresponds to a solenoid in which a fluctuating current flows as a result of a voltage gradient between the two ends. This solenoidal flow geometry leads to an equivalent electrical element possessing self-inductance. From Faraday's law we can derive an effective inductance for the actin filament in solution by

$$L = \frac{\mu N^2 A}{l} \tag{2}$$

where l is the length of the F-actin and A is the cross-sectional area of the effective coil given by

$$A = \pi(r_{\text{actin}} + \lambda_{\text{B}})^2 \tag{3}$$

The number of turns is approximated by simply working out how many ions could be lined up along the length of a monomer. We would then be approximating the helical turns as circular rings lined up along the axis of the F-actin. We also take the hydration shell of the ions into account in our calculation. The hydration shell is then the group of water molecules oriented around an ion. It can be shown that $L = 1.7$ pH for the length of the monomer.

The capacitive element of the electric circuit is obtained following the observation of oppositely charged layers surrounding the filament surface. We envisage the protein surface's negative charge to be distributed homogeneously on a cylinder defining the filament surface. Furthermore, positive counterionic charges in the bulk are expected to form another cylinder at a radius greater than the AF itself, approximately one

Bjerrum length, λ_B, away from the actin surface which includes the condensed ions. The permittivity, ε_0, is given by $\varepsilon = \varepsilon_0\varepsilon_r$ where ε_r is the relative permittivity which we take to be that of water, that is, $\varepsilon_r = 80$. We take the length of an actin monomer typically as $a = 5.4$ nm and the radius of the actin filament, r_{actin}, to be $r_{actin} = 2.5$ nm (Chasan *et al.*, 2002). The next step is to consider a cylindrical Gaussian surface of length a whose radius is r such that $r_{actin} < r < r_{actin} + \lambda_B$.

Application of Gauss's law for the total charge enclosed in the cylinder gives us the following expression for the capacitance

$$C_0 = \frac{2\pi\varepsilon a}{\ln(\frac{r_{actin}+\lambda_B}{r_{actin}})} \tag{4}$$

With the parameters given above we estimate that the capacitance per monomer is $C_0 = 96 \times 10^{-6}$ pF.

The resistive part is obtained from Ohm's law. Taking into account the potential difference and the current I, the magnitude of the resistance, $R = V/I$, for an actin filament is given by

$$R = \frac{\rho \ln((r_{actin} + \lambda_B)/r_{actin})}{2\pi a} \tag{5}$$

where ρ is the resistivity. Typically, K^+ and Na^+ intracellular ionic concentrations are 0.15 M and 0.02 M, respectively (Tuszynski and Dixon, 2001). Kohlrausch's law states that the molar conductance of a salt solution is the sum of the conductivities of the ions comprising the salt solution. Thus

$$\sigma \wedge_0^{K+} c_{K^+} + \wedge_0^{Na+} c_{Na^+} = 1.21(\Omega m)^{-1} \tag{6}$$

Using Eq. (6) with $\rho = \sigma^{-1}$, the resistance estimate becomes $R = 6.11$ MΩ which is much lower than pure water since $R_{water} = 1.8 \times 10^6$ MΩ.

To describe the properties of the whole filament, we simply connect n subcircuits as described earlier to obtain an effective resistance, inductance, and capacitance, respectively, such that:

$$R_{eff} = \left(\sum_{i=1}^{n} \frac{1}{R_{2,i}}\right)^{-1} + \sum_{i=1}^{n} R_{1,i} \tag{7}$$

$$L_{eff} = \sum_{i=1}^{n} L_i \tag{8}$$

and

$$C_{eff} = \sum_{i=1}^{n} C_{0,i} \tag{9}$$

where $R_{1,i} = 6.11 \times 10^6\ \Omega$, $R_{2,I} = 0.9 \times 10^6\ \Omega$, such that $R_{1,i} = 7R_{2,i}$. Note that we have used $R_{1,i} = R_1$, $R_{2,i} = R_2$, $L_i = L$, and $C_{0,i} = C_0$. For a 1 μm of the AF we find therefore

$$R_{\text{eff}} = 1.2 \times 10^9\ \Omega$$
$$L_{\text{eff}} = 340 \times 10^{-12}\ \text{H}$$
$$C_{\text{eff}} = 0.02 \times 10^{-12}\ \text{F}$$

From Kirchhoff's laws, one can derive an equation governing the propagation of voltage along the filament (Tuszynski *et al.*, 2004). One of the key aspects in this model is the nonlinearity of the associated capacitance (Ma *et al.*, 1999; Wang *et al.*, 1999) that eventually gives rise to the self-focusing of the ionic waves. The equations developed for the model originate from the application of Kirchhoff's laws to the RLC resonant circuit of a model actin monomer in a filament. Perhaps the most important finding is the existence of the traveling wave which describes a moving transition region between a high and low ionic concentration due to the corresponding intermonomeric voltage gradient. The velocity of propagation was estimated to range between 1 and 100 m/s depending on the characteristic properties of the electrical circuit model. It is noteworthy that these values overlap with action potential velocities in excitable tissues (Hille, 1992).

Analysis of the model described earlier reveals the possibility of stationary waves in time that may lead to the establishment of spatial periodic patterns of ionic concentration. Lader *et al.* (2000) applied an input voltage pulse with amplitude of approximately 200 mV and duration of 800 μs to an AF, and measured electrical signals at the opposite end of the AF indicating that AFs support ionic waves in the form of axial nonlinear currents. In an earlier experiment (Lin and Cantiello, 1993), the wave patterns observed in electrically stimulated single AFs were remarkably similar to recorded solitary waveforms for electrically stimulated nonlinear transmission lines (Lonngren, 1978). Considering the AF's highly nonlinear physical structure and thermal fluctuations of the counterionic cloud (Oosawa, 1971), the observation of soliton-like ionic waves is consistent with the idea of AFs functioning as biological transmission lines.

This section only provides an indication as to a realistic model of actin that can support soliton-like ionic traveling waves. Modeling relies on data constrained by experimental conditions, and/or assumptions made, including the charge density, which is calculated based on the net surface charges of actin. It should also be considered that soliton velocity is directly proportional to the magnitude of the stimulus, which in a biological setting has not been formally described. Actin interacts with a number of ion channels of different ionic permeability and conductance. Thus, it is expected that channel opening, single channel currents, and other channel properties, including the resting potential of the cell may significantly modify the amplitude and velocity of the soliton-supported waves. This should correlate with the velocity of the traveling waves along channel-coupled filaments. Other parameters that may play a role in this type of electrodynamic interaction are the local ionic gradients and the regulatory role of actin-binding proteins, which can help "focus" the conductive medium or otherwise impair wave velocity.

IV. LONG-RANGE SPATIOTEMPORAL IONIC WAVES ALONG MICROTUBULES

MTs are long hollow cylinders made of $\alpha\beta$-tubulin dimers (Dustin, 1984). These structures have outer diameter measuring approximately 25 nm and inner diameter of approximately 15 nm. MTs form by polymerization of $\alpha\beta$-tubulin dimers in a GTP-dependent process. MTs typically consist of 13 protofilaments and each μm of MT length consists of approximately 1650 heterodimers. It has become apparent that neurons utilize MTs in cognitive processing. Both kinesin and MAP2 that associate with MTs have been implicated in learning and memory (Woolf *et al.*, 1999; Khuchua *et al.*, 2003). Dendritic MTs are implicated in particular, and it is highly probable that the precisely coordinated transport of critical proteins and mRNAs to the postsynaptic density via kinesin along MT tracks in dendrites is necessary for learning, as well as for long-term potentiation (LTP) (Kiebler and DesGroseillers, 2000).

The following molecular dynamics (MD) simulation results focus on the C-termini of neighboring tubulins, whose biophysical properties have a significant influence on the transport of material to activated synapses. This affects cytoskeletal signal transduction and processing as well as synaptic functioning related to LTP. Using MD modeling we calculated conformation states of the C-termini protruding from the outer surfaces of MTs and strongly interacting with other proteins, such as MAP2 and kinesin (Sackett, 1995). To elucidate the biophysical properties of C-termini and gain insight into the role they play in the functioning of dendrites, we developed a quantitative computational model based on the available biophysical and biochemical data regarding the key macromolecular structures involved, including tubulin, their C-termini, and associated MAP2. The proposed model of the C-termini microtubular network is schematically illustrated in Fig. 3, where the tubulin dimer is considered to be the basic unit. Each dimer is decorated with two C-termini that may either extend outwardly from the surface of the protofilament or bind to it in one of few possible configurations.

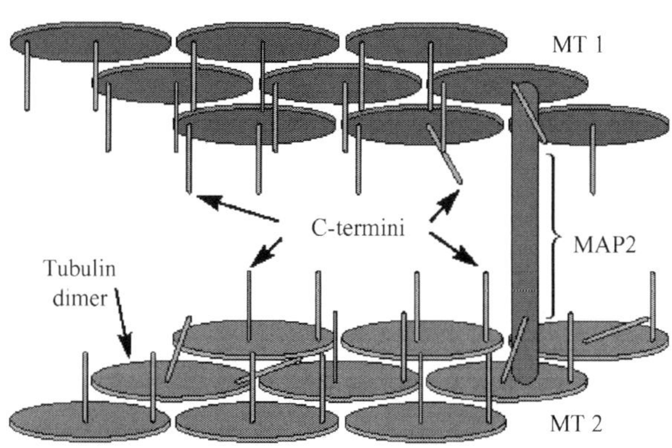

Figure 3. A schematic illustration of the C-termini microtubular network in a dendrite.

The most dynamic structural elements of the system (i.e., its elastic and electric degrees of freedom) are envisaged as conformational states of the C-termini. Each state of the unbound C-terminus evolves so as to minimize the overall interaction energy of the system. The negatively charged C-termini interact with (1) the dimer's surface, (2) neighboring C-termini, and (3) adjacent MAPs. While the surface of the dimer is highly negatively charged overall, it has positive charge regions that attract the C-termini causing them to bend and bind in a "downward" state. The energy difference between the two major metastable states is relatively small, on the order of a few $k_B T$ (Priel *et al.*, 2005).

Our simulations indicate the ability of an ionic wave to trigger a coupled wave of C-termini state changes from their upright to downward orientations. Four examples of "up" and "down" states of the C-termini obtained through bead-spring model simulations are shown in Fig. 4 (left) and results from molecular dynamics simulations

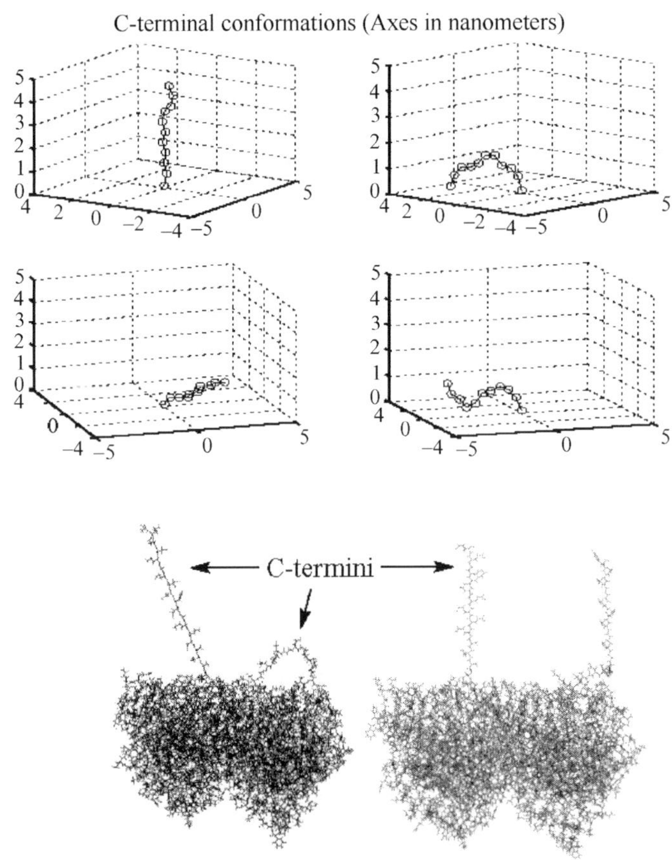

Figure 4. Examples of the conformational states of the C-termini in a tubulin dimer (right) and results from bead-spring model simulations (left).

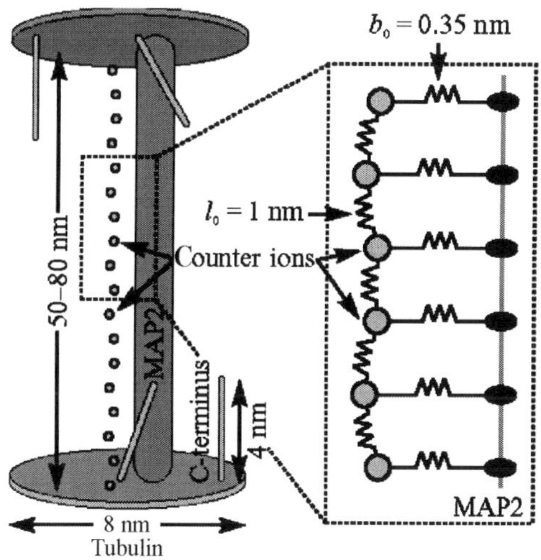

Figure 5. An illustration of a microtubule C-terminus conformational change transfer process. Ionic waves travel along an adjacent MAP via local perturbations to the bounded counterions.

for an actual tubulin dimer are shown in Fig. 4 (right). Calculations of the energy-minimized positions of the individual beads representing the amino acids of the C-terminus in two equivalent forms reveal that the probability of the down position which includes all cases of full or partial attachment is approximately 15%. This means that the system has two major states with a strong bias toward the stretched-up state.

In the remainder of this section, we present an MD simulation study of the interaction between MTs and MAP2, followed by preliminary experimental results of ionic wave conduction by MTs.

A simplified model of the interaction between MAP2 and its ionic environment via counterions was used to investigate the ability of MAP2 to function as a "biological wire" that transfers the conformational change in a C-terminus state to an adjacent MT (Fig. 5).

A perturbation applied to the counterions at one end of the MAP2 drives them out of equilibrium and initiates a wave that travels along the MAP2 (Priel *et al.*, 2005). Figure 6 depicts the main result for a localized perturbation applied for a few picoseconds to the counterions near the binding site of an MAP2. Wave propagation along the chain of counterions is demonstrated as a counterions' displacement parallel to MAP2.

To validate the proposed model we have conducted experiments on electrostimulation of MTs in solution. A detailed description of the experiment and its results is given in Priel *et al.* (2006). Here we only describe the main aspects of the preliminary work on this phenomenon. Figure 7 depicts the experimental setup used. Isolated taxol-stabilized MTs have been shown to be able to amplify an electric signal applied to

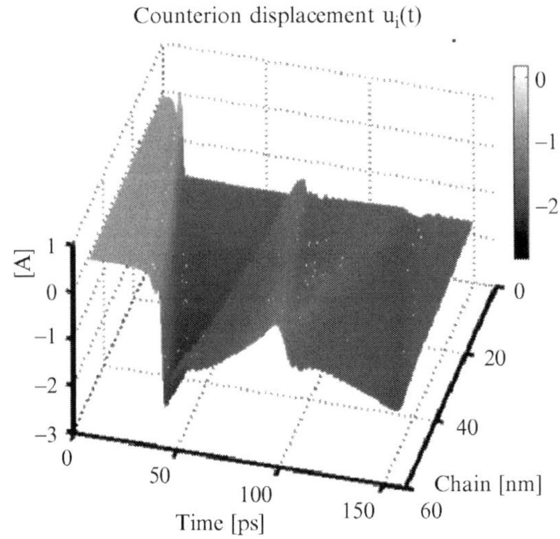

Figure 6. Propagation of a localized perturbation applied to counterions near the binding site of an MAP2.

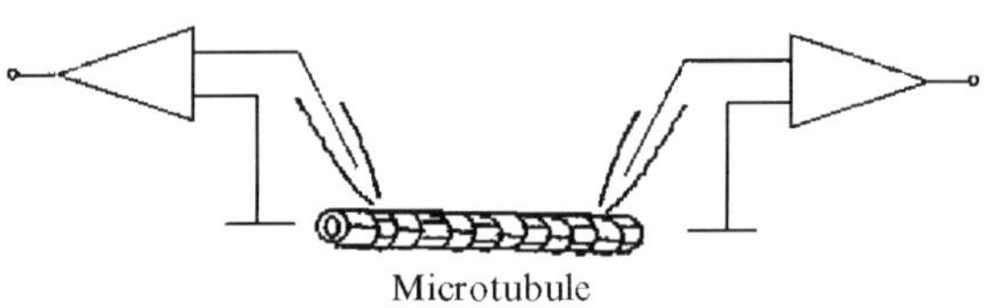

Figure 7. Schematic representation of the experimental setup used for the measurements of MT's ionic conductivity. The MT is attached to two micropipettes that are connected to signal amplifiers.

them through a micropipette. The input signal was of 5–10 ms duration with amplitude in the range of ±200 mV. The signal that arrived at the other end of the MT was more than twice higher than the signal recorded in a control experiment where the same two pipettes were immersed in a solution with no MT making contact to them. The calculated conductivity of MTs was found to be on the order of 10 nS, indicating a high level of ionic conductivity along an MT. By comparison, for a typical ion channel the corresponding value ranges between 5 and 200 ps.

V. DENDRITIC CYTOSKELETON INFORMATION PROCESSING MODEL

Our hypothesis states that the cytoskeletal biopolymers constitute the backbone for ionic wave propagation that interacts with, and regulates dendritic membrane components, such as ion channels, to effectively control synaptic connections. Figure 1 depicts a portion of the dendritic shaft where MTs are decorated by C-termini and interconnected by MAP2 (thick line). Connections between MTs and AFs are shown as well as two types of synaptic bindings. On the upper left side, actin bundles bind to the postsynaptic density (PSD) of a spineless synapse. On the lower right hand a spiny synapse is shown, where actin bundles enter the spine neck and bind to the PSD, which at the other end, is connected to the MTN.

We envision a mechanism in which a direct regulation of ion channels and thus synaptic strength by AFs and associated cytoskeletal structures controls and modifies the electrical response of the neuron. In this picture, MTs arranged in networks of mixed polarity receive signals in the form of electric perturbations, from synapses via AFs connected to MTs by MAP2 (Rodriguez *et al.*, 2003), or via direct MT connections to postsynaptic density proteins by molecules such as CRIPT (Passafaro *et al.*, 1999). As discussed later, the MTN may be viewed as a high-dimensional dynamic system where the main degrees of freedom are related to the conformational state of the C-termini. The input signals perturb the current state of the system that continues to evolve. Hypothetical integration of the ideas mentioned earlier is outlined as follows. Electrical signals arrive at the PSD via synaptic transmission, which in turn elicits ion waves along the associated AFs at the synaptic spine. These dendritic input signals propagate in the form of ionic waves through AFs to the MTN where they serve as input signals. The MTN, operating as a large high-dimensional state machine, evolves these input states, for example, by dynamically changing C-termini conformation. The output from the MTN is the state of the system that may propagate via AFs to remote ion channels. These output functions are assumed to regulate the temporal gating state of voltage-sensitive channels. This process subsequently regulates the membrane conductive properties and controls the axon hillock behavior by changing the rate, distribution, and topology of open/close channels. The overall functions of the dendrite and neuron can thus be regulated in this manner.

The attractiveness of the concept that the cytosol, with its cytoskeletal structures may behave as a large dynamical system is clear as it provides a means for real-time computation without the need for stable attractor states. Moreover, the output is relatively insensitive to small variations in either the MTN (cytoskeletal networks) or the input patterns. It should be noted that the temporal system's state evolves continuously, even without external inputs. Recent perturbations, however, have a long-term effect on the MTN trajectories, that is, there is a memory effect inherent to this system (not to be mistaken with synaptic LTP, which has a much longer timescale). The output from the MTN may converge at or near ion channels to regulate their temporal behavior. The issue of adaptation requires a feedback mechanism that will, at least locally, enable the change of the output function. In the context of neuronal function, with focus on processivity, synaptic strengthening, LTP, and memory enhancement, the output function may simply reflect an effect of the MTN on synaptic channel function, such that the desired state of the channel appears to have a higher probability of being open/close on the presentation of the associated input pattern. One possibility is a Hebbian-based response where a more frequent activity of certain subdomains of the MTN output states gives rise to higher/lower density of AFs connecting to corresponding channels.

VI. DISCUSSION

Considering the abundance of MTNs and AFs in axons and dendritic trees, the findings and theoretical models described earlier may have important consequences for our understanding of the signaling and ionic transport at an intracellular level. Extensive new information (Janmey, 1998) indicates that AFs are both directly (Chasan *et al.*, 2002) and indirectly linked to ion channels in both excitable and nonexcitable tissues, providing a potentially relevant electrical coupling between these current generators (i.e., channels), and intracellular transmission lines (i.e., AFs, MTs). Furthermore, both filaments are crucially involved in cell motility and, in this context, they are known to be able to rearrange their spatial configuration. In nerve cells AFs are mainly located in the synaptic bouton region, whereas MTs are located in both dendrites and the axon. Again, it would make sense for electrical signals supported by these filaments to help trigger neurotransmitter release through a voltage-modulated membrane deformation leading to exocytosis (Segel and Parnas, 1991). Actin is also prominent in postsynaptic dendritic spines, and its dynamics within dendritic spines has been implicated in the postsynaptic response to synaptic transmission. Kaech *et al.* (1999) have shown that general anesthetics inhibit this actin-mediated response. Among the functional roles of actin in neurons, we mention in passing glutamate receptor channels, which are implicated in long-term potentiation. It is therefore reasonable to expect ionic wave propagation along AFs and MTs to lead to a broad range of physiological effects.

The results of the MD modeling described earlier raise the possibility of transmitting electrostatic perturbations collectively among neighboring C-termini and from C-termini on one MT to those C-termini on another MT via MAP2. The potential importance of collective conformational states of C-termini and of transmitting perturbations among the C-termini as a novel information-processing mechanism operating at a subneuronal level is clear. MAP2 and kinesin bind near to the C-termini on tubulin and the electrostatic properties of C-termini affect this binding (Thorn *et al.*, 2000; Al-Bassam *et al.*, 2002). Hence the conformational states of the C-termini must at some point be taken into account in order to understand neural processing that depends on transport of synaptic proteins inside of neurons. Kim and Lisman (2001) have shown that inhibition of MT motor proteins reduces an AMPA receptor-mediated response in hippocampal slices. This means that a labile pool of AMPA-receptors depends on MT dynamics, and MT-bound motors determine the amplitude of excitatory postsynaptic currents (EPSC). A walking kinesin carries with it proteins or mRNA molecules. Since kinesin binds to an MT on a C-terminus, as it steps on it, it brings the C-terminus to the MT surface and makes it ineffective in binding for the next kinesin over a period of time that it takes the C-terminus to unbind and protrude outside. From this we can deduce that long stretches of C-termini in the upright position are going to be most efficient at transporting kinesin and kinesin cargo while C-termini that lie flat are expected to be most efficient at detaching kinesin. Thus, in considering the trafficking of many kinesins, collective electrostatic effects of C-termini become crucially important. Collective states that correspond to transport strategies that will send optimal numbers of kinesin molecules to synaptic zones are likely to occur when synapses are activated (e.g., when synapses are generating EPSCs). One type of kinesin cargo associated with learning and memory is the NMDA receptor, which is well known to be associated with LTP. The PSD might be expected to require replenishment of NMDA receptors at a later time than AMPA receptors are replenished (Kim and Lisman, 2001). Kinesin (specifically KIF17) actively transports NMDA receptor 2B subunits (NR2B) to the region of the PSD (Guillaud *et al.*, 2003). Once NR2B is in the vicinity of an active synapse it dissociates from kinesin and then becomes associated with the PSD, presumably using actin transport as an intermediary step. Hence kinesin-mediated transport of NMDA receptors along MTs has a built-in negative feedback mechanism: whenever too many NMDA receptors are transported to the synaptic site (often located on a spine head), then those NMDA receptors can initiate proteolytic breakdown of MAP2. The latter event would be expected to reduce further transport of NMDA receptors to the synaptic site. The signal transduction molecule CaMKII is also critical for learning and LTP. Similar to the NMDA receptor, CaMKII is transported to active synapses via a kinesin-mediated transport mechanism. However, it appears to be the mRNA for CaMKII to a larger extent than the protein that is transported to spines, since dendrites are enriched with polyribosomes and CaMKII mRNA (Steward and Schuman, 2001). Ribosomes are redistributed from dendrites to spines with LTP (Ostroff *et al.*, 2002). Local translation of CaMKIIα in dendrites appears to be necessary for the late phase of LTP, fear conditioning and spatial memory (Miller *et al.*, 2002). Moreover, changes in synaptic efficacy are often accompanied by changes in morphology; reorganization of underlying MTs is a fundamental factor for these morphological changes.

In summary, this chapter broadly described the physical conditions that enable cytoskeletal polymers, such as AFs and MTs, to act as electrical transmission lines for ion flow along their lengths and along MAP-mediated interconnections between adjacent filaments. In the case of AFs we propose a model in which each protein subunit is equivalent to an electric circuit oscillator. The physical parameters used in the model were evaluated based on the molecular properties of the polymer. Using the general conductivity rules that apply to electrical circuits, we analyzed the properties of ionic waves that propagate along AFs and compared these values to those observed in earlier experiments. In the context of the role played by MTs in neurons, we described the dynamics of C-termini states. We discussed both individual MTs and their networks, including the interactions with ions and signal transmission via MAPs. Experiments on ionic conductivity along AFs and MTs show the validity of the basic assumptions postulated in our models. In light of these results we conjectured a new dendritic-signaling mechanism that involves ion waves along cytoskeletal polymers, which may travel without significant decay over tens of microns, potentially affecting the function of synapses and ion channels.

ACKNOWLEDGMENTS

JT and AP were supported by grants from MITACS, NSERC, and the Allard Foundation. Additional funding from Technological Innovations, LLC of Rochester, NY is gratefully acknowledged.

REFERENCES

Anderson, C. and Record, M., Jr. (1990). Ion distributions around DNA and other cylindrical polyions: Theoretical descriptions and physical implications. *Ann. Rev. Biophys. Biophys. Chem.* **19,** 423–465.

Al-Bassam, J., Ozer, R. S., Safer, D., Halpain, S., and Milligan, R. A. (2002). MAP2 and tau bind longitudinally along the outer ridges of microtubule protofilaments. *J. Cell Biol.* **157,** 1187–1196.

Cantiello, H. F., Patenande, C., and Zaner, K. (1991). Osmotically induced electrical signals from actin filaments. *Biophys. J.* **59,** 1284–1289.

Chasan, B., Geisse, N., Pedatella, K., Wooster, D., Teintze, M., Carattino, M., Goldmann, W., and Canteillo, H. F. (2002). Evidence for direct interaction between actin and the cystic fibrosis transmembrane conductance regulator. *Eur. Biophys. J.* **30,** 617–624.

Dehmelt, L. and Halpain, S. (2004). Actin and microtubules in neurite initiation: Are MAPs the missing link? *J. Neurobiol.* **58,** 18–33.

Dustin, P. (1984). *Microtubules.* Springer-Verlag, Berlin.

Guillaud, L., Setou, M., and Hirokawa, N. (2003). KIF17 dynamics and regulation of NR2B trafficking in hippocampal neurons. *J. Neurosci.* **23,** 131–140.

Hempel, C. M., Hartman, K. H., Wang, X. J., Turrigiano, G. G., and Nelson, S. B. (2000). Multiple forms of short-term plasticity at excitatory synapses in rat medial prefrontal cortex. *J. Neurophysiol.* **83,** 3031–3041.

Hille, B. (1992). *Ionic Channels of Excitable Membranes.* Sinauer Associates, Sunderland, MA.

Janmey, P. (1998). The cytoskeleton and cell signaling: Component localization and mechanical coupling. *Physiol. Rev.* **78,** 763–781.

Kaech, S., Brinkhaus, H., and Matus, A. (1999). Volatile anesthetics block actin-based motility in dendritic spines. *Proc. Natl. Acad. Sci. USA* **96**(18), 10433–10437.

Khuchua, Z., Wozniak, D. F., Bardgett, M. E., Yue, Z., McDonald, M., Boero, J., Hartman, R. E., Sims, H., and Strauss, A. W. (2003). Deletion of the N-terminus of murine MAP2 by gene targeting disrupts hippocampal CA1 neuron architecture and alters contextual memory. *Neuroscience* **119,** 101–111.

Kiebler, M. A. and DesGroseillers, L. (2000). Molecular insights into mRNA transport and local translation in the mammalian nervous system. *Neuron* **25,** 19–28.

Kim, C. H. and Lisman, J. E. (2001). A labile component of AMPA receptor-mediated synaptic transmission is dependent on microtubule motors, actin, and N-ethylmaleimide-sensitive factor. *J. Neurosci.* **21,** 4188–4194.

Kobayasi, S. (1964). Effect of electric field on F-actin orientated by flow. *Biochim. Biophys. Acta* **88,** 541–552.

Kobayasi, S., Asai, H., and Oosawa, F. (1964). Electric birefrigence of actin. *Biochim. Biophys. Acta* **88,** 528–540.

Koch, C. and Segev, I. (2000). The role of single neurons in information processing. *Nat. Neurosci.* **3,** 1171–1177.

Lader, A., Woodward, H., Lin, E., and Cantiello, H. F. (2000). Modeling of ionic waves along actin filaments by discrete electrical transmission lines. *METMBS'00 Int. Conf.* 77–82.

Lin, E. and Cantiello, H. F. (1993). A novel method to study the electrodynamic behavior of actin filaments. Evidence for cable-like properties of actin. *Biophys. J.* **65,** 1371–1378.

Lonngren, K. (1978). Observations of solitons on non-linear dispersive transmission lines. In: *Solitons in Action*, pp. 127–152. Academic Press, New York.

Ma, Z., Wang, J., and Guo, H. (1999). Weakly nonlinear ac response: Theory and application. *Phys. Rev. B.* **59,** 7575.

Miller, S., Yasuda, M., Coats, J. K., Jones, Y., Martone, M. E., and Mayford, M. (2002). Disruption of dendritic translation of CaMKIIalpha impairs stabilization of synaptic plasticity and memory consolidation. *Neuron* **36,** 507–519.

Oosawa, F. (1970). Counterion fluctuation and dielectric dispersion in linear polyelectrolytes. *Biopolymers* **9,** 677–688.

Oosawa, F. (1971). *Polyelectrolytes.* Marcel Dekker, Inc., New York.

Ostroff, L. E., Fiala, J. C., Allwardt, B., and Harris, K. M. (2002). Polyribosomes redistribute from dendritic shafts into spines with enlarged synapses during LTP in developing rat hippocampal slices. *Neuron* **35,** 535–545.

Parodi, M., Bianco, B., and Chiabrera, A. (1985). Toward molecular electronics. Self-screening of molecular wires. *Cell Biophys.* **7,** 215–235.

Passafaro, M., Sala, C., Niethammer, M., and Sheng, M. (1999). Microtubule binding by CRIPT and its potential role in the synaptic clustering of PSD-95. *Nat. Neurosci.* **2,** 1063–1069.

Pedrotti, B. and Islam, K. (1996). Dephosphorylated but not phosphorylated microtubule associated protein MAP1B binds to microfilaments. *FEBS Lett.* **388,** 131–133.

Pedrotti, B., Colombo, R., and Islam, K. (1994). Interactions of microtubule-associated protein MAP2 with unpolymerized and polymerized tubulin and actin using a 96-well microtiter plate solid-phase immunoassay. *Biochemistry* **33,** 8798–8806.

Pollard, T. and Cooper, J. (1986). Actin and actin-binding proteins. A critical evaluation of mechanisms and functions. *Ann. Rev. Biochem.* **55,** 987–1035.

Priel, A., Tuszynski, J. A., and Woolf, N. (2005). Transitions in microtubule C-termini conformations as a possible dendritic signaling phenomenon. *Eur. Biophys. J.* **35,** 40–52.

Priel, A., Ramos, A. J., Tuszynski, J. A., and Cantiello, H. F. (2006). A bio-polymer transistor. Electrical amplification by microtubules. *Biophys. J.* **90,** 4639–4643.

Rall, W. (1959). Branching dendritic trees and motoneuron membrane resistivity. *Exp. Neurol.* **1,** 491–527.

Rodriguez, O. C., Schaefer, A. W., Mandato, C. A., Forscher, P., Bement, W. M., and Waterman-Storer, C. M. (2003). Conserved microtubule-actin interactions in cell movement and morphogenesis. *Nat. Cell Biol.* **5,** 599–609.

Sackett, D. L. (1995). Structure and function in the tubulin dimer and the role of the acid carboxyl terminus. In: *Subcellular Biochemistry – Proteins: Structure, Function and Engineering* (B. B. Biswas and S. Roy, Eds.), Vol. 24, pp. 255–302. Kluwer Academic Publishers, Dordrecht.

Sattilaro, W. (1986). Interaction of microtubule-associated protein 2 with actin filaments. *Biochemistry* **25,** 2003–2009.

Segel, L. and Parnas, H. (1991). What controls the exocytosis of neurotransmitters. In: *Biologically Inspired Physics* (L. Peliti, Ed.). Plennum Press, New York.

Segev, I. and London, M. (2000). Untangling dendrites with quantitative models. *Science* **290,** 744–750.

Selden, S. C. and Pollard, T. (1983). Phosphorylation of microtubule-associated proteins regulates their interaction with actin filaments. *J. Biol. Chem.* **258,** 7064–7071.

Sept, D., Xu, J., Pollard, T., and McCammon, J. (1999). Annealing accounts for the length of actin filaments formed by spontaneous polymerization. *Biophys. J.* **77,** 2911–2919.

Steward, O. and Schuman, E. M. (2001). Protein synthesis at synaptic sites on dendrites. *Annu. Rev. Neurosci.* **24,** 299–325.

Tang, J. and Janmey, P. (1996). The polyelectrolyte nature of F-actin and the mechanism of actin bundle formation. *J. Biol. Chem.* **271,** 8556–8563.

Thorn, K. S., Ubersax, J. A., and Vale, R. D. (2000). Engineering the processive run length of the kinesin motor. *J. Cell Biol.* **151,** 1093–1100.

Togel, M., Wiche, G., and Propst, F. (1998). Novel features of the light chain of microtubule-associated protein MAP1B: Microtubule stabilization, self interaction, actin filament binding, and regulation by the heavy chain. *J. Cell Biol.* **143,** 695–707.

Tuszynski, J. A. and Dixon, J. M. (2001). *Biomedical Applications of Introductory Physics*. John Wiley and Sons, New York.

Tuszynski, J. A., Portet, S., Dixon, J. M., Luxford, C., and Cantiello, H. F. (2004). Ionic wave propagation along actin filaments. *Biophys. J.* **86,** 1890–1903.

Wang, B., Zhao, X., and Guo, H. (1999). Nonlinear quantum capacitance. *Appl. Phys. Lett.* **74,** 2887.

Woolf, N. J., Zinnerman, M. D., and Johnson, G. V. W. (1999). Hippocampal microtubule-associated protein-2 alterations with contextual memory. *Brain Res.* **821,** 241–249.

Zimm, B. (1986). Use of the Poisson-Boltzmann equation to predict ion condensation around polyelectrolytes. In: *Coulombic Interactions in Macromolecular Systems*, pp. 212–215. American Chemical Society, Washington DC.

Chapter 9

The Functional Role of Actin Cytoskeleton Dynamics and Signaling

Christos Stournaras

Department of Biochemistry, University of Crete Medical School, GR 71110 Heraklion Voutes, Greece

In this chapter we summarize experimental evidence supporting the hypothesis that rapid modifications of actin cytoskeleton dynamics may represent cellular sensory machinery, receiving extracellular signals and mediating early cell responses. A model "from structure to function" for actin cytoskeleton dynamics and signaling is presented. Nongenomic signaling pathways triggering rapid actin reorganization are analyzed and their regulatory role in various cell functions is discussed. The emerging data thus obtained on impaired actin cytoskeleton dynamics and signaling in malignant cells are addressed and potential strategies that can be applied for the manipulation of cancer cells based on these findings are provided. It is concluded that rapid modifications of actin cytoskeleton dynamics regulated by specific signaling pathways play a pivotal role in early nongenomic cell responses. Studying and understanding

Advances in Molecular and Cell Biology, Vol. 37

ISBN: 0-444-52868-7
DOI: 10.1016/S1569-2558(06)37009-9

actin dynamics and signaling in malignant cells may represent a novel and challenging domain for cancer treatment in the future.

I. INTRODUCTION

The actin cytoskeleton is a dynamic structure providing architectural support and functional flexibility in cell morphology. The microfilament network primarily underlies the plasma membrane. The organization and dynamics of these structures in nonmuscle cells are regulated by a large number of actin-associated and/or actin-binding proteins. According to their functional role, these proteins may be divided into four major groups: (1) a group regulating actin polymerization dynamics via specific binding to G- or F-actin and barbed or pointed end of the filaments, (2) a group controlling the cross-linking and bundling of filaments, (3) a group regulating the formation of specific microfilament structures, for example, stress fibers, lamelipodia, filopodia, or membrane ruffles, and finally (4) a group regulating the attachment and interactions of actin filamentous structures to plasma membrane.

At present it is well established that rapid changes in actin cytoskeleton dynamics induce potent reorganization of actin filaments. Restructuring of actin microfilament network is modifying cell-substratum adhesion and controlling many aspects of cellular physiology. These include cell motility (Pollard and Borisy, 2003; Wozniak *et al.*, 2004), cytokinesis (Theriot and Satterwhite, 1997; Kaibuchi *et al.*, 1999), regulation of cell shape and volume (Theodoropoulos *et al.*, 1992), membrane trafficking of receptors (Mills *et al.*, 1994; Durrbach *et al.*, 1996) or ion channels (Cantiello, 1995; Papakonstanti *et al.*, 1996), secretion (Bretscher, 1991; Muallem *et al.*, 1995; Kampa *et al.*, 2002; Charalampopoulos *et al.*, 2005), and endocytosis (Durrbach *et al.*, 1996; Caron and Hall, 1998). In addition, actin cytoskeleton reorganization is reported to be involved in transcriptional regulation (Miralles *et al.*, 2003). From these and many other reports it has become widely accepted that modifications in microfilament dynamics play a pivotal role in various cell responses (Isenberg, 1996) and seem to be involved in a large number of human diseases (Carpenter, 2000).

Based on these findings, we have previously postulated that rapid modifications of actin cytoskeleton dynamics may represent cellular sensory machinery receiving extracellular signals and mediating early cell responses (Papakonstanti *et al.*, 2000a). In order to examine this complex hypothesis, three major issues should be successfully addressed: (1) How to study quantitatively actin dynamics in various cell models? (2) Which are the signaling pathways regulating the rapid changes in actin polymerization dynamics? (3) What is the biological role of rapid actin reorganization in cellular functions? The present chapter summarizes the experimental evidence addressing these questions. Nongenomic signaling pathways triggering rapid actin reorganization and their regulatory role in various cellular systems will be presented. The role of altered actin dynamics and signaling in tumor cells will be reviewed and potential applications will be discussed. Finally, a "structure to function model" for actin cytoskeleton dynamics supporting the above hypothesis will be presented and discussed.

II. STUDYING ACTIN CYTOSKELETON DYNAMICS

The last decade saw the elaboration of appropriate techniques to quantitatively measure the intracellular actin polymerization rate. This allowed a detailed, time-dependent analysis of actin dynamics in various cell models. The polymerization equilibrium of actin dynamics was expressed as the ratio of monomer (G-) to total-actin content (Theodoropoulos *et al.*, 1992). The methodological tools widely used are: (1) the DNase I inhibition assay (Blikstad *et al.*, 1978) adapted for monomer- and total-actin content determination in cell preparations (Theodoropoulos *et al.*, 1992), (2) the rhodamine-phalloidin fluorescence assay (Cable *et al.*, 1995) for filamentous actin quantification, and (3) Western blot analysis of detergent soluble and insoluble cell extracts (Nurko *et al.*, 1996) for monomer- and filamentous-actin measurements. The introduction of the G-/total- and G-/F-actin ratio (Theodoropoulos *et al.*, 1992; Papakonstanti *et al.*, 1996; Stournaras *et al.*, 1996; Jordan and Wilson, 1998) proved to be a useful tool in expressing quantitatively the time-dependent modifications of cellular actin cytoskeleton dynamics.

A. Actin Cytoskeleton Dynamics in Malignant Cells

By applying these techniques, others and ourselves have provided ample experimental evidence for the importance of actin cytoskeleton rearrangements in various cell models. One aspect of specific interest was focused on the role of actin organization in malignant cells. The sensitive dynamic equilibrium of G- to F-actin was found actively modulated during malignant cell transformation (Stournaras *et al.*, 1996; Jordan and Wilson, 1998). We found that actin polymerization dynamics may characterize malignant transformation, since systematic quantitative determinations of G-, F-, and total-actin content in various malignant cell types revealed important alterations of the actin cytoskeleton dynamics (Katsantonis *et al.*, 1994; Koukouritaki *et al.*, 1996, 1999; Stournaras *et al.*, 1996; Moustakas and Stournaras, 1999; Kampa *et al.*, 2002; Kallergi *et al.*, 2003; Papakonstanti *et al.*, 2003). Figure 1 summarizes the conceptual conclusions of these findings. In particular, in all malignant cell types investigated, a dramatic decrease of total- and F-actin content and an increase of the relative monomer-actin level were documented as compared to measurements in the corresponding marker in normal cells. Consequently, a significant increase in the G-/total-actin ratio was calculated in malignant cells, indicating that this ratio may represent a novel tumor marker for cell transformation (Jordan and Wilson, 1998). In addition, transformed cells showed impaired microfilament organization and stability, as indicated by morphological analysis in the presence of increasing concentrations of actin-destabilizing agents, for example, cytochalasins (Katsantonis *et al.*, 1994; Stournaras *et al.*, 1996). The impaired actin microfilament organization and stability was also correlated with the disassembly of focal contacts and enhanced motility. This was clearly demonstrated in LNCaP and DU145 human prostate cancer cells (Hatzoglou *et al.*, 2005), in MCF7 human breast cancer cells (Kallergi *et al.*, 2003),

Figure 1. The sensitive dynamic equilibrium of monomeric G-actin to filamentous F-actin is actively modulated during malignant cell transformation. (See Color Insert.)

and in NIH-3T3/H-Ras–transformed fibroblasts (Moustakas and Stournaras, 1999). These reports indicate a potential important role of actin dynamics in regulating cell phenotype during malignant transformation.

III. NONGENOMIC SIGNALING PATHWAYS TRIGGERING RAPID ACTIN REORGANIZATION AND REGULATING CELL RESPONSES

Rapid reorganization of the actin cytoskeleton is one of the earliest cellular responses to many extracellular signals. So far, various signals including growth factors, hormones, cytokines, ions, and opioids have been studied by our group during the last decade. Binding and activation of appropriate membrane receptors or ion channels by these factors trigger key signaling effectors via nongenomic pathways that regulate actin remodeling. Table I summarizes part of these results. In the following sections some of the most exciting and prominent ones will be given.

A. Example I: Actin Reorganization as a Predominant Event in Nongenomic Steroid Hormone Signaling

In recent years, several reports have indicated that steroid hormones may induce cell responses in cells lacking classical receptors. In addition, very rapid effects were also obtained, occurring during a time lag noncompatible with the classical genomic action of steroid hormones (Wehling, 1997; Losel and Wehling, 2003; Norman *et al.*, 2004). These findings led to the identification of estrogen-, progesterone-, glucocorticoid-, and androgen-binding sites in the plasma membrane, which are considered as new receptors for the steroid hormone action (Jensen, 1996; Koukouritaki *et al.*, 1996; Grazzini *et al.*, 1998; Nadal *et al.*, 1998; Nemere and Farach-Carson, 1998; Benten *et al.*, 1999; Kampa *et al.*, 2002).

Table I
Extracellular Signals Activating Nongenomic Signaling Toward Actin Reorganization in Different Tissues or Cell Types

Cell mode/tissue	Signal	Signaling molecules	Rapid/nongenomic	Long term	References
Liver	-Na^+ -Insulin	?	Yes	Yes	Theodoropoulos *et al.*, 1992
Kidney	Glucocorticoids	cAMP-FAK-paxillin	Yes	Yes	Koukouritaki *et al.*, 1996, 1999
Kidney	Na^+/Pi	PLC-?1-villin	Yes	No	Papakonstanti *et al.*, 1996, 2000b
Kidney	Opioids	PI-3K-PAK-Cdc42/Rac1	Yes	-	Papakonstanti and Stournaras, 2002
Kidney	Opioids	LIMK1-cofilin	-	Yes	Papakonstanti and Stournaras, 2002
Fibroblasts	TGF-β	RhoA/B-ROCK-LIMK2-cofilin	Yes	Yes	Moustakas and Stournaras, 1999; Vardouli *et al.*, 2005
Breast	Opioids	FAK-PI-3K-Cdc42/Rac1	Yes	No	Kallergi *et al.*, 2003
Prostate	Androgens	FAK-PI-3K-Cdc42/Rac1	Yes	Yes	Kampa *et al.*, 2002; Papakonstanti *et al.*, 2003; Hatzoglou *et al.*, 2005
Kidney	TNF	FAK-paxillin-PI-3K-Cdc42/PLC-?1	Yes	Yes	Papakonstanti and Stournaras, 2004
Sympathoadrenal cells	Neurosteroids	PI-3K/PLC-γ1	Yes	No	Charalampopoulos *et al.*, 2005

A detailed analysis of the involvement of actin cytoskeleton dynamics in nongenomic steroid action was initially reported for glucocorticoids (Koukouritaki *et al.*, 1996, 1997). In these reports, exposure of Ishikawa human endometrial cells to dexamethasone was shown to induce rapid actin polymerization and a marked decrease in the G-/total-actin ratio. Systematic morphological analysis revealed a clear reorganization of actin cytoskeleton and stabilization of microfilamentous structures. From these findings it was concluded that rapid restructuring of the actin network offered new perspectives for the understanding of early cellular responses to the glucocorticoids (Koukouritaki *et al.*, 1996). The key signaling molecules initiating this phenomenon were identified to be focal adhesion kinase (FAK) and paxillin (Koukouritaki *et al.*, 1999). Phosphorylation and activation of these effectors led to regulation of rapid actin reorganization. These findings provided strong evidence for the existence of a specific nongenomic signaling pathway triggered by glucocorticoids. In addition, a late increase of Rho GTPase expression was documented, suggesting that small GTPases may control the long-term actin redistribution observed in endometrial cells. The illustration in Fig. 2 summarizes the reported nongenomic glucocorticoid signaling to the actin cytoskeleton.

In a more recent study, we have reported the identification of membrane-binding sites for testosterone in intracellular androgen receptor (iAR)-positive LNCaP and in

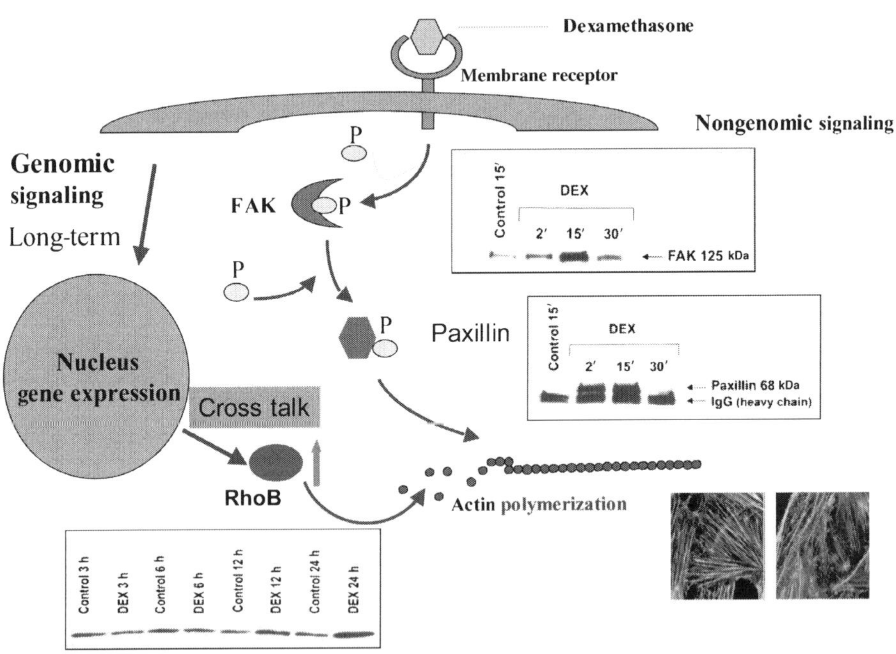

Figure 2. The nongenomic glucocorticoid signaling leading to rapid actin reorganization in Ishikawa human endometrial cells. Reprinted with permission from the article entitled "Actin Cytoskeleton and Nongenomic Glucocorticoid Action," published in *Mol. Med.*, 1999, 5:731–742.

iAR-negative DU145 human prostate cancer cells, respectively (Kampa *et al.*, 2002; Hatzoglou *et al.*, 2005). Activation of these membrane-binding sites resulted within minutes in a significant alteration of actin polymerization dynamics, as indicated by an effective decrease in the G-/total-actin ratio. This was followed by marked rearrangement of the actin network and a rapid increase in prostate-specific antigen (PSA) secretion into the culture medium (Kampa *et al.*, 2002). The molecular mechanism controlling the actin cytoskeleton modifications induced by testosterone membrane binding in LNCaP cells was analyzed (Papakonstanti *et al.*, 2003). The nongenomic pathway that transmits the signal to the actin cytoskeleton is through the focal adhesion kinase (FAK) → phosphatidylinositol-3 kinase (PI-3K) → cell division cycle 42 (Cdc42)/Rac1 → actin polymerization cascade. Blockage of this cascade by PI-3 kinase inhibitors resulted in inhibition of both actin reorganization and PSA secretion. The illustration presented in Fig. 3 summarizes the reported novel membrane-induced androgen signaling to actin cytoskeleton.

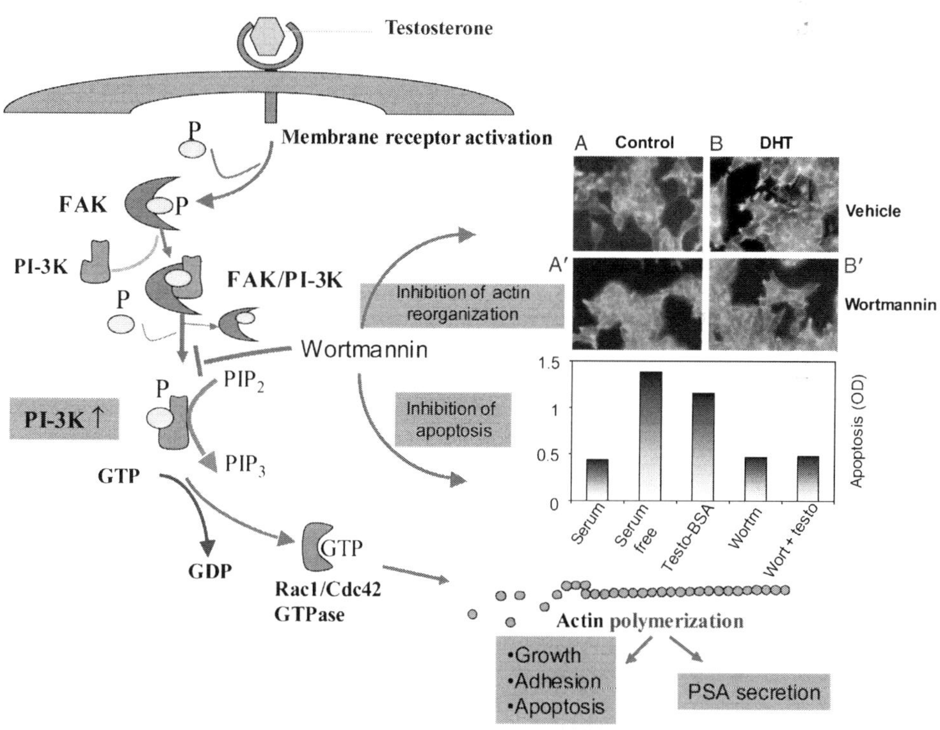

Figure 3. The nongenomic testosterone signaling leading to rapid actin reorganization regulates cell responses in LNCaP human prostate cancer cells. The four micrographs (A, B, A′, and B′) are being reprinted with permission from *The Endocrine Society, Copyright 2003* (Papakonstanti, 2003).

Biological Role of Actin Cytoskeleton Signaling in Nongenomic Steroid Hormone Action

A highly interesting aspect of these findings with potential clinical applications is focused on the biological relevance of the activation of androgen membrane-binding sites in prostate cancer cells. In a study, we have reported that activation of androgen membrane-binding sites induced inhibition of cell growth and migration in human prostate cancer cells. In addition, potent induction of apoptosis was observed in both iAR-positive and iAR-negative prostate cancer cells (Hatzoglou *et al.*, 2005). Induction of apoptosis in both cell lines was directly correlated with the actin cytoskeleton reorganization. This was clearly demonstrated by using cytoskeletal inhibitors to block actin redistribution (Papadopoulou and Stournaras, in preparation). In these experiments, induction of apoptosis was inhibited, indicating that actin reorganization is an important step in this process. This assumption was further strengthened through specific blockage of the newly identified signaling pathway. Inhibition of a key signaling step in this cascade (PI-3K activation, Fig. 3) blocked actin reorganization and caused complete inhibition of the apoptotic cellular response. These observations are in line with recent reports to be discussed momentarily, showing that actin dynamics regulated by an early, nongenomic signaling cascade play a key role in the antiapoptotic response of opossum kidney cells to tumor necrosis factor-α (TNF-α). These experimental findings provide strong evidence that rapid actin cytoskeleton reorganization may control apoptotic or antiapoptotic cell responses.

The *in vitro* effects of the activation of androgen membrane-binding sites prompted us to evaluate their potential role *in vivo*. In LNCaP cell-inoculated nude mice treatment with the nonpermeable bovine serum albumine (BSA)-coupled testosterone for 1 month resulted in 60% reduction of tumor size compared to control animals. This effect was not affected by the anti-androgen flutamide. From these findings it was postulated that activation of androgen membrane-binding sites acting via nongenomic actin cytoskeleton signaling induces apoptotic regression of prostate cancer cells both *in vitro* and *in vivo* (Hatzoglou *et al.*, 2005). It is worth noticing that in a parallel study, histological evaluation of human specimens revealed that androgen membrane-binding sites are predominantly expressed in malignant but not in benign hyperplasia or normal peritumoral cells (Stathopoulos *et al.*, 2003; Dambaki *et al.*, 2005). Taken together, these findings indicate that activators of androgen membrane receptors may represent a novel class of antitumor agents of prostate cancer (Hatzoglou *et al.*, 2005; Nifli *et al.*, 2005).

B. Example II: Actin Reorganization Regulates Nongenomic Transforming Growth Factor-β Signaling

Transforming growth factor-β (TGF-β) is a well-studied extracellular protein that regulates cell growth and differentiation and modulates cell shape (Roberts, 1998; Massagué, 2000; Lutz and Knaus, 2003). TGF-β induces motility in various cell types, is involved in wound-healing processes, and influences epithelial to mesenchymal transitions (Dumont and Arteaga, 2003; Grünert *et al.*, 2003; Roberts and Wakefield, 2003). The classical genomic TGF-β–signaling apparatus involves a complex of type I and type II receptors and the downstream signaling effectors Smad (Shi and Massagué, 2003). The TGF-β–activated type I receptor phosphorylates and activates Smad 2 and Smad 3 (regulatory, R-Smads). R-Smads oligomerize with the unique Smad 4 and translocate to the nucleus where they regulate expression of target genes (Moustakas *et al.*, 2001; Shi and Massagué, 2003). The ability of TGF-β to modulate cell morphology and the actin cytoskeleton in a variety of cell types has been reported (Moustakas and Stournaras, 1999; Piek *et al.*, 1999; Bhowmick *et al.*, 2001; Edlund *et al.*, 2002), indicating that studying TGF-β signaling in the actin cytoskeleton offers the means to evaluate regulatory pathways of malignant cell proliferation and motility.

In recent studies, we addressed the mechanistic interrelationship between TGF-β–mediated tumor regression or promotion and the modulation of cell shape via regulation of the actin microfilament network. TGF-β was reported to suppress the transformed phenotype and regulate actin reorganization via rapid and potent polymerization of actin microfilaments in oncogenic NIH-3T3/Ras fibroblasts (Moustakas and Stournaras, 1999). This effect persisted for hours; it was correlated with modulation of the organization of focal adhesions and resulted in long-term inhibition of anchorage-independent growth. In addition, it was shown that TGF-β enhances the abundance of Rho GTPases and that Rho deactivation blocks the fast restructuring effect of TGF-β on actin dynamics (Moustakas and Stournaras, 1999). These data imply the existence of a signaling pathway leading to rapid actin reorganization that involves Rho activation.

Nongenomic Signaling Pathway Regulating TGF-β–Induced Actin Reorganization and Its Biological Role

Despite our understanding of the classical Smad pathway, we have demonstrated that rapid regulation of actin dynamics downstream of TGF-β in fibroblasts appeared to be controlled by a novel, nongenomic signaling cascade (Vardouli *et al.*, 2005). As presented in the illustration in Fig. 4, the effector cascade consists of RhoA- and RhoB-GTPases, Rho-coiled-coil kinase 1 (ROCK1), LIM domain kinase 2 (LIMK2), and cofilin. Activation of Rho GTPases triggers the downstream kinases ROCK1 and LIMK2, thereby transmitting a signal to cofilin and modulating its ability to shift the actin polymerization equilibrium. Inactivation of this cascade through transfectans or inhibitors of the key signaling molecules (by small interfering RNA's (siRNA's) dominant-negative constructs or specific inhibitors) totally blocked actin reorganization. The potent actin polymerization and microfilament reorganization induced by TGF-β was shown to control cell motility, as indicated by enhanced cell migration observed in wound-healing assays in the presence of TGF-β. This phenomenon was inhibited in the presence of the specific ROCK1 inhibitor Y27632, indicating a

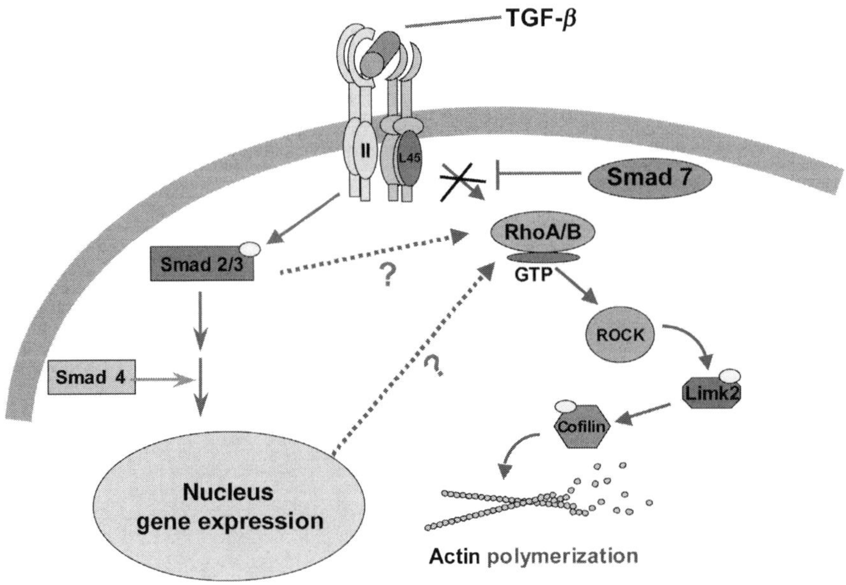

Figure 4. The nongenomic TGF-β–signaling cascade regulating actin reorganization and the possible cross talk with the classical Smad pathway in mouse fibroblasts.

regulatory role of the Rho/ROCK/cascade and early actin reorganization in cell migration (Vardouli *et al.*, in preparation).

An interesting point that remains to be elucidated is the possible cross talk between classical Smad signaling and the novel Rho/ROCK pathway (Fig. 4). Experimental evidence indicates that the type I TGF-β receptor is the activator of robust actin reorganization (Vardouli *et al.*, 2005). In addition, constitutively activated type I receptor with the mutated Smad-docking site (L45 loop) failed to activate the nongenomic signaling cascade or to trigger actin reorganization. In line with this, blocking of the classical Smad pathway by the inhibitory Smad 7 was found to inhibit TGF-β–induced Rho activation and actin reorganization. These findings suggest that the Smad pathway may be involved in the regulation of Rho GTPases and actin reorganization. Additional experiments are needed to establish a possible link between the classical genomic Smad pathway and the novel nongenomic cascade and to identify the molecule(s) recruited to bridge TGF-β receptor type I to Rho GTPases. Moreover, it would be of particular interest to elucidate the functional role of the Rho small GTPases in the observed actin cytoskeleton restructuring and regulation of cell motility. It is, in particular, to analyze the cross talk between the rapid nongenomic Rho-dependent mechanism of actin reorganization and possible Rho-dependent transcriptional mechanisms controlling the long-term alterations of actin cytoskeleton dynamics.

C. Example III: Actin Reorganization and TNF-α Signaling

TNF-α, one of the most pleiotropic proinflammatory cytokines, is produced by activated macrophages and monocytes in response to inflammation, injury, or infection, and by heart and kidney in response to ischemia and perfusion (Meldrum and Donnahoo, 1999; Baud and Karin, 2001). TNF-α may induce necrosis or apoptosis (Natoli *et al.*, 1998), although it does also elicit antiapoptotic cell signals. TNF-α–induced cell survival is dependent mostly on the nuclear factor-kappa B (NF-κB) response (Barnes and Karin, 1997; Baud and Karin, 2001; Vancurova *et al.*, 2001) and in some cases independently through signaling cascades of NF-κB (Pastorino *et al.*, 1999; Madge and Pober, 2000). This cytokine is also able to reorganize the actin cytoskeleton in various cell systems (Wojciak-Stothard *et al.*, 1998; Koukouritaki *et al.*, 1999; Puls *et al.*, 1999; Papakonstanti and Stournaras, 2004). Differential microfilament reorganization occurs in these cell systems. Analysis of actin dynamics revealed that TNF-α showed robust actin polymerization in glomerular epithelial cells (Koukouritaki *et al.*, 1999) as well as in fibroblasts (Puls *et al.*, 1999), whereas depolymerization of actin was demonstrated in proximal tubular epithelial cells (Papakonstanti and Stournaras, 2004). These observations indicate a differential mode of action of TNF-α, either due to dissimilar cellular specificity or activation of distinct key signaling molecules. The latter may also correlate with differential cell responses. These results were addressed by our group and the conclusions we drew are summarized as follows:

Exposure of glomerular epithelial cells to TNF-α resulted in marked actin polymerization, followed by microfilament reorganization and significant changes in cell volume (Koukouritaki *et al.*, 1999; Papakonstanti *et al.*, 2000a). Analysis of actin polymerization dynamics demonstrated a rapid decrease in the G-/total-actin ratio, thus substantiating the morphological observations. In this cell system, TNF-α increased the levels of vinculin and FAK phosphorylation, suggesting alterations in the membrane attachment molecules, followed by phosphorylation of paxillin. These data support the notion that rapid actin cytoskeleton polymerization and microfilament redistribution may be a crucial step in controlling the signal transduction pathway of TNF-α in response to cell volume regulation of glomerular epithelial cells (Papakonstanti *et al.*, 2000a).

Actin Reorganization Regulates the TNF-α Antiapoptotic Effect

When we used opossum kidney proximal tubular epithelial cells to test whether TNF-α exerts apoptotic or antiapoptotic cell effects, we observed rapid actin depolymerization, as reflected by actin dynamics measurements (i.e., an increased G-/total-actin ratio). In addition, TNF-α was shown to promote survival of opossum kidney cells by inhibiting caspase-3 activity, an effect that depended on actin reorganization and NF-κB activation (Papakonstanti and Stournaras, 2004). Analysis of the signaling

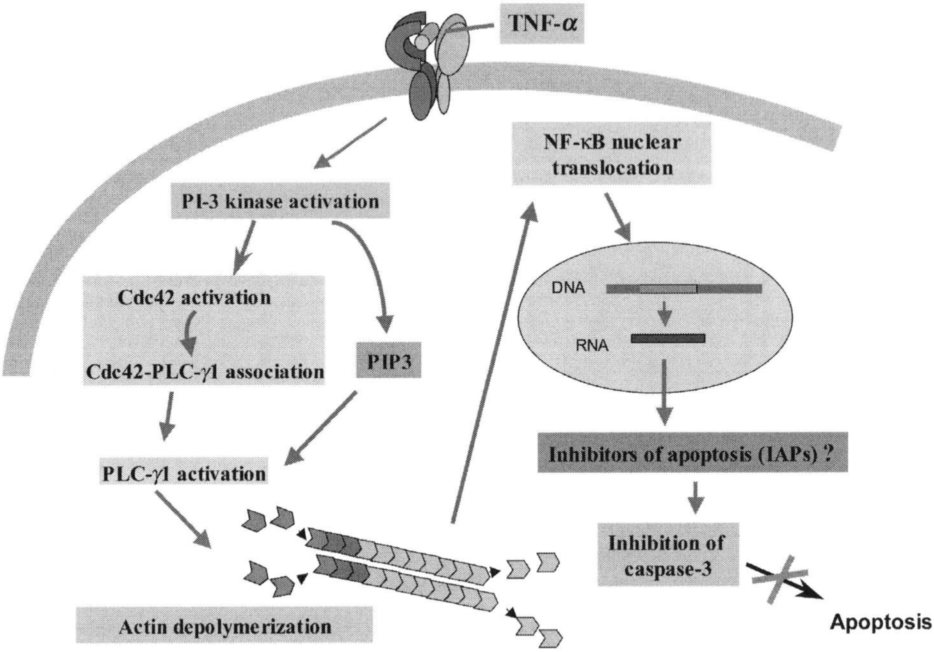

Figure 5. The TNF-α–signaling pathway leading to actin reorganization regulates the antiapoptotic mechanism in proximal tubular epithelial cells.

mechanism revealed rapid activation of PI-3 kinase, Cdc42/Rac1, and phospholipase C-γ1 (PLC-γ1). Details of the cascade of fast (nongenomic) and slow (long term) TNF-α signaling are shown in the illustration in Fig. 5. Dominant negative constructs for Cdc42, or a specific PLC-γ1 inhibitor prevented actin redistribution and depolymerization. Moreover, the actin-stabilizing agent, phallacidin, which inhibits TNF-α–induced actin depolymerization, completely blocked nuclear translocation of NF-κB and the subsequent inhibition of caspase-3. Such a result is indicative of the critical role of actin filament redistribution for NF-κB activation and its subsequent inhibition of caspase-3. It is noteworthy that in this cell system, even though upstream NF-κB–related events could occur as the result of Cdc42 and/or PLC-γ1 activation, the inhibition of early actin depolymerization is sufficient to abrogate the TNF-α–induced cell response (Papakonstanti and Stournaras, 2004). It was therefore concluded that actin dynamics regulated by an early, nongenomic signaling cascade play a key role in the antiapoptotic response of opossum kidney cells to TNF-α.

An additional critical observation, which deserves to be highlighted here, is the differential actin reorganization induced by the same signal. TNF-α activates distinct nongenomic signaling molecules, thereby exacting opposite reorganization effects on actin cytoskeleton dynamics (polymerization/depolymerization). These, in turn, regulate different biological responses. While indicative of cellular specificity to TNF-α action, it also implies that rapid actin reorganization may be an important common regulatory step for specific cell responses.

D. Example IV: Differential Actin Reorganization as Observed in Opioid Signaling

Differential actin reorganization dynamics have also been reported for opioid treatment in various normal and malignant cell lines. Generally, opioids act through membrane receptors, which belong to the seven transmembrane loop receptor superfamily, involving G regulatory elements (Reisine and Bell, 1993). Binding of these agents to their cognitive receptors decreases cell proliferation in different cell systems (Hatzoglou *et al.*, 1996a,b), arrests cells in the G2/M phase of cell cycle promoting apoptosis (Maneckjee and Minna, 1994), and induces drastic reorganization of the actin cytoskeleton (Papakonstanti *et al.*, 1998; Panagiotou *et al.*, 1999).

Analysis of the actin dynamics in nonmalignant opossum kidney tubular cells revealed rapid actin depolymerization, in fact within minutes. However, it was transient in nature, since repolymerization and restructuring of actin filaments appeared 6 h after treatment (Papakonstanti *et al.*, 1998). To account for the molecular mechanism underlying this complex effect, a novel signaling cascade was identified. That is, a cascade involving direct association of PI-3 kinase with p21-activated protein kinase 1 (PAK1), which in turn regulates PAK1 activity through a Cdc42/Rac- and Akt-independent mechanism. Finally, activated PAK1 phosphorylates actin directly, inducing actin depolymerization and restructuring of the microfilaments (Papakonstanti and Stournaras, 2002). In these studies, the long-term mechanism regulating the repolymerization of actin was defined as well: The association of PAK1 with PI-3 kinase as well as the phosphorylation of actin stops, whereas the RhoA/ROCK/LIMK1/cofilin pathway is activated resulting in repolymerization of actin and the reformation of stress fibers shown. Figure 6A illustrates the overall signaling cascade pertaining to both the early and late signaling molecules controlling actin reorganization.

Biological Significance of Differential Cytoskeleton Dynamics and Signaling Induced by Opioids

In a parallel study (Kallergi *et al.*, 2003), an analysis of actin dynamics in malignant breast epithelial cells was carried out. This was in response to opioid receptor activation, which led to opposite effects on actin reorganization. During the early phase, powerful and resistant actin polymerization and microfilament stabilization

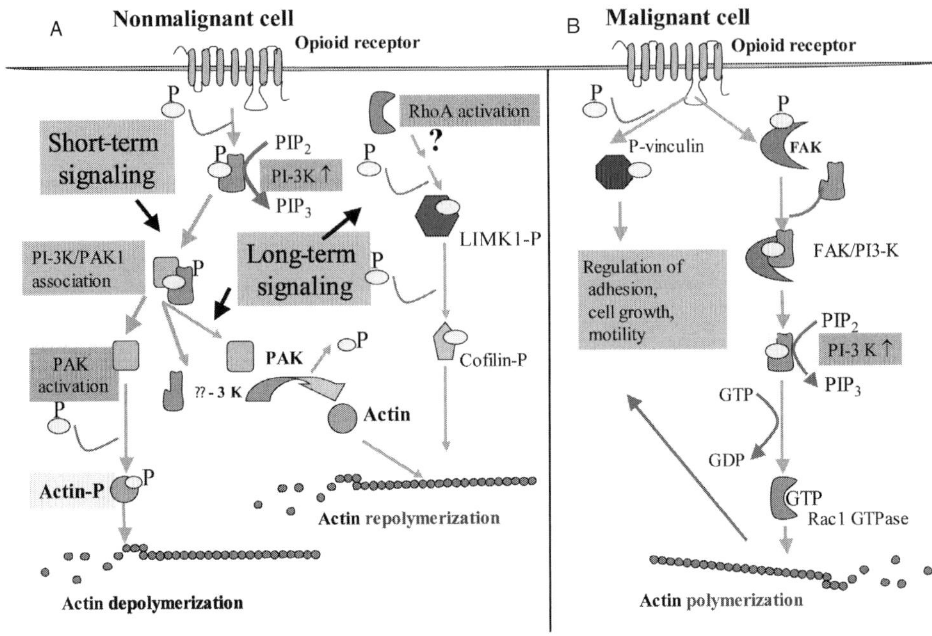

Figure 6. Distinct opioid-signaling pathways are activated in normal and malignant cells, leading to differential actin cytoskeleton reorganization.

were observed. This was regulated by a distinct signaling pathway involving the FAK → PI-3K → Cdc42/Rac1 → actin polymerization cascade (Fig. 6B). It will be recalled that the same signaling cascade was activated in human prostate cancer cells on activation of androgen membrane-binding sites. This resulted in robust actin restructuring (Fig. 3). However, in MCF12A nonmalignant breast epithelial cells, expressing opioid receptors, this pathway was inactive (Kallergi *et al.*, 2003). In such cells, opioids induced only transient actin depolymerization as described for nonmalignant tubular cells. These findings meant that there are distinct signaling pathways that regulate differential opioid effects on actin cytoskeleton in normal and malignant cells (Fig. 6A or B). The conclusion drawn here was that activation of the same membrane receptor by the same extracellular stimuli may induce differential actin cytoskeleton reorganization in malignant and nonmalignant cells. These effects are controlled by different signaling pathways, indicating cellular specificity of this phenomenon. The possibility this raises is that the key signaling molecules identified in malignant cells (breast and prostate) may represent novel targets for specific therapeutic interventions.

IV. POTENTIAL APPLICATIONS OF ALTERED ACTIN DYNAMICS AND SIGNALING IN MALIGNANT CELLS

As already indicated, impaired actin cytoskeleton dynamics in malignant cells may represent a novel and reliable tumor marker for cell transformation. In addition, an aspect of specific interest is focused on the differential signaling mechanisms shown to regulate actin dynamics in normal and malignant cells. Taken together, these results imply potential applications of those findings in cancer diagnosis and treatment. The G-/total-actin ratio may represent a useful diagnostic tool for the assessment of malignant transformation and is now under large-scale evaluation in histological preparations of human tumors.

Inhibition of cell viability, cell growth, and cell proliferation has been reported for various tumor cell types by using specific anti-actin drugs, which disrupt selectively the unstable microfilament network of malignant cells (Stournaras *et al.*, 1996; Hatzoglou *et al.*, 2005). Cell motility and invasiveness can be downregulated by using specific kinase inhibitors, siRNAs or antisense nucleotides to block distinct actin-signaling molecules activated selectively in cancer cells without affecting normal cells (Kallergi *et al.*, 2003; Hatzoglou *et al.*, 2005; Vardouli *et al.*, 2005).

Finally, inhibition of anchorage-independent growth or induction of proapoptotic mechanisms in cancer cells have also been reported using appropriate extracellular signals to selectively reorganize and stabilize the actin cytoskeleton and focal contacts (Moustakas and Stournaras, 1999; Hatzoglou *et al.*, 2005). These reports represent only a limited number of potential strategies that can be applied for the manipulation of cancer cells, based on new information about actin cytoskeleton dynamics and signaling. Accordingly, we believe that studying and understanding actin dynamics and signaling in malignant cells may provide a novel and challenging approach to cancer treatment in the future.

V. "FROM STRUCTURE TO FUNCTION" MODEL FOR ACTIN CYTOSKELETON DYNAMICS AND SIGNALING

The conceptual results presented thus far provide convincing evidence that actin cytoskeleton dynamics—in addition to their structural mechanical properties—may function as a *second messenger system*, receiving extracellular signals and mediating rapid nongenomic cell responses. A model called "from structure to function" for actin cytoskeleton dynamics and signaling is presented in Fig. 7. According to this model, extracellular signals through the binding to and activation of appropriate receptors trigger rapid, nongenomic signaling cascades through fast activation of key signaling molecules. These may control alterations in actin cytoskeleton dynamics. Actin reorganization, in turn, regulates various early cell responses. Actin modifications seem to be a crucial part of the signaling events, since inhibition of actin reorganization blocks cell responses. In parallel, receptor activation triggers genomic signaling pathways,

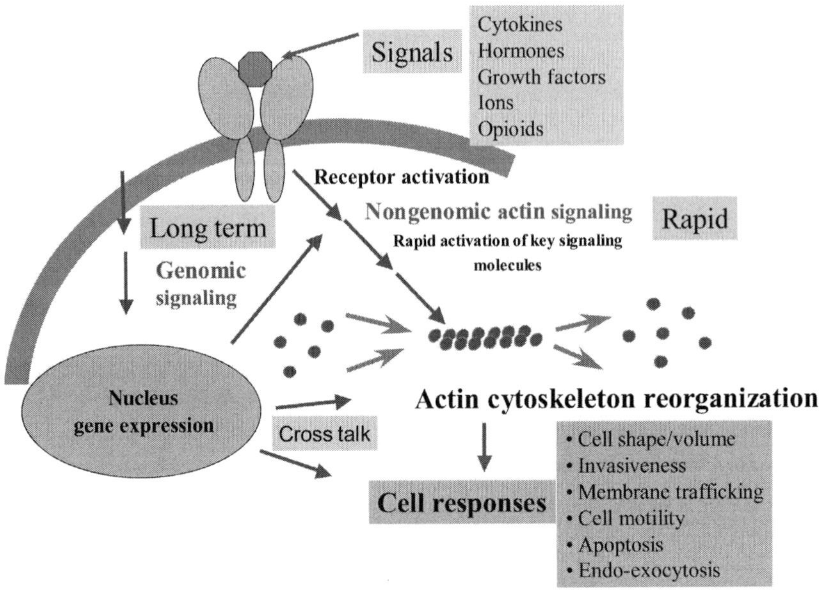

Figure 7. From structure to function model of actin cytoskeleton dynamics and signaling.

leading to gene expression and transcriptional regulation, which may control the long-term cell responses. Cross talk between transcriptional and nongenomic pathways may exist at several levels of events, including actin reorganization.

ACKNOWLEDGMENTS

The support of the Greek Secretariat for Research and Technology, the Greek Ministry of Education, the Greek Ministry of Health, and the European Union are kindly acknowledged.

REFERENCES

Barnes, P. J. and Karin, M. (1997). Nuclear factor-kappaB: A pivotal transcription factor in chronic inflammatory diseases. *N. Engl. J. Med.* **336,** 1066–1071.

Baud, V. and Karin, M. (2001). Signal transduction by tumor necrosis factor and its relatives. *Trends Cell Biol.* **11,** 372–377.

Benten, W. P., Liberherr, M., Giese, G., Wrehlke, C., Stamm, O., Sekeris, C. E., Mossmann, H., and Wunderlich, F. (1999). Functional testosterone receptors in plasma membranes of T cells. *FASEB J.* **13,** 123–133.

Bhowmick, N. A., Ghiassi, M., Bakin, A., Aakre, M., Lundquist, C. A., Engel, M. E., Arteaga, C. L., and Moses, H. L. (2001). Transforming growth factor β1 mediates epithelial to mesenchymal transdifferentiation through a RhoA dependent mechanism. *Mol. Biol. Cell* **12,** 27–36.

Blikstad, I., Markey, F., Carlsson, L., Person, T., and Lindberd, U. (1978). Selective assay of monomeric and filamentous actin in cell extracts, using inhibition of deoxyribonuclease I. *Cell* **15,** 935–943.

Bretscher, A. (1991). Microfilaments structure and function in the cortical cytoskeleton. *Annu. Rev. Cell Biol.* **7,** 337–374.

Cable, H. C., El-Mansoury, A., and Morgan, N. J. (1995). Activation of alpha-2-adrenoceptors results in an increase in F-actin formation in HIT-T15 pancreatic B-cells. *Biochem. J.* **307,** 169–174.

Cantiello, H. F. (1995). Role of the actin cytoskeleton on epithelial Na+ channel regulation. *Kidney Int.* **48**(4), 970–984.

Caron, E. and Hall, A. (1998). Identification of two distinct mechanisms of phagocytosis controlled by different Rho GTPases. *Science* **282,** 1717–1721.

Carpenter, C. L. (2000). Actin cytoskeleton and cell signaling. *Crit. Care Med.* **28,** N94–99.

Charalampopoulos, I., Dermitzaki, E., Vardouli, L., Tsatsanis, C., Stournaras, C., Margioris, N., and Gravanis, A. (2005). Dehydroepiandrosterone sulfate and allopregnanolone directly stimulate catecholamine production via induction of tyrosine hydroxylase and secretion by affecting actin polymerization. *Endocrinology* **146,** 3309–3318.

Dambaki, C., Kogia, C., Kampa, M., Darivianaki, K., Nomikos, M., Anezinis, P., Theodoropoulos, P. A., Castanas, E., and Stathopoulos, E. N. (2005). Membrane testosterone binding sites in prostate carcinoma as a potential new marker and therapeutic target: Study in paraffin tissue sections. *BMC Cancer* **5,** 148.

Dumont, N. and Arteaga, C. L. (2003). Targeting the TGF beta signaling network in human neoplasia. *Cancer Cell* **3,** 531–536.

Durrbach, A., Louvard, D., and Coudrier, E. (1996). Actin filaments facilitate two steps of endocytosis. *J. Cell Sci.* **109,** 457–465.

Edlund, S., Landstrom, M., Heldin, C. H., and Aspenström, P. (2002). Transforming growth factor-β-induced mobilization of actin cytoskeleton requires signaling by small GTPases Cdc42 and RhoA. *Mol. Biol. Cell* **13,** 902–914.

Grazzini, F., Guillon, G., Mouillae, B., and Zinjg, H. H. (1998). Inhibition of oxytocin receptor function by direct binding of progesterone. *Nature* **392,** 209–512.

Grünert, S., Jechlinger, M., and Beug, H. (2003). Diverse cellular and molecular mechanisms contribute to epithelial plasticity and metastasis. *Nat. Rev. Mol. Cell Biol.* **4,** 657–665.

Hatzoglou, A., Bakogeorgou, E., Papakonstanti, E., Stournaras, C., Emmanouel, D. S., and Castanas, E. (1996a). Identification and characterization of opioid and somatostatin binding sites in the opossum kidney (OK) cell line and their effect on growth. *J. Cell Biochem.* **63**(4), 410–421.

Hatzoglou, A., Bakogeorgou, E., Hatzoglou, C., Martin, P. M., and Castanas, E. (1996b). Antiproliferative and receptor binding properties of alpha- and beta-casomorphins in the T47D human breast cancer cell line. *Eur. J. Pharmacol.* **310**(2–3), 217–223.

Hatzoglou, A., Kampa, M., Kogia, C., Charalambopoulos, I., Theodoropoulos, P. A., Anezinis, P., Dambaki, C., Papakonstanti, E. A., Stathopoulos, E. A., Stournaras, C., Gravanis, A., and Castanas, E. (2005). Membrane androgen receptor activation induces apoptotic regression of human prostate cancer cells *in vitro* and *in vivo*. *J. Clin. Endocrinol. Metab.* **90**(2), 893–903.

Isenberg, G. (1996). New concepts for signal perception and transduction by the actin skeleton at cell boundaries. *Cell Dev. Biol.* **7,** 707–715.

Jensen, E. V. (1996). Steroid hormones, receptors and antagonists. *Ann. NY Acad. Sci.* **748,** 1–17.

Jordan, M. A. and Wilson, L. (1998). Microtubules and actin filaments: Dynamic targets for cancer chemotherapy. *Curr. Opin. Cell. Biol.* **10**(1), 123–130.

Kaibuchi, K., Kuroba, S., and Amano, M. (1999). Regulation of cytoskeleton and cell adhesion by the Rho family GTPases in mammalian cells. *Annu. Rev. Biochem.* **68,** 459–486.

Kallergi, G., Tsapara, A., Kampa, M., Krasagakis, C., Castanas, E., and Stournaras, C. (2003). Distinct signaling pathways regulate differential opioid effects on actin cytoskeleton in malignant MCF7 and non-malignant MCF12A human breast epithelial cells. *Exp. Cell Res.* **288,** 94–109.

Kampa, M., Papakonstanti, E. A., Hatzoglou, A., Stathopoulos, S., Stournaras, C., and Castanas, E. (2002). The human prostate cancer cell line LNCaP bears functional membrane testosterone receptors, which increase PSA secretion and modify actin cytoskeleton. *FASEB J.* **16,** 1429–1431.

Katsantonis, I., Tosca, N., Koukouritaki, S., Theodoropoulos, P. A., Gravanis, A., and Stournaras, C. (1994). Differential G/total actin ratio and microfilament stabilities in normal and malignant human keratinocytes. *Cell Biochem. Function* **12,** 267–274.

Koukouritaki, S. B., Theodoropoulos, P. A., Margioris, A. N., Gravanis, A., and Stournaras, C. (1996). Dexamethasone alters rapidly actin polymerization dynamics in human endometrial cells: Evidence for nongenomic actions involving cAMP turnover. *J. Cell. Biochem.* **62**(2), 251–261.

Koukouritaki, S. B., Margioris, A. N., Gravanis, A., Hartig, R., and Stournaras, C. (1997). Dexamethasone induces rapid actin assembly in human endometrial cells without affecting its synthesis. *J. Cell. Biochem.* **65**(4), 492–500.

Koukouritaki, S. B., Vardaki, E. A., Papakonstanti, E. A., Lianos, E., Stournaras, C., and Emmanouel, D. S. (1999). TNF-alpha induces actin cytoskeleton reorganization in glomerular epithelial cells involving tyrosine phosphorylation of paxillin and focal adhesion kinase. *Mol. Med.* **5**(6), 382–392.

Losel, R. and Wehling, M. (2003). Nongenomic actions of steroid hormones. *Nat. Rev. Mol. Cell. Biol.* **4,** 46–56.

Lutz, M. and Knaus, P. (2003). Integration of the TGF-β pathway into the cellular signaling network. *Cell. Signal.* **14,** 977–988.

Madge, L. A. and Pober, J. S. (2000). A phosphatidylinositol 3-kinase/Akt pathway, activated by tumor necrosis factor or interleukin-1, inhibits apoptosis but does not activate NFκB in human endothelial cells. *J. Biol. Chem.* **275,** 15458–15465.

Maneckjee, R. and Minna, J. D. (1994). Opioids induce while nicotine suppresses apoptosis in human lung cancer cells. *Cell Growth Differ.* **5**(10), 1033–1040.

Massagué, J. (2000). How cells read TGF-β signals. *Nat. Rev. Mol. Cell. Biol.* **1,** 169–178.

Meldrum, D. R. and Donnahoo, K. K. (1999). Role of TNF in mediating renal insufficiency following cardiac surgery: Evidence of a postbypass cardiorenal syndrome. *J. Surg. Res.* **85,** 185–199.

Mills, J. W., Schwiebert, E. M., and Stanton, B. A. (1994). The cytoskeleton and membrane transport. *Curr. Opin. Nephrol. Hypertens.* **3,** 529–534.

Miralles, F., Posern, G., Zaromytidou, A. T., and Treisman, R. (2003). Actin dynamics control SRF activity by regulation of its coactivator MAL. *Cell* **113,** 329–342.

Moustakas, A. and Stournaras, C. (1999). Regulation of actin organisation by TGF-beta in H-ras-transformed fibroblasts. *J. Cell. Sci.* **112**(Pt. 8), 169–179.

Moustakas, A., Souchelnytskyi, S., and Heldin, C. H. (2001). Smad regulation in TGF-β signal transduction. *J. Cell Sci.* **114,** 4359–4369.

Muallem, S., Kwiatkowska, K., Xu, X., and Yin, H. L. (1995). Actin filament disassembly is a sufficient final trigger for exocytosis in nonexcitable cells. *Cell Biol.* **128,** 589–598.

Nadal, A., Rovira, J. M., Laribi, O., Leon-quinto, T., Andrew, E., Ripoll, C., and Soria, B. (1998). Rapid insulinotropic effect of 17β-estradiol via a plasma membrane receptor. *FASEB. J.* **12,** 1341–1348.

Natoli, G., Costanzo, A., Guido, F., Moretti, F., and Levrero, M. (1998). Apoptotic, non-apoptotic, and anti-apoptotic pathways of tumor necrosis factor signalling. *Biochem. Pharmacol.* **56,** 915–920.

Nemere, I. and Farach-Carson, M. C. (1998). Breakthroughs and views. Membrane receptors for steroid hormones: A case for specific cell surface binding sites for vitamine D metabolites and estrogens. *Biochem. Biophys. Res. Commun.* **248,** 443–449.

Nifli, E. P., Bosson-Kouame, A., Papadopoulou, N., Kogia, C., Kampa, M., Castagnino, C., Stournaras, C., Vercauteren, J., and Castanas, E. (2005). Monomeric and oligomeric flavanols are agonists of membrane androgen receptors. *Exp. Cell Res.* **309**(2), 329–339.

Norman, A. W., Mizwicki, M. T., and Norman, D. P. (2004). Steroid-hormone rapid actions, membrane receptors and a conformational ensemble model. *Nat. Rev. Drug Discov.* **3,** 27–41.

Nurko, S., Sogabe, K., and Davis, J. A. (1996). Contribution of actin cytoskeleton alterations to ATP depletion and calcium-induced proximal tubule cell injury. *Am. J. Physiol.* **270,** F39–F52.

Panagiotou, S., Bakogeorgou, E., Papakonstantim, E. A., Hatzoglou, A., Walletm, F., Dussert, C., Stournaras, C., Martin, P. M., and Castanas, E. (1999). Opioid agonists modify breast cancer cell proliferation by blocking cells to the G2/M phase of the cycle: Involvement of cytoskeletal elements. *J. Cell Biochem.* **73**(2), 204–211.

Papakonstanti, E. A. (2003). A rapid, nongenomic, signaling pathway regulates the actin reorganization induced by activation of membrane testosterone receptors. **17,** 870–881.

Papakonstanti, E. A. and Stournaras, C. (2002). Association of PI-3 kinase with PAK1 leads to direct phosphorylation of actin and cytoskeletal reorganization. *Mol. Biol. Cell.* **13,** 2946–2962.

Papakonstanti, E. A. and Stournaras, C. (2004). TNFa promotes survival of proximal tubular epithelial cells via Cdc42-induced PLC-gamma 1 activation and actin filament redistribution. *Mol. Biol. Cell.* **15**(3), 1273–1286.

Papakonstanti, E. A., Emmanouel, D. S., Gravanis, A., and Stournaras, C. (1996). Na+/Pi cotransport alters rapidly cytoskeletal protein polymerization dynamics in opossum kidney cells. *Biochem. J.* **315,** 241–247.

Papakonstanti, E. A., Bakogeorgou, E., Castanas, E., Emmanouel, D. S., Hartig, R., and Stournaras, C. (1998). Early alterations of actin cytoskeleton in OK cells by opioids. *J. Cell Biochem.* **70**(1), 60–69.

Papakonstanti, E. A., Vardaki, E. A., and Stournaras, C. (2000a). Actin cytoskeleton: A signaling sensor in cell volume regulation. *Cell. Physiol. Biochem.* **10,** 257–264.

Papakonstanti, E. A., Emmanouel, D. S., Gravanis, A., and Stournaras, C. (2000b). PLC-γ1 signaling pathway and villin activation are involved in actin cytoskeletal reorganization induced by Na+/Pi cotransport upregulation. *Mol. Med.* **6**(4), 303–318.

Papakonstanti, E. A., Kampa, M., Castanas, E., and Stournaras, C. (2003). A rapid, nongenomic, signaling pathway regulates the actin reorganization and PSA secretion induced by membrane testosterone receptors' activation. *Mol. Endocrinol.* **17**(5), 870–881.

Pastorino, J. G., Tafan, M., and Farber, J. L. (1999). Tumor necrosis factor induces phosphorylation and translocation of BAD through a phosphatidylinositide-3-OH kinase-dependent pathway. *J. Biol. Chem.* **274,** 19411–19416.

Piek, E., Moustakas, A., Kurisaki, A., Heldin, C. H., and ten Dijke, P. (1999). TGF-(beta) type I receptor/ALK-5 and Smad proteins mediate epithelial to mesenchymal transdifferentiation in NMuMG breast epithelial cells. *J. Cell Sci.* **112,** 4557–4568.

Pollard, T. D. and Borisy, G. G. (2003). Cellular motility driven by assembly and disassembly of actin filaments. *Cell* **112,** 453–465.

Puls, A., Eliopoulos, A. G., Nobes, C. D., Bridges, T., Young, L. S., and Hall, A. (1999). Activation of the small GTPase Cdc42 by the inflammatory cytokines TNF(alpha) and IL-1, and by the Epstein-Barr virus transforming protein LMP1. *J. Cell Sci.* **112,** 2983–2992.

Reisine, T. and Bell, G. I. (1993). Molecular biology of opioid receptors. *Trends Neurosci.* **16,** 506–510.

Roberts, A. B. (1998). Molecular and cell biology of TGF-beta. *Miner. Electrolyte Metab.* **24,** 111–119.

Roberts, A. B. and Wakefield, L. M. (2003). The two faces of transforming growth factor beta in carcinogenesis. *Proc. Natl. Acad. Sci. USA* **100,** 8621–8623.

Shi, Y. and Massagué, J. (2003). Mechanisms of TGF-β signaling from cell membrane to the nucleus. *Cell* **113,** 685–700.

Stathopoulos, E. N., Dambaki, C., Kampa, M., Theodoropoulos, P. A., Anezinis, P., Delakas, D., Delides, G. S., and Castanas, E. (2003). Membrane androgen binding sites are preferentially expressed in human prostate carcinoma cells. *BMC Clin. Pathol.* **3,** 1.

Stournaras, C., Stiakaki, E., Koukouritaki, S., Theodoropoulos, P. A., Kalmanti, M., and Gravanis, A. (1996). Altered actin polymerization dynamics in various malignant cell types: Evidence for differential sensitivity to cytochalasin B. *Biochem. Pharmacol.* **52,** 1339–1346.

Theodoropoulos, P., Stournaras, C., Stoll, B., Markogiannakis, E., Lang, F., Gravanis, A., and Häussinger, D. (1992). Hepatocytes swelling leads to rapid decrease of the G-/total actin ratio and increases actin mRNA levels. *FEBS Lett.* **311,** 241–245.

Theriot, J. A. and Satterwhite, L. L. (1997). New wrinkles in cytokinesis. *Nature* **385,** 388–389.

Vancurova, I., Miskolci, V., and Davidson, D. (2001). NF-kappa B activation in tumor necrosis factor alpha-stimulated neutrophils is mediated by protein kinase Cdelta. Correlation to nuclear Ikappa Balpha. *J. Biol. Chem.* **276,** 19746–19752.

Vardouli, L., Moustakas, A., and Stournaras, C. (2005). LIM-kinase 2 and cofilin phosphorylation mediate actin cytoskeleton reorganization induced by transforming growth factor-beta. *J. Biol. Chem.* **280**(12), 11448–11457.

Wehling, M. (1997). Specific non-genomic effects of steroid hormones. *Annu. Rev. Physiol.* **59,** 365–393.

Wojciak-Stothard, B., Entwistle, A., Garg, R., and Ridley, A. J. (1998). Regulation of TNF-alpha-induced reorganization of the actin cytoskeleton and cell-cell junctions by Rho, Rac, and Cdc42 in human endothelial cells. *J. Cell Physiol.* **176,** 150–165.

Wozniak, M. A., Modzelewska, K., Kwong, L., and Keely, P. J. (2004). Focal adhesion regulation of cell behavior. *Biochim. Biophys. Acta* **1692,** 103–119.

Chapter 10

Regulation of the Actin Cytoskeleton by Phospholipids

Aneta Skwarek-Maruszewska, Pirta Hotulainen, and Pekka Lappalainen

Institute of Biotechnology, University of Helsinki, Finland

The actin cytoskeleton plays a central role in various motile and morphogenetic cellular and developmental processes. The assembly and disassembly of actin filaments during these processes are precisely controlled by an array of actin-binding proteins (ABPs). Studies have shown that phosphoinositides, especially $PI(4,5)P_2$ and $PI(3,4,5)P_3$, regulate the activities of many ABPs *in vitro* and have key roles in actin-dependent cellular processes such as cytokinesis, phagocytosis, endocytosis, and cell migration. These phosphoinositides typically downregulate the activities of proteins promoting actin filament disassembly and upregulate proteins promoting actin filament assembly. Thus, $PI(4,5)P_2$ and $PI(3,4,5)P_3$ induce actin filament assembly both *in vitro* and in a number of distinct cellular processes.

I. INTRODUCTION

Living cells constantly adjust the composition and size of their membrane systems to fulfill the demands for the expansion or reduction of cell size and the commitment to cell division. The dialogue between the machinery controlling cell polarization and motility involves extracellular signals and intracellular factors and requires a high degree of spatiotemporal coordination. The plasma membrane is a dynamic surface that also undergoes routine shape changes in the form of protrusions and ruffles, particularly during insertion and uptake of membrane elements by exocytic and endocytic vesicles (reviewed in Martin, 2001).

ISBN: 0-444-52868-7
DOI: 10.1016/S1569-2558(06)37010-5

The actin cytoskeleton is a highly dynamic structure that plays a central role in cellular processes involving membrane dynamics. Actin-driven processes include cell motility, cytokinesis, endocytosis, and phagocytosis. The actin cytoskeleton is also intimately involved in developmental processes of multicellular organisms because actin filaments together with myosin motors control the shape of all eukaryotic cells and generate the forces required for morphogenetic processes (reviewed in Jacinto and Baum, 2003). The three-dimensional organization and dynamics of the actin cytoskeleton are regulated by a large array of actin-binding proteins (ABPs) that interact with monomeric and/or filamentous actin. Actin filament-binding proteins regulate the nucleation, assembly, and disassembly of actin filaments, as well as bundling/crosslinking of filaments into appropriate three-dimensional networks (reviewed in Pollard and Borisy, 2003; Nicholson-Dykstra *et al.*, 2005). Actin monomer-binding proteins regulate the size and localization of the cellular actin monomer pool and monomer incorporation into filament ends (reviewed in Paavilainen *et al.*, 2004).

The dynamic nature of the actin cytoskeleton allows cells to adapt rapidly to external stimuli, and the activities of most ABPs that are downstream effectors of such stimuli are precisely regulated by various signaling pathways. Small GTPases of the Rho family are key components involved in signaling from plasma membrane receptors to the actin cytoskeleton (reviewed in Jaffe and Hall, 2005). Rho GTPases control actin dynamics by either interacting directly with central regulators of actin dynamics or by activating phosphorylation/dephosphorylation cascades that lead to activation/ inactivation of certain ABPs. For example, the active (GTP bound) form of the Cdc42 GTPase promotes actin filament assembly by directly binding to and activating Wiskott–Aldrich syndrome protein (WASP)/N-WASP proteins, which induce actin filament nucleation and assembly together with the actin-related protein 2/3 (Arp2/3) complex (reviewed in Bompard and Caron, 2004; Vartiainen and Machesky, 2004). In addition, Cdc42 can activate PAK1 kinase, which in turn phosphorylates and activates LIM kinase; phosphorylation of actin depolymerizing factor (ADF)/cofilins by LIM kinase inhibits their actin filament depolymerization activity, resulting in a local increase in F-actin concentration (Edwards *et al.*, 1999; Nishita *et al.*, 2005).

In addition to phosphorylation/dephosphorylation and direct interactions with small GTPases, the activities of many ABPs are regulated through interactions with phosphoinositides (reviewed in Hilpelä *et al.*, 2004; Janmey and Lindberg, 2004). In this chapter, we discuss the mechanisms by which membrane phospholipids regulate the activities of various ABPs and subsequently control the assembly/disassembly of actin filaments during diverse cellular processes.

II. GENERAL FEATURES OF PHOSPHOINOSITIDES AND THEIR METABOLISM

Phosphoinositides is a collective term for phosphatidylinositol (PI) and its phosphorylated derivatives. Their structural backbone consists of glycerol esterified at positions 1 and 2 with fatty acids (essentially stearic and arachidonic acids, respectively) and bearing at position 3 a *myo*-inositol head group connected to the diacylglycerol

(DAG) by a phosphodiester bond (Pendaries *et al.*, 2005). The *myo*-inositol head group of PI contains five free hydroxyl groups, but only three of them (positions D-3 D-4, and D-5) are phosphorylated *in vivo* (Fig. 1A). Reversible phosphorylation at these positions in various combinations generates seven phosphoinositides implicated

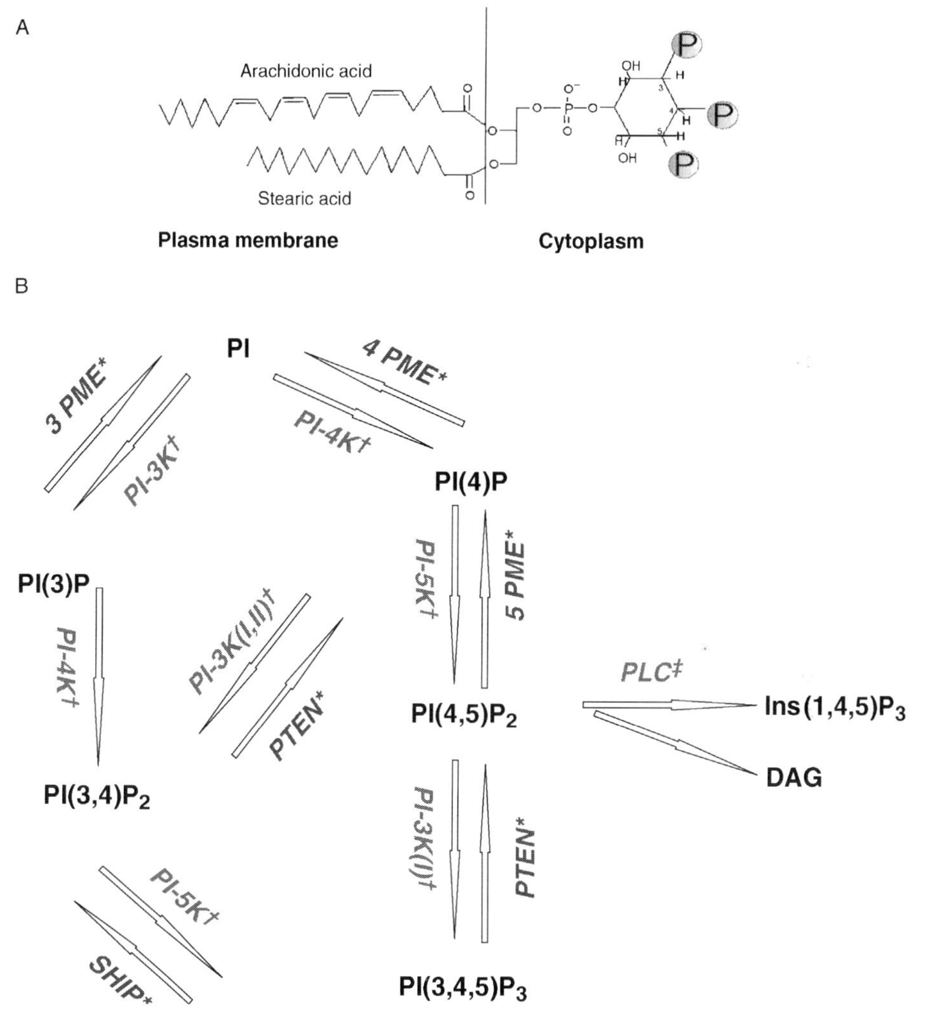

Figure 1. (A) Structure of phosphatidylinositol and its phosphorylation sites (indicated with grey-shaded "P"s). (B) The main pathways of phosphoinositide synthesis and interconversion (modified from Pendaries *et al.*, 2003). Asterisk indicates phosphatases: PTEN (phosphatase and tensin homologue), SHIP (Src homology2-containing inositol-5-phosphatase), and PMEs (phosphomonoesterases). Dagger indicates phosphoinositide kinases and double dagger indicates phospholipase C (PLC). Hydrolysis by PLCs leads to the production of the secondary messengers inositol 1,4,5-tris-phosphate ($Ins(1,4,5)P_3$) and diacylglycerol (DAG).

in the regulation of critical cellular processes (Insall and Weiner, 2001; Toker, 2002). The cellular pool of inositol phospholipids is mostly dominated by PI, whereas other phosphoinositides are less abundant. Of the total phosphoinositides, only approximately 5% are phosphorylated at position D-4 [PI(4)P], another 5% are phosphorylated at both positions D-4 and D-5 [PI(4,5)P_2], and 0.25% are phosphorylated at position D-3 (Rameh and Cantley, 1999). This explains why the cellular concentration of PI(3,4,5)P_3 is typically at least 25 times lower than that of PI(4,5)P_2 (Lemmon and Ferguson, 2000). Among the various phosphoinositides, PI(4,5)P_2 and PI(3,4,5)P_3 appear to have the most crucial roles in actin cytoskeletal reorganization in cells (reviewed in Hilpelä *et al.*, 2004; Janmey and Lindberg, 2004).

Although PI(4,5)P_2 comprises only about 1% of the total phospholipid content of the plasma membrane, it is involved in numerous cellular events. PI(4,5)P_2 is a source of two second messengers in the cell: DAG and inositol 1,4,5-trisphosphate (Ins(1,4,5)P_3). In addition, it is a substrate for the generation of PI(3,4,5)P_3, which is also an important lipid-signaling molecule. PI(4,5)P_2 is also involved in enzyme activation, regulation of ion channels, membrane trafficking (exocytosis, endocytosis), and cytoskeletal reorganization (reviewed in Niggli, 2005). The last feature is important for processes such as the formation of filopodia, lamellipodia, and membrane ruffles, as well as in chemotaxis, which requires rapid remodeling of the actin cytoskeleton. The dynamic changes in actin network structure are not only essential for cell movement but are also crucial for other dynamic membrane events such as vesicle trafficking (reviewed in Yin and Janmey, 2003; Janmey and Lindberg, 2004; Downes *et al.*, 2005).

Despite its critical role in cytoskeletal regulation, there is no evidence that PI(4,5)P_2 interacts directly with actin. Instead, numerous ABPs, which regulate the equilibrium between actin polymerization and depolymerization, have PI(4,5)P_2-binding motifs and the activities of these proteins are regulated through interactions with phosphoinositides. The PI(4,5)P_2-binding motifs vary from proper domains and more defined PI(4,5)P_2-binding pockets—like the pleckstrin homology (PH) domain and the FYVE, PX, ENTH domains—to less structured and typically less specific binding motifs (e.g., lysine-rich basic region) (reviewed in Hilpelä *et al.*, 2004; Niggli, 2005).

The cellular PI(4,5)P_2 level is regulated by lipid kinases (PIP kinases), phosphatases, and the availability of phosphoinositide precursors (Fig. 1B). The phosphoinositide kinases and the pathway that generates PI(4,5)P_2 are evolutionarily conserved. In yeast as well as in mammalian cells, PI(4,5)P_2 is synthesized via sequential phosphorylation of PI to PI(4)P and PI(4,5)P_2 through the action of PI 4-kinases and PI(4)P 5-kinases, respectively (reviewed in Fruman *et al.*, 1998; Doughman *et al.*, 2003b). Two subfamilies of PIP kinases, types I and II, allow the generation of PI(4,5)P_2 from independent pools of substrate, PI(4)P and PI(5)P, respectively. Types I and II PIP kinases show distinct subcellular localizations and are thus involved in different signaling pathways. Since PI(4)P is the major PIP in cells and its concentration is relatively high, the type I PIP kinases are thought to have a more prominent role in the production of PI(4,5)P_2. The budding yeast, *Saccharomyces cerevisiae*, expresses a single and essential PI(4)P 5-kinase termed Mss4 that is implicated in the regulation of cell cycle-dependent actin reorganization (Desrivieres *et al.*, 1998; Homma *et al.*, 1998).

In contrast, several type I isoforms, and splice variants thereof, are expressed in mammals. These enzymes are differentially localized in cells and thus appear to be involved in the synthesis of $PI(4,5)P_2$ at distinct sites. Mammalian type I PIP kinases have been found at the plasma membrane (isoform α), focal adhesions (isoform γ), nucleus (isoform α), and Golgi (isoform β) (Di Paolo *et al.*, 2002; Ling *et al.*, 2002; Doughman *et al.*, 2003a).

In contrast, the type II PIP kinases are not present in yeast but are found in multicellular organisms from *Drosophila* to mammals. Unlike type I PI(4)P 5-kinases, the type II PI(5)P 4-kinases do not appear to be central to cytoskeletal regulation in mammalian cells (reviewed in Doughman *et al.*, 2003b). Thus, it is likely that PI(5)P 4-kinases evolved to perform a specialized function in multicellular organisms and that the PI(4)P and PI(5)P pathways for $PI(4,5)P_2$ synthesis have nonredundant physiological functions in cells. Type III PIP kinases, which are homologous to yeast PIP kinase, Fab1p, have also been identified in mammalian cells. These kinases produce $PI(3,5)P_2$ *in vivo*, using PI(3)P as substrate, and are hence PI(3)P 5-kinases (Gary *et al.*, 1998; Sbrissa and Shisheva, 2005).

In addition to enzymes required for the synthesis of phosphoinositides (phosphoinositide-kinases), various phosphatases also have important roles in lipid turnover. $PI(4,5)P_2$ is a key element of the PI pathway since it is also a precursor for $PI(3,4,5)P_3$ synthesis by the $PI(4,5)P_2$ 3-kinase (PI-3K). The reverse reaction, dephosphorylation of $PI(3,4,5)P_3$ to $PI(4,5)P_2$, is catalyzed by tumor suppressor protein PTEN (phosphatase and tensin homologue deleted on chromosome 10) (Walker *et al.*, 2004). $PI(3,4,5)P_3$ is also metabolized to $PI(3,4)P_2$ by 5-phosphoinositide phosphatases such as SHIP1 and SHIP2 (SH2-domain containing inositol phosphatase-1 and -2) (Damen *et al.*, 1996). In resting *Dictyostelium* cells, PI-3Ks are found in the cytoplasm, whereas a fraction of PTEN is localized at the cell membrane. When cells are uniformly exposed to a chemoattractant, the PI-3Ks rapidly bind to the membrane and then PTEN dissociates. As stimulation continues, the PI-3Ks return to the cytoplasm and PTEN reassociates with the membrane. When cells are exposed to a gradient of a chemoattractant, the PIP-3Ks and PTEN bind to the membrane at the front and rear of the cell, respectively, thereby leading to the accumulation of $PI(3,4,5)P_3/PI(4,5)P_2$ at the leading edge (Funamoto *et al.*, 2001; Iijima and Devreotes, 2002; Chen *et al.*, 2003).

III. SUBCELLULAR DISTRIBUTION OF $PI(4,5)P_2$ AND $PI(3,4,5)P_3$

$PI(4,5)P_2$ comprises approximately 1% of the total phospholipid content at the plasma membrane, with an effective concentration of approximately 10 μM (McLaughlin *et al.*, 2002). Although $PI(4,5)P_2$ is involved in regulation of a large array of cellular processes, the concentration of this phospholipid is relatively constant in most cell types and does not change in response to various extracellular signals. In addition to the inner leaflet of the plasma membrane, $PI(4,5)P_2$ is also concentrated to "cytoplasmic leaflet" of endosomes, endoplasmic reticulum, and the Golgi compartment as well as to certain structures in the nucleus (Jones *et al.*, 2000; Watt *et al.*, 2002).

In most cell types, $PI(3,4,5)P_3$ is mainly found only from the plasma membrane (Gray *et al.*, 1999; Czech, 2000). In contrast to $PI(4,5)P_2$, the $PI(3,4,5)P_3$ levels at the plasma membrane are tightly controlled by various intracellular and extracellular signals. In resting cells, both $PI(3,4)P_2$ and $PI(3,4,5)P_3$ are almost undetectable (Auger *et al.*, 1989). After receptor stimulation, there is a transient and rapid production of $PI(3,4,5)P_3$ from $PI(4,5)P_2$ by class I PI(4,5) 3-kinases. This is followed by a dephosphorylation of $PI(3,4,5)P_3$ to $PI(3,4)P_2$ by 5-phosphoinositide phosphatases and to $PI(4,5)P_2$ by PTEN phosphatase (Auger and Cantley, 1991; Stephens *et al.*, 1991; Walker *et al.*, 2004). As a result, the accumulation of $PI(3,4,5)P_3$ is rapid and transient, while the accumulation of $PI(3,4)P_2$ is delayed and significantly more sustained (Stephens *et al.*, 1993; Franke *et al.*, 1997).

It has been proposed that the plasma membrane contains lipid-based membrane microdomains, "lipid rafts." Current models of membrane microdomains are largely based on the biochemical property of insolubility in cold, nonionic detergents. These detergent-insoluble membrane microdomains appear to be rich in cholesterol, sphingomyelin, and glycosphingolipids. In addition, many signaling molecules are also enriched in the detergent-insoluble fractions, suggesting that the membrane microdomains may be actively involved in signaling events (reviewed in Munro, 2003). Several studies in which the subcellular localization of $PI(4,5)P_2$ was probed with green fluorescent protein (GFP)-tagged PH domains in wild-type and cholesterol-depleted cells indicated that $PI(4,5)P_2$ may be concentrated to cholesterol-rich lipid microdomains at the plasma membrane. It was proposed that enrichment of $PI(4,5)P_2$ to the inner leaflet of "lipid rafts" could thus play an important role in regulation of actin-dependent processes such as cell motility and endocytosis (Botelho *et al.*, 2000; Rozelle *et al.*, 2000; Caroni, 2001; Kwik *et al.*, 2003; Aoyagi *et al.*, 2005).

However, the existence of lipid rafts is still largely unclear because results obtained by different methods are contradictory (Douglass and Vale, 2005). Furthermore, whether $PI(4,5)P_2$ localizes to specific domains at the plasma membrane is not known. The evidence that supports the presence of $PI(4,5)P_2$ in lipid microdomains is that $PI(4,5)P_2$ localizes to clusters at the plasma membrane, when localized by an anti-$PI(4,5)P_2$ antibody (Laux *et al.*, 2000), and that PI(4)P 5-kinases are recruited to rafts by the small GTPases Rho and ADP ribosylation factor (Arf) to activate local $PI(4,5)P_2$ synthesis (Chong *et al.*, 1994; Honda *et al.*, 1999). However, studies on living cells using GFP-tagged PH domains as $PI(4,5)P_2$ markers suggested that the localization of $PI(4,5)P_2$ at the plasma membrane is strictly homogenous (van Rheenen and Jalink, 2002). In addition, studies by van Rheenen *et al.* (2005) suggested that $PI(4,5)P_2$ clustering is induced by detergent treatment, thus providing an alternative explanation for the enrichment of $PI(4,5)P_2$ in detergent-insoluble membrane fractions.

Because of the contradictory results concerning the existence of $PI(4,5)P_2$-rich membrane microdomains, it is still unclear how $PI(4,5)P_2$ signals to the actin cytoskeleton. It is possible that $PI(4,5)P_2$ is distributed relatively homogenously at the plasma membrane and may act as a spatial marker at the plasma membrane. In contrast, $PI(3,4,5)P_3$ would function as a stimulus-responsive messenger at certain plasma membrane regions/domains (Insall and Weiner, 2001).

IV. REGULATION OF THE ACTIN CYTOSKELETON BY PHOSPHOINOSITIDES

Overexpression of PI(4) 5-kinases in cells produces dramatic actin phenotypes, thus establishing a tight connection between PI(4,5)P_2 and the actin cytoskeleton. PI(4) 5-kinase overexpression yields different actin phenotypes depending on the type of cells used. These phenotypes include comet-like actin tails (Rozelle *et al.*, 2000), microvilli (Matsui *et al.*, 1999), thick stress fibers (Yamamoto *et al.*, 2001), membrane ruffles (Honda *et al.*, 1999), and abnormal pine needle-like structures (Shibasaki *et al.*, 1997). One potential explanation for the large variation of phenotypes is that different cell backgrounds may favor a unique combinatorial involvement of subsets of potential players. The phenotypes may also vary within the same cell type, depending on specific environmental factors and the stage of differentiation. In preadipocytes, for example, PI(4) 5-kinase overexpression results in actin comet tail formation, whereas in differentiated adipocytes it induces multiple intracellular vacuoles coated with F-actin, N-WASP, dynamin, cortactin, and caveolin (Kanzaki *et al.*, 2004).

The small GTPases RhoA, Cdc42, and Rac1 regulate the actin cytoskeleton through direct interactions with ABPs as well as by activating protein kinase pathways (reviewed in Jaffe and Hall, 2005). In addition, these GTPases can activate the PI(4)P 5-kinases and thereby promote actin filament assembly (Weernink *et al.*, 2004). Rac directly interacts with and activates PI(4)P 5-kinase and thus increases the number of free actin filament barbed ends (Tolias *et al.*, 2000).

Studies with PTEN-deficient cells demonstrate a tight relationship between PI(3,4,5)P_3/PI(4,5)P_2 levels and actin polymerization. When exposed to uniform stimuli, PTEN-deficient cells exhibit prolonged association of the PI(3,4,5)P_3-recognizing PH domain with the cell membrane and a simultaneous increase in actin polymerization. Placed in a chemoattractant gradient, PTEN-deficient cells fail to produce sharp localization of PI(3,4,5)P_3/PI(4,5)P_2 and F-actin organization at the cell's leading edge, thus displaying severely impaired chemotaxis (Funamoto *et al.*, 2001; Iijima and Devreotes, 2002; Chen *et al.*, 2003). The mammalian PIP-3K/PTEN/Akt pathway was reconstructed in yeast (Rodriguez-Escudero *et al.*, 2005). High-level expression of the p110 catalytic subunit of the mammalian PI(4,5)P_2 3-kinase in yeast cells dramatically inhibited growth. The growth arrest by PI(4,5)P_2 3-kinase correlated with the loss of PI(4,5)P_2 and its conversion to PI(3,4,5)P_3. PI(4,5)P_2 depletion resulted in severe rearrangements of the actin and septine architecture, defects in secretion and endocytosis, and activation of the mitogen-activated protein kinase, Slt2.

A. Actin-Dependent Cellular Processes Regulated by Phospholipids

As described earlier, cell adhesion, morphogenesis, motility, chemotaxis, vesicular trafficking, and cytokinesis depend on accurate regulation of actin dynamics by membrane phospholipids and other signaling pathways (Fig. 2). The temporal and spatial regulation of actin dynamics during these processes is mediated by various

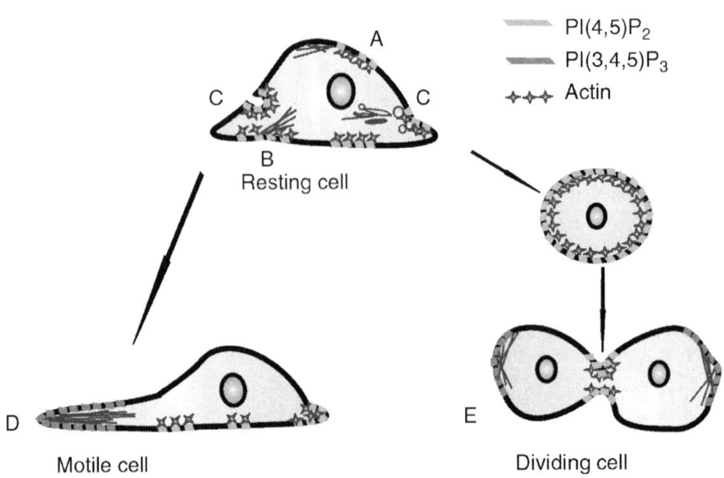

Figure 2. The actin cytoskeleton (red) is important for controlling several key cellular processes such as cell shape coordination (A), adhesion to the surface (B), vesicle transport (including endocytosis and phagocytosis) (C). Actin is also involved in cell motility (D) and cytokinesis (E). In resting cells, $PI(4,5)P_2$ is uniformly distributed, and $PI(3,4,5)P_3$ levels are almost undetectable. $PI(3,4,5)P_3$ levels are temporally and spatially increased during cell polarization and cytokinesis. Local accumulation of $PI(3,4,5)P_3$ is observed at the leading edge of migrating cells and at the cell poles during cytokinesis. $PI(4,5)P_2$ is concentrated at the cleavage furrow during cytokinesis, as well as at the sites of endocytosis/phagocytosis. It is important to note that increased actin filament accumulation occurs primarily in these phospholipid-rich areas of cells. (See Color Insert.)

ABPs that influence the equilibrium between monomeric (G) and filamentous (F) actin as well as filament nucleation and assembly. Moreover, these proteins regulate the three-dimensional organization of the actin cytoskeleton. In this chapter, we will discuss the mechanisms and significance of phospholipid–actin interactions during cell motility, cytokinesis, and vesicular trafficking.

1. Cell Motility

The actin cytoskeleton plays a crucial role in cell migration. The rapid turnover of actin filaments at the leading edge of cells provides the force-generating machinery that is required for protrusion of the leading edge and subsequent cell motility. Similarly, many morphogenetic events including neural outgrowth depend on a similar actin-dependent machinery to promote the formation of specific plasma membrane processes (reviewed in Pantaloni *et al.*, 2001; Dent and Gertler, 2003; Pollard and Borisy, 2003). Several lines of evidence suggest that phosphoinositides play a central role in cell motility. The activities of many ABPs that are linked to cell motility are regulated by $PI(4,5)P_2$ (Fig. 3). Typically, proteins that promote actin filament disassembly, such as capping protein, gelsolin, ADF/cofilin, and twinfilin, are inactivated by $PI(4,5)P_2$

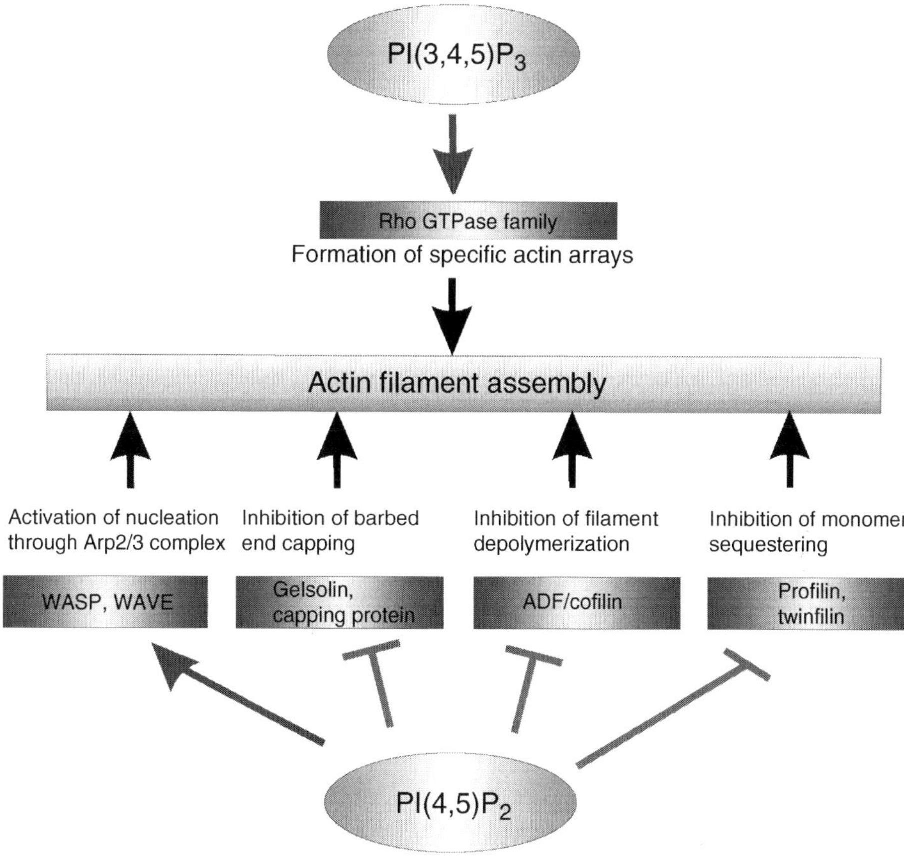

Figure 3. Regulation of actin-binding proteins (ABPs) by phosphoinositides. Phospholipids typically upregulate proteins promoting actin assembly and downregulate inhibitors of actin polymerization. PI(4,5)P_2 inhibits actin filament depolymerization and monomer sequestering by ADF/cofilin and twinfilin. Moreover, through its interactions with heterodimeric capping proteins and gelsolin, PI(4,5)P_2 promotes uncapping of actin filaments and subsequent elongation of preexisting filaments. In contrast, the WASP family proteins, which promote the formation of new actin filaments, are upregulated by PI(4,5)P_2. PI(3,4,5)P_3 appears to promote actin assembly through activation of Rho family GTPases.

in vitro, while WASP family proteins, which promote actin filament assembly in cells, are upregulated by PI(4,5)P_2 (Yonezawa *et al.*, 1990; Schafer *et al.*, 1996; Rohatgi *et al.*, 1999; Palmgren *et al.*, 2001). N-WASP activation is dependent on PI(4,5)P_2 density, and this dependence is even more pronounced than that observed for PI(4,5)P_2 binding. Thus, N-WASP responds to changes in PI(4,5)P_2 density in a switch-like manner: the transition from the fully repressed state to the fully activated state occurs above a sharp threshold (Papayannopoulos *et al.*, 2005).

In addition to indirect evidence concerning the role of PI(4,5)P_2 in cell migration, there is also direct evidence showing that depletion of cellular PI(4,5)P_2 levels impairs cell motility. Depletion of PIPIKα, or its activator Ajuba, from mouse embryonic fibroblasts resulted in a significant decrease in PI(4,5)P_2 levels at the leading edge of cells and consequent defects in cell migration (Kisseleva *et al.*, 2005).

Instead of directly interacting with ABPs, PI(3,4,5)P_3 appears mainly to be involved in cell motility through regulation of Rho family GTPases (Fig. 3). The Rho family GTPases, namely Rho, Rac, and Cdc42, are critical components that mediate various signals to the actin cytoskeleton (reviewed in Jaffe and Hall, 2005). The activity of small GTPases is determined by a cycle between inactive GDP-bound and active GTP-bound forms. This cycle is regulated by guanosine nucleotide exchange factors (GEFs), which catalyze the release of GDP and the subsequent binding of GTP, and GTPase-activating proteins (GAPs), which stimulate the usually low intrinsic GTPase activity of small GTPases. These activities convert small GTPases between their active and inactive forms (reviewed in Jaffe and Hall, 2005). Different members of the Rho family GTPases are affected by distinct sets of GEFs. Members of the Dbl family of GEFs contain a conserved domain of approximately 150 residues, designated the Dbl homology (DH) domain, which is required for GEF activity. The DH domain is invariably coupled to a PH domain, which binds certain phosphoinositides, most often PI(4,5)P_2 or PI(3,4,5)P_3 (reviewed in Erickson and Cerione, 2004). The invariant association between PH and DH domains in GEFs suggests that the PH domain may be critical for the function of DH domains. However, there is no known universal mechanism that governs the activation of Rho GEFs via the interaction of these phospholipids with the DH-PH domain pair. The PH domain can, for example, activate the DH domain by an allosteric mechanism. This activation can be regulated in a very sophisticated way: PI(4,5)P_2 binding to the PH domain downregulates GEF activity, whereas PI(3,4,5)P_3 binding upregulates the activity (Han *et al.*, 1998; Das *et al.*, 2000). Phosphoinositides can also regulate GEFs by another mechanism, in that binding of PI(4,5)P_2 or PI(3,4,5)P_3 to certain PH domains inhibits GEF activity (Russo *et al.*, 2001). It has also been proposed that PH domain–phosphoinositide interactions may regulate the subcellular localization of certain Rho GEFs (Michiels *et al.*, 1997; Snyder *et al.*, 2001).

There is also direct *in vivo* evidence supporting the role of PI(3,4,5)P_3 in cell migration. During chemotaxis of slime mold *Dictyostelium discoideum*, PI(3,4,5)P_3 accumulates at the leading edge of cells (reviewed in Postma *et al.*, 2004). Decreasing the PI(3,4,5)P_3 levels in cells by inactivation of certain PI(4,5) 3-kinases leads to defects in directed cell migration (Weiss-Haljiti *et al.*, 2004; Maffucci *et al.*, 2005). Furthermore, mice in which the gamma isoform of PI(4,5) 3-kinase was inactivated by gene targeting showed defects in dendritic cell migration and adaptive immunity.

2. Cytokinesis

Successful cell division requires coordinated regulation of the actomyosin contractile ring and the cleavage furrow. In addition to a series of proteins playing key roles at the cell poles and/or at the furrow, inositol lipids and enzymes that govern their metabolism are also intimately involved in cytokinesis. In dividing cells, F-actin is concentrated at the cell poles and cleavage furrow (Fig. 2), and proper F-actin ring formation is required for separation of the two daughter cells. Using U73122 (U7), a phospholipase C inhibitor, Saul *et al.* (2004) investigated the role of PI(4,5)P_2-dependent F-actin regulation during cytokinesis. The absence of furrow filaments and increased F-actin content at the poles in the presence of U7 suggest that actin filaments remerge at the cell poles and may be subsequently severed and transported to the cell midzone. In the absence of PI(4,5)P_2 hydrolysis, the polar filaments may be stabilized and accumulate at the cell poles.

In budding yeast, PI(4,5)P_2 is localized to the medial ring of dividing cells, and mutations in PI(4)P 5-kinase cause defects during cytokinesis (Zhang *et al.*, 2000). The subcellular localization of PI(4)P 5-kinase during cytokinesis has also been reported in various mammalian cell lines (Emoto *et al.*, 2005; Field *et al.*, 2005) and in *Drosophila melanogaster* (Brill *et al.*, 2000). Studies using two distinct and unrelated PI(4,5)P_2 reporters, PLCδ-PH-EGFP and GFP-TubbyC (Field *et al.*, 2005), show that PI(4,5)P_2 accumulates at the furrow (starting from initial furrow invagination to abscission). In contrast, other phosphoinositides, such as PtdIns(3)P, PtdIns(4)P, PtdIns(3,4)P_2, and PtdIns(3,4,5)P_3, do not appear to colocalize with PI(4,5)P_2 during cytokinesis. Thus, localized production of PI(4,5)P_2 at the cleavage furrow appears to be required for the completion of cytokinesis.

PI 4-kinases also have a role in cell division. Brill *et al.* (2000) showed that a mutation in *four wheel drive* (*fwd*), a gene required for proper organization of the contractile F-actin ring, induces the accumulation of phosphotyrosine epitopes and formation of stable intercellular bridges during male meiosis in *Drosophila*. The *fwd* gene encodes a predicted phosphatidylinositol 4-kinase (PI-4K). Additionally, the fission yeast homologue of *fwd* has been implicated in cytokinesis, indicating that the function of this PI 4-kinase during cytokinesis may be evolutionarily conserved (Desautels *et al.*, 2001).

In addition to PI(4,5)P_2, PI(3,4,5)P_3 also appears to have an important function during cytokinesis. Studies on *D. discoideum* cells lacking PI 3-kinase 1, PI 3-kinase 2 [enzymes generating PI(3,4,5)P_3], and PTEN [PI(3,4,5)P_3 phosphatase] are defective in cytokinesis and exhibit cell division delays, multinucleated cells, and problems in localization and proper function of cytoskeletal proteins (e.g., myosin II and coronin) (Janetopoulos *et al.*, 2005). Such is also the case in studies of *D. discoideum* treated with inhibitors of these enzymes. Changes in the distribution of PI(3,4,5)P_3 are temporally and spatially modulated in dividing cells. PI(3,4,5)P_3 is not detected in cells as they shift to a rounded morphology prior to cytokinesis, in contrast to the observed high levels of PTEN. This is accompanied by a loss of membrane ruffling and actin polymerization (Fig. 2). Later, when a cell spindles and elongates, the PI(3,4,5)P_3 kinases and PTEN relocalize to the poles and furrow, respectively (Janetopoulos *et al.*, 2005).

Together, these studies suggest that PI(3,4,5)P_3 accumulation at the cell poles results in the formation of actin-based projections, whereas a low concentration of PI(3,4,5)P_3 in the midzone and at the forming cleavage furrow may permit actomyosin-based contraction. $PI(4,5)P_2$, on the other hand, appears to promote actin filament assembly at the cleavage furrow during cytokinesis. In the future, it will be important to elucidate the basis for the differential roles of $PI(4,5)P_2$ and $PI(3,4,5)P_3$ during cytokinesis. Moreover, an important challenge will be to address how the accumulation of $PI(4,5)P_2$ or $PI(3,4,5)P_3$ during cytokinesis differentially affects the dynamics or three-dimensional organization of actin filaments.

3. Vesicular Trafficking

The actin cytoskeleton plays an important role in endocytosis and phagocytosis. Actin dynamics contribute to both vesicle budding and intracellular membrane transport processes (reviewed in Engqvist-Goldstein and Drubin, 2003). Endocytic uptake is driven by a modular design based on adaptors and ABPs that facilitate the timing and positioning of actin assembly and disassembly during vesicle formation and scission (Kaksonen *et al.*, 2003, 2005; Merrifield *et al.*, 2005). It is significant that the activities of many of these adaptors and ABPs are regulated by $PI(4,5)P_2$. Furthermore, several studies have provided evidence for the direct involvement of $PI(4,5)P_2$ in endocytosis and phagocytosis (Niedergang and Chavrier, 2004; Haucke, 2005).

The most detailed studies of the role of the actin cytoskeleton during endocytosis have been conducted using budding yeast. Regulated actin polymerization at the endocytic site provides the force required for membrane invagination and transport of newly formed vesicles away from the plasma membrane (Huckaba *et al.*, 2004; Kaksonen *et al.*, 2005). $PI(4,5)P_2$ regulates the function of many ABPs, including capping protein, ADF/cofilin, and WASP family proteins, which are involved in endocytosis and other motile processes (reviewed in Hilpelä *et al.*, 2004). A more specific link between phosphoinositide-regulated actin remodeling during endocytosis has come from analysis of the yeast endocytic adaptor protein, Sla2. This protein, and its mammalian homologue, Hip1R, coordinates actin polymerization to facilitate productive membrane invagination during clathrin-mediated endocytosis (Kaksonen *et al.*, 2003; Engqvist-Goldstein *et al.*, 2004). These proteins interact directly with both actin and clathrin. Sla2 binds $PI(4,5)P_2$ through its N-terminal ANTH (AP180 N-terminal homology) domain. Mutagenesis studies provided evidence that the Sla2–$PI(4,5)P_2$ interaction is important during endocytic internalization in yeast (Sun *et al.*, 2005). In addition to initiation and maintenance of actin assembly during endocytosis, $PI(4,5)P_2$ metabolism may also be important for the termination of actin assembly during postfission vesicle uncoating. Studies on the budding yeast synaptojanin-like proteins (Sjl1, Sjl2, and Sjl3), which are PIP phosphatases, provided evidence that $PI(4,5)P_2$ dephosphorylation is required for coordinating actin dynamics during endocytosis (Stefan *et al.*, 2005).

Most $PI(4,5)P_2$-regulated actin-binding endocytic proteins, as well as synaptojanins, are also conserved in mammals, suggesting that $PI(4,5)P_2$-dependent regulation of the actin cytoskeleton is universally important in endocytosis (Engqvist-Goldstein and Drubin, 2003). In addition, mammalian cells have additional links between

$PI(4,5)P_2$-regulated actin dynamics and endocytosis. Annexin 2 is one candidate that may serve as an interface between endocytic vesicles (macropinosome surface) and associated actin filaments. In addition to the F-actin–binding feature, Hayes *et al.* (2004) showed that annexin 2 binds with high selectivity to $PI(4,5)P_2$. Thus, it is an attractive candidate for a $PI(4,5)P_2$-regulated protein required for dynamic remodeling of the membrane and cytoskeleton during endocytic vesicle formation.

Actin polymerization is essential during the formation of phagosomes, and the ensuing actin depolymerization from the phagocytic cap is required for cap maturation. The first step is controlled by GTPases of the Rho family. Specifically, the roles of Rac1 and Cdc42 are well established in actin polymerization during phagocytic uptake (Hoppe and Swanson, 2004). Also, phosphoinositides are important for actin assembly during phagocytosis (Defacque *et al.*, 2002). Studies by Scott *et al.* (2005) demonstrated the accurate temporal and spatial dynamics of $PI(4,5)P_2$, and actin is essential at different stages of phagocytosis. While actin assembly is required for pseudopod formation during phagocytosis, actin disassembly proceeds asymmetrically after phagosomal closure, with the loss of actin structures occurring first at the base of the phagosome. $PI(4,5)P_2$ hydrolysis and actin disassembly show similar spatial and temporal kinetics during phagocytosis, and actin detachment from forming phagosomes in macrophages can be inhibited by impairing $PI(4,5)P_2$ hydrolysis (Scott *et al.*, 2005).

V. CONCLUSIONS AND FUTURE PERSPECTIVES

Several lines of evidence demonstrate that phosphoinositides, especially $PI(4,5)P_2$ and $PI(3,4,5)P_3$, control the activity of various actin-associated proteins and thus promote actin filament assembly at appropriate regions within the cell. These phosphoinositides typically upregulate the activities of proteins that promote actin filament assembly and downregulate those proteins that induce net actin filament disassembly (Fig. 3). It is also important to note that $PI(4,5)P_2$ may also control the orientation of actin filaments in cells, ensuring that the fast growing filament barbed ends are toward the plasma membrane.

It was discovered that some ABPs (e.g., N-WASP) interact simultaneously with multiple phosphoinositol head groups and can thus function as sensors for membrane phospholipid concentrations (Papayannopoulos *et al.*, 2005). In the future, it will be important to elucidate whether other phosphoinositide-responsive actin-regulating proteins also interact simultaneously with more than one phosphoinositide molecule and if their activities are accurately modulated by the concentration of membrane phospholipids. Furthermore, it will be important to obtain direct information about the structural mechanisms by which phospholipids interact with ABPs and modulate their activities and/or structures.

Although many actin-regulating proteins have been shown to bind phosphoinositides *in vitro*, the *in vivo* relevance of these interactions has only been demonstrated in a small number of cases. Thus, in future studies it will be essential to elucidate the possible cellular consequences of these interactions. In addition, it will be important to reveal the cellular roles of the phosphoinositide specificities [PI(4,5)P_2 binding vs PI(3,4,5)P_3 binding] of various ABPs. These studies, together with additional cell biological investigations, would be expected to provide important new insights into the molecular mechanisms by which phosphoinositides contribute to various cellular processes requiring membrane dynamics.

REFERENCES

Aoyagi, K., Sugaya, T., Umeda, M., Yamamoto, S., Terakawa, S., and Takahashi, M. (2005). The activation of exocytotic sites by the formation of phosphatidylinositol 4,5-bisphosphate microdomains at syntaxin clusters. *J. Biol. Chem.* **280,** 17346–17352.

Auger, K. R. and Cantley, L. C. (1991). Novel polyphosphoinositides in cell growth and activation. *Cancer Cell* **3,** 263–270.

Auger, K. R., Serunian, L. A., Soltoff, S. P., Libby, P., and Cantley, L. C. (1989). PDGF-dependent tyrosine phosphorylation stimulates production of novel polyphosphoinositides in intact cells. *Cell* **57,** 167–175.

Bompard, G. and Caron, E. (2004). Regulation of WASP/WAVE proteins: Making a long story short. *J. Cell Biol.* **166,** 957–962.

Botelho, R. J., Teruel, M., Dierckman, R., Anderson, R., Wells, A., York, J. D., Meyer, T., and Grinstein, S. (2000). Localized biphasic changes in phosphatidylinositol-4,5-bisphosphate at sites of phagocytosis. *J. Cell Biol.* **151,** 1353–1368.

Brill, J. A., Hime, G. R., Scharer-Schuksz, M., and Fuller, M. T. (2000). A phospholipid kinase regulates actin organization and intercellular bridge formation during germline cytokinesis. *Development* **127,** 3855–3864.

Caroni, P. (2001). New EMBO members' review: Actin cytoskeleton regulation through modulation of PI(4,5)P(2) rafts. *EMBO J.* **20,** 4332–4336.

Chen, L., Janetopoulos, C., Huang, Y. E., Iijima, M., Borleis, J., and Devreotes, P. N. (2003). Two phases of actin polymerization display different dependencies on PI(3,4,5)P3 accumulation and have unique roles during chemotaxis. *Mol. Biol. Cell* **14,** 5028–5037.

Chong, L. D., Traynor-Kaplan, A., Bokoch, G. M., and Schwartz, M. A. (1994). The small GTP-binding protein Rho regulates a phosphatidylinositol 4-phosphate 5-kinase in mammalian cells. *Cell* **79,** 507–513.

Czech, M. P. (2000). PIP2 and PIP3: Complex roles at the cell surface. *Cell* **100,** 603–606.

Damen, J. E., Liu, L., Rosten, P., Humphries, R. K., Jefferson, A. B., Majerus, P. W., and Krystal, G. (1996). The 145-kDa protein induced to associate with Shc by multiple cytokines is an inositol tetraphosphate and phosphatidylinositol 3,4,5-triphosphate 5-phosphatase. *Proc. Natl. Acad. Sci. USA* **93,** 1689 1693.

Das, B., Shu, X., Day, G. J., Han, J., Krishna, U. M., Falck, J. R., and Broek, D. (2000). Control of intramolecular interactions between the pleckstrin homology and Dbl homology domains of Vav and Sos1 regulates Rac binding. *J. Biol. Chem.* **275,** 15074–15081.

Defacque, H., Bos, E., Garvalov, B., Barret, C., Roy, C., Mangeat, P., Shin, H. W., Rybin, V., and Griffiths, G. (2002). Phosphoinositides regulate membrane-dependent actin assembly by latex bead phagosomes. *Mol. Biol. Cell* **13,** 1190–1202.

Dent, E. W. and Gertler, F. B. (2003). Cytoskeletal dynamics and transport in growth cone motility and axon guidance. *Neuron* **40,** 209–227.

Desautels, M., Den Haese, J. P., Slupsky, C. M., McIntosh, L. P., and Hemmingsen, S. M. (2001). Cdc4p, a contractile ring protein essential for cytokinesis in Schizosaccharomyces pombe, interacts with a phosphatidylinositol 4-kinase. *J. Biol. Chem.* **276,** 5932–5942.

Desrivieres, S., Cooke, F. T., Parker, P. J., and Hall, M. N. (1998). MSS4, a phosphatidylinositol-4-phosphate 5-kinase required for organization of the actin cytoskeleton in Saccharomyces cerevisiae. *J. Biol. Chem.* **273,** 15787–15793.

Di Paolo, G., Pellegrini, L., Letinic, K., Cestra, G., Zoncu, R., Voronov, S., Chang, S., Guo, J., Wenk, M. R., and De Camilli, P. (2002). Recruitment and regulation of phosphatidylinositol phosphate kinase type 1 gamma by the FERM domain of talin. *Nature* **420,** 85–89.

Doughman, R. L., Firestone, A. J., Wojtasiak, M. L., Bunce, M. W., and Anderson, R. A. (2003a). Membrane ruffling requires coordination between type I alpha phosphatidylinositol phosphate kinase and Rac signaling. *J. Biol. Chem.* **278,** 23036–23045.

Doughman, R. L., Firestone, A. J., and Anderson, R. A. (2003b). Phosphatidylinositol phosphate kinases put PI4,5P(2) in its place. *J. Membr. Biol.* **194,** 77–89.

Douglass, A. D. and Vale, R. D. (2005). Single-molecule microscopy reveals plasma membrane microdomains created by protein-protein networks that exclude or trap signaling molecules in T cells. *Cell* **121,** 937–950.

Downes, C. P., Gray, A., and Lucocq, J. M. (2005). Probing phosphoinositide functions in signaling and membrane trafficking. *Trends Cell Biol.* **15,** 259–268.

Edwards, D. C., Sanders, L. C., Bokoch, G. M., and Gill, G. N. (1999). Activation of LIM-kinase by Pak1 couples Rac/Cdc42 GTPase signaling to actin cytoskeletal dynamics. *Nat. Cell Biol.* **1,** 253–259.

Engqvist-Goldstein, A. E. and Drubin, D. G. (2003). Actin assembly and endocytosis: From yeast to mammals. *Annu. Rev. Cell Dev. Biol.* **19,** 287–332.

Engqvist-Goldstein, A. E., Zhang, C. X., Carreno, S., Barroso, C., Heuser, J. E., and Drubin, D. G. (2004). RNAi-mediated Hip1R silencing results in stable association between the endocytic machinery and the actin assembly machinery. *Mol. Biol. Cell* **15,** 1666–1679.

Emoto, K., Inadome, H., Kanaho, Y., Narumiya, S., and Umeda, M. (2005). Local change in phospholipid composition at the cleavage furrow is essential for completion of cytokinesis. *J. Biol. Chem.* **280,** 37901–37907.

Erickson, J. W. and Cerione, R. A. (2004). Structural elements, mechanism, and evolutionary convergence of Rho protein-guanine nucleotide exchange factor complexes. *Biochemistry* **43,** 837–842.

Field, S. J., Madson, N., Kerr, M. L., Galbraith, K. A., Kennedy, C. E., Tahiliani, M., Wilkins, A., and Cantley, L. C. (2005). PtdIns(4,5)P2 functions at the cleavage furrow during cytokinesis. *Curr. Biol.* **15,** 1407–1412.

Franke, T. F., Kaplan, D. R., Cantley, L. C., and Toker, A. (1997). Direct regulation of the Akt proto-oncogene product by phosphatidylinositol-3,4-bisphosphate. *Science* **275,** 665–668.

Fruman, D. A., Meyers, R. E., and Cantley, L. C. (1998). Phosphoinositide kinases. *Annu. Rev. Biochem.* **67,** 481–507.

Funamoto, S., Milan, K., Meili, R., and Firtel, R. A. (2001). Role of phosphatidylinositol 3′ kinase and a downstream pleckstrin homology domain-containing protein in controlling chemotaxis in dictyostelium. *J. Cell Biol.* **153,** 795–810.

Gary, J. D., Wurmser, A. E., Bonangelino, C. J., Weisman, L. S., and Emr, S. D. (1998). Fab1p is essential for PtdIns(3)P 5-kinase activity and the maintenance of vacuolar size and membrane homeostasis. *J. Cell Biol.* **143,** 65–79.

Gray, A., Van Der Kaay, J., and Downes, C. P. (1999). The pleckstrin homology domains of protein kinase B and GRP1 (general receptor for phosphoinositides-1) are sensitive and selective probes for the cellular detection of phosphatidylinositol 3,4-bisphosphate and/or phosphatidylinositol 3,4,5-trisphosphate *in vivo*. *Biochem. J.* **344,** 929–936.

Han, J., Luby-Phelps, K., Das, B., Shu, X., Xia, Y., Mosteller, R. D., Krishna, U. M., Falck, J. R., White, M. A., and Broek, D. (1998). Role of substrates and products of PI 3-kinase in regulating activation of Rac-related guanosine triphosphatases by Vav. *Science* **279,** 558–560.

Haucke, V. (2005). Phosphoinositide regulation of clathrin-mediated endocytosis. *Biochem. Soc. Trans.* **33,** 1285–1289.

Hayes, M. J., Merrifield, C. J., Shao, D., Ayala-Sanmartin, J., Schorey, C. D., Levine, T. P., Proust, J., Curran, J., Bailly, M., and Moss, S. E. (2004). Annexin 2 binding to phosphatidylinositol 4,5-bisphosphate on endocytic vesicles is regulated by the stress response pathway. *J. Biol. Chem.* **279,** 14157–14164.

Hilpelä, P., Vartiainen, M. K., and Lappalainen, P. (2004). Regulation of the actin cytoskeleton by PI(4,5)P2 and PI(3,4,5)P3. *Curr. Top. Microbiol. Immunol.* **282,** 117–163.

Homma, K., Terui, S., Minemura, M., Qadota, H., Anraku, Y., Kanaho, Y., and Ohya, Y. (1998). Phosphatidylinositol-4-phosphate 5-kinase localized on the plasma membrane is essential for yeast cell morphogenesis. *J. Biol. Chem.* **273,** 15779–15786.

Honda, A., Nogami, M., Yokozeki, T., Yamazaki, M., Nakamura, H., Watanabe, H., Kawamoto, K., Nakayama, K., Morris, A. J., Frohman, M. A., and Kanaho, Y. (1999). Phosphatidylinositol 4-phosphate 5-kinase alpha is a downstream effector of the small G protein ARF6 in membrane ruffle formation. *Cell* **99,** 521–532.

Hoppe, A. D. and Swanson, J. A. (2004). Cdc42, Rac1, and Rac2 display distinct patterns of activation during phagocytosis. *Mol. Biol. Cell* **15,** 3509–3519.

Huckaba, T. M., Gay, A. C., Pantalena, L. F., Yang, H. C., and Pon, L. A. (2004). Live cell imaging of the assembly, disassembly, and actin cable-dependent movement of endosomes and actin patches in the budding yeast, Saccharomyces cerevisiae. *J. Cell Biol.* **167,** 519–530.

Iijima, M. and Devreotes, P. (2002). Tumor suppressor PTEN mediates sensing of chemoattractant gradients. *Cell* **109,** 599–610.

Insall, R. H. and Weiner, O. D. (2001). PIP3, PIP2 and cell movement—similar messages, different meanings. *Dev. Cell* **1,** 743–747.

Jacinto, A. and Baum, B. (2003). Actin in development. *Mech. Dev.* **120,** 1337–1349.

Jaffe, A. B. and Hall, A. (2005). RHO GTPASES: Biochemistry and biology. *Annu. Rev. Cell Dev. Biol.* **21,** 247–269.

Janetopoulos, C., Borleis, J., Vazquez, F., Iijima, M., and Devreotes, P. (2005). Temporal and spatial regulation of phosphoinositide signaling mediates cytokinesis. *Dev. Cell* **8,** 467–477.

Janmey, P. A. and Lindberg, U. (2004). Cytoskeletal regulation: Rich in lipids. *Nat. Rev. Mol. Cell Biol.* **5,** 658–666.

Jones, D. H., Morris, J. B., Morgan, C. P., Kondo, H., Irvine, R. F., and Cockcroft, S. (2000). Type I phosphatidylinositol 4-phosphate 5-kinase directly interacts with ADP-ribosylation factor 1 and is responsible for phosphatidylinositol 4,5-bisphosphate synthesis in the Golgi compartment. *J. Biol. Chem.* **275,** 13962–13966.

Kaksonen, M., Sun, Y., and Drubin, D. G. (2003). A pathway for association of receptors, adaptors, and actin during endocytic internalization. *Cell* **115,** 475–487.

Kaksonen, M., Toret, C. P., and Drubin, D. G. (2005). A modular design for the clathrin- and actin-mediated endocytosis machinery. *Cell* **123,** 305–320.

Kanzaki, M., Furukawa, M., Raab, W., and Pessin, J. E. (2004). Phosphatidylinositol 4,5-bisphosphate regulates adipocyte actin dynamics and GLUT4 vesicle recycling. *J. Biol. Chem.* **279,** 30622–30633.

Kisseleva, M., Feng, Y., Ward, M., Song, C., Anderson, R. A., and Longmore, G. D. (2005). The LIM protein Ajuba regulates phosphatidylinositol 4,5-bisphosphate levels in migrating cells through an interaction with and activation of PIPKI alpha. *Mol. Cell. Biol.* **25,** 3956–3966.

Kwik, J., Boyle, S., Fooksman, D., Margolis, L., Sheetz, M. P., and Edidin, M. (2003). Membrane cholesterol, lateral mobility, and the phosphatidylinositol 4,5-bisphosphate-dependent organization of cell actin. *Proc. Natl. Acad. Sci. USA* **100,** 13964–13969.

Laux, T., Fukami, K., Thelen, M., Golub, T., Frey, D., and Caroni, P. (2000). GAP43, MARCKS, and CAP23 modulate PI(4,5)P(2) at plasmalemmal rafts, and regulate cell cortex actin dynamics through a common mechanism. *J. Cell Biol.* **149,** 1455–1472.

Lemmon, M. A. and Ferguson, K. M. (2000). Signal-dependent membrane targeting by pleckstrin homology (PH) domains. *Biochem. J.* **350,** 1–18.

Ling, K., Doughman, R. L., Firestone, A. J., Bunce, M. W., and Anderson, R. A. (2002). Type I gamma phosphatidylinositol phosphate kinase targets and regulates focal adhesions. *Nature* **420,** 89–93.

Maffucci, T., Cooke, F. T., Foster, F. M., Traer, C. J., Fry, M. J., and Falasca, M. (2005). Class II phosphoinositide 3-kinase defines a novel signaling pathway in cell migration. *J. Cell Biol.* **169,** 789–799.

Martin, T. F. J. (2001). PI(4,5)P2 regulation of surface membrane traffic. *Curr. Opin. Cell Biol.* **13,** 493–499.

Matsui, T., Yonemura, S., Tsukita, S., and Tsukita, S. (1999). Activation of ERM proteins *in vivo* by Rho involves phosphatidyl-inositol 4-phosphate 5-kinase and not ROCK kinases. *Curr. Biol.* **9,** 1259–1262.

McLaughlin, S., Wang, J., Gambhir, A., and Murray, D. (2002). PIP(2) and proteins: Interactions, organization, and information flow. *Annu. Rev. Biophys. Biomol. Struct.* **31,** 151–175.

Merrifield, C. J., Perrais, D., and Zenisek, D. (2005). Coupling between clathrin-coated-pit invagination, cortactin recruitment, and membrane scission observed in live cells. *Cell* **121,** 593–603.

Michiels, F., Stam, J. C., Hordijk, P. L., van der Kammen, R. A., Ruuls-Van, S. L., Feltkamp, C. A., and Collard, J. G. (1997). Regulated membrane localization of Tiam1, mediated by the NH2-terminal pleckstrin homology domain, is required for Rac-dependent membrane ruffling and C-Jun NH2-terminal kinase activation. *J. Cell Biol.* **137,** 387–398.

Munro, S. (2003). Lipid rafts: Elusive or illusive? *Cell* **115,** 377–388.

Nicholson-Dykstra, S., Higgs, H. N., and Harris, E. S. (2005). Actin dynamics: Growth from dendritic branches. *Curr. Biol.* **15,** 346–357.

Niedergang, F. and Chavrier, P. (2004). Signaling and membrane dynamics during phagocytosis: Many roads lead to the phagos(R)ome. *Curr. Opin. Cell Biol.* **16,** 422–428.

Niggli, V. (2005). Regulation of protein activities by phosphoinositide phosphates. *Annu. Rev. Cell Dev. Biol.* **21,** 57–79.

Nishita, M., Tomizawa, C., Yamamoto, M., Horita, Y., Ohashi, K., and Mizuno, K. (2005). Spatial and temporal regulation of cofilin activity by LIM kinase and Slingshot is critical for directional cell migration. *J. Cell Biol.* **171,** 349–359.

Paavilainen, V. O., Bertling, E., Falck, S., and Lappalainen, P. (2004). Regulation of cytoskeletal dynamics by actin-monomer-binding proteins. *Trends Cell Biol.* **14,** 386–394.

Palmgren, S., Ojala, P. J., Wear, M. A., Cooper, J. A., and Lappalainen, P. (2001). Interactions with PIP2, ADP-actin monomers, and capping protein regulate the activity and localization of yeast twinfilin. *J. Cell Biol.* **155,** 251–260.

Pantaloni, D., Le Clainche, C., and Carlier, M. F. (2001). Mechanism of actin-based motility. *Science* **292,** 1502–1506.

Papayannopoulos, V., Co, C., Prehoda, K. E., Snapper, S., Taunton, J., and Lim, W. A. (2005). A polybasic motif allows N-WASP to act as a sensor of PIP(2) density. *Mol. Cell* **17,** 181–191.

Pendaries, C., Tronchere, H., Plantavid, M., and Payrastre, B. (2003). Phosphoinositide signaling disorders in human diseases. *FEBS Lett.* **546,** 25–31.

Pendaries, C., Tronchere, H., Racaud-Sultan, C., Gaits-Iacovoni, F., Coronas, S., Manenti, S., Gratacap, M. P., Plantavid, M., and Payrastre, B. (2005). Emerging roles of phosphatidylinositol monophosphates in cellular signaling and trafficking. *Adv. Enzyme Regul.* **45,** 201–214.

Pollard, T. D. and Borisy, G. G. (2003). Cellular motility driven by assembly and disassembly of actin filaments. *Cell* **113,** 453–465.

Postma, M., Bosgraaf, L., Loovers, H. M., and Van Haastert, P. J. (2004). Chemotaxis: Signaling modules join hands at front and tail. *EMBO Rep.* **5,** 35–40.

Rameh, L. E. and Cantley, L. C. (1999). The role of phosphoinositide 3-kinase lipid products in cell function. *J. Biol. Chem.* **274,** 8347–8350.

Rodriguez-Escudero, I., Roelants, F. M., Thorner, J., Nombela, C., Molina, M., and Cid, V. J. (2005). Reconstitution of the mammalian PI3K/PTEN/Akt pathway in yeast. *Biochem. J.* **390,** 613–623.

Rohatgi, R., Ma, L., Miki, H., Lopez, M., Kirchhausen, T., Takenawa, T., and Kirschner, M. W. (1999). The interaction between N-WASP and the Arp2/3 complex links Cdc42-dependent signals to actin assembly. *Cell* **97,** 221–231.

Rozelle, A. L., Machesky, L. M., Yamamoto, M., Driessens, M. H., Insall, R. H., Roth, M. G., Luby-Phelps, K., Marriott, G., Hall, A., and Yin, H. L. (2000). Phosphatidylinositol 4,5-bisphosphate induces actin-based movement of raft-enriched vesicles through WASP-Arp2/3. *Curr. Biol.* **10,** 311–320.

Russo, C., Gao, Y., Mancini, P., Vanni, C., Porotto, M., Falasca, M., Torrisi, M. R., Zheng, Y., and Eva, A. (2001). Modulation of oncogenic DBL activity by phosphoinositol phosphate binding to pleckstrin homology domain. *J. Biol. Chem.* **276,** 19524–19531.

Saul, D., Fabian, L., Forer, A., and Brill, J. A. (2004). Continuous phosphatidylinositol metabolism is required for cleavage of crane fly spermatocytes. *J. Cell Sci.* **117,** 3887–3896.

Sbrissa, D. and Shisheva, A. J. (2005). Acquisition of unprecedented phosphatidylinositol 3,5-bisphosphate rise in hyperosmotically stressed 3T3-L1 adipocytes, mediated by ArPIKfyve-PIKfyve pathway. *J. Biol. Chem.* **280,** 7883–7889.

Schafer, D. A., Jennings, P. B., and Cooper, J. A. (1996). Dynamics of capping protein and actin assembly *in vitro*: Uncapping barbed ends by polyphosphoinositides. *J. Cell Biol.* **135,** 169–179.

Scott, C. C., Dobson, W., Botelho, R. J., Coady-Osberg, N., Chavrier, P., Knecht, D. A., Heath, C., Stahl, P., and Grinstein, S. (2005). Phosphatidylinositol-4,5-bisphosphate hydrolysis directs actin remodeling during phagocytosis. *J. Cell Biol.* **169,** 139–149.

Shibasaki, Y., Ishihara, H., Kizuki, N., Asano, T., Oka, Y., and Yazaki, Y. (1997). Massive actin polymerization induced by phosphatidylinositol-4-phosphate 5-kinase *in vivo*. *J. Biol. Chem.* **272,** 7578–7581.

Snyder, J. T., Rossman, K. L., Baumeister, M. A., Pruitt, W. M., Siderowski, D. P., Der, C. J., Lemmon, M. A., and Sondek, J. (2001). Quantitative analysis of the effect of phosphoinositide interactions on the function of Dbl family proteins. *J. Biol. Chem.* **276,** 45868–45875.

Stefan, C. J., Padilla, S. M., Audhya, A., and Emr, S. D. (2005). The phosphoinositide phosphatase Sjl2 is recruited to cortical actin patches in the control of vesicle formation and fission during endocytosis. *Mol. Cell. Biol.* **25,** 2910–2923.

Stephens, L. R., Hughes, K. T., and Irvine, R. F. (1991). Pathway of phosphatidylinositol(3,4,5)-trisphosphate synthesis in activated neutrophils. *Nature* **351,** 33–39.

Stephens, L. R., Jackson, T. R., and Hawkins, P. T. (1993). Agonist-stimulated synthesis of phosphatidylinositol(3,4,5)-trisphosphate: A new intracellular signaling system? *Biochim. Biophys. Acta* **1179,** 27–75.

Sun, Y., Kaksonen, M., Madden, D. T., Schekman, R., and Drubin, D. G. (2005). Interaction of Sla2p's ANTH domain with PtdIns(4,5)P2 is important for actin-dependent endocytic internalization. *Mol. Biol. Cell* **16,** 717–730.

Toker, A. (2002). Phosphoinositides and signal transduction. *Cell Mol. Life Sci.* **59,** 761–779.

Tolias, K. F., Hartwig, J. H., Ishihara, H., Shibasaki, Y., Cantley, L. C., and Carpenter, C. L. (2000). Type I alpha phosphatidylinositol-4-phosphate 5-kinase mediates Rac-dependent actin assembly. *Curr. Biol.* **10,** 153–156.

van Rheenen, J. and Jalink, K. (2002). Agonist-induced PIP(2) hydrolysis inhibits cortical actin dynamics: Regulation at a global but not at a micrometer scale. *Mol. Biol. Cell* **13,** 3257–3267.

van Rheenen, J., Achame, E. M., Janssen, H., Calafat, J., and Jalink, K. (2005). PIP2 signaling in lipid domains: A critical re-evaluation. *EMBO J.* **24,** 1664–1673.

Vartiainen, M. K. and Machesky, L. M. (2004). The WASP-Arp2/3 pathway: Genetic insights. *Curr. Opin. Cell Biol.* **16,** 174–181.

Walker, S. M., Leslie, N. R., Perera, N. M., Batty, I. H., and Downes, C. P. (2004). The tumour-suppressor function of PTEN requires an N-terminal lipid-binding motif. *Biochem. J.* **379,** 301–307.

Watt, S. A., Kular, G., Fleming, I. N., Downes, C. P., and Lucocq, J. M. (2002). Subcellular localization of phosphatidylinositol 4,5-bisphosphate using the pleckstrin homology domain of phospholipase C delta1. *Biochem. J.* **365,** 657–666.

Weernink, P. A., Meletiadis, K., Hommeltenberg, S., Hinz, M., Ishihara, H., Schmidt, M., and Jakobs, K. H. (2004). Activation of type I phosphatidylinositol 4-phosphate 5-kinase isoforms by the Rho GTPases, RhoA, Rac1, and Cdc42. *J. Biol. Chem.* **279,** 7840–7849.

Weiss-Haljiti, C., Pasquali, C., Ji, H., Gillieron, C., Chabert, C., Curchod, M. L., Hirsch, E., Ridley, A. J., van Huijsduijnen, R. H., Camps, M., and Rommel, C. (2004). Involvement of phosphoinositide 3-kinase gamma, Rac, and PAK signaling in chemokine-induced macrophage migration. *J. Biol. Chem.* **279,** 43273–43284.

Yamamoto, M., Hilgemann, D. H., Feng, S., Bito, H., Ishihara, H., Shibasaki, Y., and Yin, H. L. (2001). Phosphatidylinositol 4,5-bisphosphate induces actin stress-fiber formation and inhibits membrane ruffling in CV1 cells. *J. Cell Biol.* **152,** 867–876.

Yin, H. L. and Janmey, P. A. (2003). Phosphoinositide regulation of the actin cytoskeleton. *Annu. Rev. Physiol.* **65,** 761–789.

Yonezawa, N., Nishida, E., Iida, K., Yahara, I., and Sakai, H. (1990). Inhibition of the interactions of cofilin, destrin, and deoxyribonuclease I with actin by phosphoinositides. *J. Biol. Chem.* **265,** 8382–8386.

Zhang, Y., Sugiura, R., Lu, Y., Asami, M., Maeda, T., Itoh, T., Takenawa, T., Shuntoh, H., and Kuno, T. (2000). Phosphatidylinositol 4-phosphate 5-kinase Its3 and calcineurin Ppb1 coordinately regulate cytokinesis in fission yeast. *J. Biol. Chem.* **275,** 35600–35606.

Chapter 11

Lipid Interactions of Cytoskeletal Proteins

Verena Niggli

Department of Pathology, University of Bern, Murtenstr. 31, CH-3010 Bern, Switzerland

At least 18 different cytoskeletal proteins have been shown to interact *in vitro* with, and in some cases are regulated by, specific membrane lipids, mostly phosphoinositides, whose synthesis is themselves regulated by extracellular signals. These lipid interactions are mediated by structurally diverse specific binding sites. Lipid interactions can serve to target proteins, such as spectrin, myosin X, dynamin, or annexin 2, to the plasma membrane or inactivate processes counteracting actin polymerization and stimulate proteins, such as Wiskott–Aldrich syndrome protein (WASP), involved in nucleating actin polymerization. In other cases (ezrin, talin), phosphoinositides serve to activate proteins involved in actin-membrane linkage. Such interactions may occur highly localized in lipid rafts.

Advances in Molecular and Cell Biology, Vol. 37

ISBN: 0-444-52868-7
DOI: 10.1016/S1569-2558(06)37011-7

I. INTRODUCTION

The cytoskeleton, including actin filaments, intermediate filaments, microtubules, and associated proteins, is crucial for cell shape changes, cell migration, cell adhesion, mitosis, and intracellular transport of organelles. In order to convert changes in cytoskeletal organization into shape changes required for cell motility, and for inside-out or outside-in signaling via the cytoskeleton, reversible and regulated contacts of these structural proteins with the plasma membrane are required. Anchoring of cytoskeletal proteins may occur via direct or indirect association with transmembrane proteins, such as integrins, but may also involve phospholipids. Strikingly, especially an increasing number of actin-associated proteins have been shown to interact with membrane phospholipids (Table I). Structures of important phospholipids and their abbreviations are shown in Fig. 1. Protein–protein and protein–lipid interactions may cooperate in regulating dynamics and localized anchoring of cytoskeletal filaments. Importantly, specific phospholipids, the phosphoinositides, which are regulated by extracellular signals, themselves regulate functions of cytoskeletal proteins. Phospholipids could thus serve both to concentrate cytoskeletal proteins at the plasma membrane and also to modulate the activity of these proteins. Especially the highly negatively charged lipids $PI(4,5)P_2$, $PI(3,4)P_2$, and $PI(3,4,5)P_3$ are involved in regulating functions of cytoskeletal proteins. The latter two molecules are products of the enzyme phosphatidylinositol 3-kinase (PI 3-kinase) whose activity is tightly controlled by signaling. $PI(4,5)P_2$ is present in appreciable amounts in resting cells and is increased or decreased locally depending on activation of phospholipase C or phosphatidylinositol-4-phosphate 5-kinase (PIP-5K) (Doughman *et al.*, 2003; Oude Weernink *et al.*, 2004).

Modulation of levels of $PI(4,5)P_2$ affect cortical actin organization, membrane ruffling, endocytosis, synaptic vesicle recycling, whereas products of PI 3-kinase are thought to be involved in cell migration and exocytosis (Downes *et al.*, 2005). These processes could potentially occur in a highly localized manner in so-called lipid rafts. Rafts are plasma membrane microdomains enriched in cholesterol and glycosphingolipids. They are thought to function as organizing platforms for signaling molecules (Pike, 2004). Raft markers and $PI(4,5)P_2$ colocalize in prominent patches near the cell edge of living PC 12 cells and raft disruption results in disappearance of these $PI(4,5)P_2$-containing patches (Golub and Caroni, 2005). Moreover, Bodin *et al.* (2005) provided evidence for specific association of $PI(4,5)P_2$-containing rafts with the actin cytoskeleton, which is dependent on activation in platelets. Rafts rich in $PI(4,5)P_2$ may thus provide platforms for localized actin assembly. However, this contention is still controversial and the presence of $PI(4,5)P_2$ in rafts has been questioned (van Rheenen *et al.*, 2005). Another way of localizing $PI(4,5)P_2$ in membranes could be sequestration by proteins, such as MARCKS, which then would release these lipid molecules in response to increased cellular calcium (McLaughlin and Murray, 2005).

This chapter focuses on lipid-binding site architecture and on the functional regulation of cytosketal proteins by phospholipids, considering mainly those examples in which specific lipid-binding sites have been identified and some evidence of the functional *in situ* relevance of these interactions has been obtained by site-directed

Table I
Overview over Cytoskeletal Proteins Interacting with Specific Phospholipids

Protein	Lipid specificity	Binding site	Functional impact of lipid binding
Gelsolin	(a) PI-4,5-P_2 ~ PI-3,4,5-P_3	Clusters of basic residues	Dissociation of gelsolin from plus ends of F-actin
	(b) Acidic phospholipids	Not identified	Actin-membrane linkage?
Cofilin	PI-4,5-P_2 ~ PI-3,4,5-P_3	Clusters of basic residues	Inhibition of functions
Villin	PI-4,5-P_2	Clusters of basic residues	Inhibition of actin severing and capping; enhancement of actin bundling
Profilin	PI-3,4,5-P_3 >PI-4,5-P_2	Clusters of basic residues	Inhibition of profilin–G-actin interaction
N-WASP, WASP	PI-4,5-P_2 ~ PI-3,4,5-P_3	Clusters of basic residues	Relieve of autoinhibition
WAVE	PI-3,4,5-P_3 > PI-4,5-P_2	Clusters of basic residues	Membrane targeting
ERM	PI-4,5-P_2 ~ PI-3,4,5-P_3	Clusters of basic residues in the FERM domain	Relieve of autoinhibition
Vinculin	PS, PI, PI-4,5-P_2	Amphipathic helices	Relieve of autoinhibition? Modulation of interaction with binding partners? Inhibition of interaction with F-actin?
Talin	PI-4,5-P_2	Amphipathic helix in the FERM domain?	Relieve of autoinhibition? Membrane targeting?

(*continued*)

Table I (*continued*)

Protein	Lipid specificity	Binding site	Functional impact of lipid binding
Spectrin	(a) PI-4,5-P_2 ~ PI-3,4,5-P_3	PH domain	Membrane targeting
	(b) PI-4,5-P_2	Calponin homology domain 2	Stabilization of membrane interaction
	(c) PS	N-terminus of β-chain	
Dynamin	PI-4,5-P_2 ~ PI-3,4,5-P_3	PH domain	Membrane targeting
Annexin 2	(a) Acidic phospholipids	Endonexin folds	Membrane targeting
	(b) PI-4,5-P_2 > PI-4,5-P_2	Clusters of basic residues	Membrane targeting, clustering of lipids?
α-Actinin	PI-4,5-P_2 ~ PI-3,4,5-P_3	Calponin homology domains 1,2	Modulation of interaction with binding partners
Myosin X	PI-3,4,5-P_3 > PI-4,5-P_2	PH domains	Membrane targeting
α-Tubulin and β-tubulin	PI-4,5-P_2	Not yet identified	Modulation of phosphoinositide hydrolysis? Raft targeting?
MAP1B	PS, PI, PA	KKE repeats	Modulation of interaction with microtubules?
Kinesin (UNC-104)	PI-4 5-P_2 > PI-3,4,5-P_3	PH domain	Enhancement of vesicles binding and processive transport
Vimentin	(a) Acidic phospholipids	Basic N-terminal domain	Regulation of filament assembly?
	(b) Neutral phospholipids	α-Helical central rod domain	Lipid metabolism?

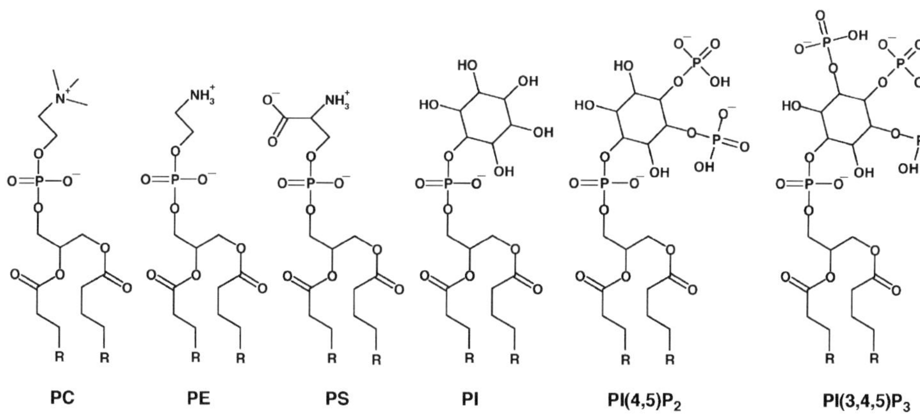

Figure 1. Structures of selected phospholipids. Schematic representations of phosphatidylcholine (PC), phosphatidylethanolamine (PE), phosphatidylserine (PS), phosphatidylinositol (PI), phosphatidylinositol 4,5-bisphosphate (PI(4,5)P_2), and phosphatidylinositol 3,4,5-bisphosphate (PI(3,4,5)P_3). R corresponds to the remainder of the fatty acid side chain. PI(3,4,5)P_3 can be dephosphorylated by 5-phosphatases resulting in the product phosphatidylinositol 3,4-bisphosphate (PI(3,4)P_2) (Downes *et al.*, 2005).

mutagenesis of lipid-binding domains. It should be mentioned in the beginning that the lipid-binding sites in cytoskeletal proteins are very diverse, featuring, for example, clusters of basic amino acid residues, classical pleckstrin homology (PH) domains, endonexin folds, or amphipathic helices (Table I, see also Niggli, 2001). Lipid interactions of actin-associated proteins involved in F-actin assembly, F-actin-membrane linkage, and actin cross-linking, of myosin isoforms, microtubules, microtubule-associated proteins, and intermediate filaments will be considered. First, the structure and function of the protein in question will be briefly summarized followed by discussion of the properties of the lipid-binding site and the physiological relevance of the lipid interactions.

II. ACTIN-ASSOCIATED PROTEINS

A. Proteins Involved in F-Actin Assembly

1. Gelsolin

The 84-kDa protein gelsolin is a calcium-activated regulator of scaffolding and motile functions of actin. Gelsolin caps the fast growing barbed ends of actin filaments, nucleates actin polymerization from the pointed end, and also severs F-actin. Fibroblasts lacking gelsolin feature excessive actin stress fibers and migrate more slowly than wild-type fibroblasts, indicating a role of gelsolin in maintaining dynamics of F-actin in motile cells

(Witke *et al.*, 1995; Silacci *et al.*, 2004). Gelsolin consists of six compactly folded homologous 100–125 amino acid modules connected by linkers (G1–G6). G1 and G4 bind actin monomers, whereas G2 mediates F-actin binding (McGough *et al.*, 2003).

a. *Lipid-Binding Site Properties*

Gelsolin binds *in vitro* $PI(4,5)P_2$ and $PI(3,4,5)P_3$ with equal affinity. Gelsolin interacts with these lipids via two regions in G1 and G2 (residues 135–149 and 150–169), which contain clusters of basic amino acids. Two sites in G6 (residues 620–634 and 732–755) may additionally contribute to phosphoinositide binding (Feng *et al.*, 2001). These data were obtained using photoaffinity labeling with modified phosphoinositides. The lipid-binding domains overlap with G-actin- and F-actin-binding sites (Xian and Janmey, 2002). Gelsolin appears not only to bind to the charged lipid head group but also interacts with fatty acid side chains and thus may pull the lipids out of the bilayer (Feng *et al.*, 2001; Liepina *et al.*, 2003). Gelsolin could therefore, besides being regulated by these lipids, also modulate phosphoinositide density in the plasma membrane. The binding of phosphoinositides leads to a conformational change of gelsolin. This conformational change very likely induces dissociation of gelsolin from the barbed ends of actin filaments and allows filament elongation (Xian and Janmey, 2002).

Méré *et al.* (2005) additionally provided evidence for a calcium-independent interaction of gelsolin with acidic phospholipids, such as phosphatidylglycerol, which is optimal at acidic pH (pH 5.5) and does not interfere with its interaction with actin filaments but impairs its severing activity. This interaction may involve the extended linker region between the two halves of gelsolin. It may serve to allow gelsolin to bind actin filaments to the membrane.

b. *Physiological Relevance of Gelsolin–Lipid Interactions*

Whether these lipid interactions are physiologically relevant is so far unclear. At least part of gelsolin appears to be in contact with lipids in intact cells and phosphoinositides modulate the location of gelsolin. In osteoclasts, for example, gelsolin immunoprecipitates obtained from cellular lysates contain $PI(4,5)P_2$ and $PI(3,4,5)P_3$ (Chellaiah and Hruska, 1996). Stimulation of osteoclasts with the bone matrix protein osteopontin increases the amount of phosphoinositides associated with gelsolin, correlating with increased actin polymerization (Chellaiah and Hruska, 1996). A synthetic peptide corresponding to the $PI(4,5)P_2$-binding region (amino acids 160–169) of gelsolin displaces gelsolin and actin from $PI(4,5)P_2$-rich adhesion complexes (Guttman *et al.*, 2002). Gelsolin also coimmunoprecipitates with PI 3-kinase, an interaction that is disrupted by the above-mentioned peptide (Biswas *et al.*, 2004). The notions envisaging a specific functional regulation of gelsolin by lipids or a role of gelsolin in modulating availability of phosphoinositides at the plasma membrane in intact cells will have to be substantiated.

2. Cofilin

The 18- to 19-kDa protein cofilin is structurally and functionally related to gelsolin. It is a globular protein, consisting of a central β-sheet surrounded by four α-helices, strikingly similar to one of the gelsolin modules (Hatanaka *et al.*, 1996). Cofilin promotes disassembly of actin filaments at the slow growing pointed ends, severs actin filaments creating new barbed ends for polymerization, and binds also to actin monomers (preferably ADP-actin). It is thought to be a key protein regulating actin dynamics at the leading edge of migrating cells and is also required for cytokinesis. Lack of nonmuscle cofilin in mice results in defects in neural tube morphogenesis and neural crest cell migration (DesMarais *et al.*, 2005; Gurniak *et al.*, 2005).

a. *Lipid-Binding Site Properties*

In vitro both PI(4,5)P_2 and PI(3,4,5)P_3 interact equally well with yeast cofilin and inhibit its activity (Ojala *et al.*, 2001). Phosphoinositides induce *in vitro* oligomerization of cofilin (Pfannstiel *et al.*, 2001). The binding sites for F-actin and phosphoinositides, located in a positively charged surface at one side of the molecule, overlap but are not identical. The lipid-binding sites are scattered over this surface, involving basic residues 23–26, 34, 35, 38, 80, 82, 96, 98, 109, and 110, based on site-directed mutagenesis. This implicates binding of cofilin to multiple lipid molecules. Arginines 109 and 110 are required for interaction with phosphoinositides but are not involved in interaction with F-actin and play a minor role in G-actin binding (Ojala *et al.*, 2001).

b. *Physiological Relevance of Cofilin–Lipid Interactions*

The *in situ* evidence on this point is scarce. Replacement of wild-type cofilin by cofilin with arginines 109 and 110 mutated to alanines in yeast resulted in abnormal actin organization, but the cells were viable, indicating that this lipid interaction is involved in actin reorganization but is not crucial for yeast survival (Ojala *et al.*, 2001).

3. Villin

Villin is structurally related to cofilin and gelsolin, consists of six gelsolin-like domains and caps, nucleates or cross-links actin filaments (Silacci *et al.*, 2004). Its regulation by calcium, phosphorylation, and phosphoinositides is extensively discussed elsewhere in this volume by S. Khurana. Evidence for its self-association has been provided *in vitro* and in living cells (George *et al.*, submitted for publication).

a. *Lipid-Binding Site Properties*

Villin interacts *in vitro* preferentially with PI(4,5)P_2 as compared to phosphatidylinositol monophosphate (PIP), PI, or phosphatidylserine (PS). This interaction impairs its actin severing and capping function but enhances its actin-bundling capacity. Villin contains three putative $PI(4,5)P_2$-binding sites consisting of clusters of basic residues similar to those present in gelsolin. Kumar *et al.* (2004) showed that especially regions 138–146 and 816–824 are required for binding to this lipid. These sites overlap with the actin-binding sites.

b. *Physiological Relevance of Villin–Lipid Interactions*

Data using fluorescence resonance energy transfer (FRET) and coimmunoprecipitation show association of villin with $PI(4,5)P_2$ in living cells. Overexpression of PIP-5Kγ in Caco-2 cells enhanced formation of villin oligomers, which are located in microvilli, filopodia, membrane ruffles, and lamellipodia. These oligomers are required for actin bundling. Thus, $PI(4,5)P_2$ appears to regulate villin self-association and villin-mediated actin bundling (George *et al.*, submitted for publication).

4. Profilin

The 14- to 17-kDa protein profilin folds into a single compact globular domain with an extended β-sheet flanked on the convex face by two α-helices and on the concave face by three small α-helical regions, as determined using NMR spectroscopy and crystallization (Metzler *et al.*, 1995). Profilin is a key regulator of actin dynamics and its function is indispensable for embryonic development in insects and mice (Verheyen and Cooley, 1994; Witke, 2004). Profilin binds monomeric actin and accelerates the ADP–ATP exchange on G-actin 1000-fold. Actin–profilin complexes bind to free barbed ends of actin filaments, followed by dissociation of profilin and incorporation of actin into the filament. Profilin also interacts with proteins containing poly L-proline (PLP) stretches such as the enabled/vasodilator-stimulated phosphoprotein Ena/VASP (Witke, 2004).

a. *Lipid-Binding Site Properties*

Profilin shows *in vitro* some preference for products of PI 3-kinase ($PI(3,4)P_2$, $PI(3,4,5)P_3$) as compared to $PI(4,5)P_2$ (Lu *et al.*, 1996), but the physiological profilin lipid ligand has not yet been identified. Two domains (residues 83–92 and 126–136) located on one side of the molecule containing clusters of basic residues form a lipid-binding pocket in profilin, involving N- and C-terminal helices. Site-directed mutagenesis showed that arginines 88 and 136 contribute importantly to interaction with $PI(4,5)P_2$ micelles. The first site (residues 83–92) overlaps with the actin-binding site and the second site (residues 126–136) with the PLP-binding site. Phosphoinositides prevent association of profilin with actin, thus keeping the protein in an inactive state and also interfere with PLP binding (Lambrechts *et al.*, 2002). Covalently linked profilin–actin complexes still interact with phosphoinositides confirming the presence of a second lipid-binding site clearly separated from the actin-binding site (Skare and Karlsson, 2002). Profilin may also control $PI(4,5)P_2$ turnover (Goldschmidt-Clermont *et al.*, 1990, 1991), similar to gelsolin, but this notion has not yet been verified *in situ*.

Gareus *et al.* (2006) provided evidence for another interaction of profilin that is controlled by $PI(4,5)P_2$. They showed that profilin 2 interacts *in vitro* with dynamin and precludes recruitment of Src homology 3 (SH3) ligands to dynamin, thus suppressing its function. This interaction can be abrogated by $PI(4,5)P_2$, suggesting a role of profilin in controlling endocytosis. Overexpression of profilin in HeLa cells inhibits endocytosis. $PI(4,5)P_2$ could thus simultaneously activate dynamin and induce actin polymerization by releasing G-actin from profilin 2 (Gareus *et al.*, 2006).

b. *Physiological Relevance of Profilin–Lipid Interactions*

Wittenmayer *et al.* (2004) studied the impact of expressing human profilin I wild type and mutants on induction of a nontumorigenic phenotype in a human breast cancer cell line expressing low levels of endogenous profilin. The following mutants were expressed: mutants Y59A (significantly lower affinity for actin; normal binding of lipids and PLP), H133S (loss of PLP binding; normal binding to actin and lipids), and R88L (less than 30% of wild-type binding to lipids; normal PLP binding, somewhat reduced actin interaction). According to this study, the actin-binding site, but not the interaction of profilin with phosphoinositides and PLP, is required for tumor suppression.

5. N-WASP, WASP, WAVE

Proteins of the Wiskott–Aldrich syndrome (WAS) protein (WASP) family and the related WAVE (WASP family verprolin homologous protein) proteins are approximately 5000 amino acid residue multidomain molecules. They interact with signaling molecules, profilin, actin-related protein 2/3 (Arp2/3), and actin filaments, activating actin assembly via Arp2/3 in response to intracellular signals. WASP isoforms are implicated in movement of endocytic vesicles and formation of filopodia, whereas WAVE proteins are involved in lamellipodia formation in the leading edge of motile cells (Takenawa and Miki, 2001). The small GTP-binding proteins Cdc42 and Rac are required for activation of WASP and WAVE, respectively. In the absence of stimuli, neural WASP (N-WASP) and WASP are locked in an autoinhibited conformation due to intramolecular interactions between C- and N-terminus. These interactions are relieved by cooperative binding of upstream activators. In contrast to N-WASP, WAVE proteins are probably constitutively active *in vitro* and occur in complexes with other proteins. WAVE activation requires proper localization of these complexes as well as the presence of F-actin and/or the release of additional proteins from the complex. Rac induces translocation of the WAVE complex to lamellipodia (Bompard and Caron, 2004).

a. Lipid-Binding Site Properties

Both WASP and WAVE proteins contain related highly basic domains thought to be involved in binding of phosphoinositides that modulate their function. Both PI(4,5)P_2 and PI(3,4,5)P_3 bind to N-WASP, thereby reducing the affinity between C- and N-termini and activating the protein (Oikawa *et al.*, 2004). Elimination of residues 186–200 in N-WASP or mutation of at least three or more basic residues to glutamic acid or alanine results in abolishment of binding to PI(4,5)P_2 (Rohatgi *et al.*, 2000; Papayannopoulos *et al.*, 2005). Interaction of the basic domain of N-WASP with phosphoinositides occurs in a multivalent, highly cooperative manner. N-WASP could thus respond in a switch-like manner to small increases in the density of PI(4,5)P_2. Reversal of the basic sequence 183–197 does not impair binding of a fragment of WASP to this lipid, suggesting that the high positive charge density of the motif, rather than any precise structure, is critical for binding (Papayannopoulos *et al.*, 2005). However, this notion contrasts with the finding that the very similar basic domain in WAVE appears to preferentially interact with PI(3,4,5)P_3 (Oikawa *et al.*, 2004), whereas that of N-WASP has been reported to show some preference for PI(4,5)P_2 (Papayannopolous *et al.*, 2005).

b. Physiological Role of WASP–Lipid Interactions and WAVE–Lipid Interactions

The role of phosphoinositides in *in situ* activation of WASP proteins is presently under debate. A "mini" version of N-WASP, containing the basic domain, initiates actin-based motility of endosomal vesicles in *Xenopus* egg extracts depleted of endogenous N-WASP. Constructs lacking residues 178–204 of the basic domain are inactive in this assay. A "gain of function mutation," that is, increasing the number of basic residues increases affinity of interaction with PI(4,5)P_2 and actin comet tail length. N-WASP full-length constructs with additional lysines, when expressed in N-WASP$^{-/-}$ mouse fibroblasts, induce vesicle rocketing even in the absence of increases in PI(4,5)P_2 (Papayannopoulos *et al.*, 2005). These data support an *in situ* role of this lipid interaction. Benesch *et al.* (2002) reported that the actin comet–dependent movement of endosomes in fibroblasts is abolished in N-WASP–defective cells. However, somewhat in contradiction to the report by Papayannopoulos *et al.* (2005), N-WASP constructs lacking the basic domain and the GTPase-binding (GBD) domain (residues 159–226) restore formation of actin comet tails as efficiently as wild-type protein in these cells (Benesch *et al.*, 2002). Possibly the construct lacking both the basic domain and GBD domain (but not the protein lacking only the basic domain) is partially activated.

Oikawa *et al.* (2004) have provided evidence for an important and selective role of PI(3,4,5)P_3 in WAVE activation. Suppression of synthesis of this lipid in intact cells inhibits growth factor-induced translocation of WAVE2 to lamellipodia. WAVE2 mutants with mutated basic residues that *in vitro* lack the capacity of interacting with PI(3,4,5)P_3 are not anymore recruited to the plasma membrane in stimulated cells and overexpression of these mutants reduces Rac-dependent formation of lamellipodia (Oikawa *et al.*, 2004). PI(3,4,5)P_3 thus may localize activated WAVE2 at the membrane together with Rac. In summary, these findings implicate that the basic domains of WAVE and WASP are more than just clusters of positively charged molecules unspecifically mediating interactions with negatively charged lipids.

B. Proteins Involved in F-Actin-Membrane Linkage

1. Ezrin/Radixin/Moesin

Ezrin, radixin, and moesin (ERM) proteins are closely related molecules attaching actin filaments to the membrane, directly via lipid interaction or indirectly to cytoplasmic tails of transmembrane proteins or scaffolding proteins linked to transmembrane proteins. The lipid- and membrane protein-binding sites are located in a conserved N-terminal membrane-binding domain of approximately 300 amino acids called the FERM (4.1-ezrin-radixin-moesin) domain that adopts a trilobed structure (Hamada *et al.*, 2000; Smith *et al.*, 2003), whereas a major F-actin-binding site is located in the C-terminus (Bretscher *et al.*, 2002; Hoeflich and Ikura, 2004). Intramolecular association between the FERM domain and the C-terminal domain keep the molecule in an inactive, "closed" state. These interactions are relieved by phosphoinositide interaction and subsequent phosphorylation (Bretscher *et al.*, 2002; Fievet *et al.*, 2004).

a. *Lipid-Binding Site Properties*

In vitro, the FERM domain of ezrin interacts preferentially in buffers of physiological ionic strength with liposomes containing $PI(4,5)P_2$ or $PI(3,4,5)P_3$ as compared to other acidic phospholipids, such as PS or PI, and does not discriminate between the first two lipids (Niggli *et al.*, 1995; Niggli, Bahloul, and Houdusse, unpublished data). Analysis of crystals of the radixin FERM domain complexed to inositol-1,4,5-trisphosphate (IP_3) showed that the binding site for the latter molecule is located in a basic groove between subdomains F1 and F3 (Hamada *et al.*, 2000). Based on site-directed mutagenesis, interaction with $PI(4,5)P_2$ involves additional basic residues (lysines 253, 254, 262, 263) contributed by two protruding loops of a PH-domain-like structure located in domain F3 (Barret *et al.*, 2000). Several distinct sites on ezrin may thus cooperate in binding to a bilayer containing phosphoinositides.

b. *Physiological Relevance of ERM–Lipid Interaction*

Phosphoinositide interaction of ezrin is physiologically relevant, as mutation of lysines 253, 254, 262, 263 to Asn precludes recruitment of ezrin to the plasma membrane in living cells (Barret *et al.*, 2000). Importantly, no other tested interactions were significantly affected by these mutations. According to the compelling study by Fievet *et al.* (2004), phosphoinositide interaction is a prerequisite for subsequent phosphorylation and stable activation of ezrin but is not required for its membrane targeting. Mutations in ezrin abolishing its interaction with $PI(4,5)P_2$ also preclude its localization in apical microvilli and its retention in the detergent-insoluble cytoskeleton. This ezrin mutant is not anymore phosphorylated on Thr-567. Phosphomimetic mutation of ezrin defective in lipid binding at Thr-567 overcame this inhibition and resulted in its membrane association (Fievet *et al.*, 2004). Locally enriched phosphoinositides may thus locally activate ezrin resulting in its association, for example, with cytosplasmic tail of transmembrane proteins, such as intercellular adhesion molecule (ICAM)-2, which also interacts with phosphoinositides (Hamada *et al.*, 2003).

2. Vinculin

The actin-membrane linker 116-kDa protein vinculin may couple cell adhesion and membrane protrusion by inducing actin assembly in focal adhesions (DeMali, 2004). Its function is absolutely essential for normal development as homozygous inactivation of the vinculin gene in mice causes embryonic lethality (Zemljic-Harpf *et al.*, 2004).

In the resting state, vinculin is a globular molecule (dimensions of $100 \times 100 \times 50$ Å), corresponding to a collection of loosely packed eight antiparallel α-helical bundles held together by head–tail interactions. Similar to ezrin, unfolding of vinculin is required for unmasking binding sites and for its full activation. Lipid and protein interaction may contribute to this process. The tail domain (residues 879–1066) contains binding sites for acidic phospholipids, actin and paxillin, whereas the head domain interacts with talin and α-actinin (Bakolitsa *et al.*, 2004; Borgon *et al.*, 2004; Cohen *et al.*, 2005).

a. Lipid-Binding Site Properties

Vinculin interacts *in vitro* with bilayers containing acidic phospholipids such as PS, PI, or phosphoinositides. Whether a specific binding site for $PI(4,5)P_2$ exists is as yet unclear. Interaction of vinculin with lipid bilayers involves a collar of basic residues in the tail domain including amino acids 910, 911, 1049, 1060, 1061, a basic ladder centered on helix 3 (residues 945–973), as well as a hydrophobic hairpin (residues 1062–1066). This C-terminal hairpin inserts into the bilayer, triggering unfurling of the helical bundle and resulting in closer association of especially helices H2 and H3 with the bilayer (Johnson *et al.*, 1998; Bakolitsa *et al.*, 1999, 2004).

Full activation of vinculin requires simultaneous binding of several partners. A kinetic pathway to activation has been proposed in which lipid binding transiently releases the tail from the head, allowing interaction with other ligands which then stabilize the active conformation. Spatial colocalization of several binding partners thus would be required for vinculin activation (Bakolitsa *et al.*, 2004; Cohen *et al.*, 2005). In support of this notion, lipid bilayer insertion of intact vinculin is enhanced in the presence of its binding partner α-actinin (Niggli and Gimona, 1993), and the interaction of vinculin with Arp2/3 in cell lysates is synergistically enhanced by $PI(4,5)P_2$ (DeMali *et al.*, 2002).

b. Physiological Relevance of Vinculin–Lipid Interactions

Vinculin is present in nascent focal contacts in its active open conformation (Chen *et al.*, 2005), but whether lipid interactions are required for physiological activation of vinculin is unclear. According to a report by Chandrasekar *et al.* (2005), lipid interaction of vinculin is not required for its activation but may regulate its interaction with actin and control its release from focal contacts. Chandrasekar *et al.* (2005) mutated the basic residues K952, K956, R963, K966, R1060, and K1061 in the tail domain to glutamines. This resulted in almost complete abolishment of interaction of the tail domain with bilayers containing 48% PC, 45% PS, and 7% PI(4,5)P_2. This lipid composition was chosen to mimic lipid concentrations in the inner leaflet of the plasma membrane assuming a local increase in phosphoinositide. The mutations did not affect interaction of this domain with F-actin or with the head domain. Expression of mutated full-length vinculin in melanoma cells resulted in abnormally long adhesive extensions containing the mutated vinculin, loss of cell polarization, and abolishment of cell migration. Expression of wild-type protein in contrast had no effect. Mutated vinculin was not anymore sensitive to increased production of PI(4,5)P_2 induced by expression of PIP-5K, which resulted in displacement of wild-type vinculin from focal adhesions and cell detachment.

In conclusion, phosphoinositide interaction, rather than being required for its activation, may negatively regulate vinculin function by precluding its interaction with F-actin. PI(4,5)P_2, by inducing dissociation of vinculin from actin filaments, may contribute to focal adhesion turnover.

3. Talin

Talin, another member of the protein 4.1 family, plays a crucial role in integrin activation by disrupting the interaction of α- and β-subunits of these adhesion receptors (Cram and Schwarzbauer, 2004). Disruption of one of the two mammalian talin genes causes embryonic lethality in mice (Critchley, 2005). Talin is an elongated 60-nm-long flexible antiparallel dimer with a small globular head connected to an extended rodlike tail domain. The globular head shows about 24% sequence homology with the ERM family FERM domain and contains binding sites for phosphoinositides, focal adhesion kinase (FAK), cytosolic tails of β1- and β3-integrins, F-actin and type I neuronal isoform of PIP-5Kγ. The rod domain contains a highly conserved actin-binding site and at least three vinculin-binding sites (Lee *et al.*, 2004; Nayal *et al.*, 2004; Critchley, 2005).

a. Lipid-Binding Site Properties

Talin interacts *in vitro* with some selectivity with liposomes containing 20% PI(4,5) P_2. This lipid also induces a conformational change in talin and enhances its interaction with integrins (Martel *et al.*, 2001). The phosphoinositide-binding site in talin has not

yet been precisely located. Most of the basic amino acid residues crucial for lipid binding of ezrin and radixin (Barret *et al.*, 2000; Hamada *et al.*, 2000) are not conserved in the talin FERM domain. Evidence for acidic phospholipid lipid bilayer insertion has been provided for an amphipathic talin peptide (residues 385–406) (Seelig *et al.*, 2000). This sequence overlaps with that identified to interact with integrin cytoplasmic β-tails (Garcia-Alvarez *et al.*, 2003) and with PIP-5Kγ (Barsukov *et al.*, 2003).

b. Physiological Relevance of Talin–Lipid Interaction

Indirect evidence suggests that the interaction with phosphoinositides occurs also in cells and relieves intramolecular inhibitory interactions in talin. For example, sequestration of PI(4,5)P_2 delocalizes talin from focal adhesions in fibroblasts (Martel *et al.*, 2001) and prevents talin-mediated integrin clustering in melanoma cells (Cluzel *et al.*, 2005).

Talin may not only be activated by phosphoinositides but it also interacts with and activates PIP-5Kγ, an enzyme involved in production of PI(4,5)P_2. This interaction involves a phosphotyrosine binding (PTB)-like domain in the talin FERM domain that also binds to the tails of β-integrins (Di Paolo *et al.*, 2002; Ling *et al.*, 2002; Morgan *et al.*, 2004). Talin may thus be locally activated by PI(4,5)P_2 at the plasma membrane in a first step, and in the next promotes highly localized production of PI(4,5)P_2 in a positive feedback loop. The interaction of talin with PIP-5Kγ appears to be physiologically relevant, as a peptide that disrupts interaction between these two proteins interferes with accumulation of F-actin at synapses and impairs clathrin-mediated synaptic vesicle endocytosis (Morgan *et al.*, 2004).

4. Spectrin

Spectrin, a rodlike protein related to α-actinin and dystrophin, consists of antiparallel heterodimers of α- and β-spectrin, which form head-to-head tetramers linked at distal ends to F-actin, protein 4.1, and so on. The α- and β-spectrin subunits consist of a series of tandem antiparallel-coiled coil repeats (comprising three helices each). α-Spectrin features 21 repeats, a C-terminal calcium-binding domain, and an SH3 domain; β-spectrin features 17 repeats and, for nonerythrocyte isoforms, a C-terminal PH domain (Broderick and Winder, 2005).

a. Lipid-Binding Site Properties

Three different types of lipid-binding sites have so far been detected in spectrin. Several repeats of erythrocyte and nonerythrocyte α- and β-spectrin, when expressed as peptides, show extensive cosedimentation with PS liposomes *in vitro* at physiological ionic strength (An *et al.*, 2004, 2005b). This interaction is very likely specific as a number of other repeats did not cosediment with PS. The PS-binding sites cluster near the N-terminus of the β-chain and are close to binding sites of ankyrin and protein 4.1,

suggesting coupling of lipid and protein interactions (An *et al.*, 2004). Furthermore, the spectrin PH domain interacts with moderate affinities with $PI(4,5)P_2$ and $PI(3,4,5)P_3$ *in vitro* (Hyvönen *et al.*, 1995; Kavran *et al.*, 1998). Finally, results suggest the presence of another $PI(4,5)P_2$-binding site in the calponin homology 2 (CH2) domain of βI- and βII-spectrin (residues 173–280) with strong homology to that of α-actinin (An *et al.*, 2005a).

b. Physiological Role of Spectrin–Lipid Interactions

PS is enriched in the inner leaflet of erythrocyte plasma membranes. Membranes with lipid asymmetry show increased stability and reduced extractability of the spectrin meshwork in low ionic strength buffers as compared to membranes with "scrambled" phospholipids, indicating that the spectrin-membrane association is strengthened by interactions with PS (Manno *et al.*, 2002).

The PH domain of nonerythrocyte human spectrin appears to be required for targeting of spectrin to, for example, $PI(4,5)P_2$-rich domains in Golgi membranes (Lorra and Huttner, 1999). Inhibition of $PI(4,5)P_2$ synthesis results in fragmentation of the Golgi apparatus and concomitant phosphorylation and redistribution of spectrin to the cytoplasm (Siddhanta *et al.*, 2003). Small amounts of the nonerythrocyte β1-spectrin isoform have been detected in human erythrocytes. Part of this isoform is located in distinct plasma membrane patches (Pradhan *et al.*, 2004). Possibly this is due to its differential lipid interactions as compared to the erythrocyte protein.

5. Dynamin

Dynamin family proteins are large GTPases that are involved in vesicle scission reactions. These mechanochemical enzymes feature a large GTPase domain followed by a middle domain, a PH domain, and a GTPase effector domain involved in oligomerization. During endocytosis, a dynamin collar assembles around the neck of invaginated pits, resulting in cleavage of the vesicle from the parent membrane (Praefcke and McMahon, 2004). Dynamin has also been implicated to coordinate membrane remodeling and actin reorganization. Dynamin interacts *in vitro* with cortactin and profilin and affects actin polymerization (Schafer, 2004). Whether its *in vitro* interaction with microtubules is physiologically relevant is unclear.

a. Lipid-Binding Site Properties

Dynamin, comparable to spectrin, binds *in vitro* preferentially with similar affinity to $PI(4,5)P_2$, $PI(3,4)P_2$, and $PI(3,4,5)P_3$. These phosphoinositides stimulate the GTPase activity of dynamin. The lipid-binding site has been located in the PH domain of dynamin. Monomeric dynamin binds with low affinity to these lipids, but oligomerization increases the affinity markedly (Klein *et al.*, 1998). Dynamin also appears to be controlled indirectly by $PI(4,5)P_2$, via its interaction with profilin 2. Profilin interacts with and inhibits dynamin function. $PI(4,5)P_2$ releases profilin from dynamin, thereby allowing its interactions with SH3 ligands (Gareus *et al.*, 2006).

b. Physiological Relevance of Dynamin–Lipid Interaction

Phosphoinositides play a key role in endocytosis and may target dynamin and other proteins to the bilayer, as a prerequisite for their function in promoting endocytosis. Moreover, they may dissociate profilin from dynamin, thereby allowing its interaction with SH3 ligands. At endocytotic sites, dynamin molecules assemble into a stable helix and constrain the lipid membrane (Praefcke and McMahon, 2004). Expression of dynamin carrying a single point mutation (K535A) in the PH domain of dynamin that abolishes *in vitro* interaction with phosphoinositides blocks endocytosis in cells (Achiriloaie *et al.*, 1999).

6. Annexins

Annexins are a family of membrane-binding proteins characterized by non-EF hand calcium-binding sites. They consist of a conserved α-helical C-terminal membrane-binding core domain and a separate N-terminal domain variable in sequence and length. Annexin 2 has been implicated in endosome trafficking. Annexins 1, 2, 5, and 6 also interact with F-actin and thus may function as reversible actin-membrane linkers (Babiychuk and Draeger, 2000; Hayes *et al.*, 2004; Gerke *et al.*, 2005).

a. Lipid-Binding Site Properties

Annexins feature a unique type of reversible calcium-dependent interaction with acidic phospholipid bilayers. Binding of these proteins to the phospholipid head groups is thought to be mediated by a calcium bridge, the calcium ions binding simultaneously to amino acid side chains and to the negatively charged head groups of acidic phospholipids. This binding site, consisting of "endonexin folds," is located near the C-terminus of the protein. Annexin 2 induces clustering of PS molecules in planar lipid bilayers (Menke *et al.*, 2005). Annexin 2 heterotetramers interact *in vitro* with tenfold higher affinity with $PI(4,5)P_2$ as compared to PI or PS. This interaction depends only partially on calcium. The affinity of this interaction is comparable to that of the PH domain of phospholipase δ (Hayes *et al.*, 2004; Rescher *et al.*, 2004; Gokhale *et al.*, 2005). The binding site for $PI(4,5)P_2$ has been located to a cluster of basic residues, similar to the phosphoinositide-binding sites in gelsolin or villin, in a C-terminal calcium-binding loop. Annexin 2 heterotetramers (but not the PH domain of phospholipase δ) cluster $PI(4,5)P_2$ in model membranes (Gokhale *et al.*, 2005).

b. Physiological Relevance of Annexin–Lipid Interaction

Phosphoinositide interaction may also occur in intact cells, as coexpression of annexin 2 and active Arf6, a GTPase that activates PIP-5K, results in relocalization of annexin 2 to intracellular membranes (Rescher *et al.*, 2004). Annexin 2 could thus be involved in formation of endocytic vesicles depending on phosphoinositides. Data suggest that annexin 2 may induce formation of phosphoinositide-containing lipid rafts, which are essential for exocytosis (Chasserot-Golaz *et al.*, 2005).

C. Proteins Involved in Actin Cross-Linking

1. α-Actinin

The rodlike protein α-actinin belongs to the spectrin protein superfamily. It contains an N-terminal actin-binding domain comprising two consecutive calponin homology domains (CH1 and CH2), followed by four spectrin-like repeats, and a C-terminal calmodulin-like domain. It forms an antiparallel homodimer. α-Actinin links actin filaments to integrins, bundles actin filaments, and can serve as a scaffold to integrate signaling molecules at adhesion sites (MacArthur and North, 2004; Otey and Carpen, 2004).

a. Lipid-Binding Site Properties

A binding site for phosphoinositides has been located in the CH1 and CH2 domains, close to the actin-binding site, involving residues 158–174 in human α-actinin-1 (Fukami *et al.*, 1996; Fraley *et al.*, 2003). Mutation of lysines 162 and 174 to isoleucine and of histidine 170 to leucine in full-length α-actinin-1 reduced phosphoinositide binding by 70% (Fraley *et al.*, 2003). Human α-actinin-3 contains a highly homologous domain (residues 172–188). The crystal structure of residues 26–273, comprising CH1 and CH2, has been determined for human α-actinin-3 (Franzot *et al.*, 2005). According to this study, lysine 188 is almost buried in the protein. The authors therefore suggest involvement of the surface-exposed basic residues 170, 176, and 199 in lipid binding without, however, providing experimental evidence for this notion.

PI(4,5)P_2 and PI(3,4,5)P_3 bind with equal affinity to α-actinin. *In vitro* studies suggest that PI(4,5)P_2 stabilizes α-actinin, whereas PI(3,4,5)P_3 increases its flexibility (Corgan *et al.*, 2004). Phosphoinositides differentially affect the interaction of α-actinin with other proteins. Whereas PI(3,4,5)P_3 disrupts the interaction of α-actinin with the integrin β-chain (Greenwood *et al.*, 2000), PI(4,5)P_2 enhances the binding of titin to the calmodulin-like domain of full-length α-actinin possibly by disrupting intramolecular interactions (Young and Gautel, 2000) and inhibits binding to CapZ (MacArthur and North, 2004). The effects of phosphoinositides on *in vitro* interactions of α-actinin with F-actin are controversial. According to Fukami *et al.* (1992), this interaction is enhanced by these lipids, whereas Fraley *et al.* (2003) and Corgan *et al.* (2004) observed attenuation. Mutations that preclude interaction with PI(4,5)P_2 appear not to affect F-actin interactions and even enhance interactions in cell lysates (Fraley *et al.*, 2003). Interactions with phosphoinositides thus do not generally activate α-actinin, but, depending on the interaction partner, inhibit or enhance interactions.

b. Physiological Relevance of α-Actinin Lipid Interactions

Expression of GFP-tagged α-actinin defective in binding to phosphoinositides in rat embryo fibroblasts suggests that lipid binding enhances reversibility of interaction of this cross-linker with actin stress fibers: Moreover, PI(3,4,5)P_3 appears to disrupt the link between actin and integrins, thus allowing restructuring of focal contacts in motile cells (Fraley *et al.*, 2005).

D. Myosin Isoforms

Lipid interactions have been characterized for several unconventional myosin isoforms such as myosin 1c, neither inactivation nor after potential C (NINAC), the eye-enriched myosin 3, and myosin 10 (Hirono *et al.*, 2004; Lee and Montell, 2004; Strissel and Arshavsky, 2004; Tacon *et al.*, 2004). These interactions may be involved in localizing these motor proteins at the plasma membrane near sites of actin assembly or could be involved in binding of myosin to their cargo. The lipid interaction of myosin 10 is best understood.

1. Myosin 10

This isoform features an N-terminal motor domain, a neck with IQ motifs, a coiled coil region, and a unique tail with three PH domains, a MyTH4 (myosin tail homology), and a FERM domain. Myosin 10 is enriched in membrane ruffles and has been implicated in filopodia formation and phagocytosis (Cox *et al.*, 2003; Tacon *et al.*, 2004).

a. Lipid-Binding Site Properties

Myosin 10 interacts *in vitro* preferentially with PI(3,5)P_2 and PI(3,4,5)P_3 and with lower affinity with PI(4,5)P_2. This interaction appears to be selectively mediated by the PH domains, whereas the myosin 10 FERM domain lacks the conserved lysines present in the ezrin FERM domain required for lipid interaction (Mashanov *et al.*, 2004; Tacon *et al.*, 2004). It is not clear which of PH domains is responsible for binding to PI(3,4,5)P_3, but structural considerations implicate mainly the PH2 domain (Tacon *et al.*, 2004).

b. Physiological Role of Myosin 10–Lipid Interaction

In intact cells, only the PH domains showed any membrane localization, and all three PH domains appear to be required for tight membrane association of myosin 10 (Yonezawa *et al.*, 2003; Mashanov *et al.*, 2004). Myosin 10 is recruited to phagocytic cups in macrophages dependent on activity of PI 3-kinase, suggesting that the interaction of this motor protein with $PI(3,4,5)P_3$ is physiologically relevant. Moreover, a point mutation in the PH domain, predicted to abolish $PI(3,4,5)P_3$ binding, suppresses the inhibitory effect of a tail fragment of myosin 10 on phagocytosis (Cox *et al.*, 2003). Myosin 10 could thus couple movement of actin filaments with that of the plasma membrane at phagocytotic cups.

III. MICROTUBULES AND MICROTUBULE-ASSOCIATED PROTEINS

A. Tubulin

α-Tubulin and β-tubulin are homologous compact globular molecules featuring a core of two β-sheets surrounded by α-helices. They contain an N-terminal nucleotide-binding region, an intermediate domain, and a C-terminal domain probably involved in binding motor proteins (Nogales *et al.*, 1998). Besides being building blocks for microtubules, tubulin molecules are also involved in the control of G-protein–mediated signal transduction and modulation of activity of enzymes hydrolyzing $PI(4,5)P_2$, that is, phospholipases Cβ1 and Cγ1 (Popova *et al.*, 2002; Chang *et al.*, 2005).

a. Lipid-Binding Site Properties

Tubulin itself appears to bind phosphoinositides, as *in vitro* polymerization of tubulin free of microtubule-associated proteins is impaired by preincubation with specifically $PI(4,5)P_2$, but not $PI(3,4,5)P_3$, phosphatidylinositol 3-phosphate, or the nonphospholipid IP_3, suggesting a specific interaction involving both the charged head group and the fatty acid chains. $PI(4,5)P_2$ does, however, not affect binding and hydrolysis of GTP by tubulin (Popova *et al.*, 2002). No information is as yet available on the putative phosphoinositide-binding site in tubulin.

b. Physiological Relevance of Tubulin–Lipid Interactions

Whether tubulin dimers, by binding of tubulin to sites rich in phospholipase Cβ and $PI(4,5)P_2$, modulate activation of phospholipase Cβ *in situ* requires further studies. Tubulin has also been suggested to activate phospholipase Cγ1, possibly by facilitating transportation of a $PI(4,5)P_2$ pool to the membrane-localized enzyme (Chang *et al.*, 2005). Palmitoylated tubulin has been reported to be enriched in detergent-resistant membrane domains (Palestini *et al.*, 2000), where it could modulate signal transduction events. Putatively raft-associated $PI(4,5)P_2$ could promote this tubulin-raft association.

B. MAP1B

The microtubule-associated protein MAP1B is a multimeric protein consisting of a 250-kDa heavy chain and two light chains of 30 and 15 kDa. The heavy chain contains a basic amphipathic tubulin-binding site (amino acids 524–848) featuring

KKEE and KKE motifs (Noble *et al.*, 1989). MAP1B, a major neuronal protein, localizes both in the cytosol and in particulate fractions. It also interacts with F-actin *in vitro*. Knockout experiments suggest important roles in neurite outgrowth (Dehmelt and Halpain, 2004).

a. *Lipid-Binding Site Properties*

MAP1B interacts *in vitro* specifically with liposomes containing acidic phospholipids, such as PS, PI, or phosphatidic acid (PA), but not with phosphatidylinositol 4-monophosphate (PIP) or PI(4,5)P_2. These acidic phospholipids also prevent association of MAP1B with microtubules *in vitro*. The interaction is very likely specific, as a basic amphiphilic lipid-binding site could be located in the C-terminal half of the tubulin-binding domain of MAP1B, which contains multiple KKE repeats (Yamauchi *et al.*, 1997). This explains the inhibitory effect of phospholipids on MAP1B–microtubule interaction.

b. *Physiological Relevance of MAP1B–Lipid Interactions*

Whether the interaction of this site with lipids is physiologically relevant has not yet been shown.

C. Kinesin

The UNC-104/KIF1 kinesins are an important class of motor proteins transporting vesicles anterogradely along microtubules in the nervous system. Kinesins feature two major functional domains, a motor domain responsible for motion along the microtubule and a cargo-binding tail domain, which in the case of UNC-104 contains a PH domain. In contrast to most other kinesins that are dimers, UNC-104/KIF1 kinesins purify as monomers. Knockout of UNC-104 in *Caenorhabditis elegans* results in reduced numbers of synaptic vesicles at nerve endings. In mice, ablation of the KIF1-gene results similarly in a defect in presynaptic vesicle transport and death of the animals shortly after birth (Yonekawa *et al.*, 1998; Klopfenstein and Vale, 2004).

a. *Lipid-Binding Site Properties*

UNC-104 from *Dictyostelium* and *C. elegans* can interact with and transport with some selectivity PI(4,5)P_2-containing liposomes via their PH domains. Mutations of the PH domain that impair lipid binding cause *in vitro* vesicle transport defects. Conditions that promote raft formation in liposomes enhance motility at low concentrations of PI(4,5)P_2. The data suggest that clustering of UNC-104 in PI(4,5)P_2-containing rafts facilitates dimerization of this kinesin isoform and thus promotes processive hand-over-hand movement along microtubules comparable to that of conventional kinesins (Klopfenstein *et al.*, 2002; Klopfenstein and Vale, 2004).

b. *Physiological Relevance of Kinesin–Lipid Interactions*

The study by Klopfenstein and Vale (2004) clearly shows that the lipid interaction of this motor also is relevant *in situ.* They analyzed the rescue of a *C. elegans* strain expressing mutated UNC-104 that shows an uncoordinated, nearly paralyzed phenotype, with different UNC-104 mutants as compared to wild-type protein. They showed that the PH domain of the protein is essential for complete rescue and that point mutations in this PH domain that impair phosphoinositide binding also result in incomplete rescue. Mutations in UNC-104 that impair lipid interaction reduce velocity and processivity of GFP-tagged UNC-104 punctae observed in cultured neurons. This finding substantiates the notion that phosphoinositide-containing rafts concentrate UNC-104 molecules at the membrane and induce formation of processive dimers.

IV. INTERMEDIATE FILAMENTS

Intermediate filaments comprise a large family of related structural proteins that are expressed in a tissue-specific manner and are involved in stabilization of cells and tissues. They consist of a central α-helical rod domain flanked by non-α-helical N-terminal and C-terminal domains of varying size and sequence (Herrmann and Aebi, 2004).

a. *Lipid-Binding Site Properties*

Vimentin, desmin, neurofilaments, and glial fibrillary protein all interact with acidic phospholipids (PI, PS) *in vitro*. These interactions appear to involve both electrostatic and hydrophobic interactions. The purified arginine-rich N-terminal head domain (amino acids 1–96) of vimentin fully retains binding properties of the full-length protein. Vimentin contains in addition a second binding site for neutral lipids, such as cholesterol, mono-, di-, and triglycerides, in its α-helical rod domain (Traub *et al.*, 1986, 1987; Perides *et al.*, 1987; Schweitzer and Evans, 1998).

PI(4,5)P_2 and PIP rather specifically block vimentin filament assembly *in vitro* and disrupt preformed filaments (Perides *et al.*, 1987; Shah *et al.*, 1998). As the basic N-terminal head domains are essential for tetramer formation via interactions with the acidic residues in the helical rod domain (Herrmann and Aebi, 2004), phosphoinositides may, by binding to the basic residues in the head domain, preclude these interactions.

b. *Physiological Relevance of Vimentin–Lipid Interactions*

It is not clear whether phosphoinositide binding plays a role in vimentin assembly and membrane association *in situ*. Certainly vimentin filaments are closely associated with lipid droplets in adipose and steroidogenic cells. Cells lacking vimentin show decreased esterification of cholesterol. Possibly delivery of cholesterol to the endoplasmatic reticulum is dependent on vimentin (Sarria *et al.*, 1992). It is not yet clear whether vimentin participates in cholesterol metabolism via direct lipid interaction or by indirectly modulating enzymes involved in these processes (Schweitzer and Evans, 1998).

V. CONCLUDING REMARKS

A. Selectivity of Cytoskeletal Protein–Lipid Interactions

With a few exceptions, most phosphoinositide-binding cytoskeletal proteins studied so far cannot discriminate between $PI(4,5)P_2$ and $PI(3,4,5)P_3$ (Fig. 2, Table I). As the cellular levels of $PI(3,4,5)P_3$ are much lower than those of $PI(4,5)P_2$, especially in resting cells, these proteins very likely interact *in situ* with $PI(4,5)P_2$. $PI(3,4,5)P_3$ may indirectly regulate cytoskeletal functions by modulating activities of signaling proteins such as small GTP-binding proteins (Niggli, 2005). A few proteins, such as myosin 10,

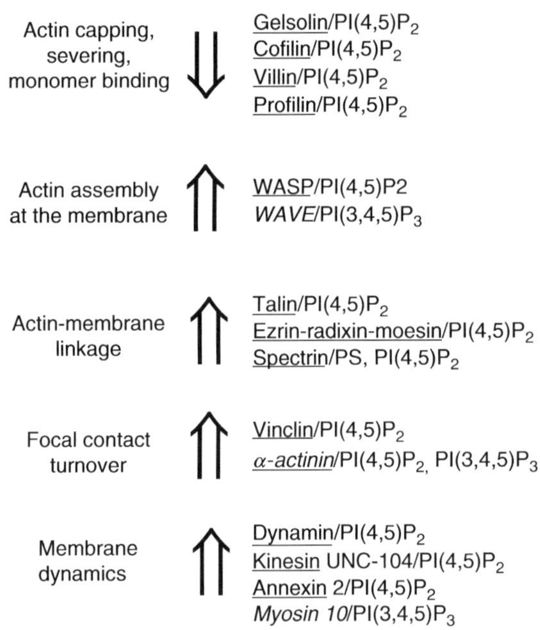

Figure 2. Regulation of cellular processes by phosphoinositides and the proteins involved. Actin assembly at the membrane, actin-membrane linkage, focal adhesion turnover, and membrane dynamics are positively regulated by phosphoinositides via the indicated proteins (upward arrows), whereas actin capping, severing, and monomer binding are negatively affected (downward arrow). Most of the proteins interact mainly with $PI(4,5)P_2$ (underlined), some proteins show selectivity for $PI(3,4,5)P_3$ (italics), and α-actinin (italics/underlined) may be regulated by both types of lipids.

WAVE, and profilin, interact *in vitro* preferentially with PI(3,4,5)P_3 and *in situ* evidence supports a physiological role of this interaction, at least for the first two proteins. For α-actinin, a protein involved both in actin cross-linking and in membrane attachment, which interacts *in vitro* equally well with PI(4,5)P_2 and PI(3,4,5)P_3, data nevertheless indicate an *in situ* role for the latter lipid. The affinities of the lipid interactions are quite low in some cases (e.g., dynamin). Clustering of phosphoinositides in rafts thus could markedly promote binding.

Interactions with acidic phospholipids other than phosphoinositides have additionally been documented for annexins, spectrin, gelsolin, and vinculin. They may serve to concentrate these proteins at the plasma membrane, facilitating their regulation by phosphoinositides.

B. Functional Impact of Cytoskeletal Protein–Lipid Interactions

PI(4,5)P_2 promotes actin assembly by removing capping and severing proteins from F-actin and by dissociating proteins from actin monomers. PI(4,5)P_2 also activates actin-membrane linker proteins. As this lipid is also present in appreciable amounts in resting cells, it may thus be involved in maintaining a cortical membrane-associated actin network in these cells. PI(4,5)P_2 moreover promotes actin assembly via WASP and Arp2/3, whereas PI(3,4,5)P_3, whose levels are tightly controlled by signaling, localizes WAVE at the plasma membrane where it may be activated by Rac, thus initiating actin polymerization in motile lamellipods. Both lipids may thus cooperate in inducing formation of F-actin–rich protrusions during cell migration.

Regulation of focal adhesions by phosphoinositides is complex. On one hand, PI(4,5) P_2 is required for focal contact localization of talin. On the other hand, phosphoinositides appear to displace vinculin from focal contacts and to disrupt the interaction of α-actinin with integrins, thus ensuring turnover of these structures. Finally, processes involving membrane dynamics, such as endocytosis, exocytosis, and phagocytosis, are regulated by phosphoinositides that target dynamin, myosin 10, kinesin, and annexin 2 to specific membrane locations and also enhance functions of kinesin and dynamin.

C. Open Questions

In most discussed cases, the *in vitro* binding site properties and lipid specificities have been well described, but compelling evidence on *in situ* interaction of these proteins with phosphoinositides and functional roles of these lipid interactions is yet scarce with a few exceptions such as ezrin, kinesin (UNC-104), WAVE, or villin. In the first three cases, data have been obtained for cellular expression of proteins with mutations in the lipid-binding domain, whereas for villin, a direct *in situ* interaction with PI(4,5)P_2 has been demonstrated. Another open question is the involvement of lipid rafts in regulation of localized assembly, turnover, and membrane linkage of cytoskeletal filaments, especially as the presence of phosphoinositides in lipid rafts has been disputed. This topic merits further research.

ACKNOWLEDGMENTS

I thank E. Sigel for preparing the figures and for careful reading of the chapter. My work is supported by the Swiss National Science Foundation (grant No. 3100A0-103708/1) and the Bernese Cancer League.

REFERENCES

Achiriloaie, M., Barylko, B., and Albanesi, J. P. (1999). Essential role of the dynamin pleckstrin homology domain in receptor-mediated endocytosis. *Mol. Cell. Biol.* **19,** 1410–1415.

An, X., Guo, X., Sum, H., Morrow, J., Gratzer, W., and Mohandas, N. (2004). Phosphatidylserine binding sites in erythroid spectrin: Location and implications for membrane stability. *Biochemistry* **43,** 310–315.

An, X., Debnath, G., Guo, X., Liu, S., Lux, S. E., Baines, A., Gratzer, W., and Mohandas, N. (2005a). Identification and functional characterization of protein 4.1R and actin-binding sites in erythrocyte beta spectrin: Regulation of the interactions by phosphatidylinositol-4,5-bisphosphate. *Biochemistry* **44,** 10681–10688.

An, X., Guo, X., Gratzer, W., and Mohandas, N. (2005b). Phospholipid binding by proteins of the spectrin family: A comparative study. *Biochem. Biophys. Res. Commun.* **327,** 794–800.

Babiychuk, E. B. and Draeger, A. (2000). Annexins in cell membrane dynamics. Ca^{2+}-regulated association of lipid microdomains. *J. Cell Biol.* **150,** 1113–1124.

Bakolitsa, C., de Pereda, J. M., Bagshaw, C. R., Critchley, D. R., and Liddington, R. C. (1999). Crystal structure of the vinculin tail suggests a pathway for activation. *Cell* **99,** 603–613.

Bakolitsa, C., Cohen, D. M., Bankston, L. A., Bobkov, A. A., Cadwell, G. W., Jennings, L., Critchley, D. R., Craig, S. W., and Liddington, R. C. (2004). Structural basis for vinculin activation at sites of cell adhesion. *Nature* **430,** 583–586.

Barret, C., Roy, C., Montcourrier, P., Mangeat, P., and Niggli, V. (2000). Mutagenesis of the phosphatidylinositol 4,5-bisphosphate (PIP_2) binding site in the NH_2-terminal domain of ezrin correlates with its altered cellular distribution. *J. Cell Biol.* **151,** 1067–1080.

Barsukov, I. L., Prescot, A., Bate, N., Patel, B., Floyd, D. N., Bhanji, N., Bagshaw, C. R., Letinic, K., Di Paolo, G., De Camilli, P., Roberts, G. C., and Critchley, D. R. (2003). Phosphatidylinositol phosphate kinase type 1γ and β1-integrin cytoplasmic domain bind to the same region in the talin FERM domain. *J. Biol. Chem.* **278,** 31202–31209.

Benesch, S., Lommel, S., Steffen, A., Stradal, T. E., Scaplehorn, N., Way, M., Wehland, J., and Rottner, K. (2002). Phosphatidylinositol 4,5-biphosphate (PIP_2)-induced vesicle movement depends on N-WASP and involves Nck, WIP, and Grb2. *J. Biol. Chem.* **277,** 37771–37776.

Biswas, R. S., Baker, D., Hruska, K. A., and Chellaiah, M. A. (2004). Polyphosphoinositides-dependent regulation of the osteoclast actin cytoskeleton and bone resorption. *BMC Cell Biol.* **5,** 19.

Bodin, S., Soulet, C., Tronchere, H., Sie, P., Gachet, C., Plantavid, M., and Payrastre, B. (2005). Integrin-dependent interaction of lipid rafts with the actin cytoskeleton in activated human platelets. *J. Cell Sci.* **118,** 759–769.

Bompard, G. and Caron, E. (2004). Regulation of WASP/WAVE proteins: Making a long story short. *J. Cell Biol.* **166,** 957–962.

Borgon, R. A., Vonrhein, C., Bricogne, G., Bois, P. R., and Izard, T. (2004). Crystal structure of human vinculin. *Structure* **12,** 1189–1197.

Bretscher, A., Edwards, K., and Fehon, R. G. (2002). ERM proteins and merlin: Integrators at the cell cortex. *Nat. Rev. Mol. Cell. Biol.* **3,** 586–599.

Broderick, M. J. and Winder, S. J. (2005). Spectrin, α-actinin, and dystrophin. *Adv. Protein Chem.* **70,** 203–246.

Chandrasekar, I., Stradal, T. E., Holt, M. R., Entschladen, F., Jockusch, B. M., and Ziegler, W. H. (2005). Vinculin acts as a sensor in lipid regulation of adhesion-site turnover. *J. Cell Sci.* **118,** 1461–1472.

Chang, J.-S., Kim, S.-K., Kwon, T.-K., Bae, S. S., Min, D. S., Lee, Y. H., Kim, S.-O., Seo, J.-K., Choi, J. H., and Suh, P.-G. (2005). Pleckstrin homology domains of phospholipase C-γ1 directly interact with β-tubulin for activation of phospholipase C-γ1 and reciprocal modulation of β-tubulin function in microtubule assembly. *J. Biol. Chem.* **280,** 6897–6905.

Chasserot-Golaz, S., Vitale, N., Umbrecht-Jenck, E., Knight, D., Gerke, V., and Bader, M. F. (2005). Annexin 2 promotes the formation of lipid microdomains required for calcium-regulated exocytosis of dense-core vesicles. *Mol. Biol. Cell* **16,** 1108–1119.

Chellaiah, M. and Hruska, K. (1996). Osteopontin stimulates gelsolin-associated phosphoinositide levels and phosphatidylinositol triphosphate-hydroxyl kinase. *Mol. Biol. Cell* **7,** 743–753.

Chen, H., Cohen, D. M., Choudhury, D. M., Kioka, N., and Craig, S. W. (2005). Spatial distribution and functional significance of activated vinculin in living cells. *J. Cell Biol.* **169,** 459–470.

Cluzel, C., Saltel, F., Lussi, J., Paulhe, F., Imhof, B. A., and Wehrle-Haller, B. (2005). The mechanisms and dynamics of $\alpha v\beta 3$ integrin clustering in living cells. *J. Cell Biol.* **171,** 383–392.

Cohen, D. M., Chen, H., Johnson, R. P., Choudhury, B., and Craig, S. W. (2005). Two distinct head-tail interfaces cooperate to suppress activation of vinculin by talin. *J. Biol. Chem.* **280,** 17109–17117.

Corgan, A. M., Singleton, C., Santoso, C. B., and Greenwood, J. A. (2004). Phosphoinositides differentially regulate α-actinin flexibility and function. *Biochem. J.* **378,** 1067–1072.

Cox, D., Berg, J. S., Cammer, M., Chinegwundoh, J. O., Dale, B. M., Cheney, R. E., and Greenberg, S. (2003). Myosin X is a downstream effector of PI(3)K during phagocytosis. *Nat. Cell Biol.* **4,** 469–477.

Cram, E. J. and Schwarzbauer, J. E. (2004). The talin wags the dog: New insights into integrin activation. *Trends Cell Biol.* **14,** 55–57.

Critchley, D. R. (2005). Genetic, biochemical and structural approaches to talin function. *Biochem. Soc. Trans.* **33,** 1308–1312.

Dehmelt, L. and Halpain, S. (2004). Actin and microtubules in neurite initiation: Are MAPs the missing link? *J. Neurobiol.* **58,** 18–33.

DeMali, K. A. (2004). Vinculin—a dynamic regulator of cell adhesion. *Trends Biochem. Sci.* **29,** 565–567.

DeMali, K. A., Barlow, C. A., and Burridge, K. (2002). Recruitment of the Arp2/3 complex to vinculin: Coupling membrane protrusion to matrix adhesion. *J. Cell Biol.* **159,** 881–891.

DesMarais, V., Ghosh, M., Eddy, R., and Condeelis, J. (2005). Cofilin takes the lead. *J. Cell Sci.* **118,** 19–26.

Di Paolo, G., Pellegrini, L., Letinic, K., Cestra, G., Zoncu, R., Voronov, S., Chang, S., Guo, J., Wenk, M. R., and De Camilli, P. (2002). Recruitment and regulation of phosphatidylinositol phosphate kinase type 1γ by the FERM domain of talin. *Nature* **420,** 85–89.

Doughman, R. L., Firestone, A. J., and Anderson, R. A. (2003). Phosphatidylinositol phosphate kinases put $PI4,5P_2$ in its place. *J. Membr. Biol.* **194,** 77–89.

Downes, C. P., Gray, A., and Lucocq, J. M. (2005). Probing phosphoinositide functions in signaling and membrane trafficking. *Trends Cell Biol.* **15,** 259–268.

Feng, L., Mejillano, M., Yin, H. L., Chen, J., and Prestwich, G. D. (2001). Full-contact domain labeling: Identification of a novel phosphoinositide binding site on gelsolin that requires the complete protein. *Biochemistry* **40,** 904–913.

Fievet, B. T., Gautreau, A., Roy, C., Del Maestro, L., Mangeat, P., Louvard, D., and Arpin, M. (2004). Phosphoinositide binding and phosphorylation act sequentially in the activation mechanism of ezrin. *J. Cell Biol.* **164,** 653–659.

Fraley, T. S., Tran, T. C., Corgan, A. M., Nash, C. A., Hao, J., Critchley, D. R., and Greenwood, J. A. (2003). Phosphoinositide binding inhibits α-actinin bundling activity. *J. Biol. Chem.* **278,** 24039–24045.

Fraley, T. S., Pereira, C. B., Tran, T. C., Singleton, C. A., and Greenwood, J. A. (2005). Phosphoinositide binding regulates α-actinin dynamics. *J. Biol. Chem.* **280,** 15479–15482.

Franzot, G., Sjoblom, B., Gautel, M., and Djinovic Carugo, K. (2005). The crystal structure of the actin binding domain from α-actinin in its closed conformation: Structural insight into phospholipid regulation of α-actinin. *J. Mol. Biol.* **348,** 151–165.

Fukami, K., Furuhashi, K., Inagaki, M., Endo, T., Hatano, S., and Takenawa, T. (1992). Requirement of phosphatidylinositol 4,5-bisphosphate for α-actinin function. *Nature* **359,** 150–152.

Fukami, K., Sawada, N., Endo, T., and Takenawa, T. (1996). Identification of a phosphatidylinositol 4,5-bisphosphate-binding site in chicken skeletal muscle α-actinin. *J. Biol. Chem.* **271,** 2646–2650.

George, S., Ramagiri, S., Chatman, L., Tomar, A., and Khurana, S. (submitted for publication). Localization and function of villin oligomers in living cells.

Garcia-Alvarez, B., de Pereda, J. M., Calderwood, D. A., Ulmer, T. S., Critchley, D., Campbell, I. D., Ginsberg, M. H., and Liddington, R. C. (2003). Structural determinants of integrin recognition by talin. *Mol. Cell* **11,** 49–58.

Gareus, R., Di Nardo, A., Rybin, V., and Witke, W. (2006). Mouse profilin 2 regulates endocytosis and competes with SH3-ligand binding to dynamin 1. *J. Biol. Chem.* **281,** 2803–2811.

Gerke, V., Creutz, C. E., and Moss, S. E. (2005). Annexins: Linking Ca^{2+}signalling to membrane dynamics. *Nat. Rev. Mol. Cell Biol.* **6,** 449–461.

Gokhale, N. A., Abraham, A., Digman, M. A., Gratton, E., and Cho, W. (2005). Phosphoinositide specificity of and mechanism of lipid domain formation by annexin A2-p11 heterotetramer. *J. Biol. Chem.* **280,** 42831–42840.

Goldschmidt-Clermont, P. J., Machesky, L. M., Baldassare, J. J., and Pollard, T. D. (1990). The actin-binding protein profilin binds to PIP_2 and inhibits its hydrolysis by phospholipase C. *Science* **247,** 1575–1578.

Goldschmidt-Clermont, P. J., Kim, J. W., Machesky, L. M., Rhee, S. G., and Pollard, T. D. (1991). Regulation of phospholipase C-γ1 by profilin and tyrosine phosphorylation. *Science* **251,** 1231–1233.

Golub, T. and Caroni, C. (2005). Spatial control of actin-based motility through plasmalemmal PtdIns(4,5) P_2-rich raft assemblies. *Biochem. Soc. Symp.* **72,** 119–127.

Greenwood, J. A., Theibert, A. B., Prestwich, G. D., and Murphy-Ullrich, J. E. (2000). Restructuring of focal adhesion plaques by PI 3-kinase. Regulation by PtdIns (3,4,5)-P_3 binding to α-actinin. *J. Cell Biol.* **150,** 627–642.

Gurniak, C. B., Perlas, E., and Witke, W. (2005). The actin depolymerizing factor n-cofilin is essential for neural tube morphogenesis and neural crest cell migration. *Dev. Biol.* **278,** 231–241.

Guttman, J. A., Janmey, P., and Vogl, A. W. (2002). Gelsolin—evidence for a role in turnover of junction-related actin filaments in Sertoli cells. *J. Cell Sci.* **115,** 499–505.

Hamada, K., Shimizu, T., Matsui, T., Tsukita, S., and Hakoshima, T. (2000). Structural basis of the membrane-targeting and unmasking mechanisms of the radixin FERM domain. *EMBO J.* **19,** 4449–4462.

Hamada, K., Shimizu, T., Yonemura, S., Tsukita, S., and Hakoshima, T. (2003). Structural basis of adhesion-molecule recognition by ERM proteins revealed by the crystal structure of the radixin-ICAM-2 complex. *EMBO J.* **22,** 502–514.

Hatanaka, H., Ogura, K., Moriyama, K., Ichikawa, S., Yahara, I., and Inagaki, F. (1996). Tertiary structure of destrin and structural similarity between two actin-regulating protein families. *Cell* **85,** 1047–1055.

Hayes, M. J., Merrifield, C. J., Shao, D., Ayala-Sanmartin, J., Schorey, C. D., Levine, T. P., Proust, J., Curran, J., Bailly, M., and Moss, S. E. (2004). Annexin 2 binding to phosphatidylinositol 4,5-bisphosphate on endocytic vesicles is regulated by the stress response pathway. *J. Biol. Chem.* **279,** 14157–14164.

Herrmann, H. and Aebi, U. (2004). Intermediate filaments: Molecular structure, assembly mechanism, and integration into functionally distinct intracellular Scaffolds. *Annu. Rev. Biochem.* **73,** 749–789.

Hirono, M., Denis, C. S., Richardson, G. P., and Gillespie, P. G. (2004). Hair cells require phosphatidylinositol 4,5-bisphosphate for mechanical transduction and adaptation. *Neuron* **44,** 309–320.

Hoeflich, K. P. and Ikura, M. (2004). Radixin: Cytoskeletal adopter and signaling protein. *Int. J. Biochem. Cell Biol.* **36,** 2131–2136.

Hyvönen, M., Macias, M. J., Nilges, M., Oschkinat, H., Saraste, M., and Wilmanns, M. (1995). Structure of the binding site for inositol phosphates in a PH domain. *EMBO J.* **14,** 4676–4685.

Johnson, R. P., Niggli, V., Durrer, P., and Craig, S. W. (1998). A conserved motif in the tail domain of vinculin mediates association with and insertion into acidic phospholipid bilayers. *Biochemistry* **37,** 10211–10222.

Kavran, J. M., Klein, D. E., Lee, A., Falasca, M., Isakoff, S. J., Skolnik, E. Y., and Lemmon, M. A. (1998). Specificity and promiscuity in phosphoinositide binding by pleckstrin homology domains. *J. Biol. Chem.* **273,** 30497–30508.

Klein, D. E., Lee, A., Frank, D. W., Marks, M. S., and Lemmon, M. A. (1998). The pleckstrin homology domains of dynamin isoforms require oligomerization for high affinity phosphoinositide binding. *J. Biol. Chem.* **273,** 27725–27733.

Klopfenstein, D. R. and Vale, R. D. (2004). The lipid binding pleckstrin homology domain in UNC-104 kinesin is necessary for synaptic vesicle transport in *Caenorhabditis elegans*. *Mol. Biol. Cell* **15,** 3729–3739.

Klopfenstein, D. R., Tomishige, M., Stuurman, N., and Vale, R. D. (2002). Role of phosphatidylinositol(4,5) bisphosphate organization in membrane transport by the Unc104 kinesin motor. *Cell* **109,** 347–358.

Kumar, N., Zhao, P., Tomar, A., Galea, C. A., and Khurana, S. (2004). Association of villin with phosphatidylinositol 4,5-bisphosphate regulates the actin cytoskeleton. *J. Biol. Chem.* **279,** 3096–3110.

Lambrechts, A., Jonckheere, V., Dewitte, D., Vandekerckhove, J., and Ampe, C. (2002). Mutational analysis of human profilin I reveals a second PI(4,5)-P_2 binding site neighbouring the poly(L-proline) binding site. *BMC Biochem.* **3,** 12.

Lee, H. S., Bellin, R. M., Walker, D. L., Patel, B., Powers, P., Liu, H., Garcia-Alvarez, B., de Pereda, J. M., Liddington, R. C., Volkmann, N., Hanein, D., Critchley, D. R., *et al.* (2004). Characterization of an actin-binding site within the talin FERM domain. *J. Mol. Biol.* **343,** 771–784.

Lee, S. J. and Montell, C. (2004). Light-dependent translocation of visual arrestin regulated by the NINAC myosin III. *Neuron* **43,** 95–103.

Liepina, I., Czaplewski, C., Janmey, P., and Liwo, A. (2003). Molecular dynamics study of a gelsolin-derived peptide binding to a lipid bilayer containing phosphatidylinositol 4,5-bisphosphate. *Biopolymers* **71,** 49–70.

Ling, K., Doughman, R. L., Firestone, A. J., Bunce, M. W., and Anderson, R. A. (2002). Type I gamma phosphatidylinositol phosphate kinase targets and regulates focal adhesions. *Nature* **420,** 89–93.

Lorra, C. and Huttner, W. B. (1999). The mesh hypothesis of Golgi dynamics. *Nat. Cell Biol.* **1,** E113–E115.

Lu, P. J., Shieh, W. R., Rhee, S. G., Yin, H. L., and Chen, C. S. (1996). Lipid products of phosphoinositide 3-kinase bind human profilin with high affinity. *Biochemistry* **35,** 14027–14034.

MacArthur, D. G. and North, K. N. (2004). A gene for speed? The evolution and function of α-actinin-3. *Bioessays* **26,** 786–795.

Manno, S., Takakuwa, Y., and Mohandas, N. (2002). Identification of a functional role for lipid asymmetry in biological membranes: Phosphatidylserine-skeletal protein interactions modulate membrane stability. *Proc. Natl. Acad. Sci. USA* **99,** 1943–1948.

Martel, V., Racaud-Sultan, C., Dupe, S., Marie, C., Paulhe, F., Galmiche, A., Block, M. R., and Albiges-Rizo, C. (2001). Conformation, localization, and integrin binding of talin depend on its interaction with phosphoinositides. *J. Biol. Chem.* **276,** 21217–21227.

Mashanov, G. I., Tacon, D., Peckham, M., and Molloy, J. E. (2004). The spatial and temporal dynamics of pleckstrin homology domain binding at the plasma membrane measured by imaging single molecules in live mouse myoblasts. *J. Biol. Chem.* **279,** 15274–15280.

McGough, A. M., Staiger, C. J., Min, J. K., and Simonetti, K. D. (2003). The gelsolin family of actin regulatory proteins: Modular structures, versatile functions. *FEBS Lett.* **552,** 75–81.

McLaughlin, S. and Murray, D. (2005). Plasma membrane phosphoinositide organization by protein electrostatics. *Nature* **438,** 605–611.

Menke, M., Gerke, V., and Steinem, C. (2005). Phosphatidylserine membrane domain clustering induced by annexin A2/S100A10 heterotetramer. *Biochemistry* **44,** 15296–15303.

Méré, J., Chahinian, A., Maciver, S. K., Fattoum, A., Bettache, N., Benyamin, Y., and Roustan, C. (2005). Gelsolin binds to polyphosphoinositide free lipid vesicles and simultaneously to actin microfilaments. *Biochem. J.* **386,** 47–56.

Metzler, W. J., Farmer, B. T., II, Constantine, K. L., Friedrichs, M. S., Lavoie, T., and Mueller, L. (1995). Refined solution structure of human profilin I. *Protein Sci.* **4,** 450–459.

Morgan, J. R., Di Paolo, G., Werner, H., Shchedrina, V. A., Pypaert, M., Pieribone, V. A., and De Camilli, P. (2004). A role for talin in presynaptic function. *J. Cell Biol.* **167,** 43–50.

Nayal, A., Webb, D. J., and Horwitz, A. F. (2004). Talin: An emerging focal point of adhesion dynamics. *Curr. Opin. Cell Biol.* **16,** 94–98.

Niggli, V. (2001). Structural properties of lipid-binding sites in cytoskeletal proteins. *Trends Biochem. Sci.* **26,** 604–611.

Niggli, V. (2005). Regulation of protein activities by phosphoinositide phosphates. *Annu. Rev. Cell Dev. Biol.* **21,** 57–79.

Niggli, V. and Gimona, M. (1993). Evidence for a ternary interaction between alpha-actinin, (meta)vinculin and acidic-phospholipid bilayers. *Eur. J. Biochem.* **213,** 1009–1015.

Niggli, V., Andreoli, C., Roy, C., and Mangeat, P. (1995). Identification of a phosphatidylinositol-4,5-bisphosphate-binding domain in the N-terminal region of ezrin. *FEBS Lett.* **376,** 172–176.

Noble, M., Lewis, S. A., and Cowan, N. J. (1989). The microtubule binding domain of microtubule-associated protein MAP1B contains a repeated sequence motif unrelated to that of MAP2 and tau. *J. Cell Biol.* **109,** 3367–3376.

Nogales, E., Wolf, S. G., and Downing, K. H. (1998). Structure of the $\alpha\beta$ tubulin dimer by electron crystallography. *Nature* **391,** 199–203.

Oikawa, T., Yamaguchi, H., Itoh, T., Kato, M., Ijuin, T., Yamazaki, D., Suetsugu, S., and Takenawa, T. (2004). PtdIns(3,4,5)P_3 binding is necessary for WAVE2-induced formation of lamellipodia. *Nat. Cell Biol.* **6,** 420–426.

Ojala, P. J., Paavilainen, V., and Lappalainen, P. (2001). Identification of yeast cofilin residues specific for actin monomer and PIP_2 binding. *Biochemistry* **40,** 15562–15569.

Otey, C. A. and Carpen, O. (2004). α-Actinin revisited: A fresh look at an old player. *Cell Motil. Cytoskel.* **58,** 104–111.

Oude Weernink, P. A., Schmidt, M., and Jakobs, K. H. (2004). Regulation and cellular roles of phosphoinositide 5-kinases. *Eur. J. Pharmacol.* **500,** 87–99.

Palestini, P., Pitto, M., Tedeschi, G., Ferraretto, A., Parenti, M., Brunner, J., and Masserini, M. (2000). Tubulin anchoring to glycolipid-enriched, detergent-resistant domains of the neuronal plasma membrane. *J. Biol. Chem.* **275,** 9978–9985.

Papayannopoulos, V., Co, C., Prehoda, K. E., Snapper, S., Taunton, J., and Lim, W. A. (2005). A polybasic motif allows N-WASP to act as a sensor of PIP_2 density. *Mol. Cell* **17,** 181–191.

Perides, G., Harter, C., and Traub, P. (1987). Electrostatic and hydrophobic interactions of the intermediate filament protein vimentin and its amino terminus with lipid bilayers. *J. Biol. Chem.* **262,** 13742–13749.

Pfannstiel, J., Cyrklaff, M., Habermann, A., Stoeva, S., Griffiths, G., Shoeman, R., and Faulstich, H. (2001). Human cofilin forms oligomers exhibiting actin bundling activity. *J. Biol. Chem.* **276,** 49476–49484.

Pike, L. (2004). Lipid rafts: Heterogeneitiy on the high seas. *Biochem. J.* **378,** 281–292.

Popova, J. S., Greene, A. K., Wang, J., and Rasenick, M. M. (2002). Phosphatidylinositol 4,5-bisphosphate modifies tubulin participation in phospholipase Cβ1 signaling. *J. Neurosci.* **22,** 1668–1678.

Pradhan, D., Tseng, K., Cianci, C. D., and Morrow, J. S. (2004). Antibodies to βIΣ2 spectrin identify in-homogeneities in the erythrocyte membrane skeleton. *Blood Cells Mol. Dis.* **32,** 408–410.

Praefcke, G. J. and McMahon, H. T. (2004). The dynamin superfamily: Universal membrane tubulation and fission molecules? *Nat. Rev. Mol. Cell. Biol.* **5,** 133–147.

Rescher, U., Ruhe, D., Ludwig, C., Zobiack, N., and Gerke, V. (2004). Annexin 2 is a phosphatidylinositol (4,5)-bisphosphate binding protein recruited to actin assembly sites at cellular membranes. *J. Cell Sci.* **117,** 3473–3480.

Rohatgi, R., Ho, H. Y., and Kirschner, M. W. (2000). Mechanism of N-WASP activation by CDC42 and phosphatidylinositol 4, 5-bisphosphate. *J. Cell Biol.* **150,** 1299–1310.

Sarria, A. J., Panini, S. R., and Evans, R. M. (1992). A functional role for vimentin intermediate filaments in the metabolism of lipoprotein-derived cholesterol in human SW-13 cells. *J. Biol. Chem.* **267,** 19455–19463.

Schafer, D. A. (2004). Regulating actin dynamics at membranes: A focus on dynamin. *Traffic* **5,** 463–469.

Schweitzer, S. C. and Evans, R. M. (1998). Vimentin and lipid metabolism. *Subcell. Biochem.* **31,** 437–462.

Seelig, A., Blatter, X. L., Frentzel, A., and Isenberg, G. (2000). Phospholipid binding of synthetic talin peptides provides evidence for an intrinsic membrane anchor of talin. *J. Biol. Chem.* **275,** 17954–17961.

Shah, J. V., Wang, L. Z., Traub, P., and Janmey, P. A. (1998). Interaction of vimentin with actin and phospholipids. *Biol. Bull.* **194,** 402–405.

Siddhanta, A., Radulescu, A., Stankewich, M. C., Morrow, J. S., and Shields, D. (2003). Fragmentation of the Golgi apparatus. A role for βIII spectrin and synthesis of phosphatidylinositol 4,5-bisphosphate. *J. Biol. Chem.* **278,** 1957–1965.

Silacci, P., Mazzolai, L., Gauci, C., Stergiopulos, N., Yin, H. L., and Hayoz, D. (2004). Gelsolin superfamily proteins: Key regulators of cellular functions. *Cell Mol. Life Sci.* **61,** 2614–2623.

Skare, P. and Karlsson, R. (2002). Evidence for two interaction regions for phosphatidylinositol(4,5)-bisphosphate on mammalian profilin I. *FEBS Lett.* **522,** 119–124.

Smith, W. J., Nassar, N., Bretscher, A., Cerione, R. A., and Karplus, P. A. (2003). Structure of the active N-terminal domain of ezrin. Conformational and mobility changes identify keystone interactions. *J. Biol. Chem.* **278,** 4949–4956.

Strissel, K. J. and Arshavsky, V. Y. (2004). Myosin III illuminates the mechanism of arrestin translocation. *Neuron* **43,** 2–4.

Tacon, D., Knight, P. J., and Peckham, M. (2004). Imaging myosin 10 in cells. *Biochem. Soc. Trans.* **32,** 689–693.

Takenawa, T. and Miki, H. (2001). WASP and WAVE family proteins: Key molecules for rapid rearrangement of cortical actin filaments and cell movement. *J. Cell Sci.* **114,** 1801–1809.

Traub, B., Perides, G., Schimmel, H., and Scherbarth, A. (1986). Interaction *in vitro* of nonepithelial intermediate filament proteins with total cellular lipids, individual phospholipids, and a phospholipids mixture. *J. Biol. Chem.* **261,** 10558–10568.

Traub, B., Perides, G., Kühn, S., and Scherbarth, A. (1987). Efficient interaction of nonpolar lipids with intermediate filaments of the vimentin type. *Eur. J. Cell Biol.* **43,** 55–64.

van Rheenen, J., Achame, E. M., Janssen, H., Calafat, J., and Jalink, K. (2005). PIP_2 signaling in lipid domains: A critical re-evaluation. *EMBO J.* **24,** 1664–1673.

Verheyen, E. M. and Cooley, L. (1994). Profilin mutations disrupt multiple actin-dependent processes during Drosophila development. *Development* **120,** 717–728.

Witke, W. (2004). The role of profilin complexes in cell motility and other cellular processes. *Trends Cell Biol.* **14,** 461–469.

Witke, W., Sharpe, A. H., Hartwig, J. H., Azuma, T., Stossel, T. P., and Kwiatkowski, D. J. (1995). Hemostatic, inflammatory, and fibroblast responses are blunted in mice lacking gelsolin. *Cell* **81,** 41–51.

Wittenmayer, N., Jandrig, B., Rothkegel, M., Schluter, K., Arnold, W., Haensch, W., Scherneck, S., and Jockusch, B. M. (2004). Tumor suppressor activity of profilin requires a functional actin binding site. *Mol. Biol. Cell* **15,** 1600–1608.

Xian, W. and Janmey, P. A. (2002). Dissecting the gelsolin-polyphosphoinositide interaction and engineering of a polyphosphoinositide-sensitive gelsolin C-terminal half protein. *J. Mol. Biol.* **322,** 755–771.

Yamauchi, E., Titani, K., and Taniguchi, H. (1997). Specific binding of acidic phospholipids to microtubule-associated protein MAP1B regulates its interaction with tubulin. *J. Biol. Chem.* **272,** 22948–22953.

Yonekawa, Y., Harada, A., Okada, Y., Funakoshi, T., Kanai, Y., Takei, Y., Terada, S., Noda, T., and Hirokawa, N. (1998). Defect in synaptic vesicle precursor transport and neuronal cell death in KIF1A motor protein-deficient mice. *J. Cell Biol.* **141,** 431–441.

Yonezawa, S., Yoshizaki, N., Sano, M., Hanai, A., Masaki, S., Takizawa, T., Kageyama, T., and Moriyama, A. (2003). Possible involvement of myosin-X in intercellular adhesion: Importance of serial pleckstrin homology regions for intracellular localization. *Dev. Growth Differ.* **45,** 175–185.
Young, P. and Gautel, M. (2000). The interaction of titin and α-actinin is controlled by a phospholipid-regulated intramolecular pseudoligand mechanism. *EMBO J.* **19,** 6331–6340.
Zemljic-Harpf, A. E., Ponrartana, S., Avalos, R. T., Jordan, M. C., Roos, K. P., Dalton, N. D., Phan, V. Q., Adamson, E. D., and Ross, R. S. (2004). Heterozygous inactivation of the vinculin gene predisposes to stress-induced cardiomyopathy. *Am. J. Pathol.* **165,** 1033.

Chapter 12

Embryo Morphogenesis and the Role of the Actin Cytoskeleton

Sarah Woolner and Paul Martin

Departments of Physiology and Biochemistry, University of Bristol, University Walk, Bristol BS8 1TD, United Kingdom

Advances in Molecular and Cell Biology, Vol. 37

ISBN: 0-444-52868-7
DOI: 10.1016/S1569-2558(06)37012-9

During embryogenesis, the tissues of the developing embryo must be bent, tugged, and sculpted in a series of morphogenetic episodes that flow from one to another to generate the final shape of the fetus. The mechanisms underlying these events share startling similarities across a great variety of animals. A primary component of morphogenesis is the actin cytoskeleton, which drives all of the cell shape changes, the tissue bending, and the fusion events that characterize embryogenesis. In this chapter we explore how regulation of the actin cytoskeleton contributes to morphogenetic episodes in worm, fly, fish, mouse, and chick embryos. In particular, we focus on how each of the different structures formed by actin function in morphogenesis and consider how recent research has expanded our knowledge of this area.

I. INTRODUCTION

Embryonic development is punctuated by morphogenetic events, which tug, bend, and sculpt the tissues of the embryo. The actin cytoskeleton is a key element in these episodes, playing a number of vital roles that are conserved across the animal kingdom. In this chapter we will focus, in particular, on the role of actin-driven events during epithelial morphogenesis. The epithelial layers of the embryo undergo the most remarkable transformations in shape by bending, invaginating, fusing, and forming intricate networks of tubes. All of these episodes are coordinated so that precisely the right structures form at the right time and in the right place and wire into the correct tissues.

In this chapter, we will first detail how different actin filament structures are formed in the cell, drawing from research carried out in cultured cells and relating this to what is known from model organisms. In particular, we will explore how filopodia and lamellipodia are assembled and contrast this to the formation of actin cable-like structures. We will also discuss how each of these structures contributes generally to morphogenesis. With the actin "tool kit" thus defined, we then go on to explore how these actin structures perform their roles during specific morphogenetic episodes. A wealth of research data has provided many clues as to the key ways in which actin regulation underpins various aspects of morphogenesis and we will detail examples during the course of the chapter. An overriding feature of morphogenesis and roles played by the actin cytoskeleton is the level of conservation seen across a huge variety of animal species. In this chapter we cover morphogenetic episodes in flies, worms, fish, mice, and chicks and the remarkable similarities between events in these diverse organisms will become apparent.

II. ACTIN STRUCTURES IN MORPHOGENESIS

The actin cytoskeleton not only provides crucial support and structure to the cell but is also key in driving dynamic cell shape change and facilitating cell motility. Actin is a globular protein, which cycles between monomeric and filamentous forms in the cell. Functional actin is filamentous (F-actin), but pools of monomeric actin exist within cells and the regulation of polymerizing filaments from these pools is a tightly regulated process involving distinct sets of actin-binding proteins (Pollard and Borisy, 2003). Actin filaments are composed of a double helical arrangement of actin monomers, sitting head-to-tail to give the filament a molecular polarity, with one end being called the "barbed" end and the other the "pointed" end. Actin filaments are built by adding monomers to the barbed end and disassembled through loss of monomers from the pointed end (Pollard and Borisy, 2003).

During embryonic morphogenesis, actin filament (F-actin) structures play key roles in bending, molding, and sticking together the epithelial tissues to shape the embryo. Crucially, actin filaments are assembled and organized into a variety of dynamic networks during morphogenesis to form cables, filopodia, and lamellipodia. These structures each contribute in unique ways to build the developing embryo. In this section, each structure will be introduced and their general roles in morphogenetic movements discussed.

A. Actin Protrusions

Actin-rich cell protrusions, in the form of sheet-like lamellipodia or finger-like filopodia, are frequently expressed by cells as they participate in key morphogenetic episodes. These structures are assembled at the leading edges or apical surfaces of cells where localized actin polymerization generates actin filaments that push at the plasma membrane to produce actin-rich extensions of the cell surface. Filopodia and lamellipodia differ morphologically due to the underlying arrangement of actin filaments in each structure. Lamellipodia are thin sheet-like extensions composed of a dendritic network of F-actin which pushes at the plasma membrane resulting in dynamic ruffles (Fig. 1). In contrast, filopodia consist of bundles of parallel actin filaments, which form a cylindrical core of actin that pushes the plasma membrane into a finger-shaped protrusion (Fig. 1). Both filopodia and lamellipodia are highly dynamic structures, extending, ruffling, and retracting rapidly as they reach out and appear to explore the cell's surroundings.

The assembly of actin protrusions is driven by changes in actin dynamics at the cell edge, with the protrusions extending as actin filaments grow by the addition of actin monomers. Growth of actin filaments occurs preferentially at the filament barbed end and filaments in filopodia and lamellipodia are orientated with their growing, barbed ends toward the exterior of the cell (Small *et al.*, 1978). To assemble the broad F-actin meshwork that makes up a lamellipodium, the cell uses the Arp2/3 complex, an actin nucleator that generates new barbed ends on which actin filaments can grow (Machesky *et al.*, 1994). Arp2/3 binds to the pointed ends of actin and initiates the growth of new actin filaments at a 70° angle to the original filament, resulting in a dense branched network of filaments which can be elegantly visualized using specialist transmission electron microscopy (TEM) techniques (Svitkina and Borisy, 1999). However, the Arp2/3 complex cannot drive actin nucleation alone and requires the

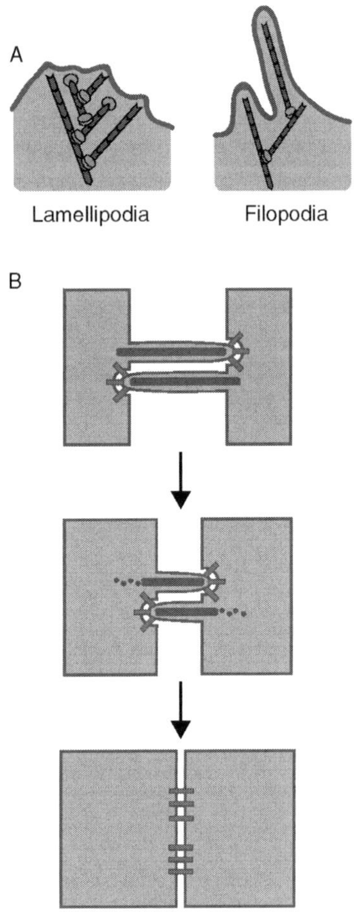

Figure 1. Actin protrusions and a filopodial interdigitation model for epithelial cell adhesion. (A) Actin protrusions are formed when localized actin polymerization generates actin filaments (red) that push against the plasma membrane. New actin filaments can be nucleated by the Arp2/3 complex, and capped at their barbed ends by capping protein (green oval). Lamellipodial protrusions are thin sheet-like protrusions composed of a network of actin filaments, whereas finger-like filopodia contain bundles of parallel actin filaments. (B) In a number of different systems, fusion of epithelial cells is preceded by the assembly of filopodia. These actin (red) protrusions interdigitate and provide initial points of contact between two cells and sites for cell–cell adhesion (blue). The cells can then pull closer together and over time the initial contacts will develop into mature adherens junctions, forming a fused epithelial surface. (See Color Insert.)

input of additional activators, primarily provided by members of the WASP/Scar family of proteins (Machesky and Insall, 1998).

The branched network formed by the nucleation of actin by Arp2/3 and its activators nicely explains how lamellipodia are assembled at the leading edge of cultured motile cells and also during *in vivo* situations such as the morphogenesis of embryonic tissues. However, it does not so easily account for the formation of filopodia because these protrusions appear to consist of unbranched bundles of aligned actin filaments (Svitkina *et al.*, 2003). These filaments are rooted in the branched network but Arp2/3 is excluded from the length of the filopodium. An *in vitro* study using WASP-coated beads revealed that a lamellipodia-like branched network could be switched to an array of bundled actin structures analogous to filopodia simply by depleting capping protein, and this may be how filopodial assembly is generally regulated *in vivo* (Vignjevic *et al.*, 2003). This is, at least in part, borne out by work in cultured cells where depletion of capping protein by short hairpin RNA caused an "explosive" increase in filopodia formation (Mejillano *et al.*, 2004). However, since cells maintain high concentrations of capping protein, it is more likely that regulation is indirect, via antagonists of capping action, such as the Ena/VASP family of proteins. If capping protein levels are depleted in Ena/VASP-deficient cells, no switch to filopodial production is seen (Mejillano *et al.*, 2004). These ideas have yet to be explored in detail during *in vivo* morphogenetic episodes.

Tissue culture experiments have shown that the Rho GTPases, Cdc42 and Rac sit upstream of the actin dynamics to mediate assembly of protrusions. Activation of Cdc42 drives the formation of filopodia, while Rac activation leads to the assembly of lamellipodia (Ridley *et al.*, 1992; Nobes and Hall, 1999). This control also appears to hold true during *in vivo* morphogenetic episodes and is described in relation to specific tissue movements later in this chapter. Downstream of Rac and Cdc42, the assembly of filopodia versus lamellipodia had been thought to run through two parallel pathways. Key to both pathways are members of the WASP family of Arp2/3 activators. These can be organized into two subfamilies, one containing WASPs and the other containing Scar/WAVE proteins. Cdc42 specifically activates WASPs (Higgs and Pollard, 2000; Rohatgi *et al.*, 2000) while Scar/WAVE proteins act downstream of Rac1 (Eden *et al.*, 2002). However, recent work in *Drosophila* cells, which carry only a single *WASP* and single *SCAR* gene, found that SCAR is the major regulator for both lamellipodia and filopodia assembly, while WASP appears less important (Rogers *et al.*, 2003; Biyasheva *et al.*, 2004). The speculation is that, in *Drosophila* cells at least, a cascade, rather than parallel pathways, may be functioning (Fig. 8). In this scheme, Rac activates SCAR leading to the assembly of lamellipodia, while the assembly of filopodia also lies downstream of SCAR and is controlled by the input of further factors, such as the recruitment of the anticapping protein Ena/VASP by the Cdc42 effector IRSp53 (Krugmann *et al.*, 2001). These ideas have yet to be explored specifically *in vivo* during embryonic morphogenesis, although it is the case that in the morphogenetic movement of *Drosophila* dorsal closure (described later) loss of either Rac1 or Cdc42 leads to a loss of both lamellipodia and filopodia (Jacinto *et al.*, 2000; Woolner *et al.*, 2005), which might endorse a cascade model of actin protrusion assembly.

Due to their dynamic, transient nature actin protrusions have not always been easy to image *in vivo*. The standard fixation processes used for the study of embryonic tissues often destroy these fragile structures. However, advances in live imaging have allowed researchers to uncover more and more examples of actin dynamics, particularly of protrusion assemblies during embryo morphogenesis of several different organisms.

A major function of actin protrusions *in vivo* appears to be to glue sheets of epithelial cells together, which is the culmination of many different morphogenetic episodes. Actin protrusions appear to prime the adhesion of adjacent cells by extending and interdigitating, so that cells, or sheets of cells, can be zipped together (Fig. 1). This action was shown particularly clearly by *in vitro* studies of keratinocytes, which can be induced to form adherent monolayers by the addition of Ca^{2+}. After Ca^{2+} exposure, keratinocytes extend numerous filopodia and when filopodia from adjacent cells make contact they slide upon each other and "embed" themselves into the neighboring cells (Vasioukhin *et al.*, 2000). E-cadherin–catenin complexes cluster at the tips of the keratinocyte filopodia and these complexes mature into strong cell–cell adhesions between the two cells (Vasioukhin *et al.*, 2000). The interdigitation of filopodia to zip two sheets of cells together seems to be a well-conserved mechanism in epithelial fusion events, which are explored in a later section.

A further role that actin protrusions may play during morphogenesis may be to push and pull sheets of cells into new positions. This is a clear function of lamellipodia in migrating single cells, such as the *Dictyostelium* amoebae, where the assembly of lamellipodial protrusions at the leading edge of the cell is fundamental for cell motility. It is less clear that protrusions can provide the level of force required to move large tissues during morphogenesis, but growing evidence from live imaging of events, such as germ band retraction in the *Drosophila* embryo, indicate that actin protrusions may contribute to the pushing and pulling of tissues.

Finally, a major function of actin protrusions during morphogenesis appears to be as sensory organs allowing cells to observe and respond to guidance cues within their environment. This function is clearly illustrated by the filopodia of a growth cone during axon guidance in the developing nervous system (Tessier-Lavigne and Goodman, 1996) but is less well described for morphogenetic events. However, strikingly, during the fusion of two epithelial sheets in *Drosophila* dorsal closure, actin protrusions are required to correctly match the segments of the embryo.

B. Stress Fibers and Actomyosin Cables

As well as forming filopodia and lamellipodia, actin filaments can also be organized into cable-like structures, and these play vital roles in many morphogenetic episodes. In cultured cells, actin filaments arrange in cable-like structures in response to tension and are called stress fibers. They provide rigidity and structure to the cell and are also involved in cell migration, driving the retraction of the rear of the cell. Analogous cable structures form in embryos during morphogenesis, although a key difference between these actin cables and stress fibers is that they can run through many cells in a tissue, providing a coordinated contractile structure across that tissue. Actin cables are anchored and linked between the cells of a tissue through adherens junctions (Fig. 2), with the actin filaments attaching to the junction through α- and β-catenin.

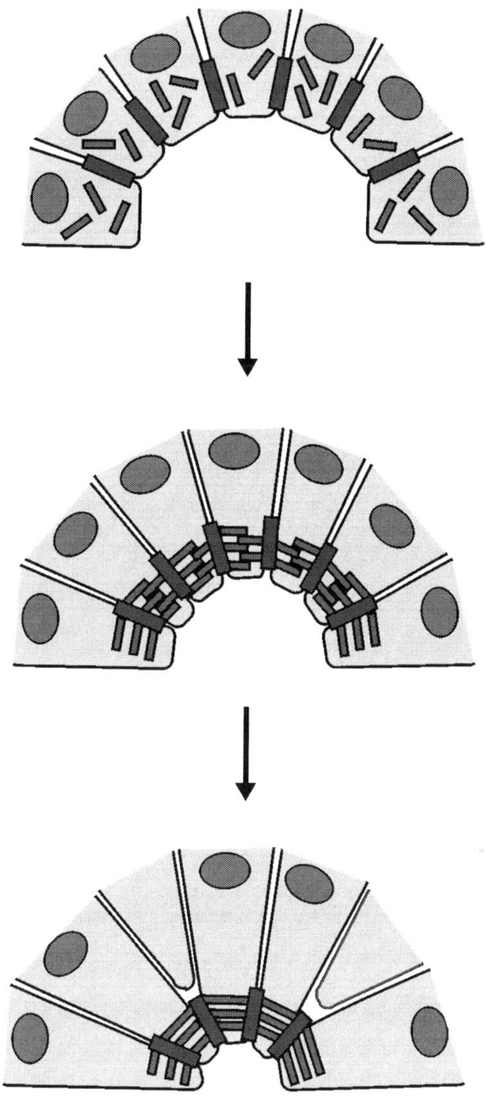

Figure 2. Purse-string contraction. Actin filaments can form contractile cables which can shape and bend epithelial layers or, as is the case in wound healing, pull cells together to close a hole. Here a temporal series illustrates how purse-string contraction can draw the edges of an epithelial hole together. The cable is made up of individual actin filaments (red bars), which are anchored to adherens junctions (blue rectangles) formed between adjacent cells. Contraction of the actin cable in each cell leads to apical cell constriction and reduces the size of the hole. As contraction continues, some of the cells are squeezed out of the front row and new adherens junctions form between the remaining cells. Sustained contraction of the actin cable draws adjacent cells together, allowing them to fuse and close the hole. (See Color Insert.)

The actin filaments in cables and stress fibers are arranged in straight, unbranched bundles, a great contrast to the branched network seen with protrusion formation. The polymerization of actin filaments for these structures cannot, therefore, be mediated by the actin nucleator, Arp2/3, which forms branched actin networks. Instead, the formin family of actin nucleators are the likely candidate for polymerizing cable structures during morphogenesis. Formins bind to the barbed end tips of actin filaments, where they can remain as the filament elongates (Higashida *et al.*, 2004), blocking the access of capping proteins but allowing the incorporation of actin monomers (Zigmond *et al.*, 2003). In this way, formins are able to drive the formation of long actin filaments. Evidence from a variety of systems indicate that formins are the primary nucleator of cable structures, with members of the family driving stress fiber formation in cultured cells (Satoh and Tominaga, 2001) and cable formation in yeast (Sagot *et al.*, 2002). Furthermore, the formins have also been implicated in number of morphogenetic episodes, including limb, kidney, and spleen morphogenesis (reviewed in Evangelista *et al.*, 2003). Upstream control of actin cable formation runs through the small GTPase, Rho. Activation of this GTPase is required for the assembly of stress fibers in cultured cells and also appears to play a similar role *in vivo* in the formation of actin cables in morphogenesis. This certainly ties in with formin nucleation of actin cables, as Rho is known to be required to activate some classes of formins, by relieving them from their usual, autoinhibited, state (Watanabe *et al.*, 1999).

A key aspect of actin cables and stress fibers is that the actin filaments in these structures can associate with myosin motors enabling the generation of tensile or contractile force. For example, if a contractile actomyosin cable forms at the apical edge of a cell, this edge will be tugged inward, giving the cell a wedge-like shape (Fig. 2). Similarly, if a cable is formed at the edge of a sheet of cells it can be used to draw cells together by "purse-string" contraction, a process that has been best described in wound-healing studies. When a wound is made in an epithelial layer, a thick cable of actin is apparent around the leading edge of the cells encompassing the hole (Martin and Lewis, 1992). If assembly of this cable is blocked by the addition of the actin polymerization blocker, cytochalasin D, or with the Rho GTPase blocker, C3 transferase, wounds fail to reepithelialize, indicating that the actin cable is providing crucial force to draw the wound edges together (McCluskey and Martin, 1995; Brock *et al.*, 1996).

Actin cables perform a number of functions during morphogenetic episodes. First, they can drive purse-string contraction, just as is seen in wound healing. This role is exhibited particularly clearly during various epithelial fusion events, such as ventral enclosure in the *Caenorhabditis elegans* embryo. Second, actin cables can act to bend and shape sheets of cells during morphogenesis such as is seen when a placode of cells invaginates to form a tube during tubulogenesis. Finally, cables can also help to restrain and organize sheets of cells as a morphogenetic episode takes place; this appears to be the case during *Drosophila* dorsal closure where an actomyosin cable helps to keep the epithelial leading edge organized and coherent (Section V.A).

III. SIGNALS REGULATING ACTIN DURING MORPHOGENESIS

A. Rho Family of Small GTPases

Rho GTPases act as molecular switches to control signal transduction pathways, cycling between an active GTP-bound and inactive GDP-bound form (Fig. 3). When in their active GTP-bound state they can interact with a wide range of downstream effectors to drive various cellular responses (reviewed in Raftopoulou and Hall, 2004; Wennerberg and Der, 2004). A major downstream effect of the Rho family is to regulate the actin cytoskeleton, with different Rho family members being responsible for the assembly of different actin structures in the cell.

The cycling of Rho GTPases between the active and inactive forms is tightly regulated by three groups of proteins (Fig. 3). Guanine nucleotide exchange factors (GEFs) activate the GTPase by promoting the exchange of GDP for GTP. In contrast, GTPase-activating proteins (GAPs) negatively regulate the cycle by enhancing the

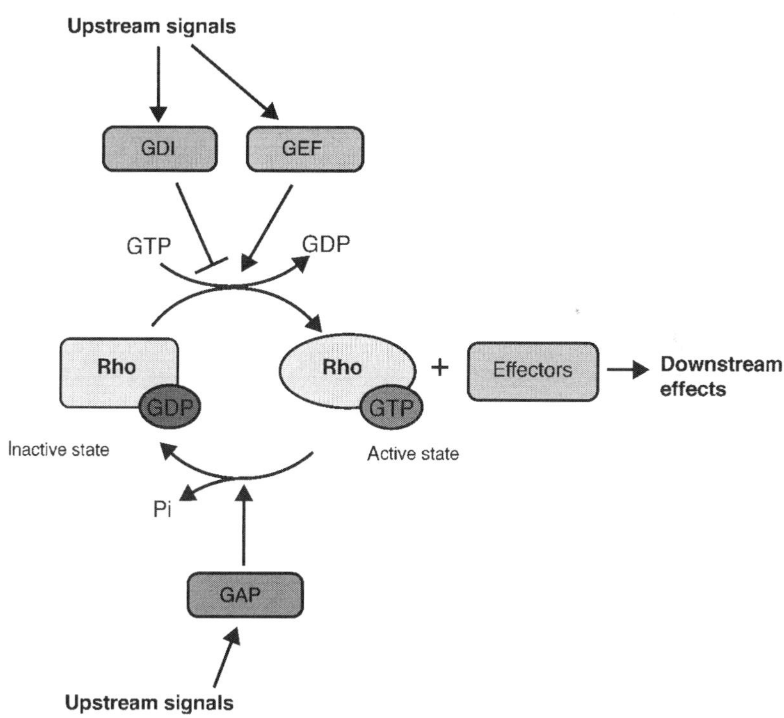

Figure 3. Cycling of Rho GTPases. The Rho family of small GTPases act as molecular switches, cycling between an inactive, GDP-bound form, and an active, GTP-bound form. Upstream signals feed into regulators of Rho GTPase activity—the GEFs, GAPs, and GDIs—and so affect the balance of the cycle. When in the active form, Rho GTPases can go on to activate numerous effectors leading to a variety of downstream effects.

GTPase activity of the Rho family member, while guanine nucleotide dissociation inhibitors (GDIs) are thought to block the cycle by sequestering the GDP-bound form (Zheng, 2001; Schmidt and Hall, 2002; Moon and Zheng, 2003). A great deal of the work carried out to investigate Rho GTPase function has used the expression of mutant forms of the GTPases which block the usual cycling. Constitutively active forms of GTPases possess mutations that reduce their intrinsic GTPase activity and do not respond to GAPs, and so, in this way, remain active irrespective of upstream signaling. Conversely, mutations which give the Rho GTPase a preferential affinity for GDP, lock the GTPase in the inactive state and are thought to act as dominant negative agents by sequestering GEFs, thus preventing the activation of their wild-type counterparts (Harden *et al.*, 1995).

Work in cultured fibroblasts has indicated that different members of the Rho family are responsible for organizing actin into specialized arrangements. Activation of Rho was found to regulate the formation of contractile actomyosin, which in cultured cells forms stress fibers (Ridley and Hall, 1992). In contrast, the Rho family members Rac and Cdc42 were found to be required for the assembly of actin protrusions, with Rac being responsible for the formation of the actin meshwork which underlies lamellipodia, and activation of Cdc42 leading to the assembly of filopodia (Ridley *et al.*, 1992; Nobes and Hall, 1995). As we will discuss, the situation is a little more complex *in vivo* but generally it appears to be true that in tissue morphogenesis Rho signaling controls events involving contractile cables, while protrusion assembly is dependent on activation of Rac and Cdc42.

B. MAP Kinase Cascades

Another important signaling component activated during some morphogenetic episodes is provided by the Jun N-terminal kinase (JNK) family of MAP kinases (MAPKs) (Xia and Karin, 2004). MAPK cascades are required for many vital cellular processes and involve the successive phosphorylation and activation of kinases. The pathway ends with the activation of transcription factors and so the transcription of a specific group of target genes (reviewed in Widmann *et al.*, 1999). At the top of the cascade is an MAPK kinase kinase (MAPKKK) which, when activated, phosphorylates and activates an MAPK kinase which in turn phosphorylates an MAPK and this then phosphorylates and activates specific transcription factors. In mammalian cells, five MAPK families have been established, which are determined by the identity of the final MAPK and therefore the transcription factors that are subsequently activated. JNK cascades end with the activation of the leucine zipper transcription factor, Jun, which then usually teams up with another leucine zipper protein, Fos, to form the transcriptionally active AP1 complex. Over recent years, JNK cascades have been implicated in a number of morphogenetic episodes involving the migration and fusion of epithelial sheets, including *Drosophila* dorsal closure and eyelid fusion in mouse embryos. In particular, the JNK pathway seems to play a role in the organization of the actin cytoskeleton during these morphogenetic events, although exactly how this is coordinated with the activity of the Rho GTPases remains unclear.

IV. GASTRULATION MOVEMENTS

Gastrulation is the first major morphogenetic event to take place during embryogenesis. While different in different organisms, gastrulation is generally characterized by large-scale cell movements and results in the establishment of the three major germ layers and the setting down of the basic body plan of the embryo. Actin is an important player in many aspects of these early shaping movements and in this section we will explore these ideas by looking at gastrulation in the *Drosophila* embryo and the related process of epiboly in the zebrafish embryo.

A. *Drosophila* Gastrulation

Gastrulation in the *Drosophila* embryo begins just after cellularization and is initiated by the formation of a furrow on the ventral surface of the embryo. The furrow forms as a consequence of mesodermal precursor cells narrowing their apices and expanding basally, thereby becoming wedge-shaped in appearance (Fig. 4). This concerted cell shape change spreads through the ventral-most domain of the embryo resulting in the formation of a furrow (Leptin, 1999). Two transcription factors, Twist and Snail, define the ventral domain and induce the expression of specific target genes. One such induced gene is *folded gastrulation* (*fog*), which encodes a secreted protein that signals to neighboring cells in the epithelium to coordinate the gastrulation cell contractions. The Fog signal operates via Concertina (Cta), a heterotrimeric G-protein, to activate Rho-GEF2, which in turn activates RhoA (Barrett *et al.*, 1997). The cells in which this signaling cascade is activated flatten and constrict their apical cortex, so driving furrow formation.

Clues from mammalian tissue culture suggested that this pathway might lead to the assembly and contraction of actomyosin cables, which could drive the cell shape changes required for gastrulation. An analogous pathway in mammalian cells, which acts through a G-protein signal to activate RhoA, induces the phosphorylation and activation of myosin II and results in the assembly of actomyosin stress fibers (Van Aelst and D'Souza-Schorey, 1997). Live imaging experiments using GFP fusions of the light or heavy chains of myosin II suggest that Fog is acting upstream of myosin II-driven contractions during *Drosophila* gastrulation (Nikolaidou and Barrett, 2004). The myosin-GFPs show a relocalization to the apical surface of the ventral cells just prior to their constriction and invagination (Fig. 4). Furthermore, this apical accumulation of myosin in the ventral cells is much reduced in embryos mutant for either *RhoGEF2* or *cta*.

B. Epiboly in Zebrafish Embryos

Zebrafish gastrulation consists of four linked cell movements and rearrangements—epiboly, involution, extension, and convergence. Epiboly is the first morphogenetic movement in this series of episodes and recent work has begun to indicate some

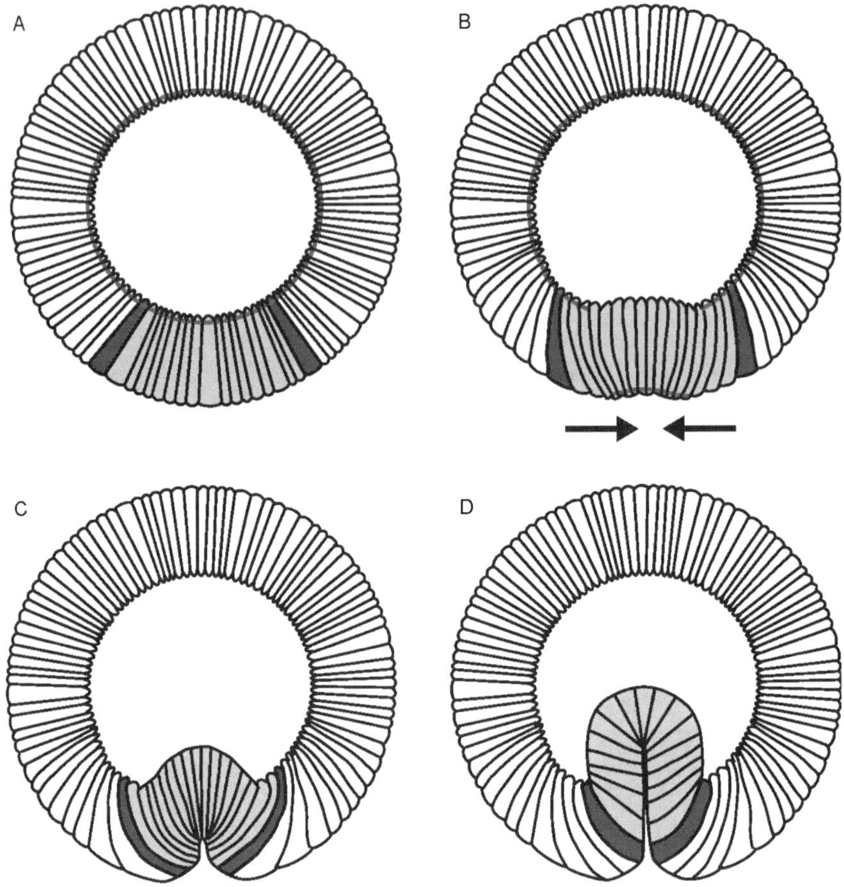

Figure 4. Ventral furrow formation during *Drosophila* gastrulation. The major steps in ventral furrow formation are shown as cross sections through the *Drosophila* embryo (top = dorsal, bottom = ventral). The domain of the gastrulation transcription factor, Twist, is shown in blue and the overlapping Twist and Snail domains are shown in yellow. The localization of myosin II is illustrated in red. (A) Immediately after cellularization, myosin II is localized to the basal surfaces of the cells. (B) The ventral furrow begins to form as myosin II relocalizes to the apical surface of the ventral cells and these cells begin to flatten and constrict their apices. (C) Continued and coordinated apical constriction drives the formation of the furrow. (D) At the end of the process, the future mesodermal cells have been completely internalized. (See Color Insert.)

intriguing roles for actin in this process (Cheng *et al.*, 2004). Epiboly begins just after the embryo has completed a series of cleavages and at this point consists of a mound of cells, the blastoderm, which sits on top of a large syncytial yolk cell. During epiboly the blastoderm spreads over and around the yolk cell, until the yolk cell is completely covered with blastodermal cells. Early studies indicated that microtubules were the major driving force for epiboly, with microtubular arrays extending from the yolk cell syncytial layer, which contacts the blastoderm, into the yolk cell cytoplasm (Solnica-Krezel and Driever, 1994). Furthermore, disruption of these arrays with the microtubule-stabilizing drug, Taxol, significantly retarded epiboly. These observations, along with the finding that the yolk cell syncytial layer continues to move toward the vegetal pole even if the blastoderm is removed, suggested that the yolk cell might act to tow the blastoderm across its surface using the microtubule connections.

However, studies of zebrafish epiboly have begun to suggest a role for blastoderm actin structures in this morphogenetic movement. First, imaging has shown that cells at the periphery of the expanding blastoderm extend actin-rich lamellipodial protrusions across the adjacent yolk cell (Zalik *et al.*, 1999). A clear possibility is that these protrusions are aiding migration of the blastoderm toward the vegetal pole. Furthermore, later in epiboly two actin rings are seen to form at the periphery of the blastoderm. These actin rings run around the circumference of the blastoderm and at least one of these rings is present until the very end of epiboly (Cheng *et al.*, 2004). If embryos in the later stages of epiboly are treated with cytochalasin B, which disrupts actin polymerization, the process slows down and the blastoderm ultimately fails to completely cover the yolk cell (Cheng *et al.*, 2004). It has therefore been postulated that actin plays two key roles in driving tissue migration during epiboly: first in the early stages of the process actin protrusions assembled by the leading blastoderm cells may help migration toward the equator of the yolk cell and then second, once the blastoderm has passed the equator, contractile actin rings may help to draw the blastodermal cells to the vegetal pole. That actin does play a vital role in epiboly now appears undisputed, since the blocking of Rho function either by injection of C3 botulinum toxin or by antagonizing Rho-associated kinase (ROCK) has been shown to prevent the epiboly process (Lai *et al.*, 2005).

V. EPITHELIAL FUSION EVENTS

The fusion of epithelial sheets is another key morphogenetic event that takes place on numerous occasions during the development of many embryos, ranging from flies to humans. In this section we will explore how the actin cytoskeleton functions to pull together and fuse epithelial sheets in a variety of examples, illustrating that many features of fusion events are highly conserved across embryogenesis.

A. *Drosophila* Dorsal Closure

Dorsal closure in the fruit fly embryo involves the closure of a large hole in the dorsal epithelium of the embryo. It provides an excellent model for the study of epithelial fusion as it takes place on the very surface of the fly embryo, making it highly amenable to live imaging; while the genetic tractability of *Drosophila* means that it can be manipulated with relative ease. The dorsal hole is formed as a result of an earlier morphogenetic movement called germ band retraction (GBR) and takes about

2–3 h to close at room temperature. Dorsal closure is one of the final tissue movements in the development of the *Drosophila* embryo, but its proper completion is required for the embryo to hatch into a larvae (reviewed in Harden, 2002; Jacinto *et al.*, 2002b). The hole left by GBR is elliptical or "eye-shaped" and closure proceeds from the most anterior and posterior ends (or canthi), zipping closed toward the center. At the completion of closure, the two epithelial edges fuse and the resulting seam resolves and vanishes, leaving a perfect, segmentally aligned, epithelial surface at the midline seam (Fig. 5). Actin is crucial to all stages of dorsal closure, with cortical actin driving key cell shape changes at the beginning of the process, and specialized actin machineries functioning to pull the two epithelial surfaces together and fuse them at the midline.

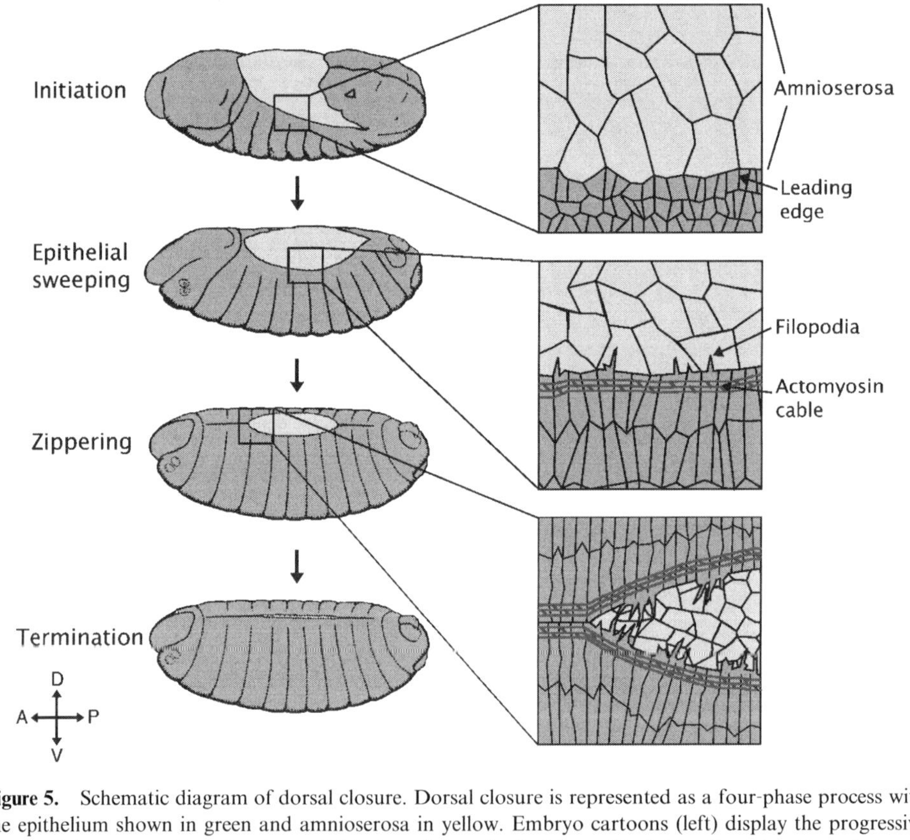

Figure 5. Schematic diagram of dorsal closure. Dorsal closure is represented as a four-phase process with the epithelium shown in green and amnioserosa in yellow. Embryo cartoons (left) display the progressive stages of the process. Boxed regions (right) show "zoom ins" of the epithelial leading edge. Dorsal closure begins as the germ band approaches its ultimate position at the posterior. Early in the process, the large cells of the amnioserosa (yellow) contract and the epithelial cells (green) begin to elongate in a dorsoventral direction (axis compass in bottom left). Slightly later in closure, the epithelial leading edge begins to assemble an actomyosin cable (red and blue) and starts to extend actin protrusions across the amnioserosa. (See Color Insert.)

Lining the eye-shaped hole is a monolayer of large flat extraembryonic cells called the amnioserosa, over which the epithelial edges must eventually travel to seal the hole. From the very start of closure, the large cells of the amnioserosa accumulate F-actin and myosin and begin to contract apically, pulling the attached epithelial edges toward each other (Fig. 5). Tight regulation of F-actin levels in the amnioserosa is required for proper closure, too much F-actin causes the amnioserosa to overcontract, distorting the embryo; while if levels are too low the epithelial surfaces are not pulled together efficiently and dorsal closure is defective (Harden *et al.*, 2002). As the amnioserosa contracts during wild-type dorsal closure, a second cell shape change takes place: the cells of the epithelium begin to elongate in a dorsoventral direction, helping to push the epithelial edges dorsally. This elongation may, in part, be driven by the assembly and contraction of an actin cable in the epithelial leading edge, which is explored next.

As dorsal closure proceeds, F-actin accumulates in cells at the leading edge of the two epithelial sheets. The accumulated actin is associated with myosin and forms a contractile cable which runs around the circumference of the hole, connected between cells at anchor points provided by cell–cell adherens junctions (Young *et al.*, 1993; Jacinto *et al.*, 2002a). The cable has contractile properties and appears to play two major roles during dorsal closure. First, the cable can act as a "purse-string," helping to pull the two epithelial surfaces together; if the cable is "cut" by laser ablation the free edges spring back reflecting a loss of tension (Kiehart, 1999; Kiehart *et al.*, 2000; Hutson *et al.*, 2003). Evidence of a second role for the actomyosin cable during dorsal closure came from studies where Rho1 or myosin II function was obliterated. In these mutants loss of the actomyosin cable results in a chaotic, no longer taut, leading edge and subsequent dorsal closure defects, indicating that the cable also acts to restrain and organize the leading edge, maintaining a coherent surface that can be closed efficiently and precisely (Jacinto *et al.*, 2002a).

As the actomyosin cable is assembled in the leading edge cells, these cells also begin to extend filopodia and lamellipodia. These actin protrusions are highly dynamic and reach out across the exposed amnioserosa. Once the epithelial edges are close enough, protrusions from opposing edges interact with one another and mediate the zipping together of the edges from the "canthi" at the most anterior and posterior ends of the hole, toward the middle until the hole is finally closed (Jacinto *et al.*, 2000). Investigations into the roles of actin protrusions during dorsal closure have inevitably highlighted the Rho GTPases, Cdc42 and Rac, which in tissue culture cells regulate the formation of filopodia and lamellipodia, respectively. Expression of either dominant negative Cdc42 or Rac in segmental stripes of the embryonic fly epithelium during dorsal closure eradicates virtually all actin protrusion assembly in affected leading edge cells and gives two key phenotypes (Fig. 6). First, leading edge cells lacking actin protrusions fail to fuse properly at the midline, indicating that these structures are required for tightly knitting the two epithelial edges together. Second, a key feature of normal dorsal closure fails in these embryos, as the segments no longer properly align as the hole is closed. This suggests that the actin protrusions play a sensing role during closure, allowing leading edge cells to find their correct partner on the opposing edge

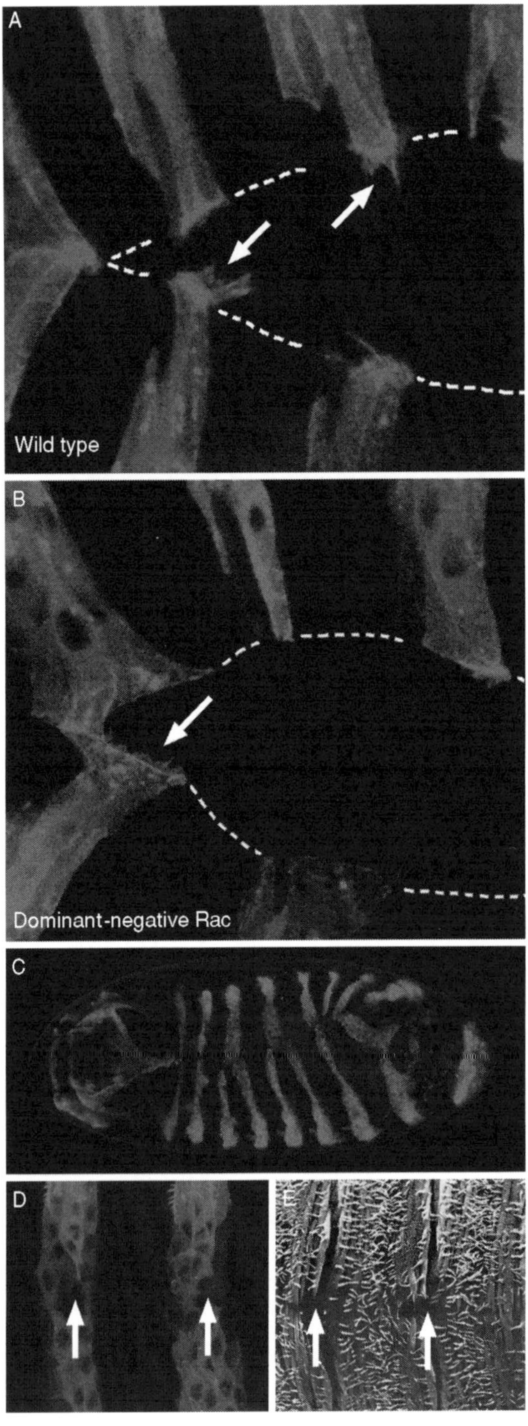
A
Wild type
B
Dominant-negative Rac
C
D
E

and thus correctly align the segments across the midline seam (Jacinto *et al.*, 2000; Woolner *et al.*, 2005).

Upstream control of actin during dorsal closure seems to feed, at least in part, through a JNK signaling pathway. This MAPK cascade appears to be the central signal transduction pathway in dorsal closure and controls a number of aspects of the process. The JNK pathway has been extensively studied in dorsal closure, and mutants in various components of the pathway exhibit similar phenotypes, characterized by cells of the leading edge losing their usual elongated state and by disruption of the actomyosin cable in these cells (Glise *et al.*, 1995; Riesgo-Escovar *et al.*, 1996; Riesgo-Escovar and Hafen, 1997; Stronach and Perrimon, 2002). Furthermore, live imaging of embryos mutant for *hemipterous* (*hep*), which encodes JNK kinase, showed that the leading edge cells of these mutants could no longer assemble actin protrusions or cable, indicating that the regulation of both these structures is mediated through the JNK pathway. The major route of action for the JNK cascade in dorsal closure appears to be through the expression and release of *Drosophila* TGF-β (Decapentaplegic) (reviewed in Harden, 2002). However, it remains unclear exactly how the JNK pathway feeds into actin assembly; some clues came from a serial analysis of gene expression (SAGE) study which identified a number of cytoskeletal regulators whose expression is upregulated in response to activation of the JNK cascade, including *Drosophila* profilin (Jasper *et al.*, 2001). Furthermore, it appears that regulation of JNK signaling is key both in the epithelial leading edge and also in the amnioserosa, where a failure to downregulate JNK leads to a failure in amnioserosa contraction.

B. *C. elegans* Ventral Enclosure

The process of ventral enclosure in the nematode worm, *C. elegans,* shares many similarities with *Drosophila* dorsal closure (Fig. 7). Like dorsal closure, ventral enclosure involves the sealing of a large hole in the epidermis of the embryo, in this case on the ventral surface. The process begins when four epithelial cells, so called "leading cells," positioned in pairs at the anterior end of the hole, begin to extend filopodia and move toward the ventral midline of the embryo. Like the filopodia of dorsal closure in flies, these protrusions are crucial for the fusion of the leading cells at the midline and thus are ultimately required for the completion of ventral enclosure

Figure 6. Actin protrusion function during *Drosophila* dorsal closure. During dorsal closure in the *Drosophila* embryo, filopodia and lamellipodia are assembled by the leading edge cells, where they function to accurately knit the two closing epithelial surfaces together. (A) A confocal image of a zippering front in a wild-type embryo, where GFP-actin is expressed in segmental stripes. Filopodia and lamellipodia assembled by the leading edge cells are indicated by arrows and the broken line highlights the edge of the dorsal hole. (B) In an embryo where a dominant negative form of the small GTPase Rac is expressed along with GFP-actin, very little actin protrusion assembly is seen. (C) Without actin protrusions, the embryos expressing dominant negative Rac are no longer able to correctly match their segments as the hole closes. In addition, these embryos also show fusion failures at the midline, shown by a confocal image in (D) and a scanning electron micrograph in (E). Reproduced with permission from Woolner *et al.* (2005). (See Color Insert.)

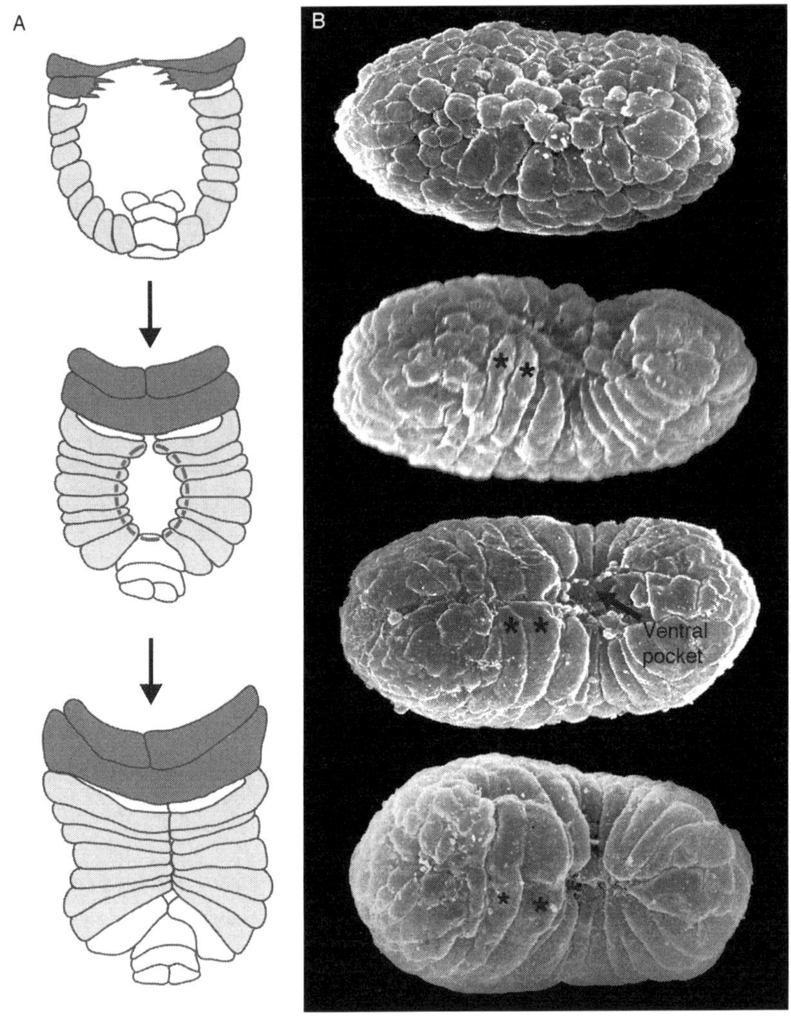

Figure 7. Ventral enclosure in the *C. elegans* embryo. *C. elegans* ventral enclosure is an epithelial fusion event which shares many similarities with dorsal closure. (A) Schematic diagrams illustrating ventral enclosure: the process starts as the four leading cells (green) assemble filopodia and begin to migrate ventrally. Once the leading cells have met at the midline, the pocket cells (yellow) assemble an actin cable and "purse-string" the hole closed. (B) A set of scanning electron micrographs of a *C. elegans* embryo undergoing ventral enclosure, leading cells are highlighted with asterisks. Reproduced with permission from Martin and Parkhurst (2004). (See Color Insert.)

(Williams-Masson *et al.*, 1997; Raich *et al.*, 1999). Studies of ventral enclosure have clearly shown that adherens junction components localize to the filopodial tips of leading cells and that filopodial contacts appear to prime cadherin-mediated adhesion and junction formation (Raich *et al.*, 1999), just as the case for keratinocyte adhesions *in vitro* (Vasioukhin *et al.*, 2000). The movement of the leading cells during ventral enclosure draws the epidermis around toward the ventral side of the embryo, but completion of the process requires the coordinated elongation and contraction of a second population of cells, called "pocket cells," which line the edges of the hole posterior to the leading cells. Once the leading cells have fused, they begin to contract their apices and elongate along the dorsoventral axis, just as *Drosophila* leading edge cells elongate at the start of dorsal closure. In addition, actin accumulates at the apical edge of the pocket cells and is thought to form a cable that aids closure of the hole by purse-string contraction. Laser ablation of regions along this edge release tension in the closing epithelium, just as is seen in similar cable ablation studies in dorsal closure (Williams-Masson *et al.*, 1997; Kiehart *et al.*, 2000).

C. Epithelial Fusion Events in Vertebrate Embryonic Development

Development of the vertebrate embryo also involves a number of epithelial fusion episodes that include neural tube closure, palate fusion, and eyelid closure. Strikingly, many of the mechanisms and signaling pathways underlying these fusions appear to be shared with the invertebrate fusion events described earlier. The vertebrate neural tube is formed from a flat layer of neural plate cells which invaginates to form a tube. In order for the neural tube to close, the lips of the folding neural plate must be drawn together and fused at the midline. This is initiated from several nucleation sites using a zippering mechanism similar to that seen in the final fusion of the epithelium in dorsal closure and ventral enclosure (reviewed in Colas and Schoenwolf, 2001). An actin network is seen just below the apical plasma membrane of the neural plate cells suggesting that the neural lips are, at least partially, drawn up to meet one another by concerted constriction of these cells. Cytochalasin blocking experiments indicate that active actin polymerization is required in the cranial region, since disassembly of the network leads to exencephaly (Morriss-Kay and Tuckett, 1985; Ybot-Gonzalez and Copp, 1999). Like dorsal closure, Rho GTPase activity is also required to close the neural tube, as the blocking of the Rho activator, Rho kinase (ROCK), leads to a failure in this process, as well as disruption of other morphogenetic movements in the mouse embryo (Wei *et al.*, 2001). *JNK1/JNK2* knockout (KO) mice also fail in neural tube closure, indicating that a JNK pathway is functioning here, just as in dorsal closure (Sabapathy *et al.*, 1999). The final fusion of the neural tube may also involve the Eph/ephrin family of adhesion signaling receptors because double *EphB2, EphB3* KO mice suffer from a failure in neural tube closure as well as related midline fusion events, leading to, for example, hypospadias of the genitalia (Dravis *et al.*, 2004).

Another midline fusion event in the mammalian embryo occurs as the secondary palatal shelves meet to form the roof of the mouth. While many of the cleft palate phenotypes seen in various KO mice are likely to be due to a failure in the machinery that pushes the shelves together early in the process, it also appears that actin protrusions are required here for the final fusion of the two surfaces. The assembly of filopodia during palate fusion appears to be dependent on TGF-β

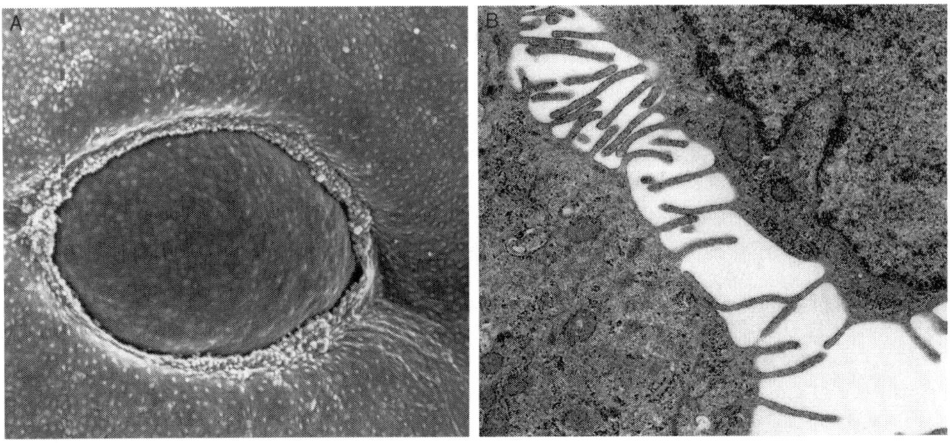

Figure 8. Eyelid fusion in the mouse embryo. Filopodia appear to assist eyelid fusion during mouse embryogenesis. (A) A scanning electron micrograph of the mouse eyelid at E15, when the eyelids are just starting to advance forward over the corneal epithelium. (B) TEM of the eye shows that filopodia are assembled by the leading edge cells and that these structures interdigitate as the two eyelids confront each other at the anterior and posterior canthi.

signaling, since TGF-β3 KO mice fail to form filopodia as the palatial shelves come together, ultimately resulting in KO mice being born with a cleft palate (Taya *et al.*, 1999).

The transient fusion of eyelids in the fetal stages of mammals is an event, which, superficially at least, bears startling resemblances to *Drosophila* dorsal closure (Fig. 8). As the eyelids approach each other they extend filopodia, allowing the edges to zip together from the nasal and lateral margins of the eye; electron microscopy shows that these protrusions interdigitate as the eyelids close, just as is seen in dorsal closure. These outward similarities extend to the signaling pathways involved in directing the two processes; the JNK pathway is clearly necessary for eyelid closure because mice in which c-Jun is conditionally inactivated in keratinocytes fail to complete eyelid closure and are born with their eyes open (Li *et al.*, 2003; Zenz *et al.*, 2003). ROCK KO mice are also born with open eyelids and this suggests that the assembly of oriented actin filaments may also operate in a purse-string-like way to aid in drawing the eyelids together, again suggesting parallels with fly dorsal closure. The other major deficit in these mice is a failure to close the ventral body wall, suggesting that this process too may be mediated by a contractile actin purse string (Shimizu *et al.*, 2005).

VI. TUGGING AND SQUEEZING MOVEMENTS

So far, we have discussed how actin is used to bend epithelial sheets during gastrulation and to knit epithelial sheets together in epithelial fusion events. However, in the morphogenetic episodes explored in this section, actin is employed in slightly different ways to drag epithelia or to squeeze tissues.

A. Germ Band Retraction in *Drosophila*

Prior to dorsal closure in the *Drosophila* embryo a pair of back-to-back morphogenetic events take place called germ band extension (GBE) and GBR. During GBE, the presumptive germ band is pushed along the dorsal side of the embryo toward the anterior end. Once it approaches the head, the germ band goes into reverse and is drawn back to its ultimate position at the posterior of the embryo, in the process known as GBR. GBE is known to be driven by intercalation of epithelial cells, causing a lengthening of the epithelial sheet and elongating the germ band (Irvine and Wieschaus, 1994). Much less is known about the mechanism behind GBR, but a live imaging study has begun to shed some light on the forces involved and indicated a key role for the actin cytoskeleton (Schock and Perrimon, 2002). As the germ band moves toward the posterior, the amnioserosa, a cell layer which will later play such an important role in dorsal closure, begins to be exposed. Live imaging has revealed that those amnioserosa cells which abut the germ band assemble lamellipodial actin protrusions that extend over the germ band as it retracts (Schock and Perrimon, 2002). It is not entirely clear exactly what role these protrusions are playing; possibly they act to help push the germ band toward the posterior, or alternatively they are tightly connecting the amnioserosa and the germ band allowing pushing forces from cell shape changes in the amnioserosa to be transmitted to the germ band. However, it is definitely the case that the regulation of actin within the amnioserosa is required for the germ band to retract, since if F-actin assembly is upset in this tissue, by the expression of either dominant negative or constitutively active versions of RhoA, the resulting embryos show severe defects in GBR (Schock and Perrimon, 2002).

B. *C. elegans* Elongation

The final stages of *C. elegans* embryonic development provide an excellent example of a stretching/constricting morphogenetic movement. Following ventral enclosure, the worm embryo undergoes an elongation event that results in a fourfold lengthening of its body (reviewed in Piekny and Mains, 2003). During elongation, the embryo is squeezed by the contraction of specialized actomyosin bundles which assemble around the circumference of the embryo (Priess and Hirsh, 1986). These bundles form in the lateral epidermal cells of the embryo and contraction causes these cells, and the embryo as a whole, to decrease in diameter and as a result dramatically increase in length. The requirement for actin in elongation is clearly shown by addition of cytochalasin D, as the circumferential bundles are disrupted and elongation fails (Priess and Hirsh, 1986). Just as actomyosin cables in other morphogenetic events are linked between cells by adherens junctions, these same junctions are required to transmit the contraction forces along the worm embryo as it elongates. The adherens junctions anchor the circumferential actin bundles to the apical surfaces of the epidermal cells but also, crucially, link the epidermal cells together allowing the contractile forces to be transmitted along the entire embryo (Priess and Hirsh, 1986). Genetic studies of elongation indicate that contraction of the actomyosin bundles bear similarities to smooth muscle contraction. Embryos mutant for myosin phosphatase hyperelongate, indicating that phosphorylation of the myosin light chain drives contraction, just as occurs in smooth muscle (Piekny *et al.*, 2003). A key player in other actomyosin contractions is ROCK; this Rho effector is required for worm elongation, with mutants in the *C. elegans* ROCK homologue, let-502, failing to elongate (Piekny *et al.*, 2003). Genetic interaction studies have highlighted several other actin regulators that are also involved in elongation, including two Rac-like GTPases and a Rho/Rac GEF (Wissmann *et al.*, 1999).

C. Heart Folding and Pharynx Development in Vertebrate Embryos

Actin is also used in novel ways to push and squeeze the developing vertebrate embryo into shape. Here we describe two morphogenetic events, heart looping and the formation of the pharyngeal pouches, which use actin in this way to sculpt embryonic tissues. The vertebrate heart begins life as single straight tube; this tube is then bent and shaped to produce the complex organ which pumps blood around the fetal and then adult body. An early step in the development of the heart is a process called tubular heart looping, in which the straight heart tube is bent over on itself toward the right. Not only does looping begin to give the heart tube some structure, it also sets up crucial left–right asymmetry within the embryo. Studies in chick embryos have revealed that the arrangement of actin filaments in the heart tube changes as the tube begins to bend. In the straight tube, all actin bundles are arranged circumferentially, while in the looping heart tube an asymmetrical pattern is seen. On the right, convex side of the tube, actin bundles are arranged in random direction, whereas the left, concave side retains a circumferential arrangement of actin bundles (Itasaki *et al.*, 1989). This asymmetry suggests that the heart tube may bend as a result of differences in the levels of pulling force provided by the different actin arrangements, with the actin bundles on the right side of the tube generating greater tension and causing the heart tube to loop toward the right. If the actin bundles on the right side of the tube are disrupted by the local application of cytochalasin B, the heart tube instead bends to the left (Itasaki *et al.*, 1991).

Other complex morphogenetic episodes in higher vertebrates are also regulated by controlled actin contractions. The pharyngeal arches are a series of bulges found on the lateral surface of the head and neck of all vertebrate embryos. Within the arches, the nerves, muscles, skeleton, and epithelia of the pharynx originate and differentiate (Graham and Smith, 2001). Separating the arches are a series of outpocketings called the pharyngeal pouches, which segregate the cell populations of each arch and act to induce the formation of arch components. As each pouch forms, a web of actin fibers assembles along the lumenal surface of the developing pouch. If this actin arrangement is disrupted by the addition of cytochalasin D, the pouches fail to develop their normal slit-like morphology and instead splay open. Disruption of the actin web slightly later in development leads to pouches that have a contorted shape (Quinlan *et al.*, 2004).

These data indicate that the actin web is involved in constraining the cells of the pharyngeal pouch and directing the expansion of the pouch in the appropriate direction. This constraining role has parallels to that played by the actin cable during *Drosophila* dorsal closure, where the cable restrains the cells of the leading edge, keeping these cells taut, organized, and coherent.

VII. TUBE FORMATION

The morphogenetic episodes described earlier have largely concerned the general shaping and sculpting of the embryo. However, several "fine tuning" aspects of organogenesis are also dependent on actin-driven morphogenetic events. One such process is tubulogenesis, which underlies the formation of many organs including the mammalian heart, lungs, and kidneys (reviewed in Hogan and Kolodziej, 2002). Here we will explore three specific aspects of tube formation: initial epithelial invagination, lumen formation, and the sensing function of tube tip cells, where vital roles for actin have been revealed (Fig. 9).

A. Epithelial Invagination

A common strategy for initiating the formation of a tube during development involves the invagination of small groups of epithelial cells (Fig. 9). This invagination scheme of tube formation is seen in a variety of tubulogenesis episodes ranging from the formation of the *Drosophila* salivary gland to the development of the complex tube systems which make up the mammalian lung, liver, and kidney. Tube formation starts when a group of cells within a polarized epithelial sheet begin to become increasingly columnar and form what is known as a placode. After elongating further along their basolateral axis, the cells in the placode then undergo a second cell shape change as they begin to constrict their apices and become wedge shaped (Hogan and Kolodziej, 2002). This shape change appears to drive the invagination of the cells in the placode, leading to the formation of a pocket of cells below the epithelial sheet, which can subsequently be further elaborated to form a tubular network.

The cell shape changes in placode cells are thought to be driven by constriction of the apical actin cytoskeleton, in much the same way as is seen during purse-string closure of a wound and also in the initial stages of *Drosophila* gastrulation. This certainly appears to be the case during the formation of the salivary gland of the *Drosophila* embryo, a structure which forms by placode invagination. Embryos mutant for *zipper,* the gene encoding the nonmuscle myosin heavy chain, fail to form normal salivary glands, and closer analysis indicates that epithelial cells in the mutants fail to adopt the columnar and wedge-like shapes required for placode invagination (Blake *et al.*, 1999). Similarly, in mammalian lung branching morphogenesis, actin is implicated as a key driver in the early invagination steps of tube formation. It is possible to culture embryonic lung epithelium and in this isolated state the tissue can still undergo "budding" with groups of cells invaginating to form tubular structures. Time-lapse

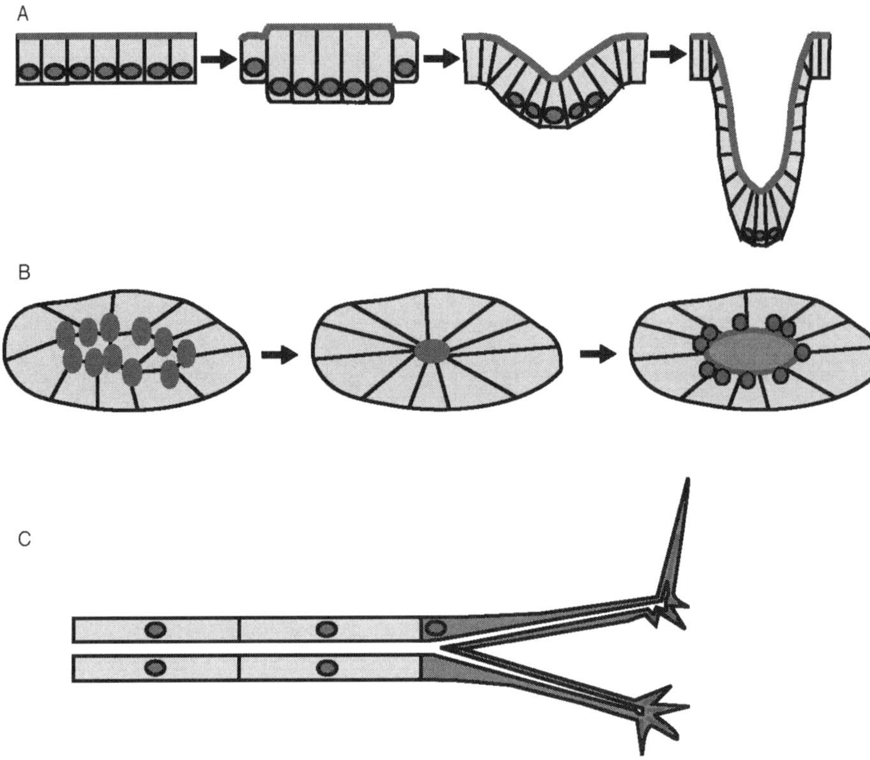

Figure 9. Mechanisms for making tubes. Actin plays a major role in three different strategies for generating tubular systems. (A) Invagination of cells from an epithelial sheet can drive tube formation. Actin (red) is found along the apical surface of the invaginating cells and it is thought that actin cables act to constrict the apices of these cells, making them wedge shaped and driving their invagination. (B) Lumen formation is an alternative tubulogenesis strategy and can occur in a single cells or a group of cells (the latter is illustrated here). Cells in a cord assemble primitive adherens contacts (green), these then cluster in the center of the cord and a lumen (pale blue) forms between the cells. It is thought that the actin (red) associated with the contacts aids in their clustering. (C) Once a tube is formed, it may migrate and elongate, often in response to external cues. An example of directed migration of tubes is found in the developing tracheal system, where secondary tracheal branches (purple) extend toward sources of fibroblast growth factor (FGF). The branches assemble actin protrusions at their distal tips, which appear to aid tube extension and sensing of FGF. (See Color Insert.)

studies have allowed this budding process to be assessed in greater detail and, crucially, actin staining of the budding epithelium has shown that F-actin accumulates along the apical side of the invaginating lung epithelial cells (Miura and Shiota, 2000). It therefore appears that, just as in the *Drosophila* salivary gland, actomyosin contraction drives the invagination of epithelial cells at the onset of tubulogenesis. It must be noted, however, that other factors, along with actin, are also likely to be required for invagination, these include changes in cell adhesion between invaginating cells and adjustments to the underlying extracellular matrix (Hogan and Kolodziej, 2002).

B. Lumen Formation

Alternative strategies for forming a tube are either to direct the formation of a lumen within a cluster of cells or to forge a hole through a single cell. There is growing evidence that lumen formation, either in multiple cells or a single cell, requires the involvement of the actin cytoskeleton. The formation and maintenance of a lumen within a single-celled tube provides an excellent paradigm for more complex multicellular lumen formation.

An especially powerful model of single cell tubulogenesis is provided by the *C. elegans* excretory cell. This cell forms two connected tubules which run down the left and right sides of the worm, stretching the entire length of the animal. These tubes act to regulate the osmolarity of the worm, and the diameter of the tube lumen must be controlled throughout the animal's life (Buechner, 2002). Soon after the excretory cell is born in the *C. elegans* embryo, a lumen begins to develop; vacuoles form within the cell and then coalesce to form the apical/lumenal surface of the cell. Actin appears to be intricately involved in the formation and maintenance of this lumen, with an F-actin "web" being seen along the new apical/lumenal surface (Buechner, 2002). Two worm mutants, *sma-1* and *exc-5*, which fail in normal lumen formation, are defective in genes that encode actin interactors. The *sma-1* gene encodes βH-spectrin, an actin cross-linking protein that may regulate lumenal dimension by anchoring the web of actin to the apical membrane. Meanwhile, the *exc-5* gene product is a GEF which is thought to activate the Rho GTPase, Cdc42 (Buechner, 2002).

A more complex, multicellular, example of lumen formation is seen during the development of the zebrafish gut. The formation of this tubular system begins when a group of cells align into a cord-like structure in the fish embryo. Cells in the center of the cord begin to adjust their adherens junctions so that the junctions cluster in the middle, giving the cells an apical surface and allowing a lumen to begin to develop (Fig. 9). Ultimately, the cells of the developing tube become wedge-shaped, possessing a tiny apical/lumenal surface (Hogan and Kolodziej, 2002). As with single-celled lumen formation, actin is key. The zebrafish mutant, *heart-and-soul* (*has*), displays defects in many epithelial organs including the heart, eye, and gut. The gut of a *has* mutant possesses several noncentralized lumina. The *has* gene encodes a member of the protein kinase C (PKC) family, an atypical PKCλ, downstream of which is the PAR3-PAR6 complex, a complex that is known to interact with members of the Rho family of GTPases, including Rho, Rac, and Cdc42 (Horne-Badovinac *et al.*, 2001).

C. Tube Elongation and Migration

Once a tube has formed it must elongate and migrate to form the complex tubular systems that are found in embryos and adult organisms. Migration of tubes is often directed by external cues, for example, in the developing tracheal system of the *Drosophila* embryo or lung buds in vertebrate embryos, branching is generally directed toward sources of an FGF. Studies of such directed branching in various systems including the *C. elegans* excretory cell, *Drosophila* salivary gland, and *Drosophila* tracheal system have found that cells at the tips of growing tubes (tip cells) extend actin-rich filopodia and lamellipodia (Nelson, 2003). These protrusions play a role in sensing external cues and guiding the tip cells to their correct location, a role analogous to that played by the filopodia in the growth cone of an axon. The best-described example of this form of tube guidance comes from the *Drosophila* tracheal system. This system begins to form by the invagination of cells within placodes spaced along the embryo; primary branches begin to develop from these internalized sacs of cells and then branch again and again to form an extensive system of tubes (Myat, 2005). The tip cells of tracheal tubes extend actin-rich filopodial protrusions and migrate toward clusters of mesodermal cells that express the FGF ligand Branchless (Bnl). The Bnl signal is transduced through its receptor, Breathless which is expressed by the tip cells (Sutherland *et al.*, 1996) (Fig. 9). This pattern of development continues with the primary branches branching to form secondary and then terminal branches which serve the respiratory needs of the organism by "sensing" oxygen tensions, again through receptors expressed on their filopodial tips (Ghabrial *et al.*, 2003).

VIII. FUTURE DIRECTIONS

In this chapter we have outlined several of the important and varied roles that actin performs during morphogenesis. While we are beginning to understand how the regulation of actin assemblies and their contractility may function to mediate morphogenetic events, vast gaps in our knowledge remain. In this section, we outline some of the exciting new areas for investigation in the study of morphogenesis and the role of the actin cytoskeleton.

A. Initiating and Linking Morphogenetic Episodes

Although it is very clear how important various actin-rich structures are in driving many morphogenetic processes during embryogenesis, there is very little known about how these machineries are initially triggered to assemble. We do know that the Rho GTPases and signaling cascades, such as the JNK pathway, regulate the formation of actin structures, but we have very little idea of what lies upstream of these to initiate and link morphogenetic episodes. Here we describe work that hints at what some of these initiating cues might be.

One possible kick start cue for morphogenetic episodes may come from hormonal signaling. Certainly, evidence has begun to suggest that in *Drosophila,* the hormone ecdysone may play a role in the morphogenetic events that characterize late embryogenesis, including dorsal closure. Ecdysone is a steroid hormone best known for its regulation of *Drosophila* metamorphosis. However, intriguingly, there is a surge of ecdysone release in the embryo that peaks toward the end of GBR (Richards, 1981), which is well timed as a possible initiator of dorsal closure. Further investigation indicates that at the end of GBR, the amnioserosa contains high levels of active ecdysteroids, since ecdysone-dependent reporter transgenes are robustly expressed in the amnioserosa at this time (Kozlova and Thummel, 2003). In keeping with a role for ecdysone in dorsal closure, embryos mutant for components of the ecdysone synthesis pathway show failures in dorsal closure, as well as GBR (Chavez *et al.*, 2000).

Another potential upstream initiator of actin machineries in morphogenesis could be Ca^{2+} waves. During wound repair, Ca^{2+} is known to play a crucial role in the assembly of the actomyosin purse string (Bement *et al.*, 1999) and the same may be true more generally in some morphogenetic episodes. Late in zebrafish epiboly, Ca^{2+} waves can be seen traversing the margin of the blastoderm at a time just after assembly of the actin rings, which help to draw the blastodermal cells to the vegetal pole (Webb and Miller, 2003). If the Ca^{2+} waves are disrupted, by the addition of a Ca^{2+} chelator, the actin rings no longer form properly and epiboly fails, indicating that the Ca^{2+} may play a role in the formation of these structures (Cheng *et al.*, 2004).

B. Mechanical Forces

During morphogenetic events, as tissues are pulled, squeezed, and sculpted, many resulting changes in tension occur across these tissues. The sensing of these changes is another powerful candidate for the regulation of morphogenesis and the assembly of actin machineries. It is certainly possible to imagine that changes in tension occurring during one morphogenetic event may help to initiate a subsequent episode, for example in the *Drosophila* embryo, it is not unreasonable to presume that the new tissue forces generated by completion of GBR may provide some of the trigger for dorsal closure to begin.

The effect of tension during development is an area that is only just beginning to be explored, but there is a precedent for mechanical forces influencing gene expression in *Drosophila* embryos. An innovative study found that expression of the dorsoventral polarity gene, *twist*, could be mechanically induced by applying pressure to the early embryo (Farge, 2003). Furthermore, this study showed good evidence that mechanical "stretching" signals may play an inductive role later in development to trigger *twist* expression in the cells of the stomodeal primordium. These cells are deformed naturally as the germ band extends and in mutant embryos where GBE fails, or in embryos where the linking tissue is cut, *twist* is no longer expressed in the stomodeal primordium, but expression can be restored by the application of a comparable mechanical compression (Farge, 2003).

The molecular mechanisms underlying tension sensing *in vivo* remain largely unknown, although a study has uncovered a means by which *Drosophila* border cells can sense stretching forces as they migrate during oogenesis (Somogyi and Rorth, 2004). As these cells begin to migrate, a transcriptional cofactor called MAL-D translocates from the cytoplasm into the nucleus. The activity of MAL-D is required in migrating border cells to somehow strengthen the actin cytoskeleton and/or cell–cell adhesions, as cells that lack its activity break apart as they initiate migration.

In mutant border cells that cannot migrate, MAL-D remains in the cytoplasm, but it does accumulate in the nucleus if the mutant cells are pulled by wild-type migratory cells, suggesting that the translocation of MAL-D is a response to cell stretching (Somogyi and Rorth, 2004).

C. Sensing Functions of Actin Protrusions

A major function of actin during morphogenesis is to assemble filopodia and lamellipodia and these protrusions exhibit some intriguing sensory properties that certainly warrant further investigation. This is seen clearly in *Drosophila* dorsal closure, where actin protrusions are required to correctly match the embryo's segments as the two epithelial surfaces close. Eradicating protrusion formation by the expression of either dominant negative *Cdc42* (Jacinto *et al.*, 2000) or *Rac1* in segmental stripes prevents cells of the leading edge from finding their correct partners on the opposing epithelial edge, resulting in frequent mismatching of segments along the dorsal midline. Furthermore, analysis of wild-type closure also shows filopodia reaching out and interacting with incorrect segments and seemingly being repulsed, and retracting back (Jacinto *et al.*, 2000). As yet, there are very few clues as to what molecules may be involved in the segmental sensing seen during dorsal closure. It would seem likely that whatever is involved would be localized at the tips of the filopodia which explore the opposing epithelial edge as the hole closes, and dissecting the molecular activities at these tips may provide a useful focus for future research in this area.

Dorsal closure in the *Drosophila* embryo is certainly not the only example of actin protrusions playing a sensing role during embryogenesis. Perhaps the most well-studied example is during the development of the nervous system, when an axon is guided along its correct path by cues that are sensed by filopodia assembled at the growth cone of the axon (Tessier-Lavigne and Goodman, 1996). A further example of filopodial sensing is seen during the development of tubular systems, where tube branches are guided by FGF signals picked up by filopodia expressed on the branch tips. Applying what is known about these systems to the less well-understood examples of filopodial sensing, such as dorsal closure, will provide many future research directions.

Another, perhaps related, function that the filopodia seem to perform during some morphogenetic episodes, is to sense, and react to, "contact inhibition" cues. This is particularly clear during epithelial fusion events, when the two closing surfaces must be prevented from overrunning and "informed" when to stop. Intriguingly, in *Drosophila* dorsal closure expression of constitutively active *Rac1* leads to the production of overly large lamellipodia, which, crucially, seem unable to respond to inhibitory cues appropriately and interact with their adjacent neighbors as if they were on the opposing epithelial edge (Woolner *et al.*, 2005). While ideas of contact inhibition in epithelial fusion are still extremely speculative, it is an exciting prospect that these events could provide a model to study this crucial cell biological question. A question which, although first outlined by Abercrombie in the 1950s (reviewed in Abercrombie, 1979), and possessing such obvious implications for cancer biology, still has no clear molecular mechanism.

In conclusion, we now have a wealth of experiments detailing the crucial, and varied, roles performed by actin during morphogenesis. The challenge now is to pinpoint exactly how these episodes are regulated and uncover the molecular mechanisms by which the actin structures perform their complex tasks.

REFERENCES

Abercrombie, M. (1979). Contact inhibition and malignancy. *Nature* **281,** 259–262.

Barrett, K., Leptin, M., and Settleman, J. (1997). The Rho GTPase and a putative RhoGEF mediate a signaling pathway for the cell shape changes in Drosophila gastrulation. *Cell* **91,** 905–915.

Bement, W. M., Mandato, C. A., and Kirsch, M. N. (1999). Wound-induced assembly and closure of an actomyosin purse string in Xenopus oocytes. *Curr. Biol.* **9,** 579–587.

Biyasheva, A., Svitkina, T., Kunda, P., Baum, B., and Borisy, G. (2004). Cascade pathway of filopodia formation downstream of SCAR. *J. Cell Sci.* **117,** 837–848.

Blake, K. J., Myette, G., and Jack, J. (1999). Ribbon, raw, and zipper have distinct functions in reshaping the Drosophila cytoskeleton. *Dev. Genes Evol.* **209,** 555–559.

Brock, J., Midwinter, K., Lewis, J., and Martin, P. (1996). Healing of incisional wounds in the embryonic chick wing bud: Characterization of the actin purse-string and demonstration of a requirement for Rho activation. *J. Cell Biol.* **135,** 1097–1107.

Buechner, M. (2002). Tubes and the single *C. elegans* excretory cell. *Trends Cell Biol.* **12,** 479–484.

Chavez, V. M., Marques, G., Delbecque, J. P., Kobayashi, K., Hollingsworth, M., Burr, J., Natzle, J. E., and O'Connor, M. B. (2000). The *Drosophila* disembodied gene controls late embryonic morphogenesis and codes for a cytochrome P450 enzyme that regulates embryonic ecdysone levels. *Development* **127,** 4115–4126.

Cheng, J. C., Miller, A. L., and Webb, S. E. (2004). Organization and function of microfilaments during late epiboly in zebrafish embryos. *Dev. Dyn.* **231,** 313–323.

Colas, J. F. and Schoenwolf, G. C. (2001). Towards a cellular and molecular understanding of neurulation. *Dev. Dyn.* **221,** 117–145.

Dravis, C., Yokoyama, N., Chumley, M. J., Cowan, C. A., Silvany, R. E., Shay, J., Baker, L. A., and Henkemeyer, M. (2004). Bidirectional signaling mediated by ephrin-B2 and EphB2 controls urorectal development. *Dev. Biol.* **271,** 272–290.

Eden, S., Rohatgi, R., Podtelejnikov, A. V., Mann, M., and Kirschner, M. W. (2002). Mechanism of regulation of WAVE1-induced actin nucleation by Rac1 and Nck. *Nature* **418,** 790–793.

Evangelista, M., Zigmond, S., and Boone, C. (2003). Formins: Signaling effectors for assembly and polarization of actin filaments. *J. Cell Sci.* **116,** 2603–2611.

Farge, E. (2003). Mechanical induction of twist in the Drosophila foregut/stomodeal primordium. *Curr. Biol.* **13,** 1365–1377.

Ghabrial, A., Luschnig, S., Metzstein, M. M., and Krasnow, M. A. (2003). Branching morphogenesis of the *Drosophila* tracheal system. *Annu. Rev. Cell Dev. Biol.* **19,** 623–647.

Glise, B., Bourbon, H., and Noselli, S. (1995). Hemipterous encodes a novel Drosophila MAP kinase kinase, required for epithelial cell sheet movement. *Cell* **83,** 451–461.

Graham, A. and Smith, A. (2001). Patterning the pharyngeal arches. *Bioessays* **23,** 54–61.

Harden, N. (2002). Signaling pathways directing the movement and fusion of epithelial sheets: Lessons from dorsal closure in *Drosophila*. *Differentiation* **70,** 181–203.

Harden, N., Loh, H. Y., Chia, W., and Lim, L. (1995). A dominant inhibitory version of the small GTP-binding protein Rac disrupts cytoskeletal structures and inhibits developmental cell shape changes in *Drosophila*. *Development* **121,** 903–914.

Harden, N., Ricos, M., Yee, K., Sanny, J., Langmann, C., Yu, H., Chia, W., and Lim, L. (2002). Dracl and Crumbs participate in amnioserosa morphogenesis during dorsal closure in *Drosophila. J. Cell Sci.* **115,** 2119–2129.

Higashida, C., Miyoshi, T., Fujita, A., Oceguera-Yanez, F., Monypenny, J., Andou, Y., Narumiya, S., and Watanabe, N. (2004). Actin polymerization-driven molecular movement of mDia1 in living cells. *Science* **303,** 2007–2010.

Higgs, H. N. and Pollard, T. D. (2000). Activation by Cdc42 and PIP(2) of Wiskott-Aldrich syndrome protein (WASp) stimulates actin nucleation by Arp2/3 complex. *J. Cell Biol.* **150,** 1311–1320.

Hogan, B. L. and Kolodziej, P. A. (2002). Organogenesis: Molecular mechanisms of tubulogenesis. *Nat. Rev. Genet.* **3,** 513–523.

Horne-Badovinac, S., Lin, D., Waldron, S., Schwarz, M., Mbamalu, G., Pawson, T., Jan, Y., Stainier, D. Y., and Abdelilah-Seyfried, S. (2001). Positional cloning of heart and soul reveals multiple roles for PKC lambda in zebrafish organogenesis. *Curr. Biol.* **11,** 1492–1502.

Hutson, M. S., Tokutake, Y., Chang, M. S., Bloor, J. W., Venakides, S., Kiehart, D. P., and Edwards, G. S. (2003). Forces for morphogenesis investigated with laser microsurgery and quantitative modeling. *Science* **300,** 145–149.

Irvine, K. D. and Wieschaus, E. (1994). Cell intercalation during *Drosophila* germband extension and its regulation by pair-rule segmentation genes. *Development* **120,** 827–841.

Itasaki, N., Nakamura, H., and Yasuda, M. (1989). Changes in the arrangement of actin bundles during heart looping in the chick embryo. *Anat. Embryol. (Berl)* **180,** 413–420.

Itasaki, N., Nakamura, H., Sumida, H., and Yasuda, M. (1991). Actin bundles on the right side in the caudal part of the heart tube play a role in dextro-looping in the embryonic chick heart. *Anat. Embryol. (Berl)* **183,** 29–39.

Jacinto, A. and Baum, B. (2003). Actin in development. *Mech. Dev.* **120,** 1337–1349.

Jacinto, A., Wood, W., Balayo, T., Turmaine, M., Martinez-Arias, A., and Martin, P. (2000). Dynamic actin-based epithelial adhesion and cell matching during *Drosophila* dorsal closure. *Curr. Biol.* **10,** 1420–1426.

Jacinto, A., Wood, W., Woolner, S., Hiley, C., Turner, L., Wilson, C., Martinez-Arias, A., and Martin, P. (2002a). Dynamic analysis of actin cable function during Drosophila dorsal closure. *Curr. Biol.* **12,** 1245–1250.

Jacinto, A., Woolner, S., and Martin, P. (2002b). Dynamic analysis of dorsal closure in *Drosophila*: From genetics to cell biology. *Dev. Cell* **3,** 9–19.

Jasper, H., Benes, V., Schwager, C., Sauer, S., Clauder-Munster, S., Ansorge, W., and Bohmann, D. (2001). The genomic response of the *Drosophila* embryo to JNK signaling. *Dev. Cell* **1,** 579–586.

Kiehart, D. P. (1999). The power of the purse string. *Curr. Biol.* **9,** R602–R605.

Kiehart, D. P., Galbraith, C. G., Edwards, K. A., Rickoll, W. L., and Montague, R. A. (2000). Multiple forces contribute to cell sheet morphogenesis for dorsal closure in *Drosophila. J. Cell Biol.* **149,** 471–490.

Kozlova, T. and Thummel, C. S. (2003). Essential roles for ecdysone signaling during Drosophila mid-embryonic development. *Science* **301,** 1911–1914.

Krugmann, S., Jordens, I., Gevaert, K., Driessens, M., Vandekerckhove, J., and Hall, A. (2001). Cdc42 induces filopodia by promoting the formation of an IRSp53: Mena complex. *Curr. Biol.* **11,** 1645–1655.

Lai, S. L., Chang, C. N., Wang, P. J., and Lee, S. J. (2005). Rho mediates cytokinesis and epiboly via ROCK in zebrafish. *Mol. Reprod. Dev.* **71,** 186–196.

Leptin, M. (1999). Gastrulation in *Drosophila*: The logic and the cellular mechanisms. *EMBO J.* **18,** 3187–3192.

Li, G., Gustafson-Brown, C., Hanks, S. K., Nason, K., Arbeit, J. M., Pogliano, K., Wisdom, R. M., and Johnson, R. S. (2003). C-Jun is essential for organization of the epidermal leading edge. *Dev. Cell* **4,** 865–877.

Machesky, L. M. and Insall, R. H. (1998). Scar1 and the related Wiskott-Aldrich syndrome protein, WASP, regulate the actin cytoskeleton through the Arp2/3 complex. *Curr. Biol.* **8,** 1347–1356.

Machesky, L. M., Atkinson, S. J., Ampe, C., Vandekerckhove, J., and Pollard, T. D. (1994). Purification of a cortical complex containing two unconventional actins from Acanthamoeba by affinity chromatography on profilin-agarose. *J. Cell Biol.* **127,** 107–115.

Martin, P. and Lewis, J. (1992). Actin cables and epidermal movement in embryonic wound healing. *Nature* **360,** 179–183.

Martin, P. and Parkhurst, S. M. (2004). Parallels between tissue repair and embryo morphogenesis. *Development* **131,** 3021–3034.

McCluskey, J. and Martin, P. (1995). Analysis of the tissue movements of embryonic wound healing—DiI studies in the limb bud stage mouse embryo. *Dev. Biol.* **170,** 102–114.

Mejillano, M. R., Kojima, S., Applewhite, D. A., Gertler, F. B., Svitkina, T. M., and Borisy, G. G. (2004). Lamellipodial versus filopodial mode of the actin nanomachinery: Pivotal role of the filament barbed end. *Cell* **118,** 363–373.

Miura, T. and Shiota, K. (2000). Time-lapse observation of branching morphogenesis of the lung bud epithelium in mesenchyme-free culture and its relationship with the localization of actin filaments. *Int. J. Dev. Biol.* **44,** 899–902.

Moon, S. Y. and Zheng, Y. (2003). Rho GTPase-activating proteins in cell regulation. *Trends Cell Biol.* **13,** 13–22.

Morriss-Kay, G. and Tuckett, F. (1985). The role of microfilaments in cranial neurulation in rat embryos: Effects of short-term exposure to cytochalasin D. *J. Embryol. Exp. Morphol.* **88,** 333–348.

Myat, M. M. (2005). Making tubes in the Drosophila embryo. *Dev. Dyn.* **232,** 617–632.

Nelson, W. J. (2003). Tube morphogenesis: Closure, but many openings remain. *Trends Cell Biol.* **13,** 615–621.

Nikolaidou, K. K. and Barrett, K. (2004). A Rho GTPase signaling pathway is used reiteratively in epithelial folding and potentially selects the outcome of Rho activation. *Curr. Biol.* **14,** 1822–1826.

Nobes, C. D. and Hall, A. (1995). Rho, rac, and cdc42 GTPases regulate the assembly of multimolecular focal complexes associated with actin stress fibers, lamellipodia, and filopodia. *Cell* **81,** 53–62.

Nobes, C. D. and Hall, A. (1999). Rho GTPases control polarity, protrusion, and adhesion during cell movement. *J. Cell Biol.* **144,** 1235–1244.

Piekny, A. J. and Mains, P. E. (2003). Squeezing an egg into a worm: *C. elegans* embryonic morphogenesis. *Scientific World J.* **3,** 1370–1381.

Piekny, A. J., Johnson, J. L., Cham, G. D., and Mains, P. E. (2003). The Caenorhabditis elegans nonmuscle myosin genes nmy-1 and nmy-2 function as redundant components of the let-502/Rho-binding kinase and mel-11/myosin phosphatase pathway during embryonic morphogenesis. *Development* **130,** 5695–5704.

Pollard, T. D. and Borisy, G. G. (2003). Cellular motility driven by assembly and disassembly of actin filaments. *Cell* **112,** 453–465.

Priess, J. R. and Hirsh, D. I. (1986). *Caenorhabditis elegans* morphogenesis: The role of the cytoskeleton in elongation of the embryo. *Dev. Biol.* **117,** 156–173.

Quinlan, R., Martin, P., and Graham, A. (2004). The role of actin cables in directing the morphogenesis of the pharyngeal pouches. *Development* **131,** 593–599.

Raftopoulou, M. and Hall, A. (2004). Cell migration: Rho GTPases lead the way. *Dev. Biol.* **265,** 23–32.

Raich, W. B., Agbunag, C., and Hardin, J. (1999). Rapid epithelial-sheet sealing in the Caenorhabditis elegans embryo requires cadherin-dependent filopodial priming. *Curr. Biol.* **9,** 1139–1146.

Richards, G. (1981). The radioimmune assay of ecdysteroid titres in Drosophila melanogaster. *Mol. Cell Endocrinol.* **21,** 181–197.

Ridley, A. J. and Hall, A. (1992). The small GTP-binding protein rho regulates the assembly of focal adhesions and actin stress fibers in response to growth factors. *Cell* **70,** 389–399.

Ridley, A. J., Paterson, H. F., Johnston, C. L., Diekmann, D., and Hall, A. (1992). The small GTP-binding protein rac regulates growth factor-induced membrane ruffling. *Cell* **70,** 401–410.

Riesgo-Escovar, J. R. and Hafen, E. (1997). Common and distinct roles of DFos and DJun during Drosophila development. *Science* **278,** 669–672.

Riesgo-Escovar, J. R., Jenni, M., Fritz, A., and Hafen, E. (1996). The Drosophila Jun-N-terminal kinase is required for cell morphogenesis but not for DJun-dependent cell fate specification in the eye. *Genes Dev.* **10,** 2759–2768.

Rogers, S. L., Wiedemann, U., Stuurman, N., and Vale, R. D. (2003). Molecular requirements for actin-based lamella formation in Drosophila S2 cells. *J. Cell Biol.* **162,** 1079–1088.

Rohatgi, R., Ho, H. Y., and Kirschner, M. W. (2000). Mechanism of N-WASP activation by CDC42 and phosphatidylinositol 4, 5-bisphosphate. *J. Cell Biol.* **150,** 1299–1310.

Sabapathy, K., Jochum, W., Hochedlinger, K., Chang, L., Karin, M., and Wagner, E. F. (1999). Defective neural tube morphogenesis and altered apoptosis in the absence of both JNK1 and JNK2. *Mech. Dev.* **89,** 115–124.

Sagot, I., Klee, S. K., and Pellman, D. (2002). Yeast formins regulate cell polarity by controlling the assembly of actin cables. *Nat. Cell Biol.* **4,** 42–50.

Satoh, S. and Tominaga, T. (2001). mDia-interacting protein acts downstream of Rho-mDia and modifies Src activation and stress fiber formation. *J. Biol. Chem.* **276,** 39290–39294.

Schmidt, A. and Hall, A. (2002). Guanine nucleotide exchange factors for Rho GTPases: Turning on the switch. *Genes Dev.* **16,** 1587–1609.

Schock, F. and Perrimon, N. (2002). Cellular processes associated with germ band retraction in Drosophila. *Dev. Biol.* **248,** 29–39.

Shimizu, Y., Thumkeo, D., Keel, J., Ishizaki, T., Oshima, H., Oshima, M., Noda, Y., Matsumura, F., Taketo, M. M., and Narumiya, S. (2005). ROCK-I regulates closure of the eyelids and ventral body wall by inducing assembly of actomyosin bundles. *J. Cell Biol.* **168,** 941–953.

Small, J. V., Isenberg, G., and Celis, J. E. (1978). Polarity of actin at the leading edge of cultured cells. *Nature* **272,** 638–639.

Solnica-Krezel, L. and Driever, W. (1994). Microtubule arrays of the zebrafish yolk cell: Organization and function during epiboly. *Development* **120,** 2443–2455.

Somogyi, K. and Rorth, P. (2004). Evidence for tension-based regulation of Drosophila MAL and SRF during invasive cell migration. *Dev. Cell* **7,** 85–93.

Stronach, B. and Perrimon, N. (2002). Activation of the JNK pathway during dorsal closure in Drosophila requires the mixed lineage kinase, slipper. *Genes Dev.* **16,** 377–387.

Sutherland, D., Samakovlis, C., and Krasnow, M. A. (1996). Branchless encodes a Drosophila FGF homolog that controls tracheal cell migration and the pattern of branching. *Cell* **87,** 1091–1101.

Svitkina, T. M. and Borisy, G. G. (1999). Arp2/3 complex and actin depolymerizing factor/cofilin in dendritic organization and treadmilling of actin filament array in lamellipodia. *J. Cell Biol.* **145,** 1009–1026.

Svitkina, T. M., Bulanova, E. A., Chaga, O. Y., Vignjevic, D. M., Kojima, S., Vasiliev, J. M., and Borisy, G. G. (2003). Mechanism of filopodia initiation by reorganization of a dendritic network. *J. Cell Biol.* **160,** 409–421.

Taya, Y., O'Kane, S., and Ferguson, M. W. (1999). Pathogenesis of cleft palate in TGF-beta3 knockout mice. *Development* **126,** 3869–3879.

Tessier Lavigne, M. and Goodman, C. S. (1996). The molecular biology of axon guidance. *Science* **274,** 1123–1133.

Van Aelst, L. and D'Souza-Schorey, C. (1997). Rho GTPases and signaling networks. *Genes Dev.* **11,** 2295–2322.

Vasioukhin, V., Bauer, C., Yin, M., and Fuchs, E. (2000). Directed actin polymerization is the driving force for epithelial cell-cell adhesion. *Cell* **100,** 209–219.

Vignjevic, D., Yarar, D., Welch, M. D., Peloquin, J., Svitkina, T., and Borisy, G. G. (2003). Formation of filopodia-like bundles in vitro from a dendritic network. *J. Cell Biol.* **160,** 951–962.

Watanabe, N., Kato, T., Fujita, A., Ishizaki, T., and Narumiya, S. (1999). Cooperation between mDia1 and ROCK in Rho-induced actin reorganization. *Nat. Cell. Biol.* **1,** 136–143.

Webb, S. E. and Miller, A. L. (2003). Imaging intercellular calcium waves during late epiboly in intact zebrafish embryos. *Zygote* **11,** 175–182.

Wei, L., Roberts, W., Wang, L., Yamada, M., Zhang, S., Zhao, Z., Rivkees, S. A., Schwartz, R. J., and Imanaka-Yoshida, K. (2001). Rho kinases play an obligatory role in vertebrate embryonic organogenesis. *Development* **128,** 2953–2962.

Wennerberg, K. and Der, C. J. (2004). Rho-family GTPases: It's not only Rac and Rho (and I like it). *J. Cell Sci.* **117,** 1301–1312.

Widmann, C., Gibson, S., Jarpe, M. B., and Johnson, G. L. (1999). Mitogen-activated protein kinase: Conservation of a three-kinase module from yeast to human. *Physiol. Rev.* **79,** 143–180.

Williams-Masson, E. M., Malik, A. N., and Hardin, J. (1997). An actin-mediated two-step mechanism is required for ventral enclosure of the C. elegans hypodermis. *Development* **124,** 2889–2901.

Wissmann, A., Ingles, J., and Mains, P. E. (1999). The Caenorhabditis elegans mel-11 myosin phosphatase regulatory subunit affects tissue contraction in the somatic gonad and the embryonic epidermis and genetically interacts with the Rac signaling pathway. *Dev. Biol.* **209,** 111–127.

Woolner, S., Jacinto, A., and Martin, P. (2005). The small GTPase Rac plays multiple roles in epithelial sheet fusion—dynamic studies of Drosophila dorsal closure. *Dev. Biol.* **282,** 163–173.

Xia, Y. and Karin, M. (2004). The control of cell motility and epithelial morphogenesis by Jun kinases. *Trends Cell Biol.* **14,** 94–101.

Ybot-Gonzalez, P. and Copp, A. J. (1999). Bending of the neural plate during mouse spinal neurulation is independent of actin microfilaments. *Dev. Dyn.* **215,** 273–283.

Young, P. E., Richman, A. M., Ketchum, A. S., and Kiehart, D. P. (1993). Morphogenesis in Drosophila requires nonmuscle myosin heavy chain function. *Genes Dev.* **7,** 29–41.

Zalik, S. E., Lewandowski, E., Kam, Z., and Geiger, B. (1999). Cell adhesion and the actin cytoskeleton of the enveloping layer in the zebrafish embryo during epiboly. *Biochem. Cell Biol.* **77,** 527–542.

Zenz, R., Scheuch, H., Martin, P., Frank, C., Eferl, R., Kenner, L., Sibilia, M., and Wagner, E. F. (2003). c-Jun regulates eyelid closure and skin tumor development through EGFR signaling. *Dev. Cell* **4,** 879–889.

Zheng, Y. (2001). Dbl family guanine nucleotide exchange factors. *Trends Biochem. Sci.* **26,** 724–732.

Zigmond, S. H., Evangelista, M., Boone, C., Yang, C., Dar, A. C., Sicheri, F., Forkey, J., and Pring, M. (2003). Formin leaky cap allows elongation in the presence of tight capping proteins. *Curr. Biol.* **13,** 1820–1823.

Chapter 13

Mechanisms of Ion Transport Regulation by Microfilaments

Zhihong Hu, Daniel R. Clayburgh, and Jerrold R. Turner

Department of Pathology, The University of Chicago, 5841 South Maryland, Chicago, Illinois 60637

A major function of transporting epithelia is vectorial ion movement. This is accomplished by ion channels, pumps, antiporters, and symporters. The activity of these transport proteins is regulated by diverse signaling transduction and membrane trafficking events, many of which involve the actin cytoskeleton. This chapter will focus on the role of actin microfilaments and associated proteins in regulation of ion transport. The apical Na^+H^+antiporter NHE3 will be examined in detail in this context due to its general biological importance as well as recent advances that have helped to elucidate the mechanisms of regulation of this protein. Appreciation of the mechanisms of NHE3 regulation should serve as a framework with which to understand cytoskeletal regulation of other transport proteins as well.

Advances in Molecular and Cell Biology, Vol. 37

ISBN: 0-444-52868-7

DOI: 10.1016/S1569-2558(06)37013-0

I. INTRODUCTION

Cells lining the intestine and renal tubules represent the major ion transporting epithelia. The ability of these epithelia to modify ion transport rates rapidly and reversibly is critical to the maintenance of osmotic, acid–base, and volume homeostasis. Theoretically, the kinetic behavior of these transport proteins could be modified by at least three molecular mechanisms: regulation of the transporter activity at the plasma membrane, removal of transport proteins from the plasma membrane, and insertion of transport proteins into the plasma membrane. Each of these mechanisms has been demonstrated in transporting epithelia and, in many cases, this regulation is modulated by actin and actin-binding proteins. Thus, this chapter will explore these three mechanisms by which the actin cytoskeleton participates in the regulation of ion transport. While many transport proteins including $Na^+K^+2Cl^-$ cotransporter NKCC1 and the Na^+K^+ ATPase have been shown to be regulated by microfilaments (Table I) (Matthews *et al.*, 1997), actin-dependent regulation of the apical Na^+H^+ antiporter isoform 3, NHE3, has been the subject of intense investigation. Moreover, NHE3 has been shown to be regulated by each of the three mechanisms to be discussed via a wide variety of stimuli, including hormones, inflammatory mediators, and nutrient transport (Table II). The important physiological role of NHE3, a critical antiporter that absorbs luminal Na^+ in exchange for cytoplasmic H^+, is exemplified by NHE3 knockout mice that have chronic diarrhea, as a result of intestinal malabsorption, and mild acidosis, secondary to absence of renal tubular NHE3 (Schultheis *et al.*, 1998). Thus, while many transporters are regulated by microfilaments NHE3 will, therefore, be used as the primary example throughout this chapter.

Table I
Ion Transporter Regulation by Microfilaments

Transporter	Mechanism	Effect on activity	References
CFTR	cAMP-mediated protein kinase A signaling (increased surface activity)	Increase	Prat *et al.*, 1995; Cantiello, 1996; Naren *et al.*, 2003
H^+K^+ ATPase	Ezrin-dependent translocation (exocytosis)	Increase	Mangeat *et al.*, 1990; Agnew *et al.*, 1999; Zhou *et al.*, 2003
Na^+K^+ ATPase	PKC-induced endocytosis	Decrease	Matthews *et al.*, 1993; Suzuki *et al.*, 2001
NKCC1	Internalization by PKC signaling	Decrease	Liedtke *et al.*, 2003; Gimenez and Forbush, 2005

Table II
Mechanisms of NHE3 Transporter Regulation

Stimulus	Effect on NHE3 activity	Mechanism	References
Adenosine	Decrease	Phospholipase C-mediated PKC signaling	Di Sole *et al.*, 2004
Aldosterone	Increase	Increased expression	Cho *et al.*, 1998
Angiotensin II	Increase	c-Src	Tsuganezawa *et al.*, 2002
Caffeine	Decrease	Decreased expression	Lee *et al.*, 2002
Carbachol	Decrease	PKCα signaling	Lee-Kwon *et al.*, 2003b
COX	Decrease	Decreased expression	Norregaard *et al.*, 2005
Epidermal growth factor	Increase	PI3K signaling	Khurana *et al.*, 1996
EPEC	Decrease	Mediated via type III secretion	Hecht *et al.*, 2004
Forskolin	Decrease	cAMP-mediated PKA signaling	Cabado *et al.*, 1996
Glucocorticoids	Increase	Increased transcription, SGK1-mediated phosphorylation of NHE3	Kandasamy *et al.*, 1995; Yun *et al.*, 2002
IFNγ	Decrease	Decreased expression	Rocha *et al.*, 2001
Insulin	Increase	Unknown, but requires glucocorticoids	Klisic *et al.*, 2002
Na^+-glucose transport	Increase	Ezrin-mediated translocation	Zhao *et al.*, 2004
Nitric oxide	Increase	Increased expression; cGMP-mediated protein kinase G phosphorylation of NHE3	Gill *et al.*, 2002; Turban *et al.*, 2003
Parathyroid hormone	Decrease	cAMP-mediated PKA signaling; PKC signaling	Azarani *et al.*, 1995
Serotonin	Decrease	Calcium-mediated PKC signaling	Gill *et al.*, 2005

II. ROLE OF ACTIN IN REGULATION OF TRANSPORTERS AT THE PLASMA MEMBRANE

A. NHE3 Regulation at the Cell Surface Requires Interactions with Actin and Other Cytoskeletal and Signaling Proteins

NHE3 can be found within the brush border of the apical plasma membrane in detergent-soluble membranes and in two detergent-insoluble forms: within cholesterol-rich detergent-resistant membranes and in association with actin microfilaments (Li *et al.*, 2001). As the presence of NHE3 within cholesterol-rich membranes requires the presence of intact microfilaments, both pools of detergent-insoluble NHE3 depend on actin structure (Li *et al.*, 2001). The partitioning of NHE3 into these two detergent-insoluble plasma membrane pools, particularly the cholesterol-rich membranes, may be one mechanism of NHE3 regulation. For example, epidermal growth factor (EGF)-induced increases in NHE3 activity are associated with increased partitioning of NHE3 into detergent-insoluble cholesterol-rich membranes (Li *et al.*, 2001). In contrast, actin-depolymerizing drugs cause marked decreases in NHE3 activity, with NHE3 preferentially localized to apical actin aggregates (Kurashima *et al.*, 1999). In the renal tubule, luminal flow increases NHE3 activity within microvillous membranes by mechanisms that are also sensitive to actin-depolymerizing drugs (Du *et al.*, 2006). Thus, data from many experimental systems demonstrate a critical role for actin microfilaments in regulation of NHE3 activity at the plasma membrane.

NHE3 is thought to span the membrane 10–12 times, giving it ample potential sites to associate with membrane subdomains and actin filaments (Zizak *et al.*, 2000). The cytoplasmic C-terminus includes regions that bind to many intracellular mediators including NHE regulatory factor (NHERF)-1, NHERF-2, PDZK1, ezrin, Hsc70, megalin, calmodulin kinase II, and protein phosphatase 2A (Fig. 1). In addition, NHERF proteins can interact with ezrin, potentially forming a bridge between NHE3 and ezrin (Reczek and Bretscher, 1998; Yun *et al.*, 1998). Ezrin, in turn, interacts directly with microfilaments (Algrain *et al.*, 1993). The roles of each of these complex binding interactions remain to be established, but it is clear that specific regions of the cytoplasmic tail play distinct roles in NHE3 regulation (Levine *et al.*, 1995). These cytoplasmic tail domains also contribute to the formation of large NHE3-containing complexes at the apical plasma membrane (Li *et al.*, 2004). For example, a correlation has been reported between acute NHE3 downregulation and increased complex size in rabbit ileal brush border membranes (Li *et al.*, 2004). The larger complexes contain NHERF-2, α-actinin-4, and activated protein kinase C, suggesting the assembly of an inhibitory signaling aggregate.

NHE3 regulation was initially thought to occur primarily by changes in the maximal rate of Na^+H^+ exchange (Vmax). This could represent complete activation or inactivation of NHE3 at the surface or changes in NHE3 copy number at the surface. For example, as noted earlier, pharmacological actin depolymerization markedly inhibits NHE3 activity (Kurashima *et al.*, 1999). This effect of actin depolymerization requires the presence of a portion of the NHE3 C-terminal cytoplasmic domain

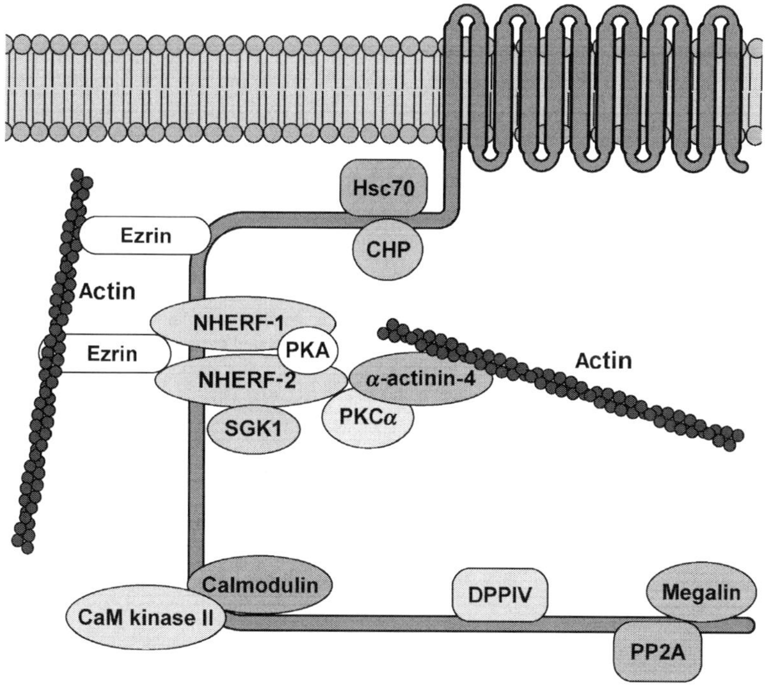

Figure 1. NHE3 protein–protein interactions. NHE3 is composed of 12 transmembrane domains, which are responsible for ion transport, and a cytoplasmic C-terminal tail. This C-terminal tail interacts with many proteins, which help to regulate NHE3 activity, including protein kinase A (PKA), protein kinase C α (PKCα), calmodulin-dependent kinase II (CaM kinase II), adaptor proteins (NHERF-1 and NHERF-2), and cytoskeletal proteins ezrin and α-actinin-4.

(Kurashima *et al.*, 1999). This region overlaps with regions required for interactions with NHERF, raising the possibility that NHE3–NHERF interactions may mediate this inhibition.

To assess the dynamic nature of interactions between NHE3 and cytoskeletal structures, an NHE3-enhanced green fluorescent protein (EGFP) fusion protein was developed (Janecki *et al.*, 2000b). Mobility of surface NHE3-EGFP was then assessed through fluorescence recovery after photobleaching (FRAP) studies (Cha *et al.*, 2004). The data showed that approximately 50% of NHE3-EGFP is mobile at the apical membrane (Cha *et al.*, 2004). This mobile fraction was markedly reduced when NHERF-2 was overexpressed, suggesting that NHERF-2 contributes to NHE3 tethering. Consistent with this, an NHE3-EGFP mutant lacking the NHERF-binding domain but retaining the ezrin-binding domain within the C-terminal cytoplasmic tail demonstrated an increased mobile fraction (Cha *et al.*, 2004). One plausible model for these data might be that the NHERF-2 tethers NHE3 to cortical actin. However, the mobile fraction of NHE3-EGFP was reduced by actin disruption (Cha *et al.*, 2004). These data therefore suggest that NHE3 mobility at the apical membrane requires an intact actin cytoskeleton and is limited by NHERF-2 (Cha *et al.*, 2004). The role of ezrin in NHE3 mobility remains to be explored.

B. Actomyosin Contraction Can Both Regulate and Be Regulated by NHE3

As described earlier, global actin depolymerization can dramatically inhibit NHE3 activity. However, such actin depolymerization represents a model pharmacological event that is not seen in response to physiological stimuli. To determine if physiological regulators of actin polymerization and actomyosin contraction, for example, small GTPases, could regulate NHE3 activity fibroblast-like CHO cells lacking $Na^{+}H^{+}$ exchange activity were studied. These were stably transfected with NHE3 and then transiently transfected with dominant-negative mutants of Rho, Rac, and Cdc42. While neither dominant-negative Rac1 nor Cdc42 affected NHE3 activity, dominant-negative RhoA markedly reduced NHE3 activity (Szaszi *et al.*, 2000). This NHE3 inhibition was not due to changes in NHE3 distribution and could be reproduced by toxin-mediated Rho inactivation (Hayashi *et al.*, 2004) and either pharmacological or genetic inhibition of Rho-associated kinase (ROCK) (Szaszi *et al.*, 2000). This ROCK inhibition was associated with reduced myosin II regulatory light chain phosphorylation. Both myosin light chain dephosphorylation and NHE3 inhibition could be reproduced by a compound that inhibits myosin light chain kinase, protein kinase A, and protein kinase C, suggesting that myosin light chain dephosphorylation can inhibit NHE3 activity (Szaszi *et al.*, 2000). Conversely, studies of cultured intestinal epithelial monolayers have shown that NHE3 inhibition can reduce myosin light chain phosphorylation (Turner *et al.*, 2000). Thus, it may be that a feedback loop exists to regulate myosin light chain phosphorylation, cortical actomyosin structure, and NHE3 activity.

C. Regulation of NHE3 Activity by Protein Kinases

Numerous protein kinases have been implicated in NHE3 regulation. Perhaps the most well characterized of these is the NHE3 inhibition that follows activation of protein kinase A (Azarani *et al.*, 1995; Moe *et al.*, 1995). This NHE3 inhibition has been reported to be independent of NHE3 removal from the plasma membrane (Moe *et al.*, 1995) but does require an intact C-terminal cytoplasmic domain (Moe *et al.*, 1995; Yun *et al.*, 1995). The initial observation that protein kinase A-dependent NHE3 inhibition was not observed in all cell types expressing NHE3 suggested a requirement for an accessory regulatory protein (Yun *et al.*, 1997). The search for this protein led to the discovery of NHERF-2, also known as NHE3 kinase A regulatory protein (E3KARP) (Yun *et al.*, 1997). Consistent with NHERF-1 and NHERF-2 representing these accessory proteins, NHE3 inhibition by protein kinase A activation occurred in the presence, but not the absence, of NHERF-1 or NHERF-2 (Yun *et al.*, 1997; Zizak *et al.*, 1999). Subsequent work showed that intracellular cAMP elevation could induce NHE3 phosphorylation in the presence, but not the absence, of NHERF-1 or NHERF-2, suggesting that NHERF proteins are scaffolds that facilitate protein kinase A-mediated NHE3 phosphorylation (Zizak *et al.*, 1999). Thus, it may be that protein kinase A-mediated NHERF-dependent NHE3 phosphorylation event can trigger NHE3 inhibition (Zhao *et al.*, 1999). When considered the observations that NHERF-1 and NHERF-2 bind ezrin, that ezrin binds microfilaments and that microfilament stabilization prevents protein kinase A-mediated NHE3 inhibition (Szaszi *et al.*, 2001), one might propose that a physical association between NHE3 and microfilaments may be necessary for protein kinase A activation to trigger NHE3 phosphorylation and inhibition.

In addition to this inhibition at the plasma membrane, one intriguing report has suggested protein kinase A may cause redistribution of NHE3 from microvilli to intermicrovillous membrane domains, both at the cell surface (Yang *et al.*, 2004). Although further investigation is necessary, this study showed that parathyroid hormone, which acts primarily through adenylate cyclase, both inhibits and redistributes NHE3 at the plasma membrane without evidence of endocytosis. This suggests that NHE3 may be inactive when sequestered within certain membrane domains.

In addition to modifying NHE3 activity at the plasma membrane and, perhaps, targeting NHE3 to surface domains where it is inactive, some data suggest that protein kinase A may also trigger NHE3 endocytosis (Zhang *et al.*, 1999; Kocinsky *et al.*, 2005). The most striking example of this comes from analyses of NHE3 localization following dopamine treatment (Kocinsky *et al.*, 2005). Dopamine inhibits NHE3 by a protein kinase A-dependent mechanism that includes phosphorylation of the NHE3 cytoplasmic tail. When antisera specific for particular NHE3 phosphorylation sites were used to track NHE3 after dopamine treatment, the phosphorylated NHE3 was specifically localized to clathrin-coated pits (Kocinsky *et al.*, 2005). Like the effects of parathyroid hormone discussed earlier, this may simply represent NHE3 trafficking to a plasma membrane domain where it is inactive. However, this targeting to coated pits may also represent the first step in NHE3 endocytosis. NHE3 endocytosis has also been reported to be induced by protein kinase A activation in renal tubular brush border membranes (Weinman *et al.*, 2003). In this case, endocytosis does not occur in brush border membranes obtained from NHERF-1 knockout mice, despite the presence of abundant NHERF-2 in these cells. Thus, although NHERF-1 and NHERF-2 are homologous, they may serve nonidentical roles in NHE3 regulation. For example, although protein kinase G-mediated inhibition of NHE3 is less well described than protein kinase A-mediated NHE3 inhibition, it also appears to require NHERF-2 (Cha *et al.*, 2005). However, in this case, NHERF-1 is insufficient, perhaps because unlike NHERF-2, NHERF-1 cannot binds to cGMP-dependent protein kinase II (Cha *et al.*, 2005). This, protein kinase A, and potentially protein kinase G, may inhibit NHE3 by as many as three distinct mechanisms: direct phosphorylation at the plasma membrane, trafficking to plasma membrane domains that limit activity, and endocytic removal from the plasma membrane.

D. Regulation of CFTR by Protein Kinase A

In contrast to protein kinase A-mediated NHE3 inhibition, CFTR, the cystic fibrosis transmembrane conductance regulator, is stimulated by cAMP and protein kinase A activation (Anderson *et al.*, 1991). While early work suggested that this might solely reflect binding of cAMP to nucleotide-binding domains within CFTR, subsequent work suggested that microfilaments were also necessary for this regulation (Prat *et al.*, 1995, 1999; Cantiello, 1996). For example, although incomplete microfilament depolymerization with cytochalasin D initially opened CFTR, extended cytochalasin D treatment prevented CFTR activation by cAMP analogues in whole cells or direct protein kinase A addition in excised membrane patches (Prat *et al.*, 1995). This was not due to some other effect of cytochalasin D, as CFTR activation by direct protein kinase A addition to excised membrane patches was restored by addition of exogenous actin filaments (Prat *et al.*, 1995; Cantiello, 1996). This effect is not due to CFTR insertion into the plasma membrane (Moyer *et al.*, 1998). Thus, CFTR at the surface is regulated by cAMP via a microfilament-dependent mechanism. CFTR can, in turn, also regulate ENaC, the epithelial sodium channel, via a mechanism that is enhanced by microfilaments (Ismailov *et al.*, 1997).

III. MECHANISMS OF ION TRANSPORTER REMOVAL (ENDOCYTOSIS) FROM THE PLASMA MEMBRANE

A. Protein Kinase C Inhibits NHE3 by Inducing Endocytosis

Early studies of acute NHE3 regulation noted that phorbol ester treatment, which increases in intracellular Ca^{2+} and activates protein kinase C, markedly inhibited NHE3 activity (Levine *et al.*, 1993, 1995; Tse *et al.*, 1993; Azarani *et al.*, 1995; Kandasamy *et al.*, 1995; Bookstein *et al.*, 1999). While some data conflict, most agree that this inhibition is due to reduced Vmax with no change in K_{Na} (Levine *et al.*, 1993, 1995; Tse *et al.*, 1993; Azarani *et al.*, 1995; Kandasamy *et al.*, 1995; Bookstein *et al.*, 1999). This suggests either complete inactivation of a subset of plasma membrane NHE3, for example, by phosphorylation or sequestration, or NHE3 removal from the plasma membrane. While most studies of protein kinase C-dependent NHE3 inhibition have relied on the nonphysiological stimuli provided by phorbol ester or calcium ionophore (Lee-Kwon *et al.*, 2003b), this does appear to be a physiologically relevant mechanism of NHE3 regulation. For example, the NHE3 inhibition induced by serotonin requires protein kinase C α activation (Gill *et al.*, 2005).

In contrast to protein kinase A, protein kinase C does not appear to inhibit NHE3 by direct phosphorylation; detailed analyses have failed to identify any phorbol ester-induced NHE3 phosphorylation (Yip *et al.*, 1997). Protein kinase C α may phosphorylate NHERF-1 and NHERF-2, but the relevance of this phosphorylation to NHE3 endocytosis is unknown. This has led some to speculate that the critical targets of protein kinase C α may be components of the endocytic machinery. This remains to be determined.

Calcium ionophore activates protein kinase C α and causes it to bind to NHERF-2 (Lee-Kwon *et al.*, 2003b). This appears to trigger assembly of a complex that includes NHE3, NHERF-2, α-actinin-4, and activated protein kinase C α (Kim *et al.*, 2002). Thus, like protein kinase A-dependent inhibition, protein kinase C-dependent NHE3 inhibition is associated with protein complex assembly (Fig. 2). Some data suggest that assembly of this complex also requires Src family kinases, although specific kinase targets have not been defined (Li *et al.*, 2004b). In any case, complex assembly may be a critical event that triggers protein kinase C α-dependent NHE3 endocytosis. This process requires NHERF-2 (Kim *et al.*, 2002), which cannot be replaced by NHERF-1, perhaps because NHERF-1 does not bind to α-actinin-4 (Kim *et al.*, 2002).

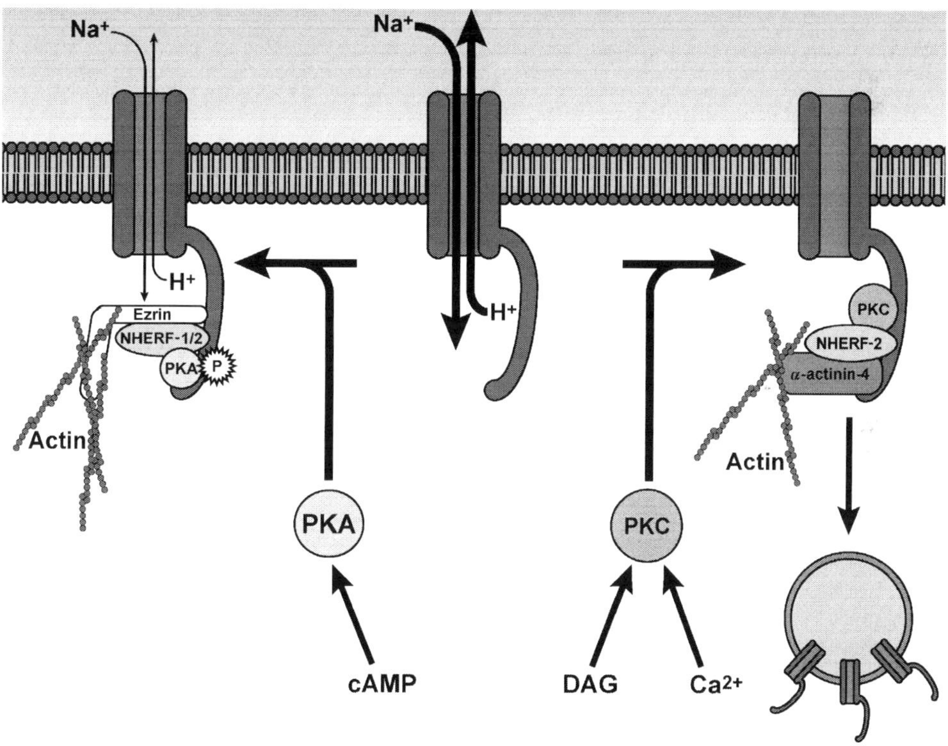

Figure 2. General mechanisms of acute NHE3 downregulation. NHE3 activity may be inhibited acutely via two distinct pathways. First, increases in intracellular cAMP lead to protein kinase A (PKA) activation and association with NHERF-1 or NHERF-2, ezrin, and NHE3. PKA phosphorylates the C-terminal tail of NHE3, which is associated with NHE3 downregulation at the apical membrane. Alternatively, diacylglycerol (DAG) or Ca^{2+} may activate protein kinase C (PKC), leading to the assembly of large complexes containing NHE3, protein kinase C, NHERF-2, and α-actinin-4 that are linked to the actin cytoskeleton. Assembly of this complex is associated with NHE3 endocytosis.

Thus, like protein kinase A-induced NHE3 internalization, protein kinase C-induced NHE3 internalization requires NHERF proteins. However, in the case of protein kinase A, NHERF-1 is uniquely required (Weinman *et al.*, 2003). Thus, NHERF-1 and NHERF-2 have distinct roles in regulating NHE3 endocytosis in response to protein kinase A and protein kinase C α. Participation of α-actinin-4 in protein kinase-C-induced complex assembly is also critical, as dominant-negative α-actinin-4 prevents Ca^{2+}- and protein kinase C α-dependent endocytosis-mediated NHE3 inhibition (Kim *et al.*, 2002).

B. NHE3 Internalization Occurs Through Clathrin-Coated Pits

Like other plasma membrane proteins, NHE3 could potentially be internalized by clathrin-mediated endocytosis, caveolae-mediated endocytosis, or macropinocytosis. Numerous approaches have been used to identify the mechanism(s) of NHE3 internalization in response to either protein kinase A or protein kinase C activation. In general, the results indicate that NHE3 is primarily internalized via clathrin-coated vesicles (Janecki *et al.*, 1998; Chow *et al.*, 1999; Hu *et al.*, 2001; Gekle *et al.*, 2002; Kocinsky *et al.*, 2005). This stimulated NHE3 internalization may reflect an increase in the rate of the basal NHE3 endocytosis (D'souza *et al.*, 1998; Chow *et al.*, 1999), which also occurs through clathrin-coated pits (Chow *et al.*, 1999). Basal NHE3 endocytosis may serve purposes beyond simple removal from the plasma membrane. For example, some data suggest that NHE3 is active when localized within recycling endosomes and may even contribute to the acidification of some endosomes subsets (D'souza *et al.*, 1998).

C. Na^+K^+ ATPase Can Be Regulated by a Clathrin-Mediated Endocytosis

Like NHE3, Na^+K^+ ATPase function can also be regulated by microfilaments. This may be due to direct association of the ATPase with actin filaments (Cantiello, 1995; Ebner *et al.*, 2005) but may also reflect microfilament-dependent regulation of Na^+K^+ ATPase expression at the plasma membrane (Matthews *et al.*, 1993; Chibalin *et al.*, 1997; Suzuki *et al.*, 2001; Beltowski *et al.*, 2004). For example, dopamine-dependent protein kinase C activation has been shown to cause clathrin-mediated Na^+K^+ ATPase endocytosis (Chibalin *et al.*, 1997; Beltowski *et al.*, 2004).

IV. MECHANISMS OF ION TRANSPORTER DELIVERY TO THE CELL SURFACE (EXOCYTOSIS)

Like NHE3 inhibition, NHE3 activation has also been tied to membrane trafficking. In this case, the mechanism of activation involves exocytic insertion of NHE3 into the plasma membrane, resulting in an increase in Vmax (Levine *et al.*, 1993).

The stimuli that can lead to exocytic insertion of NHE3 are diverse and include growth factors, glucocorticoids, hormones, cytoplasmic acidification, and Na^+-glucose cotransport (Peng *et al.*, 1999; Turner and Black, 2001; Klisic *et al.*, 2002; Yun *et al.*, 2002; du Cheyron *et al.*, 2003; Zhao *et al.*, 2004). While the regulatory mechanisms that direct this NHE3 exocytosis are not well defined, several important clues have been collected. First, a number of kinase pathways, including p38 MAP kinase (Turner and Black, 2001; Zhao *et al.*, 2004), phosphatidylinositol 3-kinase (Khurana *et al.*, 1996; Janecki *et al.*, 2000a), and serum- and glucocorticoid-inducible kinase-1 (SGK1) (Yun *et al.*, 2002) are involved in NHE3 exocytosis by various stimuli. Second, NHE3 insertion into the plasma membrane requires an intact actin cytoskeleton (du Cheyron *et al.*, 2003). Interactions between NHE3 and actin may also be required for NHE3 retention at the plasma membrane (Alexander *et al.*, 2005). Finally, NHE3 can interact with actin through NHERF proteins, which bind to both NHE3 and the actin-binding protein ezrin, or by direct NHE3 binding to ezrin (Yun *et al.*, 1998; Kurashima *et al.*, 1999).

A. Ezrin Regulates NHE3 Translocation from Endosomal Pools to the Plasma Membrane

To better understand the role of ezrin in NHE3 activation, we used the model of NHE3 activation following initiation of Na^+-glucose cotransport (Turner and Black, 2001). This model stimulus is physiologically relevant as, in the small intestine, it may allow the Na^+-glucose cotransporter, SGLT1, to "sense" luminal nutrients and, in turn, activate NHE3-mediated Na^+ absorption (Turner and Black, 2001). Given the harsh luminal environment, one possibility is that this allows NHE3 to be sequestered in a protected endosomal pool and only translocated to the plasma membrane once luminal nutrients are present (Zhao *et al.*, 2004). This SGLT1-dependent NHE3 translocation can easily be detected as an increase in cytoplasmic pH that is a direct result of increased NHE3 activity and does not require carbohydrate metabolism (Fig. 3) (Turner and Black, 2001; Zhao *et al.*, 2004).

To determine if ezrin–NHE3 interactions, either direct or indirect, could be involved in NHE3 translocation, we first assessed ezrin activation after initiation of Na^+-glucose cotransport. Ezrin was redistributed to the detergent-insoluble cytoskeletal fraction with kinetics similar to the cytoplasmic alkalinization that followed initiation of Na^+-glucose cotransport (Zhao *et al.*, 2004). Ezrin association with F-actin is typically associated with ezrin phosphorylation at Thr-567. This allows inactive soluble ezrin to unfold and link target molecules, such as NHE3 and NHERF, to the actin cytoskeleton. We therefore assessed ezrin phosphorylation and found that, like cytoskeletal association, phosphorylation at Thr-567 increased with kinetics similar to the cytoplasmic alkalinization (Zhao *et al.*, 2004).

To further explore the role of ezrin in NHE3 translocation, we developed intestinal epithelial cell lines expressing dominant-negative ezrin (Zhao *et al.*, 2004). These cells formed polarized monolayers and were capable of basal SGLT1-mediated

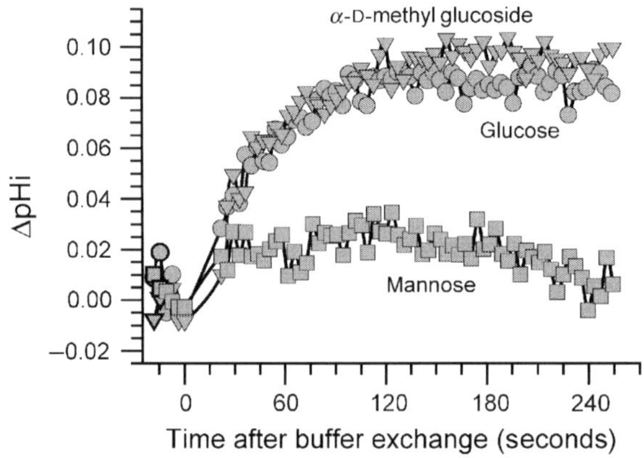

Figure 3. Cytoplasmic alkalinization requires SGLT1-mediated Na^+-glucose cotransport. Caco-2 cells were incubated in medium containing 25 mM mannose and 0.5 mM phloridzin. Exchange of mannose-phloridzin medium for medium containing 25 mM glucose (circles) resulted in a rapid and sustained rise in pHi. Exchange into medium with 25 mM α-D-methyl glucoside, a nonmetabolizable glucose analogue, caused a similar rapid and sustained rise in pHi (inverted triangles). In contrast, medium containing 25 mM mannose, resulted in only a small transient pH change (squares). Reprinted with permission from the American Journal of Physiology (Turner and Black, 2001).

Na^+-glucose cotransport and NHE3-mediated Na^+H^+ exchange that were comparable to control cells. However, unlike cells expressing only endogenous ezrin or those transfected to overexpress wild-type ezrin, those expressing dominant-negative ezrin failed to alkalinize after initiation of Na^+-glucose cotransport (Zhao *et al.*, 2004). Translocation of NHE3 to the apical membrane in response to Na^+-glucose cotransport was also blocked by dominant-negative ezrin, suggesting that functional ezrin is required for NHE3 recruitment to the apical membrane. This suggests that ezrin activation is a critical regulatory step controlling NHE3 trafficking flux from endosomal pools to the plasma membrane (Fig. 4).

B. Ezrin Activation Represents a Point of Convergence in NHE3 Regulation

The studies discussed above suggest a critical role for phosphorylation-dependent ezrin activation in NHE3 translocation but do not provide insight into the regulation of ezrin. Several kinases have been implicated in ezrin phosphorylation at Thr-567, including protein kinase C α and Rho kinase (Matsui *et al.*, 1998, 1999; Ng *et al.*, 2001). However, as discussed earlier, protein kinase C α activation causes NHE3 endocytosis and would, therefore, not be expected to be involved in NHE3 translocation to the surface. Moreover, Rho kinase inhibition has no effect on the cytoplasmic alkalinization induced followed initiation of Na^+-glucose cotransport (Turner and Black, 2001). We therefore used an *in silico* approach to identify potential kinases that

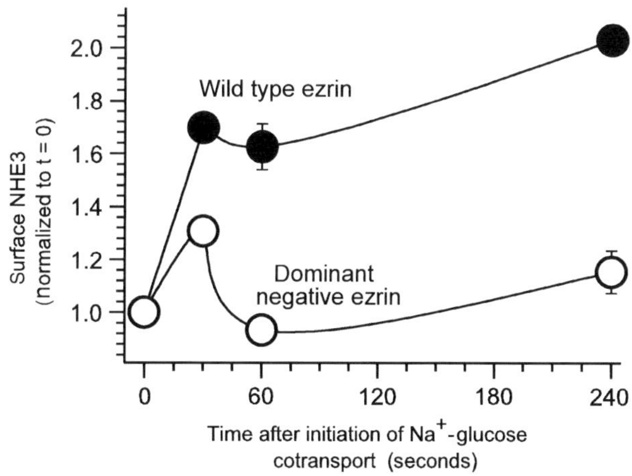

Figure 4. Na^+-glucose cotransport activates ezrin-dependent translocation of NHE3 to the apical membrane. Initiation of Na^+-glucose cotransport markedly increased the quantity of NHE3 detectable at the apical surface of Caco-2 cells expressing wild-type (closed circles) but not dominant-negative (open circles) ezrin. Reprinted with permission from Proceedings of the National Academy of Science, USA (Zhao *et al.*, 2004).

could phosphorylate ezrin at Thr-567 (Shiue *et al.*, 2005). This identified protein kinase A and Akt as candidate ezrin Thr-567 kinases. We excluded protein kinase A from consideration as it inhibits, rather than activates, NHE3. In contrast, Akt seemed plausible, as it had been previously associated with NHE3 translocation (Lee-Kwon *et al.*, 2001; Li *et al.*, 2004a).

We found that Akt could phosphorylate ezrin at Thr-567 *in vitro* (Shiue *et al.*, 2005). Moreover, Akt was activated followed initiation of Na^+-glucose cotransport with kinetics similar to ezrin phosphorylation. Finally, either pharmacological Akt inhibition or small interfering RNA (siRNA) knockdown of Akt expression blocked ezrin phosphorylation, NHE3 translocation, and cytoplasmic alkalinization after initiation of Na^+-glucose cotransport (Shiue *et al.*, 2005). We therefore conclude that ezrin phosphorylation and the NHE3 translocation and activation that follow are triggered by Akt (Fig. 5).

The observation that ezrin can be phosphorylated by Akt has profound implications given the widespread involvement of ezrin family members in diverse cellular functions. It is also striking that this observation provides an explanation that unifies what were previously thought to be unrelated means of triggering NHE3 translocation. For example, EGF can cause acute NHE3 translocation to the plasma membrane through activation of phosphatidylinositol 3-kinase (Khurana *et al.*, 1996). In turn, phosphatidylinositol 3-kinase can activate Akt, and this Akt activation appears to be necessary for NHE3 translocation (Lee-Kwon *et al.*, 2001). The targets of Akt in this pathway have not been defined, but it now seems likely that ezrin is among them. Lysophosphatidic acid may also use this pathway, as phosphatidylinositol 3-kinase

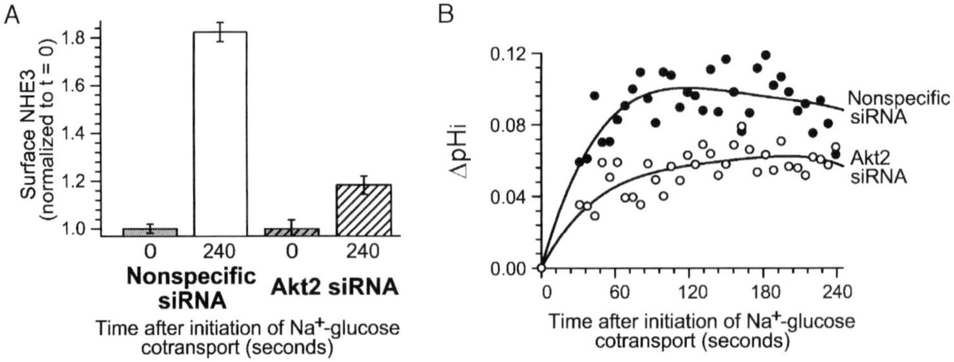

Figure 5. Akt2 expression is required for NHE3 translocation and cytoplasmic alkalinization after Na^+-glucose cotransport. (A) SiRNA-mediated knockdown of Akt2 expression prevents NHE3 translocation to the plasma membrane after initiation of Na^+-glucose cotransport. Akt2 knockdown prevented 78 ± 3% of the increase in surface NHE3 expression detected in monolayers transfected with nonspecific siRNA. (B) NHE3-dependent cytoplasmic alkalization is inhibited by Akt2 knockdown. Akt2 knockdown (white circles) inhibited 41 ± 3% of NHE3-dependent cytoplasmic alkalinization after initiation of Na^+-glucose cotransport, relative to monolayers transfected with nonspecific siRNA (black circles), paralleling the 55 ± 9% knockdown of Akt2 expression in these monolayers. Reprinted with permission from the Journal of Biological Chemistry (Shiue *et al.*, 2005).

activity is necessary for lysophosphatidic acid-induced NHE3 translocation (Lee-Kwon *et al.*, 2003a). It is also noteworthy that the SGK1, which mediates NHE3 translocation to the surface following glucocorticoid stimulation, is related to Akt (Yun *et al.*, 2002). This may therefore represent yet another stimulus that results in NHE3 translocation by an ezrin-dependent process. Thus, ezrin phosphorylation may represent a point where multiple signal transduction pathways converge, all leading to NHE3 translocation. No data are available to suggest a role for NHERF proteins in acute NHE3 translocation to the surface. Thus, it may be that, in contrast to endocytic pathways, NHE3 exocytosis is mediated by direct NHE3–ezrin interactions (Fig. 6). This hypothesis remains to be tested experimentally.

C. Ezrin Regulates H^+K^+ ATPase Translocation

Like the translocation of NHE3, the gastric parietal cell H^+K^+ ATPase, or proton pump, is exocytically inserted into and endocytically retrieved from the plasma membrane in response to stimuli. For example, histamine stimulation provokes dramatic cortical actin reorganization (Rosenfeld *et al.*, 1981; Forte *et al.*, 1998). This actin reorganization is functionally related to histamine-induced proton pump translocation to the canalicular surface, as actin disruption with cytochalasin D induced pump translocation and increased H^+K^+ ATPase activity (Forte *et al.*, 1998). Moreover, as shown for NHE3 (Zhao *et al.*, 2004), the cytoskeletal linker protein ezrin plays a

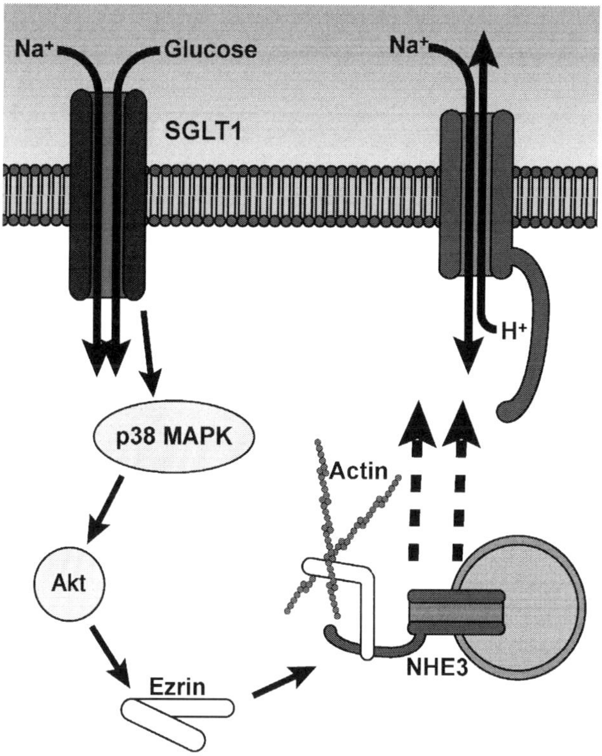

Figure 6. Na^+-glucose cotransport mediated upregulation of NHE3 activity. Increased luminal glucose and Na^+ triggers SGLT1-mediated absorption. This activates a signaling cascade wherein p38 MAPK activates Akt2, which phosphorylates ezrin. This activates ezrin, allowing it to link endosomal NHE3 pools to F-actin and cause exocytic insertion of NHE3 into the apical membrane. This increases Na^+H^+ exchange by increasing Vmax.

critical role in this process (Mangeat *et al.*, 1990; Ingraffea *et al.*, 2002; Zhou *et al.*, 2003; Tamura *et al.*, 2005). Thus, translocation-dependent activation of the proton pump may occur by mechanisms similar to translocation-dependent NHE3 activation. There is also some data to suggest that CFTR can be regulated by interactions with ezrin (Naren *et al.*, 2003).

V. CONCLUSIONS

Microfilaments play many critical roles in cellular function. They regulate cell shape, migration, and membrane organization. In transporting epithelia, microfilaments also regulate ion transport proteins. In the case of NHE3, this can occur by at least three distinct mechanisms. First, microfilaments can play a role in regulating the localization and activity of NHE3 within the plasma membrane. This can alter both the Km and Vmax of Na^+H^+ exchange. The second and third mechanisms of cytoskeletally mediated NHE3 regulation involve trafficking between the surface and intracellular compartments. This can result in inhibition or stimulation of Na^+H^+ exchange, both by changes in Vmax, due to NHE3 endocytosis or exocytosis, respectively. Studies have shed much light on the signaling events that regulate NHE3 trafficking allowing previously separate observations to be mechanistically unified.

ACKNOWLEDGMENTS

This work was supported by grants from the NIH (R01DK61931 and R01DK68271) and the Crohn's and Colitis Foundation of America to J.R.T. D.R.C. is a predoctoral fellow of the NIH (The University of Chicago Medical Scientist Training Program T32 GM07281).

REFERENCES

Agnew, B. J., Duman, J. G., Watson, C. L., Coling, D. E., and Forte, J. G. (1999). Cytological transformations associated with parietal cell stimulation: Critical steps in the activation cascade. *J. Cell. Sci.* **112** (Pt. 16), 2639–2646.

Alexander, R. T., Furuya, W., Szaszi, K., Orlowski, J., and Grinstein, S. (2005). Rho GTPases dictate the mobility of the Na/H exchanger NHE3 in epithelia: Role in apical retention and targeting. *Proc. Natl. Acad. Sci. USA* **102,** 12253–12258.

Algrain, M., Turunen, O., Vaheri, A., Louvard, D., and Arpin, M. (1993). Ezrin contains cytoskeleton and membrane binding domains accounting for its proposed role as a membrane-cytoskeletal linker. *J. Cell Biol.* **120,** 129–139.

Anderson, M. P., Berger, H. A., Rich, D. P., Gregory, R. J., Smith, A. E., and Welsh, M. J. (1991). Nucleoside triphosphates are required to open the CFTR chloride channel. *Cell* **67,** 775–784.

Azarani, A., Goltzman, D., and Orlowski, J. (1995). Parathyroid hormone and parathyroid hormone-related peptide inhibit the apical Na^+/H^+ exchanger NHE-3 isoform in renal cells (OK) via a dual signaling cascade involving protein kinase A and C. *J. Biol. Chem.* **270,** 20004–20010.

Beltowski, J., Marciniak, A., Jamroz-Wisniewska, A., Borkowska, E., and Wojcicka, G. (2004). Bidirectional regulation of renal cortical Na^+,K^+-ATPase by protein kinase C. *Acta Biochim. Pol.* **51,** 757–772.

Bookstein, C., Musch, M. W., Xie, Y., Rao, M. C., and Chang, E. B. (1999). Regulation of intestinal epithelial brush border Na^+/H^+ exchanger isoforms, NHE2 and NHE3, in c2bbe cells. *J. Membr. Biol.* **171,** 87–95.

Cabado, A. G., Yu, F. H., Kapus, A., Lukacs, G., Grinstein, S., and Orlowski, J. (1996). Distinct structural domains confer cAMP sensitivity and ATP dependence to the Na^+/H^+ exchanger NHE3 isoform. *J. Biol. Chem.* **271,** 3590–3599.

Cantiello, H. F. (1995). Actin filaments stimulate the $Na(^+)$-$K(^+)$-ATPase. *Am. J. Physiol.* **269,** F637–F643.

Cantiello, H. F. (1996). Role of the actin cytoskeleton in the regulation of the cystic fibrosis transmembrane conductance regulator. *Exp. Physiol.* **81,** 505–514.

Cha, B., Kenworthy, A., Murtazina, R., and Donowitz, M. (2004). The lateral mobility of NHE3 on the apical membrane of renal epithelial OK cells is limited by the PDZ domain proteins NHERF1/2, but is dependent on an intact actin cytoskeleton as determined by FRAP. *J. Cell Sci.* **117,** 3353–3365.

Cha, B., Kim, J. H., Hut, H., Hogema, B. M., Nadarja, J., Zizak, M., Cavet, M., Lee-Kwon, W., Lohmann, S. M., Smolenski, A., Tse, C. M., Yun, C., *et al.* (2005). cGMP inhibition of Na^+/H^+ antiporter 3 (NHE3) requires PDZ domain adapter NHERF2, a broad specificity protein kinase G-anchoring protein. *J. Biol. Chem.* **280,** 16642–16650.

Chibalin, A. V., Katz, A. I., Berggren, P. O., and Bertorello, A. M. (1997). Receptor-mediated inhibition of renal $Na(^+)$-$K(^+)$-ATPase is associated with endocytosis of its alpha- and beta-subunits. *Am. J. Physiol.* **273,** C1458–C1465.

Cho, J. H., Musch, M. W., Bookstein, C. M., McSwine, R. L., Rabenau, K., and Chang, E. B. (1998). Aldosterone stimulates intestinal Na^+ absorption in rats by increasing NHE3 expression of the proximal colon. *Am. J. Physiol.* **274,** C586–C594.

Chow, C. W., Khurana, S., Woodside, M., Grinstein, S., and Orlowski, J. (1999). The epithelial Na^+/H^+ exchanger, NHE3, is internalized through a clathrin-mediated pathway. *J. Biol. Chem.* **274,** 37551–37558.

Di Sole, F., Cerull, R., Babich, V., Quinones, H., Gisler, S. M., Biber, J., Murer, H., Burckhardt, G., Helmle-Kolb, C., and Moe, O. W. (2004). Acute regulation of Na/H exchanger NHE3 by adenosine A(1) receptors is mediated by calcineurin homologous protein. *J. Biol. Chem.* **279,** 2962–2974.

D'Souza, S., Garcia-Cabado, A., Yu, F., Teter, K., Lukacs, G., Skorecki, K., Moore, H. P., Orlowski, J., and Grinstein, S. (1998). The epithelial sodium-hydrogen antiporter Na^+/H^+ exchanger 3 accumulates and is functional in recycling endosomes. *J. Biol. Chem.* **273,** 2035–2043.

du Cheyron, D., Chalumeau, C., Defontaine, N., Klein, C., Kellermann, O., Paillard, M., and Poggioli, J. (2003). Angiotensin II stimulates NHE3 activity by exocytic insertion of the transporter: Role of PI 3-kinase. *Kidney Int.* **64,** 939–949.

Du, Z., Yan, Q., Duan, Y., Weinbaum, S., Weinstein, A. M., and Wang, T. (2006). Axial flow modulates proximal tubule NHE3 and H-ATPase activities by changing microvillus bending moments. *Am. J. Physiol. Renal Physiol.* **290,** F289–F296.

Ebner, H. L., Cordas, A., Pafundo, D. E., Schwarzbaum, P. J., Pelster, B., and Krumschnabel, G. (2005). Importance of cytoskeletal elements in volume regulatory responses of trout hepatocytes. *Am. J. Physiol. Regul. Integr. Comp. Physiol.* **289,** R877–R890.

Forte, J. G., Ly, B., Rong, Q., Ogihara, S., Ramilo, M., Agnew, B., and Yao, X. (1998). State of actin in gastric parietal cells. *Am. J. Physiol.* **274,** C97–C104.

Gekle, M., Serrano, O. K., Drumm, K., Mildenberger, S., Freudinger, R., Gassner, B., Jansen, H. W., and Christensen, E. I. (2002). NHE3 serves as a molecular tool for cAMP-mediated regulation of receptor-mediated endocytosis. *Am. J. Physiol. Renal Physiol.* **283,** F549–F558.

Gill, R. K., Saksena, S., Syed, I. A., Tyagi, S., Alrefai, W. A., Malakooti, J., Ramaswamy, K., and Dudeja, P. K. (2002). Regulation of NHE3 by nitric oxide in Caco-2 cells. *Am. J. Physiol. Gastrointest. Liver Physiol.* **283,** G747–G756.

Gill, R. K., Saksena, S., Tyagi, S., Alrefai, W. A., Malakooti, J., Sarwar, Z., Turner, J. R., Ramaswamy, K., and Dudeja, P. K. (2005). Serotonin inhibits Na^+/H^+ exchange activity via 5-HT4 receptors and activation of PKC alpha in human intestinal epithelial cells. *Gastroenterology* **128,** 962–974.

Gimenez, I. and Forbush, B. (2005). Regulatory phosphorylation sites in the NH2 terminus of the renal Na-K-Cl cotransporter (NKCC2). *Am. J. Physiol. Renal Physiol.* **289,** F1341–F1345.

Hayashi, H., Szaszi, K., Coady-Osberg, N., Furuya, W., Bretscher, A. P., Orlowski, J., and Grinstein, S. (2004). Inhibition and redistribution of NHE3, the apical Na^+/H^+ exchanger, by Clostridium difficile toxin B. *J. Gen. Physiol.* **123,** 491–504.

Hecht, G., Hodges, K., Gill, R. K., Kear, F., Tyagi, S., Malakooti, J., Ramaswamy, K., and Dudeja, P. K. (2004). Differential regulation of Na^+/H^+ exchange isoform activities by enteropathogenic E. coli in human intestinal epithelial cells. *Am. J. Physiol. Gastrointest. Liver Physiol.* **287,** G370–G378.

Hu, M. C., Fan, L., Crowder, L. A., Karim-Jimenez, Z., Murer, H., and Moe, O. W. (2001). Dopamine acutely stimulates Na^+/H^+ exchanger (NHE3) endocytosis via clathrin-coated vesicles: Dependence on protein kinase A-mediated NHE3 phosphorylation. *J. Biol. Chem.* **276,** 26906–26915.

Ingraffea, J., Reczek, D., and Bretscher, A. (2002). Distinct cell type-specific expression of scaffolding proteins EBP50 and E3KARP: EBP50 is generally expressed with ezrin in specific epithelia, whereas E3KARP is not. *Eur. J. Cell Biol.* **81,** 61–68.

Ismailov, I. I., Berdiev, B. K., Shlyonsky, V. G., Fuller, C. M., Prat, A. G., Jovov, B., Cantiello, H. F., Ausiello, D. A., and Benos, D. J. (1997). Role of actin in regulation of epithelial sodium channels by CFTR. *Am. J. Physiol.* **272,** C1077–C1086.

Janecki, A. J., Montrose, M. H., Zimniak, P., Zweibaum, A., Tse, C. M., Khurana, S., and Donowitz, M. (1998). Subcellular redistribution is involved in acute regulation of the brush border Na^+/H^+ exchanger isoform 3 in human colon adenocarcinoma cell line Caco-2. Protein kinase C-mediated inhibition of the exchanger. *J. Biol. Chem.* **273,** 8790–8798.

Janecki, A. J., Janecki, M., Akhter, S., and Donowitz, M. (2000a). Basic fibroblast growth factor stimulates surface expression and activity of Na^+/H^+ exchanger NHE3 via mechanism involving phosphatidylinositol 3-kinase. *J. Biol. Chem.* **275,** 8133–8142.

Janecki, A. J., Janecki, M., Akhter, S., and Donowitz, M. (2000b). Quantitation of plasma membrane expression of a fusion protein of Na/H exchanger NHE3 and green fluorescence protein (GFP) in living PS120 fibroblasts. *J. Histochem. Cytochem.* **48,** 1479–1492.

Kandasamy, R. A., Yu, F. H., Harris, R., Boucher, A., Hanrahan, J. W., and Orlowski, J. (1995). Plasma membrane Na^+/H^+ exchanger isoforms (NHE-1, -2, and -3) are differentially responsive to second messenger agonists of the protein kinase A and C pathways. *J. Biol. Chem.* **270,** 29209–29216.

Khurana, S., Nath, S. K., Levine, S. A., Bowser, J. M., Tse, C. M., Cohen, M. E., and Donowitz, M. (1996). Brush border phosphatidylinositol 3-kinase mediates epidermal growth factor stimulation of intestinal NaCl absorption and Na^+/H^+ exchange. *J. Biol. Chem.* **271,** 9919–9927.

Kim, J. H., Lee-Kwon, W., Park, J. B., Ryu, S. H., Yun, C. H., and Donowitz, M. (2002). $Ca2^+$-dependent inhibition of Na^+/H^+ exchanger 3 requires a NHE3/E3KARP/alpha-actinin-4 complex for oligomerization and endocytosis. *J. Biol. Chem.* **277,** 23714–23724.

Klisic, J., Hu, M. C., Nief, V., Reyes, L., Fuster, D., Moe, O. W., and Ambuhl, P. M. (2002). Insulin activates $Na(^+)/H(^+)$ exchanger 3: Biphasic response and glucocorticoid dependence. *Am. J. Physiol. Renal Physiol.* **283,** F532–F539.

Kocinsky, H. S., Girardi, A. C., Biemesderfer, D., Nguyen, T., Mentone, S., Orlowski, J., and Aronson, P. S. (2005). Use of phospho-specific antibodies to determine the phosphorylation of endogenous Na^+/H^+ exchanger NHE3 at PKA consensus sites. *Am. J. Physiol. Renal Physiol.* **289,** F249–F258.

Kurashima, K., D'Souza, S., Szaszi, K., Ramjeesingh, R., Orlowski, J., and Grinstein, S. (1999). The apical Na^+/H^+ exchanger isoform NHE3 is regulated by the actin cytoskeleton. *J. Biol. Chem.* **274,** 29843–29849.

Lee, J., Ha, J. H., Kim, S., Oh, Y., and Kim, S. W. (2002). Caffeine decreases the expression of Na^+/K^+-ATPase and the type 3 Na^+/H^+ exchanger in rat kidney. *Clin. Exp. Pharmacol. Physiol.* **29,** 559–563.

Lee-Kwon, W., Johns, D. C., Cha, B., Cavet, M., Park, J., Tsichlis, P., and Donowitz, M. (2001). Constitutively active phosphatidylinositol 3-kinase and AKT are sufficient to stimulate the epithelial Na^+/H^+ exchanger 3. *J. Biol. Chem.* **276,** 31296–31304.

Lee-Kwon, W., Kawano, K., Choi, J. W., Kim, J. H., and Donowitz, M. (2003a). Lysophosphatidic acid stimulates brush border Na^+/H^+ exchanger 3 (NHE3) activity by increasing its exocytosis by an NHE3 kinase A regulatory protein-dependent mechanism. *J. Biol. Chem.* **278,** 16494–16501.

Lee-Kwon, W., Kim, J. H., Choi, J. W., Kawano, K., Cha, B., Dartt, D. A., Zoukhri, D., and Donowitz, M. (2003b). $Ca2^+$-dependent inhibition of NHE3 requires PKC alpha which binds to E3KARP to decrease surface NHE3 containing plasma membrane complexes. *Am. J. Physiol. Cell Physiol.* **285,** C1527–C1536.

Levine, S. A., Montrose, M. H., Tse, C. M., and Donowitz, M. (1993). Kinetics and regulation of three cloned mammalian Na^+/H^+ exchangers stably expressed in a fibroblast cell line. *J. Biol. Chem.* **268,** 25527–25535.

Levine, S. A., Nath, S. K., Yun, C. H., Yip, J. W., Montrose, M., Donowitz, M., and Tse, C. M. (1995). Separate C-terminal domains of the epithelial specific brush border Na^+/H^+ exchanger isoform NHE3 are involved in stimulation and inhibition by protein kinases/growth factors. *J. Biol. Chem.* **270,** 13716–13725.

Li, X., Galli, T., Leu, S., Wade, J. B., Weinman, E. J., Leung, G., Cheong, A., Louvard, D., and Donowitz, M. (2001). Na^+-H^+ exchanger 3 (NHE3) is present in lipid rafts in the rabbit ileal brush border: A role for rafts in trafficking and rapid stimulation of NHE3. *J. Physiol.* **537,** 537–552.

Li, X., Leu, S., Cheong, A., Zhang, H., Baibakov, B., Shih, C., Birnbaum, M. J., and Donowitz, M. (2004a). Akt2, phosphatidylinositol 3-kinase, and PTEN are in lipid rafts of intestinal cells: Role in absorption and differentiation. *Gastroenterology* **126,** 122–135.

Li, X., Zhang, H., Cheong, A., Leu, S., Chen, Y., Elowsky, C. G., and Donowitz, M. (2004b). Carbachol regulation of rabbit ileal brush border Na^+-H^+ exchanger 3 (NHE3) occurs through changes in NHE3 trafficking and complex formation and is Src dependent. *J. Physiol.* **556,** 791–804.

Liedtke, C. M., Hubbard, M., and Wang, X. (2003). Stability of actin cytoskeleton and PKC-delta binding to actin regulate NKCC1 function in airway epithelial cells. *Am. J. Physiol. Cell Physiol.* **284,** C487–C496.

Mangeat, P., Gusdinar, T., Sahuquet, A., Hanzel, D. K., Forte, J. G., and Magous, R. (1990). Acid secretion and membrane reorganization in single gastric parietal cell in primary culture. *Biol. Cell* **69,** 223–231.

Matsui, T., Maeda, M., Doi, Y., Yonemura, S., Amano, M., Kaibuchi, K., and Tsukita, S. (1998). Rho-kinase phosphorylates COOH-terminal threonines of ezrin/radixin/moesin (ERM) proteins and regulates their head-to-tail association. *J. Cell Biol.* **140,** 647–657.

Matsui, T., Yonemura, S., and Tsukita, S. (1999). Activation of ERM proteins *in vivo* by Rho involves phosphatidyl-inositol 4-phosphate 5-kinase and not ROCK kinases. *Curr. Biol.* **9,** 1259–1262.

Matthews, J. B., Awtrey, C. S., Hecht, G., Tally, K. J., Thompson, R. S., and Madara, J. L. (1993). Phorbol ester sequentially downregulates cAMP-regulated basolateral and apical Cl- transport pathways in T84 cells. *Am. J. Physiol.* **265,** C1109–C1117.

Matthews, J. B., Smith, J. A., and Hrnjez, B. J. (1997). Effects of F-actin stabilization or disassembly on epithelial Cl- secretion and Na-K-2Cl cotransport. *Am. J. Physiol.* **272,** C254–C262.

Moe, O. W., Amemiya, M., and Yamaji, Y. (1995). Activation of protein kinase A acutely inhibits and phosphorylates Na/H exchanger NHE-3. *J. Clin. Invest.* **96,** 2187–2194.

Moyer, B. D., Loffing, J., Schwiebert, E. M., Loffing-Cueni, D., Halpin, P. A., Karlson, K. H., Ismailov, I. I., Guggino, W. B., Langford, G. M., and Stanton, B. A. (1998). Membrane trafficking of the cystic fibrosis gene product, cystic fibrosis transmembrane conductance regulator, tagged with green fluorescent protein in madin-darby canine kidney cells. *J. Biol. Chem.* **273,** 21759–21768.

Naren, A. P., Cobb, B., Li, C., Roy, K., Nelson, D., Heda, G. D., Liao, J., Kirk, K. L., Sorscher, E. J., Hanrahan, J., and Clancy, J. P. (2003). A macromolecular complex of beta 2 adrenergic receptor, CFTR, and ezrin/radixin/moesin-binding phosphoprotein 50 is regulated by PKA. *Proc. Natl. Acad. Sci. USA* **100,** 342–346.

Ng, T., Parsons, M., Hughes, W. E., Monypenny, J., Zicha, D., Gautreau, A., Arpin, M., Gschmeissner, S., Verveer, P. J., Bastiaens, P. I., and Parker, P. J. (2001). Ezrin is a downstream effector of trafficking PKC-integrin complexes involved in the control of cell motility. *EMBO J.* **20,** 2723–2741.

Norregaard, R., Jensen, B. L., Li, C., Wang, W., Knepper, M. A., Nielsen, S., and Frokiaer, J. (2005). COX-2 inhibition prevents downregulation of key renal water and sodium transport proteins in response to bilateral ureteral obstruction. *Am. J. Physiol. Renal Physiol.* **289,** F322–F333.

Peng, Y., Moe, O. W., Chu, T., Preisig, P. A., Yanagisawa, M., and Alpern, R. J. (1999). ETB receptor activation leads to activation and phosphorylation of NHE3. *Am. J. Physiol.* **276,** C938–C945.

Prat, A. G., Xiao, Y. F., Ausiello, D. A., and Cantiello, H. F. (1995). cAMP-independent regulation of CFTR by the actin cytoskeleton. *Am. J. Physiol.* **268,** C1552–C1561.

Prat, A. G., Cunningham, C. C., Jackson, G. R., Jr., Borkan, S. C., Wang, Y., Ausiello, D. A., and Cantiello, H. F. (1999). Actin filament organization is required for proper cAMP-dependent activation of CFTR. *Am. J. Physiol.* **277,** C1160–C1169.

Reczek, D. and Bretscher, A. (1998). The carboxyl-terminal region of EBP50 binds to a site in the amino-terminal domain of ezrin that is masked in the dormant molecule. *J. Biol. Chem.* **273,** 18452–18458.

Rocha, F., Musch, M. W., Lishanskiy, L., Bookstein, C., Sugi, K., Xie, Y., and Chang, E. B. (2001). IFN-gamma downregulates expression of $Na(^+)/H(^+)$ exchangers NHE2 and NHE3 in rat intestine and human Caco-2/bbe cells. *Am. J. Physiol. Cell Physiol.* **280,** C1224–C1232.

Rosenfeld, G. C., McAllister, E., and Thompson, W. J. (1981). Cytochalasin inhibition of isolated rat gastric parietal cell function. *J. Cell Physiol.* **109,** 53–57.

Schultheis, P. J., Clarke, L. L., Meneton, P., Miller, M. L., Soleimani, M., Gawenis, L. R., Riddle, T. M., Duffy, J. J., Doetschman, T., Wang, T., Giebisch, G., Aronson, P. S., *et al.* (1998). Renal and intestinal absorptive defects in mice lacking the NHE3 Na^+/H^+ exchanger. *Nat. Genet.* **19,** 282–285.

Shiue, H., Musch, M. W., Wang, Y., Chang, E. B., and Turner, J. R. (2005). Akt2 phosphorylates ezrin to trigger NHE3 translocation and activation. *J. Biol. Chem.* **280,** 1688–1695.

Suzuki, A., Yamanaka, T., Hirose, T., Manabe, N., Mizuno, K., Shimizu, M., Akimoto, K., Izumi, Y., Ohnishi, T., and Ohno, S. (2001). Atypical protein kinase C is involved in the evolutionarily conserved par protein complex and plays a critical role in establishing epithelia-specific junctional structures. *J. Cell Biol.* **152,** 1183–1196.

Szaszi, K., Kurashima, K., Kapus, A., Paulsen, A., Kaibuchi, K., Grinstein, S., and Orlowski, J. (2000). RhoA and rho kinase regulate the epithelial Na^+/H^+ exchanger NHE3. Role of myosin light chain phosphorylation. *J. Biol. Chem.* **275,** 28599–28606.

Szaszi, K., Kurashima, K., Kaibuchi, K., Grinstein, S., and Orlowski, J. (2001). Role of the cytoskeleton in mediating cAMP-dependent protein kinase inhibition of the epithelial Na^+/H^+ exchanger NHE3. *J. Biol. Chem.* **276,** 40761–40768.

Tamura, A., Kikuchi, S., Hata, M., Katsuno, T., Matsui, T., Hayashi, H., Suzuki, Y., Noda, T., and Tsukita, S. (2005). Achlorhydria by ezrin knockdown: Defects in the formation/expansion of apical canaliculi in gastric parietal cells. *J. Cell Biol.* **169,** 21–28.

Tse, C. M., Levine, S. A., Yun, C. H., Montrose, M. H., Little, P. J., Pouyssegur, J., and Donowitz, M. (1993). Cloning and expression of a rabbit cDNA encoding a serum-activated ethylisopropylamiloride-resistant epithelial Na^+/H^+ exchanger isoform (NHE-2). *J. Biol. Chem.* **268,** 11917–11924.

Tsuganezawa, H., Sato, S., Yamaji, Y., Preisig, P. A., Moe, O. W., and Alpern, R. J. (2002). Role of c-SRC and ERK in acid-induced activation of NHE3. *Kidney Int.* **62,** 41–50.

Turban, S., Wang, X. Y., and Knepper, M. A. (2003). Regulation of NHE3, NKCC2, and NCC abundance in kidney during aldosterone escape phenomenon: Role of NO. *Am. J. Physiol. Renal Physiol.* **285,** F843–F851.

Turner, J. R. and Black, E. D. (2001). NHE3-dependent cytoplasmic alkalinization is triggered by Na^+-glucose cotransport in intestinal epithelia. *Am. J. Physiol. Cell Physiol.* **281,** C1533–C1541.

Turner, J. R., Black, E. D., Ward, J., Tse, C. M., Uchwat, F. A., Alli, H. A., Donowitz, M., Madara, J. L., and Angle, J. M. (2000). Transepithelial resistance can be regulated by the intestinal brush border Na^+-H^+ exchanger NHE3. *Am. J. Physiol. Cell Physiol.* **279,** C1918–C1924.

Weinman, E. J., Steplock, D., and Shenolikar, S. (2003). NHERF-1 uniquely transduces the cAMP signals that inhibit sodium-hydrogen exchange in mouse renal apical membranes. *FEBS Lett.* **536,** 141–144.

Yang, L. E., Maunsbach, A. B., Leong, P. K., and McDonough, A. A. (2004). Differential traffic of proximal tubule Na^+ transporters during hypertension or PTH: NHE3 to base of microvilli vs. NaPi2 to endosomes. *Am. J. Physiol. Renal Physiol.* **287,** F896–F906.

Yip, J. W., Ko, W. H., Viberti, G., Huganir, R. L., Donowitz, M., and Tse, C. M. (1997). Regulation of the epithelial brush border Na^+/H^+ exchanger isoform 3 stably expressed in fibroblasts by fibroblast growth factor and phorbol esters is not through changes in phosphorylation of the exchanger. *J. Biol. Chem.* **272,** 18473–18480.

Yun, C. C., Chen, Y., and Lang, F. (2002). Glucocorticoid activation of $Na(^+)/H(^+)$ exchanger isoform 3 revisited. The roles of SGK1 and NHERF2. *J. Biol. Chem.* **277,** 7676–7683.

Yun, C. H., Tse, C. M., Nath, S. K., Levine, S. A., Brant, S. R., and Donowitz, M. (1995). Mammalian Na^+/H^+ exchanger gene family: Structure and function studies. *Am. J. Physiol.* **269,** G1–G11.

Yun, C. H., Oh, S., Zizak, M., Steplock, D., Tsao, S., Tse, C. M., Weinman, E. J., and Donowitz, M. (1997). cAMP-mediated inhibition of the epithelial brush border Na^+/H^+ exchanger, NHE3, requires an associated regulatory protein. *Proc. Natl. Acad. Sci. USA* **94,** 3010–3015.

Yun, C. H., Lamprecht, G., Forster, D. V., and Sidor, A. (1998). NHE3 kinase A regulatory protein E3KARP binds the epithelial brush border Na^+/H^+ exchanger NHE3 and the cytoskeletal protein ezrin. *J. Biol. Chem.* **273,** 25856–25863.

Zhang, Y., Norian, J. M., Magyar, C. E., Holstein-Rathlou, N. H., Mircheff, A. K., and McDonough, A. A. (1999). *In vivo* PTH provokes apical NHE3 and NaPi2 redistribution and Na-K-ATPase inhibition. *Am. J. Physiol.* **276,** F711–F719.

Zhao, H., Wiederkehr, M. R., Fan, L., Collazo, R. L., Crowder, L. A., and Moe, O. W. (1999). Acute inhibition of Na/H exchanger NHE-3 by cAMP. Role of protein kinase a and NHE-3 phosphoserines 552 and 605. *J. Biol. Chem.* **274,** 3978–3987.

Zhao, H., Shiue, H., Palkon, S., Wang, Y., Cullinan, P., Burkhardt, J. K., Musch, M. W., Chang, E. B., and Turner, J. R. (2004). Ezrin regulates NHE3 translocation and activation after Na^+-glucose cotransport. *Proc. Natl. Acad. Sci. USA* **101,** 9485–9490.

Zhou, R., Cao, X., Watson, C., Miao, Y., Guo, Z., Forte, J. G., and Yao, X. (2003). Characterization of protein kinase A-mediated phosphorylation of ezrin in gastric parietal cell activation. *J. Biol. Chem.* **278,** 35651–35659.

Zizak, M., Lamprecht, G., Steplock, D., Tariq, N., Shenolikar, S., Donowitz, M., Yun, C. H., and Weinman, E. J. (1999). cAMP-induced phosphorylation and inhibition of Na^+/H^+ exchanger 3 (NHE3) are dependent on the presence but not the phosphorylation of NHE regulatory factor. *J. Biol. Chem.* **274,** 24753–24758.

Zizak, M., Cavet, M. E., Bayle, D., Tse, C. M., Hallen, S., Sachs, G., and Donowitz, M. (2000). $Na(^+)/H(^+)$ exchanger NHE3 has 11 membrane spanning domains and a cleaved signal peptide: Topology analysis using *in vitro* transcription/translation. *Biochemistry* **39,** 8102–8112.

Chapter 14

Domain-Specific Phosphorylation as a Regulator of Intermediate Filaments

Hanna-Mari Pallari[1,2] and John E. Eriksson[1,2]

[1]*Turku Centre for Biotechnology, University of Turku and Åbo Akademi University, FIN-20521 Turku, Finland*
[2]*Department of Biology, Åbo Akademi University, FI-20520 Turku, Finland*

Intermediate filaments (IFs) are cytoskeletal structures that maintain cell and tissue integrity. In addition, they have been implicated in a number of other functions, including maintenance of cell shape, subcellular organelle positioning, support to cell migration, radial growth of axons, and scaffolding of signaling molecules. Many of the recognized tasks are affected by phosphorylation, which is the key mechanism to regulate both the organization of IFs and the association of IFs with interacting molecules. Another important function of IF phosphorylation is the disassembly of cytoplasmic and karyoskeletal IF networks during mitosis. An intriguing feature of IFs is the tripartite structure of their building blocks, with a highly conserved central rod domain, which is flanked by variable length N- and C-terminal regions. While the rod domain is crucial for the assembly of IFs, a common feature of the distal domains is that they represent the main targets of phosphorylation. Since phosphate targeting seems to have domain-specific functions, this chapter aims at summarizing the different roles of N- and C-terminal phosphorylation of IFs.

Advances in Molecular and Cell Biology, Vol. 37

ISBN: 0-444-52868-7

DOI: 10.1016/S1569-2558(06)37014-2

I. INTRODUCTION

The vertebrate cytoskeleton is composed of three different filament systems, actin filaments, microtubules, and intermediate filaments (IFs). Contrary to the highly conserved, globular protein subunits in the actin filaments and microtubules, IFs are composed of 65 different fibrous proteins that vary considerably with respect to their sequences, expression patterns, and abundance in different tissues (reviewed in Herrmann *et al.*, 2003; Coulombe and Wong, 2004). Based on their characteristic molecular features, IF proteins are subdivided into five different types. Type I–IV IF proteins form cytoplasmic IFs, whereas type V IF proteins, lamins, are located in the nucleus, where they form a filamentous network inside the nuclear membrane (Table I) (Herrmann *et al.*, 2003; Coulombe and Wong, 2004).

The major function of IFs is to provide cells with a supporting framework against mechanical stresses (Herrmann *et al.*, 2003). With the assumption that structural stability is a prerequisite for mechanical integrity, IFs were long considered as extremely stable and static structures. This assumption was supported by the exceptional insolubility of IFs in buffers of physiological ionic strength, the small amounts of intracellular soluble, unpolymerized IF proteins, and the seemingly stationary complex networks reaching from the plasma membrane to the nucleus. However, as these filamentous structures need to be continuously modified throughout different cellular processes, such as mitosis and differentiation, and as they need to adapt to rapidly changing environmental conditions, the IF polymers need to be in a dynamic state of continuous exchange between soluble subunits and assembled polymers. Phosphorylation is the key regulator of IF assembly, organization, subcellular distribution, and their association with partner proteins (Table I) (Eriksson *et al.*, 1992a; Inagaki *et al.*, 1996; Ku *et al.*, 1996b; Helfand *et al.*, 2005; Toivola *et al.*, 2005). During the past decades, an increasing number of cell-specific functions of IFs have been established. These functions require dynamic regulation of the interactions between individual IF proteins and the interactions of IFs with different IF-associated proteins (Coulombe and Wong, 2004; Toivola *et al.*, 2005). Phosphorylation is emerging as a versatile mechanism to regulate these interactions and, thereby, to influence the cellular processes that are affected by these interactions. Hence, phosphorylation-based modification of the building blocks of IF networks plays a key role when cells are adapting to the extreme changes in their environment that they frequently need to encounter during their lifespan.

II. REGULATION OF IFs BY PHOSPHORYLATION

Protein phosphorylation is the major posttranslational regulatory mechanism in eukaryotic cells, with extremely broad ramifications in multiple cellular processes, ranging from cell division to cell death. Phosphorylation was suggested to be a regulator of IFs more than 20 years ago, when keratins (Sun and Green, 1978), neurofilaments (Pant *et al.*, 1978), vimentin, and desmin (O'Connor *et al.*, 1981) were

Table I
The Intermediate Filament Protein Family and Examples of Situations When They Become Phosphorylated

Intermediate filament type	Protein name	Tissue distribution	Examples of situations when phosphorylation induced
I	Keratins (acidic), K9–K20	Epithelial cells	Mitosis,[a] stress, and apoptosis[b]
II	Keratins (basic), K1–K8	Epithelial cells	Mitosis,[c] stress and apoptosis,[d] EGF stimulation[e]
III	Vimentin	Mesenchymal cells	Mitosis,[f] myogenesis,[g] PDGF-stimulation,[h] viral infection[i]
	Desmin	Muscle cells	Mitosis,[j] myogenesis[k]
	Glial fibrillary acidic protein (GFAP)	Astrocytes, glia cells	Mitosis,[l] glutamate stimulation[m]
IV	NF-L	Neurons	Neuronal differentiation,[n] neurodegenerative disorders[o]
	NF-M	Neurons	"
	NF-H	Neurons	"
	Nestin	Neuroepithelial stem cells, myoblasts	Mitosis,[p] myogenesis[q]
V	Lamin A/C	Differentiated cells	Mitosis,[r] apoptosis[s]
	Lamin B1	All cell types	"
	Lamin B2	All cell types	"

[a]Liao *et al.*, 1995b.
[b]Liao *et al.*, 1995a; Ku *et al.*, 1996a, 1998b; Toivola *et al.*, 2004.
[c]Liao *et al.*, 1997; Toivola *et al.*, 2002.
[d]Liao *et al.*, 1995a, 1997; Ku *et al.*, 1996a; Toivola *et al.*, 2002, 2004; Ridge *et al.*, 2005.
[e]Ku and Omary, 1997.
[f]Evans and Fink, 1982; Chou *et al.*, 1990, 1991; Takai *et al.*, 1996; Goto *et al.*, 1998, 2003.
[g]Gard and Lazarides, 1982.
[h]Valgeirsdottir *et al.*, 1998.
[i]Stefanovic *et al.*, 2005.
[j]Inada *et al.*, 1998, 1999; Kawajiri *et al.*, 2003.
[k]Gard and Lazarides, 1982.
[l]Matsuoka *et al.*, 1992; Kawajiri *et al.*, 2003.
[m]Kommers *et al.*, 2002.
[n]Ackerley *et al.*, 2000, 2003; Grant and Pant, 2000; Jung *et al.*, 2000; Yabe *et al.*, 2001; Garcia *et al.*, 2003; Rao *et al.*, 2003.
[o]Nguyen *et al.*, 2001; Ackerley *et al.*, 2004.
[p]Sahlgren *et al.*, 2001.
[q]Sahlgren *et al.*, 2003.
[r]Heald and McKeon, 1990; Peter *et al.*, 1990; Haas and Jost, 1993.
[s]Shimizu *et al.*, 1998; Cross *et al.*, 2000.

shown to undergo this modification. Since these early observations, many different research laboratories have made significant contribution toward understanding the purpose and regulation of IF phosphorylation, as well as the mechanism underlying this modification.

An early observation regarding the relationship between IF phosphorylation and structural modifications was attained when the amount of phosphorylated vimentin was noticed to increase during cell division concurrently with the characteristic mitotic reorganization of the filamentous networks (Evans and Fink, 1982). It has become increasingly evident that phosphorylation alters the filament structure. This modification plays a crucial role during mitosis, when the cytosolic filamentous network is unraveled and reorganized to enable cell division (Foisner, 1997). The filament disassembly is especially important for lamins that form the nuclear lamina underneath the nuclear membrane, as the nuclear envelope is disassembled during mitosis (Foisner, 1997; Moir *et al.*, 2000).

In addition to cell division, changes in IF phosphorylation have been observed during a number of different cellular processes. Notably, IF phosphorylation is a key feature of many different types of cell stresses, induced by heat, viral infection, drugs, or disease (Liao *et al.*, 1995a; Ku *et al.*, 1996a; Toivola *et al.*, 2004, 2005). The role of stress-induced IF phosphorylation is not entirely clear, but it appears to have a protective effect (Ku *et al.*, 1998b; He *et al.*, 2002). Extensive studies on neurofilaments (NFs, the IFs of neuronal tissue) have shown that phosphorylation of NFs plays a role in neuronal development and differentiation, and contributes to axonal structure and especially axonal caliber (Grant and Pant, 2000; Garcia *et al.*, 2003; Rao *et al.*, 2003). Moreover, IFs, especially desmin, vimentin, and nestin, are phosphorylated during myogenesis (Gard and Lazarides, 1982; Sahlgren *et al.*, 2003). Gard and Lazarides speculated early in 1982 that phosphorylation might be involved in filament reorganization observed during myogenesis. In addition, as mentioned in the introduction, accumulated evidence implicates IF phosphorylation as an important regulator of the association of IFs with other cellular proteins, such as signaling determinants (Liao and Omary, 1996; Sin *et al.*, 1998; Tzivion *et al.*, 2000; Sahlgren *et al.*, 2003). Finally, recent results provide new insight into the previously reported role of IFs in cell migration (Eckes *et al.*, 2000; Lepekhin *et al.*, 2001). Especially interesting is the recently demonstrated role of vimentin IFs in the transcellular migration of lymphocytes (Nieminen *et al.*, 2006). It is likely that phosphorylation is instrumental in regulating these functions. In fact, it has been demonstrated that the phosphorylation of vimentin controls the trafficking of integrins to the plasma membrane, thereby, determining the directional motility of the cell (Ivaska *et al.*, 2005).

III. IF STRUCTURE AND THE TARGETING OF PHOSPHORYLATION

The structure of IF proteins is highly conserved. They all share a central, α-helical rod domain flanked by nonhelical N- and C-terminal regions (Herrmann *et al.*, 2003; Strelkov *et al.*, 2003). The rod domains of all IF proteins display pronounced repeats of heptads in which the first and the fourth positions are occupied by hydrophobic amino acids. These repeats are responsible for the generation of a coiled-coiled dimer, the initial assembly step of IF assembly initiated by two IF proteins wound around each other (Herrmann *et al.*, 2003; Strelkov *et al.*, 2003). Contrary to the conserved rod domains, the distal regions show considerable variation in both size and amino acid sequence (Herrmann *et al.*, 2003; Strelkov *et al.*, 2003; Coulombe and Wong, 2004). Therefore, these distal areas are considered to be responsible for the cell type-specific functions of IFs. Phosphorylation seems to be exclusively limited to these regions, as exemplified by many *in vitro* and *in vivo* studies (reviewed in Inagaki *et al.*, 1996; Ku *et al.*, 1996b). Although the conformation of the flexible distal domains in fully polymerized filaments has not yet been solved, it has been suggested that some of these domains would reside outside the filament core (Strelkov *et al.*, 2003), thus, being more easily accessible to protein kinases. To date, several phosphorylation sites have been revealed in almost every IF protein (Inagaki *et al.*, 1996). The majority of them are serines and threonines, although a few instances of tyrosine phosphorylation have also been reported (Valgeirsdottir *et al.*, 1998; Feng *et al.*, 1999).

The structural features and the assembly of IFs comprise an important model for structural biologists and have, therefore, been characterized in detail. The IF proteins are classified into three different assembly groups based on their special characteristics in filament formation (for review see Herrmann and Aebi, 2004). It has been demonstrated that in addition to the rod domain, which is crucial for assembly, the distal regions affect the polymerization process. N-terminally truncated, headless vimentin, and keratin have been shown to be unable to assemble into filaments (Hatzfeld and Burba, 1994; Herrmann *et al.*, 1996). In addition, the removal of the N-terminal domain also interferes with the filament-assembling function of desmin (Raats *et al.*, 1992). In vimentin, amino acids 20–42 of head domain are known to be involved in the emergence of a tetramer formed by two laterally interacting dimers and assist also at the later stages when tetramers form short, uniform assembly intermediates, termed unit-length filaments (ULFs; Herrmann and Aebi, 2004). The intermolecular interactions between dimers, the formation of which is the basis for IF assembly, are based on salt bridges formed between the arginines in the head domain and acidic residues in the central rod domain (Herrmann and Aebi, 2004). In fact, the rod domains of IFs are highly charged. For example, the rod domain of vimentin contains 116 charged amino acids, 70 of them being acidic and in principle capable of ionic interactions with the 12 arginines within the vimentin head region (Herrmann and Aebi, 2004).

Based on the findings about the importance of the head domain for filament assembly (Raats *et al.*, 1992; Hatzfeld and Burba, 1994; Herrmann *et al.*, 1996), the role of the N-terminal arginines in polymerization (Herrmann and Aebi, 2004), and the ability of phosphorylation to alter the filament structure (see preceding paragraph), it is reasonable to ask whether the phosphorylation of the head domain could by the introduction of negative charges influence the arginine-mediated interactions and, thereby, the structure of the filaments. In the 1980s, Inagaki and collaborators demonstrated that phosphorylation causes the disassembly of IFs by using purified protein kinase C (PKC) and protein kinase A (PKA) to phosphorylate reconstituted vimentin filaments *in vitro* (Inagaki *et al.*, 1987, 1988). The specific sites found to be responsible for the phosphorylation-mediated depolymerization were localized on the N-terminal domain of vimentin (Ando *et al.*, 1989). Subsequently, the preference of N-terminal phosphorylation in disassembly of IFs and in reduced capability to form filaments *in vitro* has been demonstrated with several IF proteins, for example, vimentin (Evans, 1988b; Chou *et al.*, 1990; Matsuzawa *et al.*, 1997; Goto *et al.*, 1998, 2002, 2003; Eriksson *et al.*, 2004), desmin (Geisler and Weber, 1988; Kusubata *et al.*, 1993; Inada *et al.*, 1998; Kawajiri *et al.*, 2003), GFAP (Inagaki *et al.*, 1990; Kosako *et al.*, 1997; Kawajiri *et al.*, 2003), and low-molecular weight neurofilament (NF-L) (Gonda *et al.*, 1990; Nakamura *et al.*, 1990; Mukai *et al.*, 1996). In further support of the specific role of the N-terminus in IF assembly/disassembly, studies using surface plasmon resonance have indicated that the N-terminal domain of vimentin binds to the 2B helix in the rod domain and that phosphorylation interferes with this interaction (Gohara *et al.*, 2001).

There is the distinct possibility that the ionic interactions between arginines in the vimentin head region and the charged amino acids in the rod domain (Fig. 1) could be regulated by the phosphorylation sites that reside in close proximity to these arginines (Herrmann, H., personal communication). Results from the laboratory of Harald Herrmann (personal communication) reveal that even the far N-terminal arginines of vimentin (Fig. 1) affect filament formation. While these arginines do not seem to be important for filament assembly, they are important for filament strength (Herrmann, H., personal communication). Phosphorylation of the far N-terminal cluster of serines could easily eliminate the effect of these arginine residues (Fig. 1). These serines have been shown to be putative PKC phosphorylation sites *in vivo* (Eriksson *et al.*, 2004) and were shown to participate in the trafficking of integrins to the plasma membrane (Ivaska *et al.*, 2005). Phosphorylation of these sites also regulates the association of vimentin with endocytosed vesicles containing integrins that are in the process of returning to the plasma membrane (Ivaska *et al.*, 2005). Serine to alanine mutations of this cluster of serine residues interferes with the assembly of filaments, lead to the accumulation of integrins in the endocytosed vesicles, and retards cell migration, whereas serine to aspartate mutations of these sites accelerated migration (Ivaska *et al.*, 2005). This particular cluster is located next to (and partly into) an SYRRMF motif that has been shown (by using a few amino acids longer peptide) to severely disturb the assembly of vimentin (Herrmann and Aebi, 2004). It is tempting to speculate that this would be a measure to link phosphorylation-mediated regulation to salt bridge formation at a site that could modulate both filament strength and affect an important biological role of vimentin. In fact, Ralton *et al.* (1994) have discussed similar regulation in the case of GFAP after having characterized an N-terminal motif important in assembly containing closely spaced arginines and phosphorylation sites. All these results point to a role for phosphorylation in assembly/disassembly, but the exact molecular and structural mechanisms underlying this control mechanism remain to be clarified, especially the role of phosphorylation in counteracting the arginine-based salt bridges.

The C-terminal tail, on the other hand, does not seem to have an essential role in filament formation. It has been shown that C-terminal deletions do not interfere with

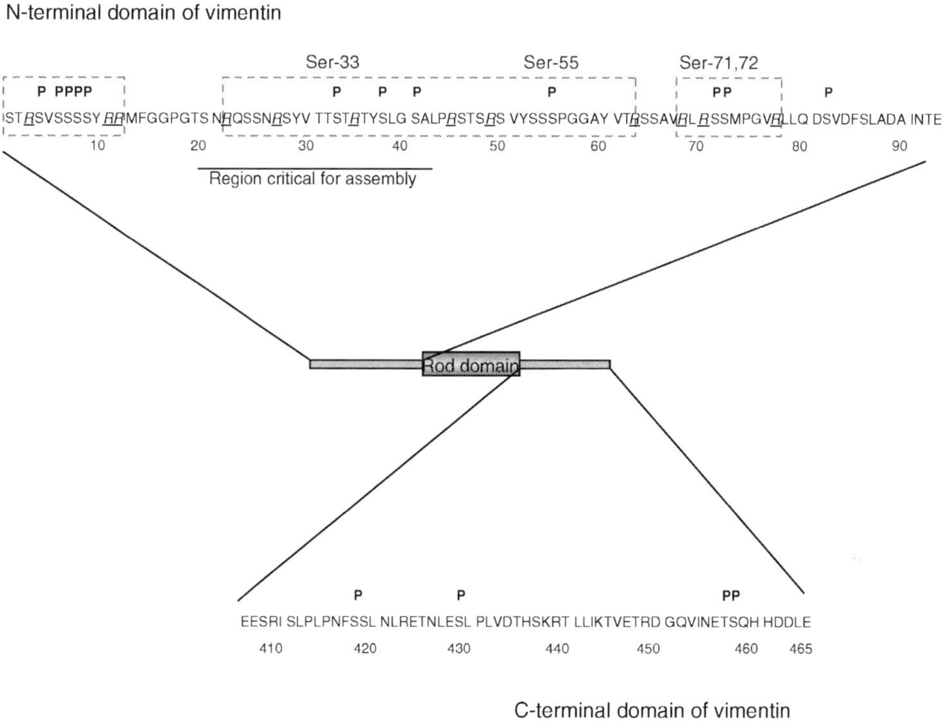

Figure 1. The distribution of known *in vivo* phosphorylation sites along the vimentin sequence. Summary of known *in vivo* phosphorylation sites on vimentin. Mitosis-specific sites are specially indicated and the dashed line highlights the arginines surrounding these phosphorylation sites. Arginines around the far N-terminal PKC-phosphorylation sites are marked in a similar way.

the polymerization process, (Bader *et al.*, 1991; Eckelt *et al.*, 1992), but rather participate in the regulation of the filament width (Herrmann *et al.*, 1996). C-terminal phosphorylation sites regulating IF structure *in vitro* have been detected mainly in lamins (Peter *et al.*, 1990; Eggert *et al.*, 1993), although C-terminal phosphorylation sites have been identified also in cytosolic IF proteins (Ando *et al.*, 1996; Eriksson *et al.*, 2004). Nevertheless, the C-terminal domains of NFs have been shown to play a critical role in the formation of the cytoskeletal network in axons by interacting with other axonal components, thus, exemplifying the importance of C-terminal regions in neurons (Grant and Pant, 2000).

IV. PHOSPHORYLATION OF IFs DURING MITOSIS

IFs undergo marked structural changes during mitosis. These changes in IF organization have been shown to be mediated by phosphorylation. The mitotic phosphorylation has been indicated to be targeted at both N- and C-termini, although the importance of the N-terminus seems to be more prominent. In the first study addressing the question of mitotic IF phosphorylation (Evans, 1988a), the increased phosphorylation of vimentin and desmin during cell division was shown to occur within their N-terminal domains. After identification of the mitotic cyclin-dependent kinase 1, Cdk1 ($p34^{cdc2}$), as the crucial kinase of the maturation-promoting factor (Dunphy *et al.*, 1988), IFs were among the first cytoskeletal substrates to be identified as major targets for this kinase. The Cdk1-mediated phosphorylation of nuclear lamins established the concept of specific Cdk1 sites as determinants of mitosis-specific disassembly of IFs (Heald and McKeon, 1990). The disassembly of nuclear lamina is the most dramatic IF-targeted structural alteration during mitosis. The collapse of lamin network is almost total compared to less extensive reorganization observed with other IFs. Type A lamins disassemble into soluble particles, whereas B-type lamins seem to remain associated with the membrane fragments of the dissociated nuclear envelope (Gerace and Blobel, 1980). The Cdk1-mediated phosphorylation has been shown to occur in both N- and C-terminal domains of lamins, both regions contain sites that affect filament formation. B-type lamins are phosphorylated *in vitro* by Cdk1 on the same residues that are phosphorylated *in vivo* during cell division (Peter *et al.*, 1990). One of the target sites was identified as a conserved SPTR motif situated in the N-terminal domain of all lamin proteins (Peter *et al.*, 1990). On lamin B2 this site is Ser-16 (Peter *et al.*, 1990). In lamin A the Cdk1-specific sites are Ser-22 in the N-terminal domain and Ser-392 in the C-terminal domain (Heald and McKeon, 1990). Mutational analysis has revealed that Ser-22 alone has an effect on mitotic reorganization, but this effect is significantly increased if Ser-392 is mutated as well, although the latter mutation alone has no effect by itself (Heald and McKeon, 1990). The authors propose that phosphorylation next to the respective far ends of the rod domain would control the assembly dynamics of lamins (Heald and McKeon, 1990). Thr-19 on lamin A has also been implicated to act similarly as Ser-22, that is, the mutation on Thr-19 interfered with the behavior of lamins during mitosis whereas the double mutant Thr-19/Ser-392 produced even more aberrant mitotic phenotypes, pointing to synergistic functions between these residues namely, Thr-19, Ser-22, and Ser-392 (Haas and Jost, 1993). In support of the results that Ser-392 by itself is not likely to be important for disassembly of IF, studies investigating Cdk1-induced *in vitro* disassembly of longitudinal lamin B2 polymers have demonstrated that the C-terminal sites are not as important as the N-terminal sites for disassembly of these polymers (Peter *et al.*, 1991).

Among the cytoplasmic IFs, the N-terminal Ser-55 of vimentin was the first to be identified as a major Cdk1 target during mitosis (Chou *et al.*, 1990, 1991). More information was obtained by mutational analysis, which showed impaired disassembly of the filament network when Ser-55 on vimentin was mutated to alanine (Chou *et al.*, 1996). The latter site is not the only N-terminal residue modified on vimentin during mitosis, since vimentin has also been shown to be phosphorylated on Ser-33 by PKC (Takai *et al.*, 1996). This modification is initiated at metaphase and maintained throughout anaphase. In contrast, phosphorylation of Ser-55 is only maintained during metaphase (Takai *et al.*, 1996), indicating that various spatiotemporally regulated kinases participate in the mitotic reorganization of IFs (Foisner, 1997).

In addition to vimentin, other cytoplasmic IF proteins, including GFAP, are phosphorylated by Cdk1 (on N-terminal Ser-8) during cell division (Matsuoka *et al.*, 1992). The amount of phosphorylation at this particular site is prominent when cells enter mitosis and like vimentin, it is dephosphorylated during the later stages of cell division (Matsuoka *et al.*, 1992).

Inagaki and collaborators have described in detail the phosphorylation of type III IF proteins during the last step of mitosis namely cytokinesis. They have revealed cooperation between Rho-binding kinase (which binds to the small GTPase Rho) and Aurora-B kinase in the phosphorylation of filaments associated with the cleavage furrow. Vimentin has been reported to be phosphorylated on Ser-71 by Rho-binding kinase (Goto *et al.*, 1998) and on Ser-72 by Aurora-B (Goto *et al.*, 2003). In GFAP, three N-terminal amino acids (Ser-13, Ser-34, and Thr-7) are the preferred phosphorylation sites during cytokinesis (Matsuoka *et al.*, 1992). These sites have been shown to be phosphorylated by both Rho-binding kinase and Aurora-B (Kawajiri *et al.*, 2003). Furthermore, another type III IF protein, desmin, is modified by these kinases, and the mitosis-specific sites in desmin are partly overlapping (Thr-16, Thr-75, and Thr-76 for Rho-binding kinase [Inada *et al.*, 1998, 1999] and Ser-11, Ser-59, and Thr-16 for Aurora-B kinase [Kawajiri *et al.*, 2003]).

In addition to type III IF proteins, type I and II keratins are also known to be phosphorylated during cell division. Ser-52 is the major phosphorylation site identified in epithelial keratin 18 (K18) (Ku and Omary, 1994). The phosphorylation of this particular residue has been shown to increase during S and G2/M phases of the cell cycle and is associated with filament reorganization (Liao *et al.*, 1995b). A good candidate for the protein kinase responsible for phosphorylation of K18 is PKC since it can phosphorylate K18 *in vitro*, as well as associate with K8/K18 filaments in cells (Ku and Omary, 1994). Generation of hyperphosphorylated K8, the binding partner of K18, has been described to occur in mitotic HT29 cells, mouse intestinal crypt mitotic cells, and regenerating mouse livers after partial hepatectomy (Liao *et al.*, 1997). This phosphorylation takes place on Ser-73 in N-terminal head but its steady state levels are relatively low (Liao *et al.*, 1997). Moreover, keratins phosphorylated on Ser-73 can partly maintain the filamentous pattern, indicating that this site could participate in mitosis, for example, by modulating interactions with other cellular elements (Liao *et al.*, 1997). However, the possible effect on filament organization cannot be excluded (Liao *et al.*, 1997). The particular LLS/TPL sequence containing the Ser-73 is well conserved among group II keratins. It has been detected in epidermal K5/K6, esophageal K4, as well as in type II hair keratins, with the exception that serine is replaced by threonine (Thr-150 in K4, Thr-145 in K5, and Thr-133 in K6; Toivola *et al.*, 2002). A dramatic increase in K5/K6 phosphorylation has been observed during mitosis (Toivola *et al.*, 2002). Moreover, phosphorylated keratins are reorganized during cell division into punctuated structures indicating a role in disassembly (Toivola *et al.*, 2002). The results described earlier imply that during cell division the N-terminal phosphorylation has an important role as a facilitator of filament reorganization. In many cases, the mutation of phosphorylation sites leads to dramatic defects in filament disassembly (Chou *et al.*, 1996). Serine to alanine mutations at sites important in cytokinesis cause severe problems in separation of IFs between daughter cells and lead to a formation of unusually long, bridge-like cytoplasmic structures between the daughter cells. The latter effect has been shown with vimentin (Goto *et al.*, 2003), desmin (Kawajiri *et al.*, 2003), and GFAP (Yasui *et al.*, 1998).

While all results regarding mitosis-specific phosphorylation of IFs strengthen the assumptions of N-terminal phosphorylation being a critical modulator of IF structure, the role of mitotic phosphorylation in controlling arginine-mediated interactions, and especially the molecular mechanism underlying this event, are not completely resolved. A closer examination of the N-terminal domain of vimentin shows that the mitosis-specific phosphorylation sites, Ser-33, Ser-55, Ser-71, and Ser-72, all reside fairly close to several arginines (Fig. 1). Particularly Ser-33, which is surrounded by arginines, is located in the region known to engage in head–rod interactions (Herrmann and Aebi, 2004). Serine-71 and Ser-72 are close to Arg-68, Arg-70, and Arg-77, all residues that could potentially participate in salt bridge formation. However, the molecular details underlying the effect of these sites on filament formation remains to be clarified. Mutation of other identified phosphorylation sites (Ser-55, Ser-71, and Ser-72) have been shown to affect the mitotic IF structure, hence their role in mitosis is well defined.

Despite the remarkable predominance of N-terminal phosphorylation sites as regulators of mitotic disassembly, some IF proteins have been shown to be phosphorylated on their C-terminus on cell division. Nestin, an IF protein expressed in the developing central nervous system and muscle tissue, is phosphorylated on its C-terminal tail during cell division (Sahlgren *et al.*, 2001). Nestin has a strikingly different structure compared to all other cytoskeletal IF proteins. It has an extremely short N-terminal head, only 11 amino acids without any possible phosphorylation sites, and an exceptionally long C-terminal tail with numerous serines and threonines. The structures of nestin N- and C-termini differ so radically from the other IF proteins that the assembly and disassembly regulation of nestin is likely to be completely different from the other IFs. This hypothesis becomes even more convincing when considering the role of N-terminus in dimer formation. Nestin is unable to form filaments on its own, but polymerizes with vimentin instead (Eliasson *et al.*, 1999; Steinert *et al.*, 1999). Therefore, it could be that the disassembly of nestin during mitosis is regulated solely by the disassembly of vimentin. If that is true, the identified mitotic phosphorylation sites of nestin, Thr-316 and Thr-1495 (Sahlgren *et al.*, 2001), are likely to have functions other than regulation of filament assembly. These functions remain to be clarified as no mutational analysis of nestin phosphorylation sites has ever been done. In addition to nestin, mitotic, C-terminal phosphorylation sites have been identified in vimentin. However, Thr-457 and Ser-458 in the tail of vimentin have not turned out to be significant for filament reorganization during cell division (Chou *et al.*, 1996). These findings demonstrate that although phosphorylation is extremely important to IF reorganization, it may have additional roles during mitosis, for example, in mediating the interactions between IFs and other cellular molecules. In fact, it has been suggested that nestin could play an important role in cell division by controlling the trafficking and distribution of cellular factors to arising daughter cells (Chou *et al.*, 2003). In a highly speculative model, phosphorylation could induce the formation of such complexes and IFs could piggyback interacting molecules to the forming daughter cells.

It has been suggested that the C-terminal phosphorylation of lamins may have a crucial role in breaking the interactions between lamins and other nuclear membrane proteins (Peter *et al.*, 1991) rather than in interrupting the interactions between the lamin proteins within the filament structure. While there is no clear description of the roles of domain-specific phosphorylation of lamins, β_{II} PKC-mediated phosphorylation of the C-terminal Ser-405 in lamin B has been proposed to be involved in the regulation of lamin B dynamics during cell division (Goss *et al.*, 1994). β_{II} PKC phosphorylates and induces the solubilization of lamin B *in vitro*, and Ser-405 is among the prominent phosphorylation sites in lamin B in mitotic human cells (Hocevar *et al.*, 1993; Goss *et al.*, 1994). The differences in lamin phosphorylation during mitosis compared to the majority of cytoplasmic IFs may rise from the completely different assembly and organization of this nuclear IF network, as well as the completely different head domain structure of lamin proteins. The assembly of lamins differs from that of cytoplasmic IFs, since lamin dimers first associate longitudinally into head-to-tail filaments that later anneal laterally, whereas the cytoplasmic dimers form partly overlapping tetramers preferring lateral interactions before any longitudinal annealing occurs (Herrmann and Aebi, 2004). It is likely that the N- and C-terminal phosphorylation of lamins could affect one or more of these interactions differently (Peter *et al.*, 1991). Furthermore, the N-terminal domains of lamins are shorter than, for example, the corresponding region of vimentin and do not contain many basic residues that could mediate the dimer–dimer interactions (Herrmann and Aebi, 2004). Hence, it could be concluded that phosphorylation-mediated disassembly of IFs during mitosis is primarily regulated by N-terminal phosphorylation events, while C-terminal phosphorylation affects mitosis due to structural specificities of individual IFs.

V. STRESS-INDUCED PHOSPHORYLATION OF IFs

Mitosis is not the only physiological event during which intermediate filaments are markedly phosphorylated. Numerous studies with cultured cells and mouse models have revealed a strong correlation between IF hyperphosphorylation and cell stress (reviewed in Ku *et al.*, 1999). Different IF proteins, especially keratins (see later) and NFs (Giasson and Mushynski, 1996), have been shown to become phosphorylated either during stress or by stress-activated kinases. In addition to mouse models and cell line studies, K8/K18 hyperphosphorylation was shown to correlate with the progression of human liver disease (Toivola *et al.*, 2004). Stress-associated phosphorylation frequently results in the reorganization of IFs but could also affect the interaction of IFs with other proteins. In this context, keratins and the role of phosphorylated keratins during stress have been well examined.

Keratins 8 and 18 are phosphorylated by heat stress and rotavirus infection of human HT29 colonic cells (Liao *et al.*, 1995a). In these cells, stress alters the filament organization and increases the amount of soluble keratins (Liao *et al.*, 1995a). Phosphorylation was localized to the N-terminus of K8 namely, Ser-73 (Liao *et al.*, 1997). A significant increase in K8/K18 phosphorylation has also been detected in mice fed with a hepatotoxic drug, griseofulvin (Ku *et al.*, 1996a). In mice, the phosphorylated site on K8 was identified as Ser-79, homologous to Ser-73 in human, and the human antibody against Ser-73 cross-reacts with the mouse Ser-79 (Liao *et al.*, 1997). Results from anisomycin and etoposide-treated cells have revealed that phosphorylation of Ser-73 is also associated with apoptosis (Liao *et al.*, 1997). Moreover, it has been demonstrated that this site is phosphorylated in response to activation of the death receptor Fas (He *et al.*, 2002), and in alveolar epithelial cells in response to shear stress induced by laminar flow (Ridge *et al.*, 2005). In addition to K8, epidermal K5 and K6 have also been reported to be phosphorylated in the conserved LLS/TPL motif during UV light or anisomycin-induced apoptosis (Toivola *et al.*, 2002).

Since Ser-73 is frequently phosphorylated during different stressful conditions, it has been suggested to be a marker of physiological stress that accumulates in epithelial cells (Liao *et al.*, 1997). It also suggests that this site may be recognized by more than one kinase(s). To date, three different kinases have been suggested to be responsible for this modification. p38 stress-activated kinase phosphorylates K8 (Ku *et al.*, 2002a) as well as K5 and K6 *in vitro* (Toivola *et al.*, 2002). p38 phosphorylates K8 on Ser-73 in cells transfected with p38 and also associates with K8/K18 filaments (Ku *et al.*, 2002a). The second identified kinase is also a member of the MAP-kinase family, c-Jun kinase, which is activated during apoptosis and also phosphorylates Ser-73 both *in vitro* and *in vivo* (He *et al.*, 2002). The kinase responsible for the phosphorylation of K8 in alveolar epithelial cells in response to sheer stress has been suggested to be PKCδ, and a PKCδ peptide antagonists has been shown to attenuate the stress-induced increase in keratin phosphorylation and solubility (Ridge *et al.*, 2005).

Although stress-induced phosphorylation and reorganization of keratin filaments is well described, the significance of this modification is not clear (Toivola *et al.*, 2002). Phosphorylated keratins do not disrupt the intermediate filament organization, suggesting that phosphorylation of keratins is not required for structural modifications (Liao *et al.*, 1995a, 1997). It has been suggested that phosphorylation of keratins could instead regulate the interactions between keratins and other cellular elements (Liao *et al.*, 1997). Such interactions between phosphorylated keratins and their ligand have been shown to enhance the stress-related pathogenesis (Kirfel *et al.*, 2003). However, there is also evidence to support a protective effect of keratin phosphorylation during cell stress. For instance, transgenic mice overexpressing a mutant K18 (Ser-52 mutated to alanine) are more susceptible to microcystin- and griseofulvin-induced liver injury compared with their wild-type littermates (Ku *et al.*, 1998b). IFs have been appropriately referred to as "stress-buffering" elements of metazoan cells (Herrmann and Aebi, 2004), thus, illustrating the versatile roles of IFs in protection against both mechanical and chemical stresses. Consistent with this, keratins and NFs have been suggested to function as phosphate sinks, thus, protecting cells from excess kinase activity (Ku *et al.*, 1998b; Nguyen *et al.*, 2001). Phosphorylation also protects the keratins themselves from ubiquitination and subsequent proteasome-mediated degradation (Ku and

Omary, 2000), as well as from cleavage by caspases (Ku and Omary, 2001). In conclusion, the N-terminal phosphorylation of keratins has an effect on cell fate during stress including apoptosis. Phosphorylation seems to protect cells and, at least partially, modulate the organization of the filaments either directly or through other modifications. While the major stress-related phosphorylation sites of keratins are located in the head domain, the role of C-terminal phosphorylation during these conditions cannot be discounted.

Phosphorylation of vimentin in context of stress was reported, when vimentin was shown to become phosphorylated on Ser-82 during African swine fever virus (ASFV) infection (Stefanovic *et al.*, 2005). During the early steps of infection, vimentin was demonstrated to concentrate on virus assembly sites and on later stages to be rearranged into a cage-like structure surrounding virus factories. Vimentin within the cage was shown to be phosphorylated on Ser-82, a site recognized by calmodulin-dependent kinase II (CaMKII; Ando *et al.*, 1991), but the authors were not able to confirm whether the reorganization was due to Ser-82 phosphorylation or was the phosphorylation just a consequence of activation of CaMKII by ASFV (Stefanovic *et al.*, 2005). However, it is tempting to speculate the reasons for the observed vimentin reorganization. It may be that vimentin is needed to form a structural scaffold to recruit the proteins necessary to viral DNA replication, thus, favoring the virus, or it might serve as a cytoprotector to prevent the movement of viral components into the cytoplasm (Stefanovic *et al.*, 2005).

VI. EFFECTS OF IF PHOSPHORYLATION ON CELL SIGNALING

During the past 10 years, a number of new functions of IFs have emerged. Most importantly, numerous findings have connected IF networks to cell signaling. Phosphorylation has been shown to affect the interactions between IFs and different signaling determinants. Members of the 14-3-3 protein family are important regulators of the cell cycle and diverse signal transduction pathways. These proteins bind to their targets, such as PKC, Raf, Cdc25, and BAD, in a phosphorylation-dependent manner and affect the activity, interactions, and subcellular distribution of their target proteins (reviewed in Mackintosh, 2004). 14-3-3 proteins have been shown to associate with epithelial K8/K18 (Liao and Omary, 1996). They bind to phosphorylated K18 and this binding can be detected during S/G2/M phases of the cell cycle when keratins disassemble on hyperphosphorylation (Liao and Omary, 1996). The N-terminal Ser-33 was determined to be essential for the interaction between K18 and 14-3-3 (Ku *et al.*, 1998a). In addition to K18, 14-3-3 also binds to vimentin. The latter interaction has been induced in COS7 cells by phosphatase inhibitor calyculin A and the region responsible is defined to the N-terminal domain of vimentin (Tzivion *et al.*, 2000). Liao and Omary (1996) have suggested that 14-3-3 could function as a solubility cofactor for K8/K18, since in *in vitro* assays 14-3-3 improves the solubility of keratin filaments isolated from mitotic cells. This is consistent with studies demonstrating the significance of N-terminal phosphorylation sites in filament assembly (as discussed earlier). On the other hand, this interaction could modulate other interactions occurring between keratins and different proteins, or alternatively alter interactions between 14-3-3 and its targets (Liao and Omary, 1996). An interrelationship of the latter kind has been demonstrated with vimentin in which the association between vimentin and 14-3-3 replaces the interactions with other partners, one of them being protooncogene Raf (Tzivion *et al.*, 2000). In transgenic mice expressing the K18 mutant (Ser-33 to alanine), the mitotic progression of hepatocytes was disturbed and was accompanied by the nuclear retention of 14-3-3ζ as opposed to a more diffuse distribution in wild-type littermates (Ku *et al.*, 2002b). The same phenomenon has been seen in K8 knockout mice (Toivola *et al.*, 2001). Since the absence of K8 and K18 results in severe structural defects in cells and disrupts cell cycle, it has been suggested that the lack of filaments may lead to unregulated interactions between 14-3-3 proteins and the regulators of the cell cycle, such as Cdc25, thereby, disturbing the progression of mitosis (Toivola *et al.*, 2001).

IFs also appear to control the distribution of kinases that phosphorylate them. Activation of RhoA induces RhoA-binding kinase α-mediated phosphorylation of the N-terminal head of vimentin (Sin *et al.*, 1998). This causes a collapse of the vimentin network and a release of vimentin-associated inactive RhoA-binding kinase α because of its reduced affinity to phosphorylated vimentin. This results in a positive feedback loop in which the released kinases are reactivated, thus, allowing it to phosphorylate additional vimentin molecules (Sin *et al.*, 1998). In addition to vimentin and RhoA-binding kinase α, an interesting regulatory system has been revealed between K8/K18 filaments, 14-3-3 and Raf-1. It has been shown that Raf-1 directly associates with K8 and is capable of phosphorylating K18 preferentially on Ser-52 (Ku *et al.*, 2004). During oxidative or toxin-mediated stress, Raf-1 is activated and released from keratin filaments. The authors suggest that inactive Raf-1 associating with K8 would, on activation, phosphorylate K18 and dissociate it from the filaments (Ku *et al.*, 2004). However, it is difficult to demonstrate whether the phosphorylation is mediated by the bound kinase or by the unbound pool of Raf-1 (Ku *et al.*, 2004). The same report also demonstrated that the same sites on both Raf-1 and 14-3-3 that are essential for their interaction with each other also regulate the association of these molecules with K8/K18 indicating a role for keratins in modulating the signaling mediated by Raf-1 and 14-3-3 (Ku *et al.*, 2004).

Another example of IFs as signaling determinants is the association of nestin with the cyclin-dependent kinase 5 (Cdk5). Cdk5 interacts with nestin in its inactive state, as the association between p35, the activator protein of Cdk5, and nestin is detected only when the activity of Cdk5 is inhibited (Sahlgren *et al.*, 2003). It has been suggested that activated Cdk5 is released from filaments after it has performed its task by phosphorylating nestin (Sahlgren *et al.*, 2003). Unlike vimentin and keratin phosphorylation, the Cdk5-mediated phosphorylation sites reside in the C-terminal region of the nestin protein (Sahlgren *et al.*, 2003). These sites determine the association between the kinase and filaments and play a role in regulating the proapoptotic activity of Cdk5 (Sahlgren *et al.*, in press). A more recent observation has identified the phosphorylation of vimentin on N-terminal Ser-55 during mitosis by Cdk1, which resulted in the association of phosphorylated vimentin and yet another kinase, Plk1 (Yamaguchi *et al.*, 2005). This interaction in turn resulted in the activation of Plk1, phosphorylation of vimentin at yet another N-terminal site and segregation of filaments between the daughter cells (Yamaguchi *et al.*, 2005). Thus, phosphorylation itself may generate additional interactions between the kinase(s) and an IF protein.

VII. PHOSPHORYLATION AND NEUROFILAMENTS

Although C-terminal phosphorylation seems to be less frequent among cytoplasmic IFs, there is one well-described instance. Type IV IF proteins, NFs, are strictly regulated through phosphorylation of their C-terminal domain (Grant and Pant, 2000). The phosphorylation-mediated regulation of NFs is discussed extensively elsewhere (for review see Grant and Pant, 2000), therefore, in this chapter we will only highlight the main features of this regulation. NFs, the neuron-specific intermediate filaments, together with other cytoskeletal components form the axonal network, which includes the structure of the axons and influences the axonal transport of nerve impulses (Grant and Pant, 2000). NFs are divided into three distinct groups according to their molecular masses: NF-H (~200 kDa, high), NF-M (~160 kDa, medium), and NF-L (~68 kDa, low; Hirokawa and Takeda, 1998). When NFs assemble, NF-L forms the filament backbone to which NF-M and NF-H (both of which are incapable of self-assembly) attach (Hirokawa and Takeda, 1998). NF-H and NF-M carry long tail domains, side arms that extend from the filament core and form links to adjacent NFs as well as to other nearby cytoskeletal components (Hirokawa and Takeda, 1998; Grant and Pant, 2000). A distinctive feature of the C-terminal side arms is an enrichment of lysine–serine–proline (KSP)-repeat motifs (Grant and Pant, 2000). NF subunits are the most extensively phosphorylated proteins in the nervous system (for review see Chan *et al.*, 2004). The phosphorylation is initiated in the perikaryal region, where NFs are synthesized, and the modification becomes more apparent when NFs enter the axons (Grant and Pant, 2000).

C-terminal phosphorylation regulates NFs in multiple ways. First of all, several results indicate that phosphorylation of KSP-repeats has an effect on the transport of NFs along axons; NFs produced in neuronal cell body are delivered to axons. It is not clear if they are transported as intact filaments or disassembled into smaller complexes (Al-Chalabi and Miller, 2003). The transport of NFs is known to be assisted by microtubules, and NFs are also known to associate with the motor proteins kinesin (Yabe *et al.*, 1999). The role of phosphorylation is not completely understood, but there is evidence to show that hyperphosphorylation slows the transport of NFs down while hypophosphorylation accelerates it (Ackerley *et al.*, 2000, 2003; Jung *et al.*, 2000; Yabe *et al.*, 2001). C-terminal serine to aspartic mutations (mimicking permanent phosphorylation) of NF-H delays the movements while alanine mutants (that cannot be phosphorylated) travel faster (Ackerley *et al.*, 2003). The precise mechanism of this effect is not clear but different suggestions have been presented. One possibility is that phosphorylation causes the NFs to detach from the motor proteins (Shea, 2000; Ackerley *et al.*, 2003). It was reported that phosphorylation of NF-H induces dissociation of NFs from kinesin, thereby, retarding the axonal transport (Jung *et al.*, 2005). Alternatively, phosphorylation may induce NFs to interact with other axonal structures that could somehow modify the movements of NFs (Ackerley *et al.*, 2003). However, the exact role and the extent of phosphorylation has remained controversial, because the rate of neurofilament transport is not altered in transgenic mice in which wild-type NF-H is replaced by a deleted form of NF-H lacking the C-terminal tail (Rao *et al.*, 2002; Yuan *et al.*, 2006). A similar effect has been seen in mice expressing tailless NF-M (Rao *et al.*, 2003). There may be partially overlapping roles of NF-M and NF-H, which could explain some of the transgenic mice data (Rao *et al.*, 2003).

In addition to axonal transport, phosphorylation of NFs has also been demonstrated to regulate radial growth of axons (Grant and Pant, 2000). It has been suggested that C-terminal phosphorylation could change the charge density of side arms and, thereby, induce negative repulsions and protrusion of side arms from the filament backbone (Hirokawa and Takeda, 1998; Grant and Pant, 2000). These lateral protrusions would increase NF spacing and, therefore, increase axonal caliber and conduction velocity (Grant and Pant, 2000). In addition, phosphorylation of NF tail regions modulates the interactions of NFs with each other and with other cytoskeletal structures such as microtubules, actin, and plectin (Grant and Pant, 2000). These connections influence the formation of cytoskeletal network supporting the structure of axon. There is evidence of both attractive and forbidding interactions created by NF phosphorylation representing the dynamic nature of cytoskeletal network in axons (reviewed in Grant and Pant, 2000). Studies with mice models indicate that NF-H phosphorylation may have only a partial effect on axon radius while the overall composition of the NF network would contribute more prominently (Hirokawa and Takeda, 1998; Rao *et al.*, 1998, 2002). It has been suggested that the tail domain of NF-M could actually be more crucial in mediating radial axonal growth rather than the longer tail of NF-H, since abolition of NF-M tails eliminates the long cross-bridges between NFs that participate in strengthening the axonal structure (Garcia *et al.*, 2003; Rao *et al.*, 2003).

Various kinases have been demonstrated to phosphorylate KSP-repeats, for example, Cdk5 (Shetty *et al.*, 1993), glycogen synthase kinase (GSK)-3α and GSK-3β (Bajaj and Miller, 1997; Sasaki *et al.*, 2002), Erk1/2 (Veeranna *et al.*, 1998; Chan *et al.*, 2004), and stress-activated protein kinase 1b or JNK3 (Brownlees *et al.*, 2000). These kinases phosphorylate distinct C-terminal serines; for example, Cdk5 prefers KSPXK-repeats (Shetty *et al.*, 1993; Pant *et al.*, 1997) whereas Erk1/2 phosphorylates KSPXXXK-sequences (Veeranna *et al.*, 1998). Cdk5, which is the workhorse of the developing central nervous system, is involved in the movement of NFs into axons, and its inhibition accelerates axonal transport (Ackerley *et al.*, 2003; Shea *et al.*, 2004). Contrary to Cdk5, Erk1/2-mediated phosphorylation seems to promote anterograde NF transport (Chan *et al.*, 2004). Kinases responsible for neurofilament phosphorylation are tightly regulated as demonstrated by the fact that overexpression of Cdk5 induces perikaryal accumulation of phosphorylated NFs (Shea *et al.*, 2004).

The importance of stringent control of kinases is also well illustrated by the deleterious effects of Cdk5 in certain pathological conditions such as Alzheimers and amyotrophic lateral sclerosis (ALS; Nguyen *et al.*, 2001; Cruz and Tsai, 2004). In these diseases, Cdk5 has been shown to contribute to the formation of insoluble neurofilament aggregates that are a characteristic feature of most neurodegenerative disorders (Nguyen *et al.*, 2001). Likewise p38α, which also phosphorylates NFs, is associated with accumulation of phosphorylated NFs in ALS (Ackerley *et al.*, 2004).

Although neurons do not go through cell divisions, NFs still contain N-terminal phosphorylation sites, as discussed earlier. N-terminal phosphorylation blocks the assembly of NFs (Gonda *et al.*, 1990). It has been postulated that phosphorylation of the head domain of NF-M prevents the neurofilament assembly in cell bodies and inhibits tail domain phosphorylation (Zheng *et al.*, 2003). This could impede the premature assembly of newly synthesized NFs before they enter axons (Zheng *et al.*, 2003). Hence, there seems to be a very clear division between the roles of N- and C-terminal phosphorylation of NFs, as N-terminal phosphorylation regulates the assembly and controls the initiation of C-terminal modifications while C-terminal phosphorylation regulates neuron-specific functions of NFs.

VIII. DEPHOSPHORYLATION OF IFs

In addition to kinases, protein phosphatases carry out an important task in phosphorylation-mediated regulation of IF dynamics. Two major groups of protein phosphatases regulate eukaryotic cell functions, the serine/threonine protein phosphatases, and tyrosine phosphatases. With the help of serine/threonine phosphatase inhibitors, such as okadaic acid, calyculin A, and microcystin, it has been demonstrated that there is an accumulation of hyperphosphorylated IF proteins, suggesting that serine/threonine phosphatases are required for the rapid and continuous turnover of phosphorylated IFs, including vimentin (Eriksson *et al.*, 1992b), keratin (Kasahara *et al.*, 1993; Yatsunami *et al.*, 1993; Toivola *et al.*, 1997), NFs (Sacher *et al.*, 1994; Saito *et al.*, 1995; Gong *et al.*, 2003), and lamins (Luscher *et al.*, 1991). Moreover, phosphatase-induced inhibition hyperphosphorylation of IF proteins also alters IF organization, demonstrating the significance of these phosphatases in the maintenance of the structural integrity of IFs (Eriksson *et al.*, 1992b; Kasahara *et al.*, 1993; reviewed in Ku *et al.*, 1996b; Eriksson *et al.*, 1998). The effect of phosphatase inhibition on IF structure and dynamics has also been explored using GFP-tagged IFs and live cell imaging (Strnad *et al.*, 2001, 2002). In this study both serine/threonine and tyrosine phosphatase inhibitions were shown to induce a breakdown of keratin network and result in the formation of granular keratin aggregates instead (Strnad *et al.*, 2001, 2002). In addition to the structure and organization of IFs, phosphatases have been determined to maintain the interactions between IFs and other cellular complexes. Studies using microcystin have revealed that protein phosphatase activity is required for the regulation of the interaction between keratin IFs and desmosomes (Toivola *et al.*, 1997).

The temporary increase in IF phosphorylation on mitosis, when filaments are disassembled, indicates that phosphatase activation (as well as kinase inactivation) is likely to be responsible for the restoration of the basal IF phosphorylation levels (Ku *et al.*, 1996b) and readapting to the interphase-specific IF conformation. It has been demonstrated that protein phosphatase 1 (PP1) is a mitotic lamin phosphatase required to remove the mitotic phosphates from lamin B to enable nuclear lamina reassembly (Thompson *et al.*, 1997). In certain instances, IFs are phosphorylated at mitosis-specific sites also during interphase, although the amount of phosphorylation during this process is minor compared with the phosphorylation levels during cell division. For instance, in cells treated with phosphatase inhibitors, vimentin and desmin are phosphorylated during interphase at sites that are also phosphoryalted by Rho-kinase during cytokinesis (Inada *et al.*, 1999). However, the phosphorylation levels are diminished to almost undetectable levels due to action of PP1, indicating a continuous, tightly controlled balance of kinase and phosphatase activities (Inada *et al.*, 1999). In a similar way as IFs interact with kinases they can also bind protein phosphatases. In several studies, the neurofilaments have been described to be associated with protein phosphotases that dephosphorylate NFs (Veeranna *et al.*, 1995; Strack *et al.*, 1997; Terry-Lorenzo *et al.*, 2000). Protein phosphatase 2A (PP2A) has been shown to be associated with vimentin through a specific complex subunit (Turowski *et al.*, 1999), as well as to K8/K18 (Tao *et al.*, 2006). The association of PP2A with K8/K18 has also been shown to protect cells against hypo-osmotic stress (Tao *et al.*, 2006). This study then provides another link between phosphorylation of IFs and stress tolerance in the cell. Further, these studies suggest that IFs form a scaffold for both kinases and phosphatases (Terry-Lorenzo *et al.*, 2000).

IX. CONCLUSIONS

Site-specific phosphorylation is an important regulator of the dynamic and functional properties of IFs. N-terminal modifications seem to be most crucial for assembly/disassembly equilibria, during mitosis, differentiation, and cell stress. This is likely due to the effects of phosphorylation on the ionic interactions between the head and the rod domains, the formation of which are crucial for filament assembly. The exceptions to the rule of the N-terminus as a regulator of assembly dynamics and equilibria seem to stem from members of the IF protein family that have either an exceptionally short N-terminal domain (e.g., nestin) or different polymerization designs (lamins) (Fig. 2). While it has been well established that protein phosphatases are crucial in maintaining IF structure, recent results suggest that the association between IFs and protein phosphatases also have a role in directing signals and mediating IF-mediated stress tolerance. C-terminal phosphorylation is best characterized in NFs. The exceptionally high phosphorylation stoichiometry and the peculiar localization and pattern of the phosporylation sites are unique features of the NFs and are associated with neurospecific functions such as determining axonal caliber and mobility of the IFs and their associated proteins along the axons (Fig. 2). While

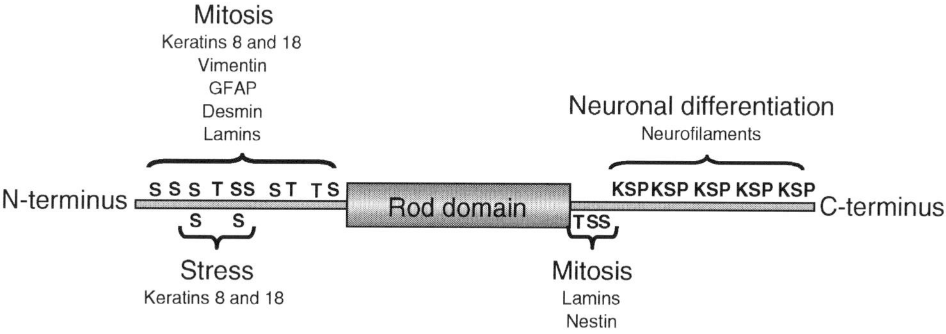

Figure 2. Domain-specific localization of phosphorylation sites with specific functions. Mitosis and stress-associated sites, which have been shown to contribute to filament assembly/disassembly, are almost exclusively located at the N-terminus, whereas the phosphorylation sites responsible for the specific functions of NFs are localized at the C-terminus.

C-terminal sites in other IFs have been shown to be phosphorylated in many situations, for example, the vimentin C-terminus during mitosis (Chou *et al.*, 1996) and the K8 C-terminus after epidermal growth factor stimulation (Ku and Omary, 1997), the precise functions of these sites remain to be elucidated. During the past few years, an increasing number of new functions have been assigned to IFs. As many of these functions are linked to interactions between IFs and different IF-associated molecules, it is likely that phosphorylation-mediated regulation orchestrates these interactions, as there are already a number of examples of phosphorylation as a regulator of IFs interaction with signaling molecules (Liao and Omary, 1996; Sin *et al.*, 1998; Tzivion *et al.*, 2000). The C-terminal phosphorylation sites that still lack an identified function could participate in regulating some of these known interactions or yet unidentified interactions.

ACKNOWLEDGMENTS

We thank Aura Kaunisto and Thomas Söderstöm for comments and critical reading of the review. This work was supported by Academy of Finland, the Sigrid Jasélium Foundation, the Finnish Cancer Associations, the Research Institute of Åbo Akademi University, and the Foundation of the Åbo Akademi University.

REFERENCES

Ackerley, S., Grierson, A. J., Brownlees, J., Thornhill, P., Anderton, B. H., Leigh, P. N., Shaw, C. E., and Miller, C. C. J. (2000). Glutamate slows axonal transport of neurofilaments in transfected neurons. *J. Cell Biol.* **150,** 165–176.

Ackerley, S., Thornhill, P., Grierson, A. J., Brownlees, J., Anderton, B. H., Leigh, P. N., Shaw, C. E., and Miller, C. C. J. (2003). Neurofilament heavy chain side arm phosphorylation regulates axonal transport of neurofilaments. *J. Cell Biol.* **161,** 489–495.

Ackerley, S., Grierson, A. J., Banner, S., Perkinton, M. S., Brownlees, J., Byers, H. L., Ward, M., Thornhill, P., Hussain, K., Waby, J. S., Anderton, B. H., *et al.* (2004). p38alpha stress-activated protein kinase phosphorylates neurofilaments and is associated with neurofilament pathology in amyotrophic lateral sclerosis. *Mol. Cell. Neurosci.* **26,** 354–364.

Al-Chalabi, A. and Miller, C. C. (2003). Neurofilaments and neurological disease. *Bioessays* **25,** 346–355.

Ando, S., Tanabe, K., Gonda, Y., Sato, C., and Inagaki, M. (1989). Domain- and sequence-specific phosphorylation of vimentin induces disassembly of the filament structure. *Biochemistry* **28,** 2974–2979.

Ando, S., Tokui, T., Yamauchi, T., Sugiura, H., Tanabe, K., and Inagaki, M. (1991). Evidence that Ser-82 is a unique phosphorylation site on vimentin for Ca2(+)-calmodulin-dependent protein kinase II. *Biochem. Biophys. Res. Com.* **175,** 955–962.

Ando, S., Tokui, T., Yano, T., and Inagaki, M. (1996). Keratin 8 phosphorylation *in vitro* by cAMP-dependent protein kinase occurs within amino- and carboxyl-terminal end domains. *Biochem. Biophys. Res. Com.* **221,** 67–71.

Bader, B. L., Magin, T. M., Freudenmann, M., Stumpp, S., and Franke, W. W. (1991). Intermediate filaments formed de novo from tail-less cytokeratins in the cytoplasm and in the nucleus. *J. Cell Biol.* **115,** 1293–1307.

Bajaj, N. P. and Miller, C. C. (1997). Phosphorylation of neurofilament heavy-chain side-arm fragments by cyclin-dependent kinase-5 and glycogen synthase kinase-3alpha in transfected cells. *J. Neurochem.* **69,** 737–743.

Brownlees, J., Bajaj, Y. N. P., Anderton, D. B. H., Leigh, P. N., Shaw, C. E., and Miller, C. C. J. (2000). Phosphorylation of neurofilament heavy chain side-arms by stress activated protein kinase-1b/Jun N-terminal kinase-3. *J. Cell Sci.* **113,** 401–407.

Chan, W. K.-H., Dickerson, A., Ortiz, D., Pimenta, A. F., Moran, C. M., Motil, J., Snyder, S. J., Malik, K., Pant, H. C., and Shea, T. B. (2004). Mitogen-activated protein kinase regulates neurofilament axonal transport. *J. Cell Sci.* **117,** 4629–4642.

Chou, Y.-H., Bischoff, J. R., Beach, D., and Goldman, R. D. (1990). Intermediate filament reorganization during mitosis is mediated by p34cdc2 phosphorylation of vimentin. *Cell* **62,** 1063–1071.

Chou, Y.-H., Ngai, K. L., and Goldman, R. D. (1991). The regulation of intermediate filament reorganization in mitosis. p34cdc2 phosphorylates vimentin at a unique N-terminal site. *J. Biol. Chem.* **266,** 7325–7328.

Chou, Y.-H., Opal, P., Quinlan, R. A., and Goldman, R. D. (1996). The relative roles of specific N- and C-terminal phosphorylation sites in the disassembly of intermediate filament in mitotic BHK-21 cells. *J. Cell Sci.* **109,** 817–826.

Chou, Y.-H., Khuon, S., Herrmann, H., and Goldman, R. D. (2003). Nestin promotes the phosphorylation-dependent disassembly of vimentin intermediate filaments during mitosis. *Mol. Biol. Cell* **14,** 1468–1478.

Coulombe, P. A. and Wong, P. (2004). Cytoplasmic intermediate filaments revealed as dynamic and multipurpose scaffolds. *Nat. Cell Biol.* **6,** 699–706.

Cross, T., Griffiths, G., Deacon, E., Sallis, R., Gough, M., Watters, D., and Lord, J. M. (2000). PKC-delta is an apoptotic lamin kinase. *Oncogene* **19,** 2331–2337.

Cruz, J. C. and Tsai, L.-H. (2004). Cdk5 deregulation in the pathogenesis of Alzheimer's disease. *Trends Mol. Med.* **10,** 452–458.

Dunphy, W. G., Brizuela, L., Beach, D., and Newport, J. (1988). The *Xenopus* cdc2 protein is a component of MPF, a cytoplasmic regulator of mitosis. *Cell* **29,** 423–431.

Eckelt, A., Herrmann, H., and Franke, W. W. (1992). Assembly of a tail-less mutant of the intermediate filament protein, vimentin, *in vitro* and *in vivo*. *Eur. J. Cell Biol.* **58,** 319–330.

Eckes, B., Colucci-Guyon, E., Smola, H., Nodder, S., Babinet, C., Krieg, T., and Martin, P. (2000). Impaired wound healing in embryonic and adult mice lacking vimentin. *J. Cell Sci.* **113,** 2455–2462.

Eggert, M., Radomski, N., Linder, D., Tripier, D., Traub, P., and Jost, E. (1993). Identification of novel phosphorylation sites in murine A-type lamins. *Eur. J. Biochem.* **213,** 659–671.

Eliasson, C., Sahlgren, C., Berthold, C. H., Stakeberg, J., Celis, J. E., Betsholtz, C., Eriksson, J. E., and Pekny, M. (1999). Intermediate filament protein partnership in astrocytes. *J. Biol. Chem.* **274,** 23996–24006.

Eriksson, J. E., Opal, P., and Goldman, R. D. (1992a). Intermediate filament dynamics. *Curr. Opin. Cell Biol.* **4,** 99–104.

Eriksson, J. E., Brautigan, D. L., Vallee, R., Olmsted, J., Fujiki, H., and Goldman, R. D. (1992b). Cytoskeletal integrity in interphase cells requires protein phosphatase activity. *Proc. Natl. Acad. Sci. USA* **89,** 11093–11097.

Eriksson, J. E., Toivola, D. M., Sahlgren, C., Mikhailov, A., and Harmala-Brasken, A.-S. (1998). Strategies to assess phosphoprotein phosphatase and protein kinase-mediated regulation of the cytoskeleton. *Meth. Enzymol.* **298,** 542–569.

Eriksson, J. E., He, T., Trejo-Skalli, A., Harmala-Braskén, A.-S., Hellman, J., Chou, Y.-H., and Goldman, R. D. (2004). Specific *in vivo* phosphorylation sites determine the assembly dynamics of vimentin intermediate filaments. *J. Cell Sci.* **117,** 919–932.

Evans, R. M. (1988a). The intermediate-filament proteins vimentin and desmin are phosphorylated in specific domains. *Eur. J. Cell Biol.* **46,** 152–160.

Evans, R. M. (1988b). Cyclic AMP-dependent protein kinase-induced vimentin filament disassembly involves modification of the N-terminal domain of intermediate filament subunits. *FEBS Lett.* **234,** 73–78.

Evans, R. M. and Fink, L. M. (1982). An alteration in the phosphorylation of vimentin-type intermediate filaments is associated with mitosis in cultured mammalian cells. *Cell* **29,** 43–52.

Feng, L., Zhou, X., Liao, J., and Omary, M. B. (1999). Pervanadate-mediated tyrosine phosphorylation of keratins 8 and 19 via a p38 mitogen-activated protein kinase-dependent pathway. *J. Cell Sci.* **112,** 2081–2090.

Foisner, R. (1997). Dynamic organisation of intermediate filaments and associated proteins during the cell cycle. *Bioessays* **19,** 297–305.

Garcia, M. L., Lobsiger, C. S., Shah, S. B., Deerinck, T. J., Crum, J., Young, D., Ward, C. M., Crawford, T. O., Gotow, T., Uchiyama, Y., Ellisman, M. H., Calcutt, N. A., *et al.* (2003). NF-M is an essential target for the myelin-directed "outside-in" signaling cascade that mediates radial axonal growth. *J. Cell Biol.* **163,** 1011–1020.

Gard, D. L. and Lazarides, E. (1982). Cyclic AMP-modulated phosphorylation of intermediate filament proteins in cultured avian myogenic cells. *Mol. Cell. Biol.* **2,** 1104–1114.

Geisler, N. and Weber, K. (1988). Phosphorylation of desmin *in vitro* inhibits formation of intermediate filaments; identification of three kinase A sites in the aminoterminal head domain. *EMBO J.* **7,** 15–20.

Gerace, L. and Blobel, G. (1980). The nuclear envelope lamina is reversibly depolymerized during mitosis. *Cell* **19,** 277–287.

Giasson, B. I. and Mushynski, W. E. (1996). Aberrant stress-induced phosphorylation of perikaryal neurofilaments. *J. Biol. Chem.* **271,** 30404–30409.

Gohara, R., Tang, D., Inada, H., Inagaki, M., Takasaki, Y., and Ando, S. (2001). Phosphorylation of vimentin head domain inhibits interaction with the carboxyl-terminal end of alpha-helical rod domain studied by surface plasmon resonance measurements. *FEBS Lett.* **489,** 182–186.

Gonda, Y., Nishizawa, K., Ando, S., Kitamura, S., Minoura, Y., Nishi, Y., and Inagaki, M. (1990). Involvement of protein kinase C in the regulation of assembly-disassembly of neurofilaments *in vitro*. *Biochem. Biophys. Res. Commun.* **167,** 1316–1325.

Gong, C.-X., Wang, J.-Z., Iqbal, K., and Grundke-Iqbal, I. (2003). Inhibition of protein phosphatase 2A induces phosphorylation and accumulation of neurofilaments in metabolically active rat brain slices. *Neurosci. Lett.* **340,** 107–110.

Goss, V. L., Hocevar, B. A., Thompson, L. J., Stratton, C. A., Burns, D. J., and Fields, A. P. (1994). Identification of nuclear beta II protein kinase C as a mitotic lamin kinase. *J. Biol. Chem.* **269,** 19074–19080.

Goto, H., Kosako, H., Tanabe, K., Yanagida, M., Sakurai, M., Amano, M., Kaibuchi, K., and Inagaki, M. (1998). Phosphorylation of vimentin by Rho-associated kinase at a unique amino-terminal site that is specifically phosphorylated during cytokinesis. *J. Biol. Chem.* **273,** 11728–11736.

Goto, H., Tanabe, K., Manser, E., Lim, L., Yasui, Y., and Inagaki, M. (2002). Phosphorylation and reorganization of vimentin by p21-activated kinase (PAK). *Genes Cells* **7,** 91–97.

Goto, H., Yasui, Y., Kawajiri, A., Nigg, E. A., Terada, Y., Tatsuka, M., Nagata, K., and Inagaki, M. (2003). Aurora-B regulates the cleavage furrow-specific vimentin phosphorylation in the cytokinetic process. *J. Biol. Chem.* **278,** 8526–8530.

Grant, P. and Pant, H. C. (2000). Neurofilament protein synthesis and phosphorylation. *J. Neurocytol.* **29,** 843–872.

Haas, M. and Jost, E. (1993). Functional analysis of phosphorylation sites in human lamin A controlling lamin disassembly, nuclear transport and assembly. *Eur. J. Cell Biol.* **62,** 237–247.

Hatzfeld, M. and Burba, M. (1994). Function of type I and type II keratin head domains: Their role in dimer, tetramer and filament formation. *J. Cell Sci.* **107,** 1959–1972.

He, T., Stepulak, A., Holmström, T. H., Omary, M. B., and Eriksson, J. E. (2002). The intermediate filament protein keratin 8 is a novel cytoplasmic substrate for c-Jun N-terminal kinase. *J. Biol. Chem.* **277,** 10767–10774.

Heald, R. and McKeon, F. (1990). Mutations of phosphorylation sites in lamin A that prevent nuclear lamina disassembly in mitosis. *Cell* **61,** 579–589.

Helfand, B. T., Chou, Y.-H., Shumaker, D. K., and Goldman, R. D. (2005). Intermediate filament proteins participate in signal transduction. *Trends Cell Biol.* **15,** 568–570.

Herrmann, H. and Aebi, U. (2004). Intermediate filaments: Molecular structure, assembly mechanism, and integration into functionally distinct intracellular scaffolds. *Annu. Rev. Biochem.* **73,** 749–789.

Herrmann, H., Häner, M., Brettel, M., Müller, S. A., Goldie, K. N., Fedtke, B., Lustig, A., Franke, W. W., and Aebi, U. (1996). Structure and assembly properties of the intermediate filament protein vimentin: The role of its head, rod and tail domains. *J. Mol. Biol.* **264,** 933–953.

Herrmann, H., Hesse, M., Reichenzeller, M., Aebi, U., and Magin, T. M. (2003). Functional complexity of intermediate filament cytoskeletons: From structure to assembly to gene ablation. *Int. Rev. Cytol.* **223,** 83–175.

Hirokawa, N. and Takeda, S. (1998). Gene targeting studies begin to reveal the function of neurofilament proteins. *J. Cell Biol.* **143,** 1–4.

Hocevar, B. A., Burns, D. J., and Fields, A. P. (1993). Identification of protein kinase C (PKC) phosphorylation sites on human lamin B. Potential role of PKC in nuclear lamina structural dynamics. *J. Biol. Chem.* **268,** 7545–7552.

Inada, H., Goto, H., Tanabe, K., Nishi, Y., Kaibuchi, K., and Inagaki, M. (1998). Rho-associated kinase phosphorylates desmin, the myogenic intermediate filament protein, at unique amino-terminal sites. *Biochem. Biophys. Res. Com.* **253,** 21–25.

Inada, H., Togashi, H., Nakamura, Y., Kaibuchi, K., Nagata, K., and Inagaki, M. (1999). Balance between activities of Rho kinase and type 1 protein phosphatase modulates turnover of phosphorylation and dynamics of desmin/vimentin filaments. *J. Biol. Chem.* **274,** 34932–34939.

Inagaki, M., Nishi, Y., Nishizawa, K., Matsuyama, M., and Sato, C. (1987). Site-specific phosphorylation induces disassembly of vimentin filaments *in vitro. Nature* **328,** 649–652.

Inagaki, M., Gonda, Y., Matsuyama, M., Nishizawa, K., Nishi, Y., and Sato, C. (1988). Intermediate filament reconstitution *in vitro*. The role of phosphorylation on the assembly-disassembly of desmin. *J. Biol. Chem.* **263,** 5970–5978.

Inagaki, M., Gonda, Y., Nishizawa, K., Kitamura, S., Sato, C., Ando, S., Tanabe, K., Kikuchi, K., Tsuiki, S., and Nishi, Y. (1990). Phosphorylation sites linked to glial filament disassembly *in vitro* locate in a non-alpha-helical head domain. *J. Biol. Chem.* **265,** 4722–4729.

Inagaki, M., Matsuoka, Y., Tsujimura, K., Ando, S., Tokui, T., Takahashi, T., and Inagaki, N. (1996). Dynamic property of intermediate filaments: Regulation by phosphorylation. *Bioessays* **18,** 481–487.

Ivaska, J., Vuoriluoto, K., Huovinen, T., Izawa, I., Inagaki, M., and Parker, P. J. (2005). PKCepsilon-mediated phosphorylation of vimentin controls integrin recycling and motility. *EMBO J.* **24,** 3834–3845.

Jung, C., Yabe, J. T., and Shea, T. B. (2000). C-terminal phosphorylation of the high molecular weight neurofilament subunit correlates with decreased neurofilament axonal transport velocity. *Brain Res.* **856,** 12–19.

Jung, C., Lee, S., Ortiz, D., Zhu, Q., Julien, J.-P., and Shea, T. B. (2005). The high and middle molecular weight neurofilament subunits regulate the association of neurofilaments with kinesin: Inhibition by phosphorylation of the high molecular weight subunit. *Brain Res. Mol. Brain Res.* **141,** 151–155.

Kasahara, K., Kartasova, T., Ren, X. Q., Ikuta, T., Chida, K., and Kuroki, T. (1993). Hyperphosphorylation of keratins by treatment with okadaic acid of BALB/MK-2 mouse keratinocytes. *J. Biol. Chem.* **268,** 23531–23537.

Kawajiri, A., Yasui, Y., Goto, H., Tatsuka, M., Takahashi, M., Nagata, K., and Inagaki, M. (2003). Functional significance of the specific sites phosphorylated in desmin at cleavage furrow: Aurora-B may phosphorylate and regulate type III intermediate filaments during cytokinesis coordinatedly with Rho-kinase. *Mol. Biol. Cell* **14,** 1489–1500.

Kirfel, J., Magin, T. M., and Reichelt, J. (2003). Keratins: A structural scaffold with emerging functions. *Cell. Mol. Life Sci.* **60,** 56–71.

Kosako, H., Amano, M., Yanagida, M., Tanabe, K., Nishi, Y., Kaibuchi, K., and Inagaki, M. (1997). Phosphorylation of glial fibrillary acidic protein at the same sites by cleavage furrow kinase and Rho-associated kinase. *J. Biol. Chem.* **272,** 10333–10336.

Ku, N.-O. and Omary, M. B. (1994). Identification of the major physiologic phosphorylation site of human keratin 18: Potential kinases and a role in filament reorganization. *J. Cell Biol.* **127,** 161–171.

Ku, N.-O. and Omary, M. B. (1997). Phosphorylation of human keratin 8 *in vivo* at conserved head domain serine 23 and at epidermal growth factor-stimulated tail domain serine 431. *J. Biol. Chem.* **272,** 7556–7564.

Ku, N.-O. and Omary, M. B. (2000). Keratins turn over by ubiquitination in a phosphorylation-modulated fashion. *J. Cell Biol.* **149,** 547–552.

Ku, N.-O. and Omary, M. B. (2001). Effect of mutation and phosphorylation of type I keratins on their caspase-mediated degradation. *J. Biol. Chem.* **276,** 26792–26798.

Ku, N.-O., Michie, S. A., Soetikno, R. M., Resurreccion, E. Z., Broome, R. L., Oshima, R. G., and Omary, M. B. (1996a). Susceptibility to hepatotoxicity in transgenic mice that express a dominant-negative human keratin 18 mutant. *J. Clin. Invest.* **98,** 1034–1046.

Ku, N.-O., Liao, J., Chou, C. F., and Omary, M. B. (1996b). Implications of intermediate filament protein phosphorylation. *Cancer Metastasis Rev.* **15,** 429–444.

Ku, N.-O., Liao, J., and Omary, M. B. (1998a). Phosphorylation of human keratin 18 serine 33 regulates binding to 14-3-3 proteins. *EMBO J.* **17,** 1892–1906.

Ku, N.-O., Michie, S. A., Soetikno, R. M., Resurreccion, E. Z., Broome, R. L., and Omary, M. B. (1998b). Mutation of a major keratin phosphorylation site predisposes to hepatotoxic injury in transgenic mice. *J. Cell Biol.* **143,** 2023–2032.

Ku, N.-O., Zhou, X., Toivola, D. M., and Omary, M. B. (1999). The cytoskeleton of digestive epithelia in health and disease. *Am. J. Physiol.* **277,** G1108–G1137.

Ku, N.-O., Azhar, S., and Omary, M. B. (2002a). Keratin 8 phosphorylation by p38 kinase regulates cellular keratin filament reorganization: Modulation by a keratin 1-like disease causing mutation. *J. Biol. Chem.* **277,** 10775–10782.

Ku, N.-O., Michie, S. A., Resurreccion, E. Z., Broome, R. L., and Omary, M. B. (2002b). Keratin binding to 14-3-3 proteins modulates keratin filaments and hepatocyte mitotic progression. *Proc. Natl. Acad. Sci. USA* **99,** 4373–4378.

Ku, N.-O., Fu, H., and Omary, M. B. (2004). Raf-1 activation disrupts its binding to keratins during cell stress. *J. Cell Biol.* **166,** 479–485.

Kusubata, M., Matsuoka, Y., Tsujimura, K., Ito, H., Ando, S., Kamijo, M., Yasuda, H., Ohba, Y., Okumura, E., Kishimoto, T., and Inagaki, M. (1993). cdc2 kinase phosphorylation of desmin at three serine/threonine residues in the amino-terminal head domain. *Biochem. Biophys. Res. Com.* **190,** 927–934.

Kommers, T., Rodnight, R., Boeck, C., Vendite, D., Oliviera, D., Horn, J., Oppelt, D., and Wofchuk, S. (2002). Phosphorylation of glial fibrillary acidic protein is stimulated by glutamate via NMDA receptors in cortical microslices and in mixed neuronal/glial cell cultures prepared from the cerebellum. *Dev. Brain Res.* **137,** 139–148.

Lepekhin, E. A., Eliasson, C., Berthold, C.-H., Berezin, V., Bock, E., and Pekny, M. (2001). Intermediate filaments regulate astrocyte motility. *J. Neurochem.* **79,** 617–625.

Liao, J. and Omary, M. B. (1996). 14-3-3 proteins associate with phosphorylated simple epithelial keratins during cell cycle progression and act as a solubility cofactor. *J. Cell Biol.* **133,** 345–357.

Liao, J., Lowthert, L. A., and Omary, M. B. (1995a). Heat stress or rotavirus infection of human epithelial cells generates a distinct hyperphosphorylated form of keratin 8. *Exp. Cell Res.* **219,** 348–357.

Liao, J., Lowthert, L. A., Ku, N.-O., Fernandez, R., and Omary, M. B. (1995b). Dynamics of human keratin 18 phosphorylation: Polarized distribution of phosphorylated keratins in simple epithelial tissues. *J. Cell Biol.* **131,** 1291–1301.

Liao, J., Ku, N.-O., and Omary, M. B. (1997). Stress, apoptosis, and mitosis induce phosphorylation of human keratin 8 at Ser-73 in tissues and cultured cells. *J. Biol. Chem.* **272,** 17565–17573.

Luscher, B., Brizuela, L., Beach, D., and Eisenman, R. N. (1991). A role for the p34cdc2 kinase and phosphatases in the regulation of phosphorylation and disassembly of lamin B2 during the cell cycle. *EMBO J.* **10,** 865–875.

Mackintosh, C. (2004). Dynamic interactions between 14-3-3 proteins and phosphoproteins regulate diverse cellular processes. *Biochem. J.* **381,** 329–342.

Matsuoka, Y., Nishizawa, K., Yano, T., Shibata, M., Ando, S., Takahashi, T., and Inagaki, M. (1992). Two different protein kinases act on a different time schedule as glial filament kinases during mitosis. *EMBO J.* **11,** 2895–2902.

Matsuzawa, K., Kosako, H., Inagaki, N., Shibata, H., Mukai, H., Ono, Y., Amano, M., Kozo, K., Matsuura, Y., Azuma, Y., and Inagaki, M. (1997). Domain-specific phosphorylation of vimentin and glial fibrillary acidic protein by PKN. *Biochem. Biophys. Res. Com.* **234,** 621–625.

Moir, R. D., Spann, T. P., Lopez-Soler, R. I., Yoon, M., Goldman, A. E., Khuon, S., and Goldman, R. D. (2000). Review: The dynamics of the nuclear lamins during the cell cycle—relationship between structure and function. *J. Struct. Biol.* **129,** 324–334.

Mukai, H., Toshimori, M., Shibata, H., Kitagawa, M., Shimakawa, M., Miyahara, M., Sunakawa, H., and Ono, Y. (1996). PKN associates and phosphorylates the head-rod domain of neurofilament protein. *J. Biol. Chem.* **271,** 9816–9822.

Nakamura, Y., Takeda, M., Angelides, K. J., Tanaka, T., Tada, K., and Nishimura, T. (1990). Effect of phosphorylation on 68 KDa neurofilament subunit protein assembly by the cyclic AMP dependent protein kinase *in vitro*. *Biochem. Biophys. Res. Commun.* **169,** 744–750.

Nieminen, M., Henttinen, T., Merinen, M., Marttila, F., Eriksson, J. E., and Jalkanen, S. (2006). Vimentin function in lymphocyte adhesion and transcellular migration. *Nat. Cell Biol.* **8,** 156–162.

Nguyen, M. D., Lariviere, R. C., and Julien, J.-P. (2001). Deregulation of Cdk5 in a mouse model of ALS: Toxicity alleviated by perikaryal neurofilament inclusions. *Neuron* **30,** 135–147.

O'Connor, C. M., Gard, D. L., and Lazarides, E. (1981). Phosphorylation of intermediate filament proteins by cAMP-dependent protein kinases. *Cell* **23,** 134–143.

Pant, A. C., Veeranna Pant, H. C., and Amin, N. (1997). Phosphorylation of human high molecular weight neurofilament protein (hNF-H) by neuronal cyclin-dependent kinase 5 (cdk5). *Brain Res.* **765,** 259–266.

Pant, H. C., Shecket, G., Gainer, H., and Lasek, R. J. (1978). Neurofilament protein is phosphorylated in the squid giant axon. *J. Cell Biol.* **78,** R23–R27.

Peter, M., Nakagawa, J., Doree, M., Labbe, J. C., and Nigg, E. A. (1990). *In vitro* disassembly of the nuclear lamina and M phase-specific phosphorylation of lamins by cdc2 kinase. *Cell* **61,** 591–602.

Peter, M., Heitlinger, E., Häner, M., Aebi, U., and Nigg, E. A. (1991). Disassembly of *in vitro* formed lamin head-to-tail polymers by CDC2 kinase. *EMBO J.* **10,** 1535–1544.

Raats, J. M., Gerards, W. L., Schreuder, M. I., Grund, C., Henderik, J. B., Hendriks, I. L., Ramaekers, F. C., and Bloemendal, H. (1992). Biochemical and structural aspects of transiently and stably expressed mutant desmin in vimentin-free and vimentin-containing cells. *Eur. J. Cell Biol.* **58,** 108–127.

Ralton, J. E., Lu, X., Hutcheson, A. M., and Quinlan, R. A. (1994). Identification of two N-terminal non-alpha-helical domain motifs important in the assembly of glial fibrillary acidic protein. *J. Cell Sci.* **107,** 1935–1948.

Rao, M. V., Houseweart, M. K., Williamson, T. L., Crawford, T. O., Folmer, J., and Cleveland, D. V. (1998). Neurofilament-dependent radial growth of motor axons and axonal organization of neurofilaments does not require the neurofilament heavy subunit (NF-H) or its phosphorylation. *J. Cell Biol.* **143,** 171–181.

Rao, M. V., Garcia, M. L., Miyazaki, Y., Gotow, T., Yuan, A., Mattina, S., Ward, C. M., Calcutt, N. A., Uchiyama, Y., Nixon, R. A., and Cleveland, D. W. (2002). Gene replacement in mice reveals that the heavily phosphorylated tail of neurofilament heavy subunit does not affect axonal caliber or the transit of cargoes in slow axonal transport. *J. Cell Biol.* **158,** 681–693.

Rao, M. V., Campbell, J., Yuan, A., Kumar, A., Gotow, T., Uchiyama, Y., and Nixon, R. A. (2003). The neurofilament middle molecular mass subunit carboxyl-terminal tail domains is essential for the radial growth and cytoskeletal architecture of axons but not for regulating neurofilament transport rate. *J. Cell Biol.* **163,** 1021–1031.

Ridge, K. M., Linz, L., Flitney, F. W., Kuczmarski, E. R., Chou, Y.-H., Omary, M. B., Sznajder, J. I., and Goldman, R. D. (2005). Keratin 8 phosphorylation by protein kinase C delta regulates shear stress-mediated disassembly of keratin intermediate filaments in alveolar epithelial cells. *J. Biol. Chem.* **280,** 30400–30405.

Sacher, M. G., Athlan, E. S., and Mushynski, W. E. (1994). Increased phosphorylation of the amino-terminal domain of the low molecular weight neurofilament subunit in okadaic acid-treated neurons. *J. Biol. Chem.* **269,** 18480–18484.

Sahlgren, C. M., Mikhailov, A., Hellman, J., Chou, Y.-H., Lendahl, U., Goldman, R. D., and Eriksson, J. E. (2001). Mitotic reorganization of the intermediate filament protein nestin involves phosphorylation by cdc2 kinase. *J. Biol. Chem.* **276,** 16456–16463.

Sahlgren, C. M., Mikhailov, A., Vaittinen, S., Pallari, H.-M., Kalimo, H., Pant, H. C., and Eriksson, J. E. (2003). Cdk5 regulates the organization of nestin and its association with p35. *Mol. Cell. Biol.* **23,** 5090–5106.

Sahlgren, C. M., Pallari, H.-M., He, T., Chou, Y.-H., Goldman, R. D., and Eriksson, J. E. (2006). A nestin scaffold links Cdk5/p35 signaling to oxidant induced cell death. *EMBO J.* (in press).

Saito, T., Shima, H., Osawa, Y., Nagao, M., Hemmings, B. A., Kishimoto, T., and Hisanaga, S. (1995). Neurofilament-associated protein phosphatase 2A: Its possible role in preserving neurofilaments in filamentous states. *Biochemistry* **34,** 7376–7384.

Sasaki, T., Taoka, M., Ishiguro, K., Uchida, A., Saito, T., Isobe, T., and Hisanaga, S. (2002). *In vivo* and *in vitro* phosphorylation at Ser-493 in the glutamate (E)-segment of neurofilament-H subunit by glycogen synthase kinase 3beta. *J. Biol. Chem.* **277,** 36032–36039.

Shea, T. B. (2000). Microtubule motors, phosphorylation and axonal transport of neurofilaments. *J. Neurocytol.* **29,** 873–887.

Shea, T. B., Yabe, J. T., Ortiz, D., Pimenta, A., Loomis, P., Goldman, R. D., Amin, N., and Pant, H. C. (2004). Cdk5 regulates axonal transport and phosphorylation of neurofilaments in cultured neurons. *J. Cell Sci.* **117,** 933–941.

Shetty, K. T., Link, W. T., and Pant, H. C. (1993). cdc2-like kinase from rat spinal cord specifically phosphorylates KSPXK motifs in neurofilament proteins: Isolation and characterization. *Proc. Natl. Acad. Sci. USA* **90,** 6844–6848.

Shimizu, T., Cao, C.-X., Shao, R.-G., and Pommier, Y. (1998). Lamin B phosphorylation by protein kinase calpha and proteolysis during apoptosis in human leukemia HL60 cells. *J. Biol. Chem.* **273,** 8669–8674.

Sin, W.-C., Chen, X.-Q., Leung, T., and Lim, L. (1998). RhoA-binding kinase alpha translocation is facilitated by the collapse of the vimentin intermediate filament network. *Mol. Cell. Biol.* **18,** 6325–6339.

Stefanovic, S., Windsor, M., Nagata, K., Inagaki, M., and Wileman, T. (2005). Vimentin rearrangement during African swine fever virus infection involves retrograde transport along microtubules and phosphorylation of vimentin by calcium calmodulin kinase II. *J. Virol.* **79,** 11766–11775.

Steinert, P. M., Chou, Y.-H., Prahlad, V., Parry, D. A., Marekov, L. N., Wu, K. C., Jang, S. I., and Goldman, R. D. (1999). A high molecular weight intermediate filament-associated protein in BHK-21 cells is nestin, a type VI intermediate filament protein. Limited co-assembly *in vitro* to form heteropolymers with type III vimentin and type IV alpha-internexin. *J. Biol. Chem.* **274,** 9881–9890.

Strelkov, S. V., Herrmann, H., and Aebi, U. (2003). Molecular architecture of intermediate filaments. *Bioessays* **25,** 243–251.

Strack, S., Westphal, R. S., Colbran, R. J., Ebner, F. F., and Wadzinski, B. E. (1997). Protein serine/threonine phosphatase 1 and 2A associate with and dephosphorylate neurofilaments. *Brain Res. Mol. Brain Res.* **49,** 15–28.

Strnad, P., Windoffer, R., and Leube, R. E. (2001). *In vivo* detection of cytokeratin filament network breakdown in cells treated with the phosphatase inhibitor okadaic acid. *Cell Tissue Res.* **306,** 277–293.

Strnad, P., Windoffer, R., and Leube, R. E. (2002). Induction of rapid and reversible cytokeratin filament network remodeling by inhibition of tyrosine phosphatases. *J. Cell Sci.* **115,** 4133–4148.

Sun, T.-T. and Green, H. (1978). Keratin filaments of cultured human epidermal cells. Formation of intermolecular disulfide bonds during terminal differentiation. *J. Biol. Chem.* **253,** 2053–2060.

Takai, Y., Ogawara, M., Tomono, Y., Moritoh, C., Imajoh-Ohmi, S., Tsutsumi, O., Taketani, Y., and Inagaki, M. (1996). Mitosis-specific phosphorylation of vimentin by protein kinase C coupled with reorganization of intracellular membranes. *J. Cell Biol.* **133,** 141–149.

Tao, G.-Z., Toivola, D. M., Zhou, Q., Strnad, P., Xu, B., Michie, S. A., and Omary, M. B. (2006). Protein phosphatase-2A associates with and dephosphorylates keratin 8 after hyposmotic stress in a site- and cell-specific manner. *J. Cell Sci.* **119,** 1425–1432.

Terry-Lorenzo, R. T., Inoue, M., Connor, J. H., Haystead, T. A. J., Armbruster, B. N., Gupta, R. P., Oliver, C. J., and Shenolikar, S. (2000). Neurofilament-L is a protein phosphatase-1-binding protein associated with neuronal plasma membrane and post-synaptic density. *J. Biol. Chem.* **275,** 2439–2446.

Thompson, L. J., Bollen, M., and Fields, A. P. (1997). Identification of protein phosphatase 1 as a mitotic lamin phosphatase. *J. Biol. Chem.* **272,** 29693–29697.

Toivola, D. M., Goldman, R. D., Garrod, D. R., and Eriksson, J. E. (1997). Protein phosphatases maintain the organization and structural interactions of hepatic keratin intermediate filaments. *J. Cell Sci.* **110,** 23–33.

Toivola, D. M., Nieminen, M. I., Hesse, M., He, T., Baribault, H., Magin, T. M., Omary, M. B., and Eriksson, J. E. (2001). Disturbances in hepatic cell-cycle regulation in mice with assembly-deficient keratins 8/18. *Hepatology* **34,** 1174–1183.

Toivola, D. M., Zhou, Q., English, L. S., and Omary, M. B. (2002). Type II keratins are phosphorylated on a unique motif during stress and mitosis in tissues and cultured cells. *Mol. Biol. Cell* **13,** 1857–1870.

Toivola, D. M., Ku, N.-O., Resurreccion, E. Z., Nelson, D. R., Wright, T. L., and Omary, M. B. (2004). Keratin 8 and 18 hyperphosphorylation is a marker of progression of human liver disease. *Hepatology* **40,** 459–466.

Toivola, D. M., Tao, G.-Z., Habtezion, A., Liao, J., and Omary, M. B. (2005). Cellular integrity plus: Organelle-related and protein-targeting functions of intermediate filaments. *Trends. Cell Biol.* **15,** 608–617.

Turowski, P., Myles, T., Hemmings, B. A., Fernandez, A., and Lamb, N. J. (1999). Vimentin dephosphorylation by protein phosphatase 2A is modulated by the targeting subunit B55. *Mol. Biol. Cell* **10,** 1997–2015.

Tzivion, G., Luo, Z.-J., and Avruch, J. (2000). Calyculin A-induced vimentin phosphorylation sequesters 14-3-3 and displaces other 14-3-3 partners *in vivo*. *J. Biol. Chem.* **275,** 29772–29778.

Valgeirsdottir, S., Claesson-Welsh, L., Bongcam-Rudloff, E., Hellman, U., Westermark, B., and Heldin, C. H. (1998). PDGF induces reorganization of vimentin filaments. *J. Cell Sci.* **111,** 1973–1980.

Veeranna Amin, N. D., Ahn, N. G., Jaffe, H., Winters, C. A., Grant, P., and Pant, H. C. (1998). Mitogen-activated protein kinases (Erk1,2) phosphorylate Lys-Ser-Pro (KSP) repeats in neurofilament proteins NF-H and NF-M. *J. Neurosci.* **18,** 4008–4021.

Veeranna Shetty, K. T., Link, W. T., Jaffe, H., Wang, J., and Pant, H. C. (1995). Neuronal cyclin-dependent kinase-5 phosphorylation sites in neurofilament protein (NF-H) are dephosphorylated by protein phosphatase 2A. *J. Neurochem.* **64,** 2681–2690.

Yamaguchi, Y., Goto, H., Yokoyama, T., Sillje, H., Hanisch, A., Uldschmid, A., Takai, Y., Oguri, T., Nigg, E. A., and Inagaki, M. (2005). Phosphorylation by Cdk1 induces Plk1-mediated vimentin phosphorylation during mitosis. *J. Cell Biol.* **171,** 431–436.

Yabe, J. T., Pimenta, A., and Shea, T. B. (1999). Kinesin-mediated transport of neurofilament protein oligomers in growing axons. *J. Cell Sci.* **112,** 3799–37814.

Yabe, J. T., Chylinski, T., Wang, F.-S., Pimenta, A., Kattar, S. D., Linsley, M.-D., Chan, W. K.-H., and Shea, T. B. (2001). Neurofilaments consist of distinct populations that can be distinguished by C-terminal phosphorylation, bundling, and axonal transport rate in growing axonal neurites. *J. Neurosci.* **21,** 2195–2205.

Yasui, Y., Amano, M., Nagata, K., Inagaki, N., Nakamura, H., Saya, H., Kaibuchi, K., and Inagaki, M. (1998). Roles of Rho-associated kinase in cytokinesis; mutations in Rho-associated kinase phosphorylation sites impair cytokinetic segregation of glial filaments. *J. Cell Biol.* **143,** 1249–1258.

Yatsunami, J., Komori, A., Ohta, T., Suganuma, M., Yuspa, S. H., and Fujiki, H. (1993). Hyperphosphorylation of cytokeratins by okadaic acid class tumor promoters in primary human keratinocytes. *Cancer Res.* **53,** 992–996.

Yuan, A., Nixon, R. A., and Rao, M. V. (2006). Deleting the phosphorylated tail domain of the neurofilament heavy subunit does not alter neurofilament transport rate *in vivo*. *Neurosci. Lett.* **393,** 264–268.

Zheng, Y., Li, B.-S., Veeranna, B.-S., and Pant, H. C. (2003). Phosphorylation of the head domain of neurofilament protein (NF-M): A factor regulating topographic phosphorylation of NF-M tail domain KSP sites in neurons. *J. Biol. Chem.* **278,** 24026–24032.

Abbreviations and Acronyms

ARF1	ADP ribosylation factor 1
Arp2/3	Actin-related protein 2/3
BDM	2,3-Butanedione monoxime
CFTR	Cystic fibrosis transmembrane conductance regulator
E3KARP	NHE3 kinase A regulatory protein
EGF	Epidermal growth factor
EGFP	Enhanced green fluorescent protein
ER	Endoplasmic reticulum
FRAP	Fluorescence recovery after photobleaching
$Ins(1,4,5)P_3$	Inositol 1,4,5-trisphosphate
MT	Microtubule
NHE3	Na^+H^+ antiporter isoform 3
NHERF	NHE regulatory factor
NKCC1	$Na^+K^+2Cl^-$ cotransporter
NRK	Normal rat kidney cells
PI	Phosphatidylinositol
PI(3)P, PI(4)P, PI(5)P	Phosphatidylinositol 3/4/5-phosphate
$PI(3,4)P_2$ $PI(3,5)P_2$ $PI(4,5)P_2$	Phosphatidylinositol 3,4/3,5/4,5-bisphosphate
$PI(3,4,5)P_3$	Phosphatidylinositol 3,$PI(4,5)P_2$-trisphosphate
PIP	All phosphorylated phosphoinositides
PIP kinases	Phosphatidylinositol phosphate kinases
PM	Plasma membrane
ROCK	Rho-associated kinase
SGK1	Serum- and glucocorticoid-inducible protein kinase-1
SGLT1	Na^+-glucose cotransporter
TGN	Trans-Golgi network
VSV-G	Vesicular stomatitis virus glycoprotein

Index

T

W

X

Y

Z

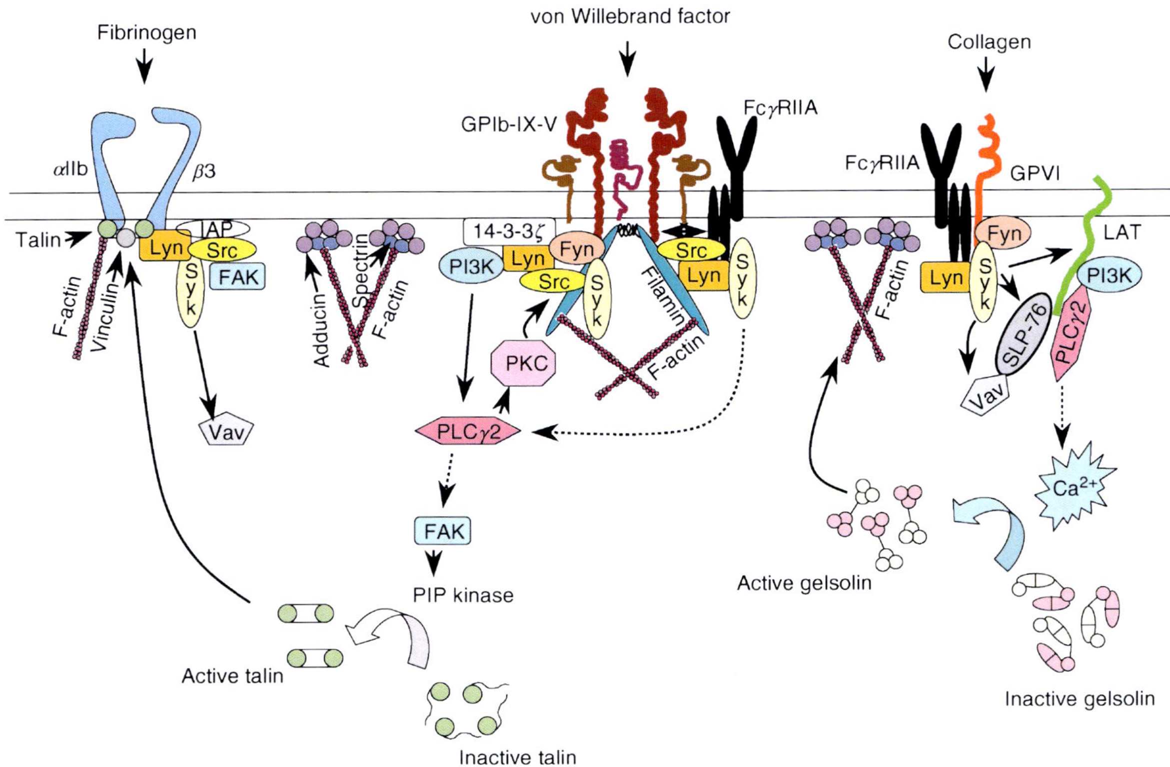

Chapter 1, Figure 1. *Continued*

Chapter 1, Figure 1. Model of platelet receptor signaling and cytoskeletal organization. Filamin attaches the GPIb-IX-V complex to actin filaments via the cytoplasmic domain of GPIbα. Binding of vWF to platelet GPIb-IX-V recruits a signaling complex including tyrosine kinases and the FcγRIIA receptor that leads to PI3-kinase-mediated activation of phospholipase Cγ2 (PLCγ2). PLCγ2 generates inositol (1,4,5)-trisphosphate and diacylglycerol, which lead to the mobilization of calcium from intracellular stores and activation of PKC, respectively. In platelets, the FcγRIIA receptor may associate with GPVI. Collagen binding leads to GPVI clustering, phosphorylation of FcγRIIA by tyrosine kinases as well as phosphorylation of the transmembrane adaptor protein LAT, which assembles a complex of signaling proteins including PI3-kinase, PLCγ2 and the adapter proteins SLP-76 and Vav. Increased cytoplasmic concentration of calcium induces a conformational change of gelsolin to expose its actin-binding sites. Active gelsolin severs actin filaments via a unique tail latch mechanism. Adductin caps the fast-growing ends of actin filaments and recruits spectrin to the filaments. Both spectrin and adductin provide the connection of the membrane skeleton to the plasma membrane.

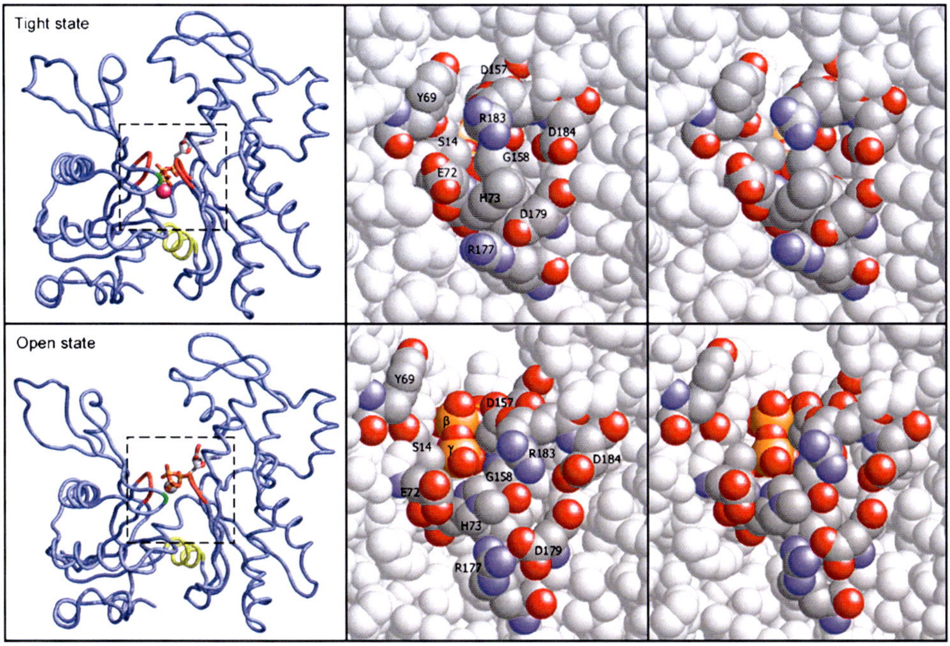

Chapter 3, Figure 1. Comparison of the tight and open states of actin in profilin:β-actin crystals. The overview to the left outlines the area shown in stereo containing the barrier residues E72, H73, R177, D179, and R183, the phosphate-binding loop residues S14, D157, and G158, and finally the hydrogen bond-forming residues Y69 and R183. In the open state (bottom row) the β- and γ-phosphates (yellow) of the ATP are exposed. Pdb accession codes 2BTF and 1HLU, respectively. Figure published in *J. Mol. Biol.* (2002) 317, 577–589.

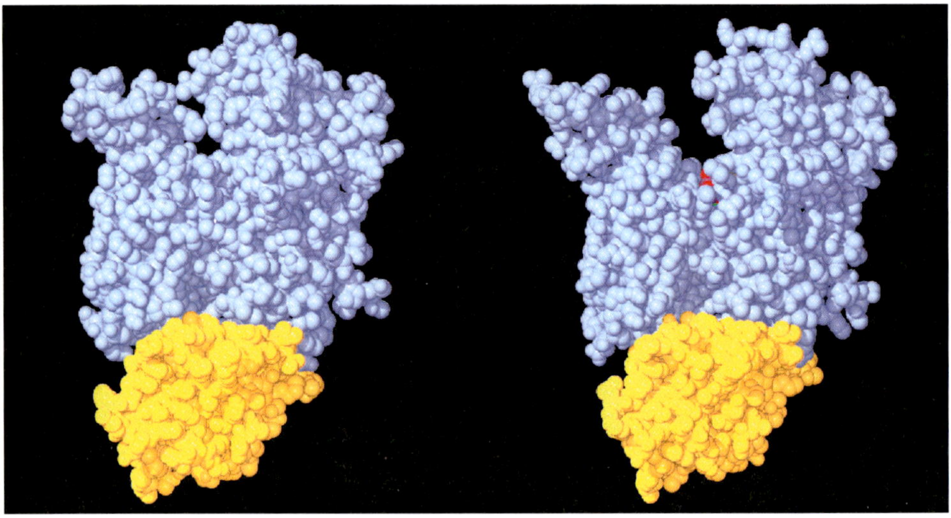

Chapter 3, Figure 3. The profilin:β-actin closed and open states (pdb accession codes 2BTF and 1HLU, respectively). Overview of the crystal structures of profilin:β-actin solved in 3.5 M NH_4SO_4 (closed state, left) and 1.8 M potassium phosphate (open state, right) in the presence of Ca^{2+} and ATP, illustrating the magnitude of the conformational difference between the two states. The polyphosphate tail of the ATP is exposed in the open state (oxygen O1beta, red dot). See also Fig. 1.

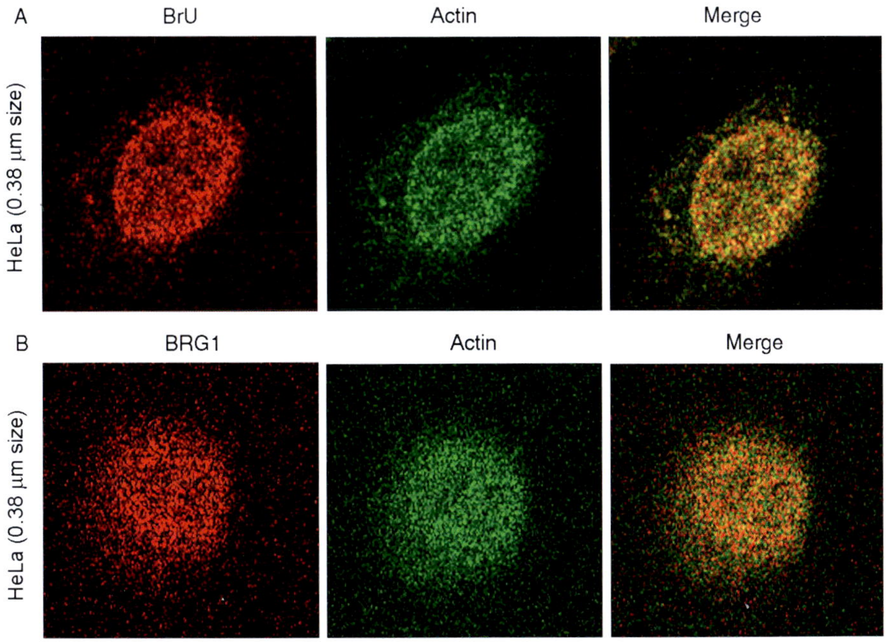

Chapter 7, Figure 3. (A) HeLa cells grown on cover slips were treated with 10 μM BrU for 10 minutes after osmotic shock, fixed, and antibodies against BrU and actin were used to visualize transcription sites and the location of actin in the nucleus. In the confocal images, BrU is red and actin green, and colocalization is shown by yellow dots in the panel labeled merge. (B) HeLa cells grown on cover slips, fixed, and stained with antibodies against BRG1 (red) and actin (red), colocalization is shown by yellow spots in the panel labeled merge, which is a merge of the two confocal images to the left.

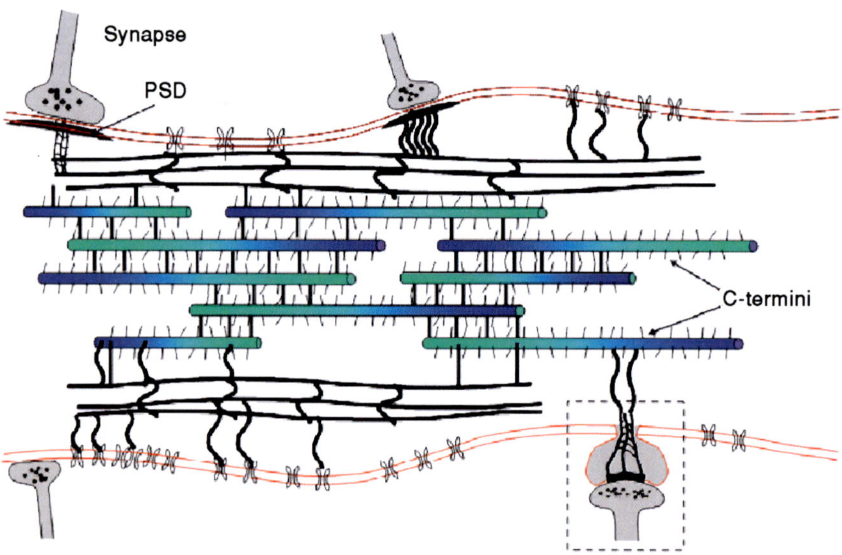

Chapter 8, Figure 1. A scheme of the dendritic shaft with MTs arranged in networks of mixed polarity which receive signals in the form of electric perturbations from synapses via AFs connected to MTs by MAP2, or via direct MT connections to postsynaptic density proteins by molecules such as CRIPT. A spiny synapse is depicted in the bottom right where an actin bundle enters the spine neck.

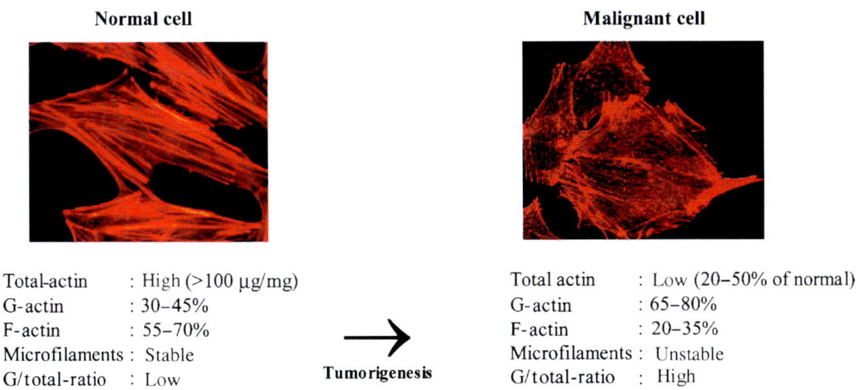

Chapter 9, Figure 1. The sensitive dynamic equilibrium of monomeric G-actin to filamentous F-actin is actively modulated during malignant cell transformation.

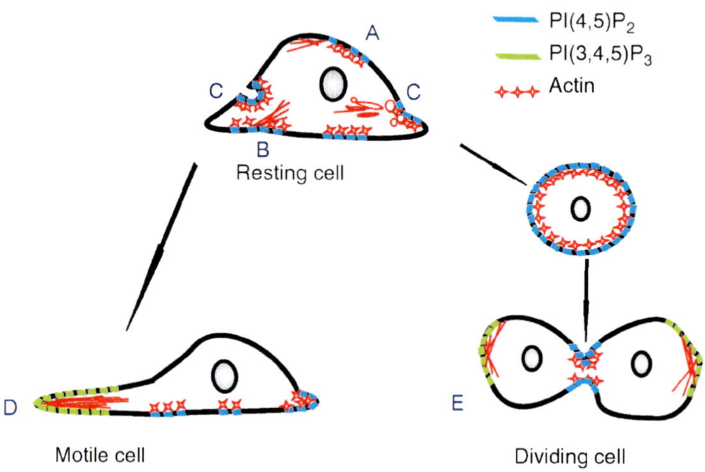

Chapter 10, Figure 2. The actin cytoskeleton (red) is important for controlling several key cellular processes such as cell shape coordination (A), adhesion to the surface (B), vesicle transport (including endocytosis and phagocytosis) (C). Actin is also involved in cell motility (D) and cytokinesis (E). In resting cells, $PI(4,5)P_2$ is uniformly distributed, and $PI(3,4,5)P_3$ levels are almost undetectable. $PI(3,4,5)P_3$ levels are temporally and spatially increased during cell polarization and cytokinesis. Local accumulation of $PI(3,4,5)P_3$ is observed at the leading edge of migrating cells and at the cell poles during cytokinesis. $PI(4,5)P_2$ is concentrated at the cleavage furrow during cytokinesis, as well as at the sites of endocytosis/phagocytosis. It is important to note that increased actin filament accumulation occurs primarily in these phospholipid-rich areas of cells.

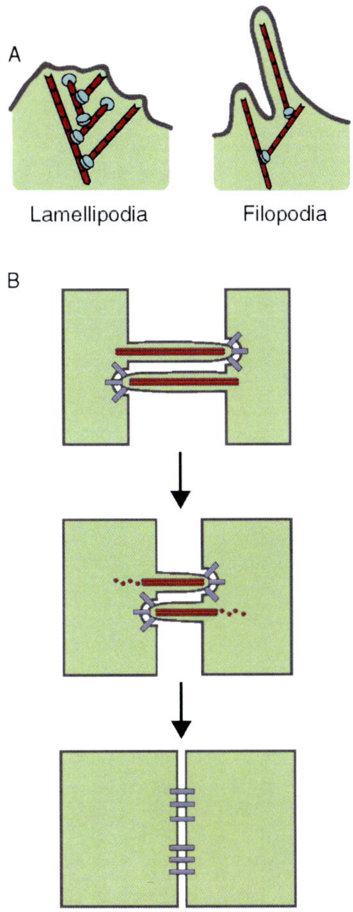

Chapter 12, Figure 1. Actin protrusions and a filopodial interdigitation model for epithelial cell adhesion. (A) Actin protrusions are formed when localized actin polymerization generates actin filaments (red) that push against the plasma membrane. New actin filaments can be nucleated by the Arp2/3 complex, and capped at their barbed ends by capping protein (green oval). Lamellipodial protrusions are thin sheet-like protrusions composed of a network of actin filaments, whereas finger-like filopodia contain bundles of parallel actin filaments. (B) In a number of different systems, fusion of epithelial cells is preceded by the assembly of filopodia. These actin (red) protrusions interdigitate and provide initial points of contact between two cells and sites for cell–cell adhesion (blue). The cells can then pull closer together and over time the initial contacts will develop into mature adherens junctions, forming a fused epithelial surface.

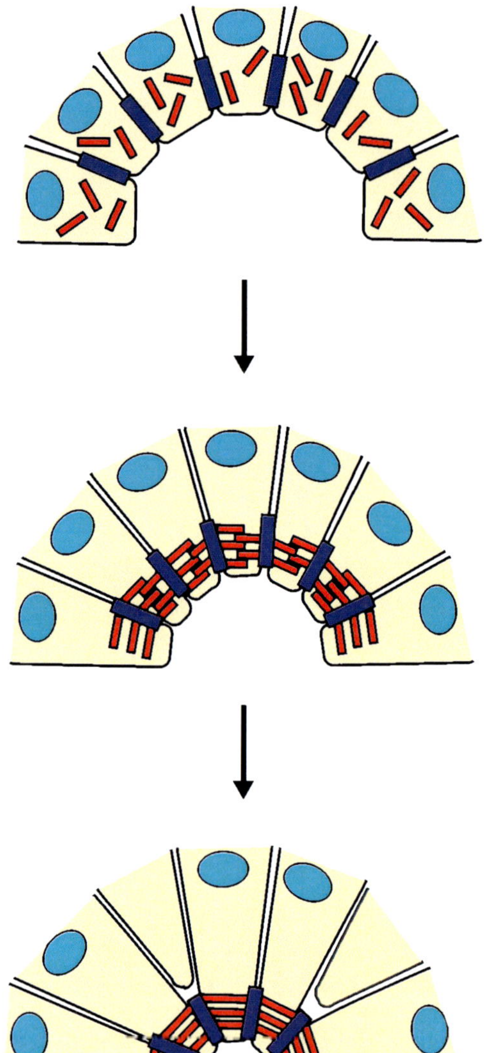

Chapter 12, Figure 2. Purse-string contraction. Actin filaments can form contractile cables which can shape and bend epithelial layers or, as is the case in wound healing, pull cells together to close a hole. Here a temporal series illustrates how purse-string contraction can draw the edges of an epithelial hole together. The cable is made up of individual actin filaments (red bars), which are anchored to adherens junctions (blue rectangles) formed between adjacent cells. Contraction of the actin cable in each cell leads to apical cell constriction and reduces the size of the hole. As contraction continues, some of the cells are squeezed out of the front row and new adherens junctions form between the remaining cells. Sustained contraction of the actin cable draws adjacent cells together, allowing them to fuse and close the hole.

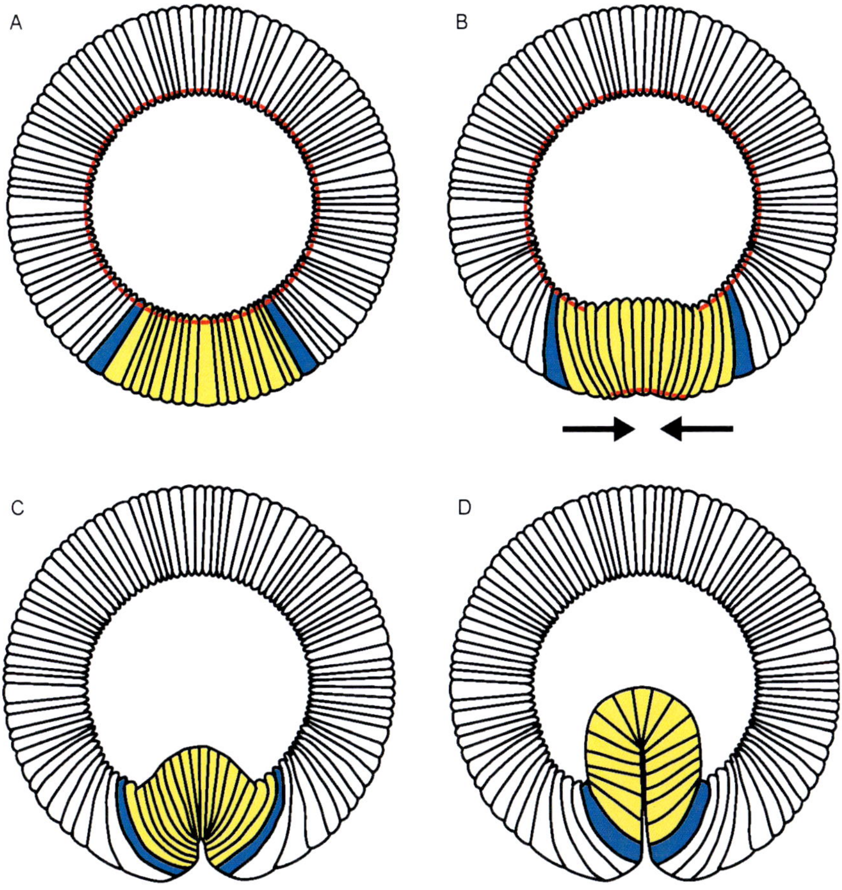

Chapter 12, Figure 4. Ventral furrow formation during *Drosophila* gastrulation. The major steps in ventral furrow formation are shown as cross sections through the *Drosophila* embryo (top = dorsal, bottom = ventral). The domain of the gastrulation transcription factor, Twist, is shown in blue and the overlapping Twist and Snail domains are shown in yellow. The localization of myosin II is illustrated in red. (A) Immediately after cellularization, myosin II is localized to the basal surfaces of the cells. (B) The ventral furrow begins to form as myosin II relocalizes to the apical surface of the ventral cells and these cells begin to flatten and constrict their apices. (C) Continued and coordinated apical constriction drives the formation of the furrow. (D) At the end of the process, the future mesodermal cells have been completely internalized.

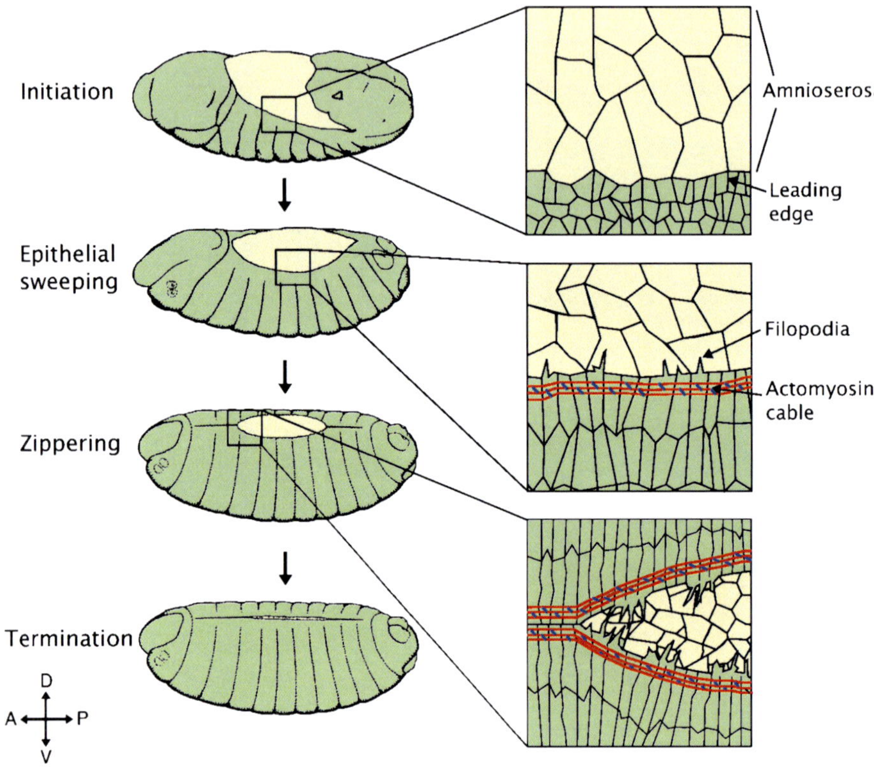

Chapter 12, Figure 5. Schematic diagram of dorsal closure. Dorsal closure is represented as a four-phase process with the epithelium shown in green and amnioserosa in yellow. Embryo cartoons (left) display the progressive stages of the process. Boxed regions (right) show "zoom ins" of the epithelial leading edge. Dorsal closure begins as the germ band approaches its ultimate position at the posterior. Early in the process, the large cells of the amnioserosa (yellow) contract and the epithelial cells (green) begin to elongate in a dorsoventral direction (axis compass in bottom left). Slightly later in closure, the epithelial leading edge begins to assemble an actomyosin cable (red and blue) and starts to extend actin protrusions across the amnioserosa.

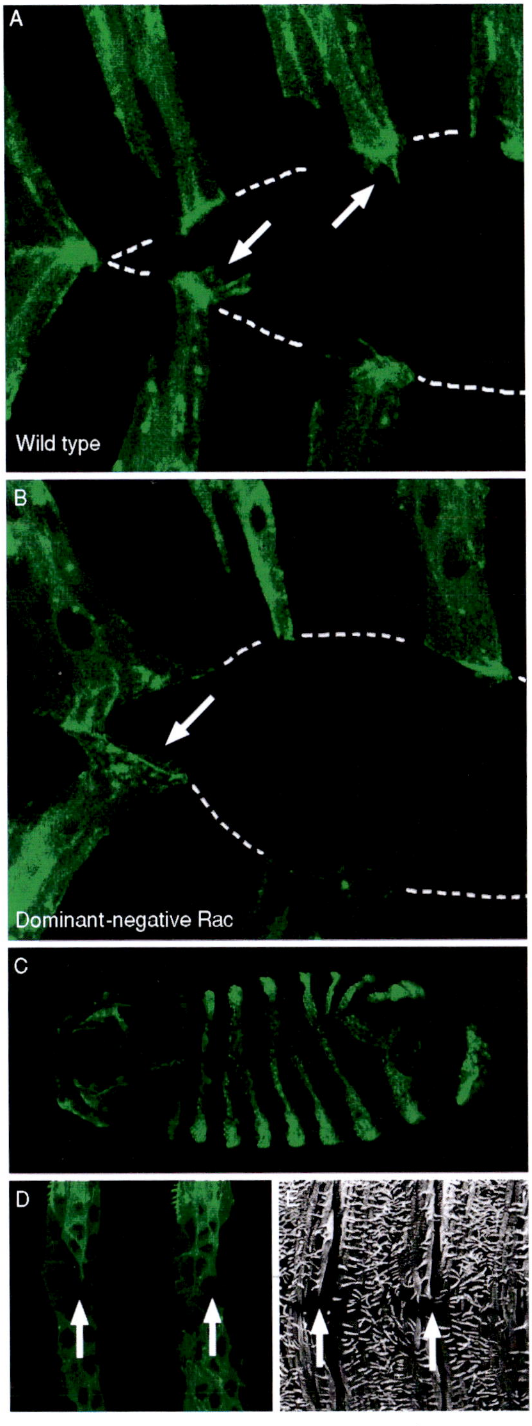

Chapter 12, Figure 6. *Continued*

Chapter 12, Figure 6. Actin protrusion function during *Drosophila* dorsal closure. During dorsal closure in the *Drosophila* embryo, filopodia and lamellipodia are assembled by the leading edge cells, where they function to accurately knit the two closing epithelial surfaces together. (A) A confocal image of a zippering front in a wild-type embryo, where GFP-actin is expressed in segmental stripes. Filopodia and lamellipodia assembled by the leading edge cells are indicated by arrows and the broken line highlights the edge of the dorsal hole. (B) In an embryo where a dominant negative form of the small GTPase Rac is expressed along with GFP-actin, very little actin protrusion assembly is seen. (C) Without actin protrusions, the embryos expressing dominant negative Rac are no longer able to correctly match their segments as the hole closes. In addition, these embryos also show fusion failures at the midline, shown by a confocal image in (D) and a scanning electron micrograph in (E). Reproduced with permission from Woolner *et al.* (2005).

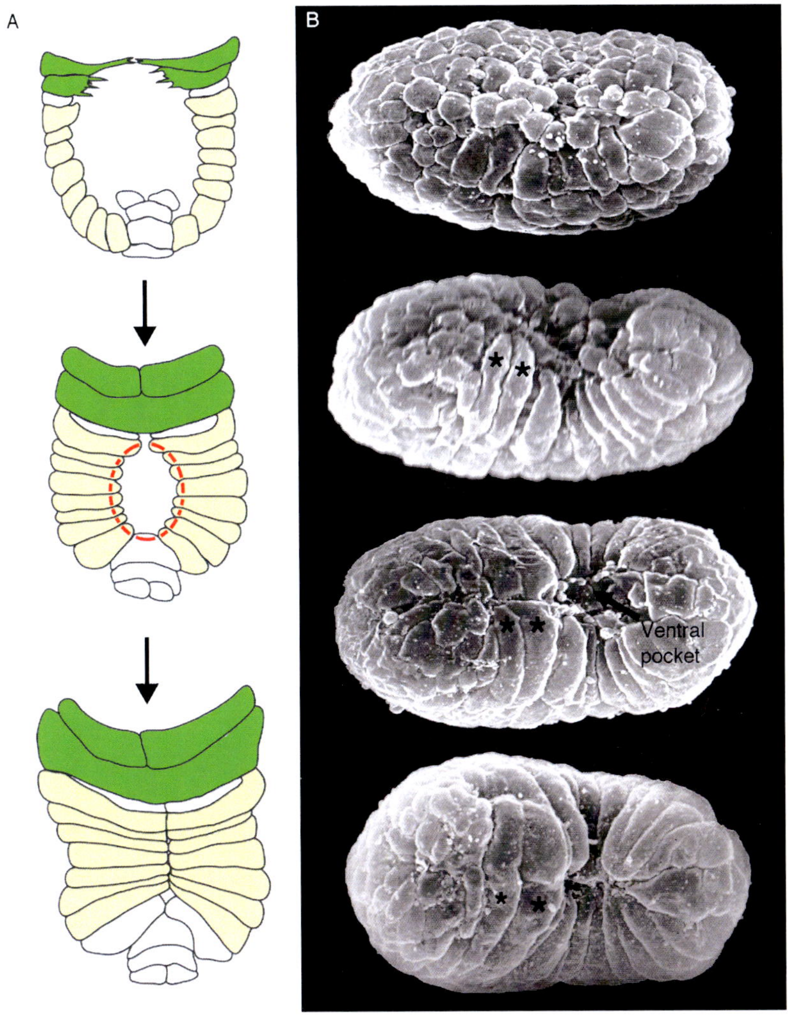

Chapter 12, Figure 7. Ventral enclosure in the *C. elegans* embryo. *C. elegans* ventral enclosure is an epithelial fusion event which shares many similarities with dorsal closure. (A) Schematic diagrams illustrating ventral enclosure: the process starts as the four leading cells (green) assemble filopodia and begin to migrate ventrally. Once the leading cells have met at the midline, the pocket cells (yellow) assemble an actin cable and "purse-string" the hole closed. (B) A set of scanning electron micrographs of a *C. elegans* embryo undergoing ventral enclosure, leading cells are highlighted with asterisks. Reproduced with permission from Martin and Parkhurst (2004).

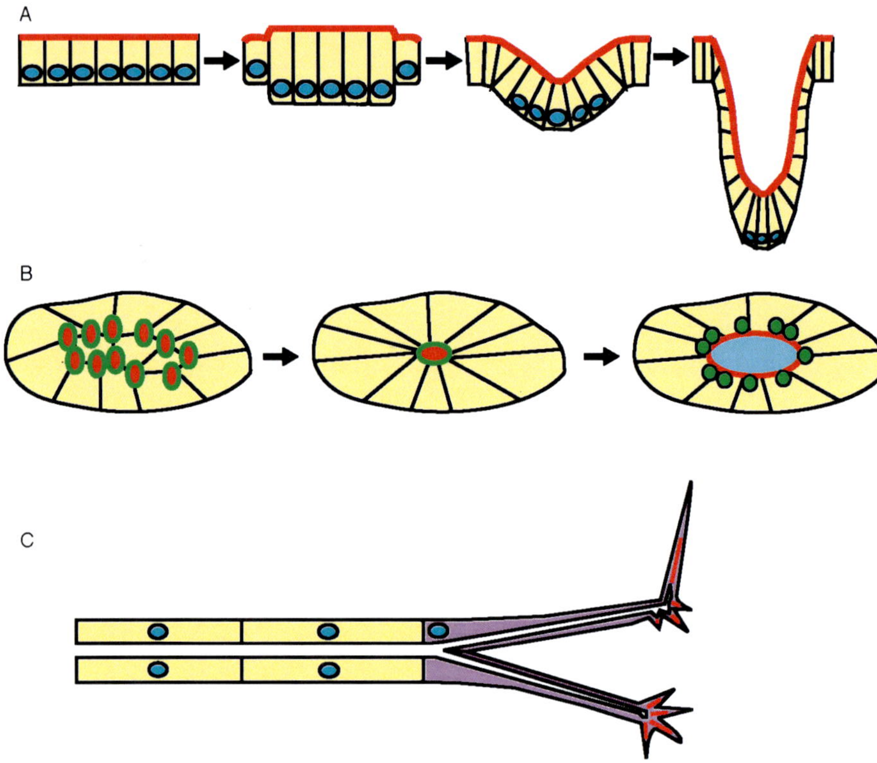

Chapter 12, Figure 9. Mechanisms for making tubes. Actin plays a major role in three different strategies for generating tubular systems. (A) Invagination of cells from an epithelial sheet can drive tube formation. Actin (red) is found along the apical surface of the invaginating cells and it is thought that actin cables act to constrict the apices of these cells, making them wedge shaped and driving their invagination. (B) Lumen formation is an alternative tubulogenesis strategy and can occur in a single cells or a group of cells (the latter is illustrated here). Cells in a cord assemble primitive adherens contacts (green), these then cluster in the center of the cord and a lumen (pale blue) forms between the cells. It is thought that the actin (red) associated with the contacts aids in their clustering. (C) Once a tube is formed, it may migrate and elongate, often in response to external cues. An example of directed migration of tubes is found in the developing tracheal system, where secondary tracheal branches (purple) extend toward sources of fibroblast growth factor (FGF). The branches assemble actin protrusions at their distal tips, which appear to aid tube extension and sensing of FGF.